KU-185-769

Contents

WITH~~~~
CHRIST~~~~
~~~~
Christ Church

T065168

BIODIVERSITY II
Understanding and Protecting Our Biological Resources

Marjorie L. Reaka-Kudla, Don E. Wilson,
and Edward O. Wilson, editors

JOSEPH HENRY PRESS
Washington, D.C.   1997

JOSEPH HENRY PRESS • 2101 Constitution Avenue, N.W. • Washington, DC 20418

The Joseph Henry Press, an imprint of the National Academy Press, was created with the goal of making books on science, technology, and health more widely available to professionals and the public. Joseph Henry was one of the founders of the National Academy of Sciences and a leader of early American science.

Library of Congress Cataloging-in-Publication Data

Biodiversity II : understanding and protecting our biological
   resources / Marjorie L. Reaka-Kudla, Don E. Wilson, and Edward O.
   Wilson, editors.
         p.   cm.
      Includes bibliographical references and index.
      ISBN 0-309-05227-0 (hardcover : alk. paper).—
      ISBN 0-309-05584-9 (paperback : alk. paper)
      1. Biological diversity conservation. 2. Biological diversity.
   I. Reaka-Kudla, Marjorie L.   II. Wilson, Don E.   III. Wilson, Edward
   Osborne, 1929-   .
   QH75.B5228   1996
   333.95'16—dc20   96-30851
                                                               CIP

 This book is printed on recycled paper.

Cover art: Center panel of the "Tree of Life," a triptych by Alfredo Arreguin.

Copyright 1997 by the National Academy of Sciences. All rights reserved.

Printed in the United States of America.

PART VI. GETTING THE JOB DONE:
INSTITUTIONAL, HUMAN, AND INFORMATIONAL INFRASTRUCTURE

PART VII. CONCLUSIONS

BIODIVERSITY II

CHAPTER

1

Introduction

EDWARD O. WILSON

*Pellegrino University Professor, Museum of Comparative Zoology,
Harvard University, Cambridge, Massachusetts*

"Biodiversity," the term and concept, has been a remarkable event in recent cultural evolution: 10 years ago the word did not exist, except perhaps through occasional idiosyncratic use. Today it is one of the most commonly used expressions in the biological sciences and has become a household word. It was born "BioDiversity" during the National Forum on BioDiversity, held in Washington, D.C., on September 21-24, 1986, under the auspices of the National Academy of Sciences and the Smithsonian Institution. The proceedings of the forum, published in 1988 under the title *BioDiversity* (later to be cited with less than bibliographical accuracy by most authors as *Biodiversity*), became a best-seller for the National Academy Press. By the summer of 1992, as a key topic of the Rio environmental summit meeting, biodiversity had moved to center stage as one of the central issues of scientific and political concern world-wide.

So what is it? Biologists are inclined to agree that it is, in one sense, everything. Biodiversity is defined as all hereditarily based variation at all levels of organization, from the genes within a single local population or species, to the species composing all or part of a local community, and finally to the communities themselves that compose the living parts of the multifarious ecosystems of the world. The key to the effective analysis of biodiversity is the precise definition of each level of organization when it is being addressed.

Even though the study of biodiversity can be traced back as far as Aristotle, what finally has given it such extraordinarily widespread attention is the realization that it is disappearing. In the late 1970s and through the 1980s, the first convincing estimates were made of the rate of tropical deforestation, which translates to the areal loss of habitat where most of living diversity is concentrated. This information led to disturbingly high estimates of the rates of loss of species in these forests. The magnitude of erosion also drew attention to ongoing extinction in other habitats, from deserts to coral reefs, at all levels of biological organization from alleles to entire local ecosystems. It became clear that the decline of Earth's biodiversity was serious. Worse, unlike toxic pollution and ozone depletion, it cannot be reversed.

Scientists who once had devoted their careers to bits and pieces of biodiversity now became holists, or at least more approving of the holistic approach, and they were energized by a new sense of mission. For the good of society as a whole, they now realized that the classification of such organisms as braconid wasps and lauraceous shrubs mattered. Moreover, the ecologists also were included: the processes by which natural communities are assembled and their constituent species maintained have central importance in both science and the real world. The study of diversity subsumed old problems in systematics and ecology, and specialists in these and in related fields of biology began to talk in common parlance as never before. Just as significantly, physical scientists, social scientists, geographers, and artists were drawn into the colloquy. The subject consequently has begun to be reshaped into a new, often surprisingly eclectic field of inquiry. Today we now hear regularly of "biodiversity science" and "biodiversity studies."

Since the 1986 National Forum on BioDiversity, there has been an exponential rise in research and technical innovation. Scientists appreciate that only a tiny fraction of biodiversity on Earth has been explored, and that its origin and maintenance pose some of the most fundamental problems of the biological sciences. These problems are also among the least technically tractable. Those who have cut into the outer surface of ecology and evolution suspect that molecular and cell biology eventually will prove simple by comparison.

The present volume is a 10-year report on the state of the art in biodiversity studies, with an emphasis on concept formation and technique. Overall, it makes a striking contrast with the original *BioDiversity*, showing how extraordinarily far we have come and at the same time mapping how far scientists yet must travel in their reinvigorated exploration of the biosphere.

Some scientists and policy-makers have worried that the magnitude of the biodiversity we now know to be present in the world's habitats is so enormous, the cost of exploring and documenting it so overwhelming, and the number of biologists who can analyze and document it so small that the goal of understanding the diversity of the world's species is unattainable. The central message of this volume is, to the contrary, that the potential benefits of knowing and conserving this biodiversity are too great and the costs of losing it are too high to take a path of least resistance. By documenting the infrastructure of knowledge and institutions that already are in place, this volume suggests that there is a cost-effective and feasible way of approaching the conservation of the world's biological resources. The key to a cost-effective solution to the biodiversity crisis lies in the collaboration of museums, research institutions, and universities; the pooling of human and financial resources; and the shared use of physical and institutional structures that are already present. Rather than building the knowledge, institutional, and physical infrastructure for documenting biodiversity from the ground up, we need to build upon the preexisting infrastructure and increase support for systematics, training, and museums.

This volume is an outgrowth of one such endeavor, the recent establishment of a Consortium for Systematics and Biodiversity between the Smithsonian Institution, the University of Maryland at College Park, the U.S. Department of Agriculture Systematics Laboratories, the University of Maryland Biotechnology Institute, and the American Type Culture Collection. The Consortium, dedicated to enhancing the conceptual understanding and documentation of biodiversity of organisms (from viruses and bacteria to invertebrate and vertebrate animals, protists, fungi, algae, and higher plants) in living and nonliving museum collections, represents the type of cooperation that will be necessary for us to understand and protect our natural resources.

THE MEANING AND
VALUE OF BIODIVERSITY

Biodiversity, the planet's most valuable resource, is on loan to us from our children.

CHAPTER

2

Biodiversity: What Is It?

THOMAS E. LOVEJOY

Counselor to the Secretary for Biodiversity and Environmental Affairs, Smithsonian Institution, Washington, D.C.

Biodiversity has to be thought of in a number of different ways. One would begin by looking at the overall perspective of evolutionary time and more specifically by looking at a great radiation (such as that of the Hawaiian honeycreepers) from a single ancestor. Another way of viewing biodiversity is as a characteristic of natural communities. For instance, a forest in Southern New England characteristically will contain about 20 or 30 tree species, but farther north or in a marshy area in Southern New England, forests are composed of only two or three species. In contrast, a tropical forest, such as in Amazonian Peru, will contain hundreds of species of trees. Therefore, natural communities each have their own characteristic biodiversity, both in terms of numbers and composition of species.

Another way of looking at biodiversity is globally and collectively. While the number of species currently described is on the order of 1.4 million, the big question is how many species are there totally? Current estimates of the total number of species run from 10-100 million. One can break it down and look at certain segments of global totals, such as the diversity of higher plants, number of species, or expressed as sheer weight (biomass). The degree of knowledge about biodiversity varies with both location and taxon (classificatory groups of organisms). For example, the British insect fauna is much better known than the Australian insect fauna. (Indeed I believe that there are so many naturalists in Britain that it is impossible for a bird to lay an egg without three people, including at least one cleric, recording it.) Within Australia, vertebrates are better known than insects. That relation is true throughout most of the world, because we have tended toward vertebrate chauvinism in the exploration of biodiversity (Wilson, 1985).

Another way to think about biodiversity is where it is most concentrated. The best known concentration, of course, is in tropical forests—those places near the equator where there is enough moisture sufficiently evenly distributed through the year to maintain a tropical forest formation. These forests comprise roughly 7% of the dry land surface of the Earth and may hold more than 50% of all species. The reality, however, is that we do not know precisely how much biodiversity is concentrated in a particular biological formation. A characteristic pattern of biodiversity is a general increase in numbers of species as one approaches the equator. For example, São Paulo, just within the tropics, holds 100 times more species of ants than Tierra del Fuego, and the number at the equator will be even higher than at São Paulo (MacArthur, 1972). This is a pattern that repeats itself again and again for many groups of organisms, but, of course, not all. The marine realm has more representations of the major groupings of life than those on land, and the realm of soil biology is so poorly explored that it is hard to tell what rich forms of life might exist there. So as we learn more about life on Earth, the relative proportions of where life on Earth is concentrated can be expected to change.

In the tropical forest, the forest canopy is an entirely different realm that can be studied with a tower or a crane like that used by the Smithsonian Tropical Research Institute. The only species that the canopy shares with the forest soil are the canopy trees themselves. All the rest are almost entirely different. The canopy is a vast, unexplored place for biodiversity. Terry Erwin, a Smithsonian entomologist, in the first major analysis of biodiversity of the canopy, suggested that the number of species with which we share this planet had been underestimated by at least a factor of three (Erwin, 1982). The larger point is that the more people look at the tropical forest in different ways, as Terry Erwin has done, the more biodiversity there seems to be. This does not mean that the same thing will not happen also in the marine realm or in the area of soils and microorganisms. It may well happen and is part of the excitement of exploring life on Earth.

Biodiversity matters to human beings in a variety of ways (Lovejoy, 1994). There are important aesthetic and ethical dimensions, but part of our existence depends on direct use, whether it is the botanical species that flavor gin or the wild relatives of a major agricultural crop such as the species of wild perennial corn found in Mexico about 20 years ago. Previously there had been only one species of perennial corn described, discovered in fact by a Smithsonian botanist early in the century. But unlike the species discovered in the 1970s, it did not have the same number of chromosomes as domestic corn. With the discovery of this second species of perennial corn, it became relatively easy to transfer some of the traits of perennial corn into corn agriculture, making the long-term dream of a perennial corn crop, as well as the more short-term one of disease resistance, an achievable goal. The importance of these kinds of contributions to agriculture can be underscored by noting that corn is the third most important grain supporting human societies. Another example in which a wild species benefits agriculture is illustrated by a species of wild potato from Peru that has a characteristic resistance to insect attacks that is being incorporated actively into potato agriculture to deal with potato beetles.

Another completely different way in which organisms can benefit humans is the rapidly growing area known as bioremediation. For example, a species of bacteria discovered in the sediments of the Potomac River by a scientist with the U.S. Geological Survey is capable of breaking down chlorofluorocarbons (CFCs) in anaerobic conditions (Lovley and Woodward, 1992). There are many organisms in nature with unusual metabolisms and appetites that could prove to be beneficial in cleaning up some pollution problems and could become of great significance in the rapidly developing field of industrial ecology.

One of the most interesting aspects of biodiversity is the evolutionary tension between species of insects that exert grazing pressure on species of plants and the chemical and biochemical defenses that these plant species produce to reduce that pressure. This "tug of war" is a source of a wide range of useful molecules for the field of medicine. The best-selling medicine of all time, aspirin (salicylic acid which derives its name from the Latin, *Salix*, for willow), is based on just one of these molecules, the discovery of which dates back to the time of Hippocrates when he prescribed mashed willow bark as a painkiller. Today, of course, the molecule is synthesized in a factory and there is no need to go back to nature for it, but the *idea* behind it came from looking at natural biodiversity.

Discoveries for the advancement of medicine and understanding of the life sciences constitute one of the most powerful ways in which biodiversity can contribute to human society. Often going unnoticed and rarely documented, extremely important discoveries have been made from organisms that were previously considered relatively peripheral to human society. For example, *Penicillium* mold at one time was valued for what it did to flavor blue cheeses, but that was dwarfed in subsequent times, when it sparked the concept of antibiotics.

There are interesting intellectual chains deriving from natural phenomena. A South American pit viper might seem of little relevance to someone living in Washington, D.C., or Chicago, yet studies on the venom of one species of these vipers led to the discovery of the angiotensin system that regulates blood pressure in human beings. Once that system was known, it became possible to devise a molecule that alters blood pressure and is the preferred prescription drug for hypertension. This compound brings the Squibb Company $1.3 billion a year in sales and contributes to the well-being and longevity of millions of Americans and others.

Yet another way to think about biodiversity is the way in which it collectively provides us with "free services." Consider the Amazon River, with its tremendous drainage area and the important fishery on which people of the

Amazon depend. The water chemistry of the Amazon basin is insufficient to support the productivity of that fishery. Instead, the productivity depends on a linkage between the terrestrial ecosystem and the aquatic one. When the rivers of the basin flood (10-15, sometimes 20 m high), they spill over into the floodplain forest. Fish then can swim into the forest and feed on the fruits, nuts, seeds, and other organic material that fall into the water. The Amazonian fishery thus depends on an ecological service in which nutrients are transferred from the terrestrial ecosystem to the aquatic ecosystem, with major benefits to people living in the area (Goulding, 1980).

Sometimes even a single species performs a vital service. In the Chesapeake Bay, the oyster plays a very critical role. Today the oyster population of the Chesapeake Bay filters a volume of water equal to that of the entire bay about once a year. Before various factors, most human-driven, led to the decline of the oyster population, it filtered a volume equal to the entire bay about once a week (Newell, 1988). That is an interesting example in which a single keystone species in an ecosystem can have profound implications for how, in this instance, a grand estuary and major source of various types of seafood actually functions.

On larger landscape scales, biodiversity provides services such as underpinning the hydrological cycle in the Amazon, in which literally half of the rainfall in the Amazon is generated within the basin. If the Amazon forest were to be replaced with grassland, a rough computer model predicts that about one-fourth of that rainfall would not occur, and would be accompanied by associated temperature increases (Nobre et al., 1991). These effects would spill over to other geographical areas, for example, central Brazil, which also depends on the Amazon as a source of moisture for rainfall. Thus, the hydrological cycle is basically dependent on those vast natural forest communities.

Another extremely valuable way in which biodiversity serves human society is as an indicator of ecological change. A few years ago, herpetologists studying amphibians, particularly frogs, began to compare incidental notes and realized that there was a major decline in populations of frogs throughout the world in patterns that still are hard to understand and explain. Something or some things are happening that appear to affect frog populations, and it would be extremely valuable to identify these vectors of change before they affect humans directly.

Stress in the biological community also reduces biodiversity. For example, extreme air pollution in Cubatão in São Paulo state (often called the valley of death) was so severe in the 1970s and 1980s that it killed most of the trees in the surrounding Atlantic forest. It reduced biodiversity and vegetative cover to the point that there were landslides. In another example, high loads of fertilizers that put stress on a pasture in England during the period 1856-1949 reduced the number of species of native plants from 49 to 3. The good news, of course, is that if you remove the stress (and the species are still in existence and nearby), characteristic diversity will recover over time.

An excellent way to understand factors affecting biodiversity is by looking at islands. They are small, confined areas, more easily managed for study, where cause and effect often appear more clearly. In the Hawaiian islands, for example, the loss of species from human activity has been taking place for centuries. A series of Hawaiian bird species became extinct at the hands of the native Polynesians prior to European arrival. Studies by Storrs Olson at the National Museum of Natural History and David Steadman at the New York State Museum have shown that this pattern repeats itself over and over again in the Caribbean and Pacific islands (Steadman and Olson, 1985). The notion that indigenous peoples were always in exquisite harmony with their environment is just not true.

One of the major problems in global biodiversity is introduced species. Exotic faunas represent a very severe problem all over the world, but particularly in island situations. The Stephen Island wren is a dramatic example. The first and last specimen of that species confined to that single island was killed by a species not indigenous to the island—the lighthouse keeper's cat. The most recent flora of Hawaii contains more alien or exotic species than native ones (Wagner et al., 1990).

By far the biggest problem in protecting the world's biodiversity is habitat destruction, whether it is southern California, where the California gnatcatcher and many other species live in declining patches of coastal sage scrub habitat, or whether it is the rapidly shrinking tropical forests in many parts of the world. The numbers of loss can be staggering. The state of Rondonia in western Amazonia lost 20% of its forest in 5 years. Today Suriname and Guyana teeter on the brink of losing much of their forest to foreign interests.

Another outcome of habitat destruction is that the available habitat is broken up into pieces. When we begin to look at what this means for biodiversity, a very disturbing picture begins to appear. For example, in 1920 there were 208 species of birds known on the Smithsonian's Barro Colorado Island, an island that was created by the flooding of the area of the Gatun Lake for the Panama Canal. Fifty years later, a number of these species were gone and, in the original analysis by Ed Willis (1974) of that loss, a significant number of the species (18) were thought to be lost simply because the isolated fragment of the island was not large enough to support them. The fragmentation of habitats leaves remnants no longer connected to a larger wilderness and hence species are lost over time. This has serious implications for conservation. Among other study sites, Smithsonian and Brazilian scientists are studying fragmentation in the middle of the Amazon (Bierregaard et al., 1992; Lovejoy et al., 1983, 1984). Fragmentation has serious implications for the use of landscapes. The good news is that if riparian habitats (vegetation along watercourses) are restored, the landscape has much more connectivity, eliminating some of the fragmentation problems.

There are also regional effects of pollution. Some examples include the air pollution of Cubatão discussed above and acid rain, which is a problem in tropical as well as northern industrial countries. Taken together, all the different ways human activity affect biodiversity have driven extinction rates to a level 1,000-10,000 times the normal rate (May et al., 1995; Wilson, 1985).

An additional and ultimate concern is global climatic change due to increasing levels of greenhouse gases. Most of these gases come from the burning of fossil fuels (essentially burning of old vegetation stored under the surface of the Earth) that represent carbon reservoirs that have been stored for thousands or millions of years, but which now are being oxidized and released into the atmosphere in a very short geological time. Burning of the forests is also a major contributor to the greenhouse gases, releasing nearly 1 billion tons of carbon as carbon dioxide (CO_2) annually.

The most famous graph of the twentieth century may well be that showing the curve of increasing CO_2 in the atmosphere as measured from the top of a volcano in Hawaii. Every year, inexorably, there is an increase in the amount of CO_2. There is also an annual decrease in atmospheric CO_2 caused by spring in the Northern Hemisphere, when the trees leaf out and, together with other plants, take up approximately 5 or 6 billion tons of CO_2. This is returned at the end of the year as the leaves drop and plant materials decay. However, a significant portion of the net annual increase in CO_2 comes from the pool of carbon that results from destroyed biomass and biodiversity.

Climatic change has serious implications for agriculture, but the resulting problems probably can be solved much easier than can serious problems with biodiversity. Biodiversity is dependent on an intricate web of factors that can be upset by rapid climatic change. Climatic change is, in fact, nothing new in the history of life on Earth. Climatic change has been extensive in the past 1 million years or so, which have been characterized by glacial and interglacial periods. During these periods, species tracked their required conditions, each migrating according to their particular dispersal rate—often moving up or down slope, up or down latitude. Today, however, most biodiversity, or at least an increasing proportion of it, is locked up in isolated patches. In the face of climatic change, even natural climatic change, human activity has created an obstacle course for the dispersal of biodiversity. This could establish one of the greatest biotic crises of all time.

Beyond the immediate causes that threaten biodiversity, there are ultimate causes, such as human population growth—which adds roughly 100 million new people to the human population every year—and the massive impact of economic activities. In addition to these activities and the per capita consumption in the industrial world, there is an enormous, complex web of interactions. When a product is purchased, there may be a long chain between that product and some other part of this country or some other part of the world which often go unnoticed. For example, had New Coke been successful, the lack of vanilla in the formula would have undercut the only element (vanilla) in the economy of one region of Madagascar, possibly forcing that island country to fall back on its only other source of income—its remaining forests—which are already in peril.

It will be a major challenge to avoid a staggering loss of biodiversity in the decades and centuries ahead. The increasing interest in ecosystem management and attempts to implement it will be very helpful. South Florida is a prime example. There, a channelized Kissimmee River feeds into Lake Okeechobee. The increasing use of phosphates from agricultural areas pollutes the drainages to the south of Lake Okeechobee. Also, ditching, diking, and draining to provide flood control and water supply for the benefit of large populated areas along the coast has led to the ultimate arrival of only one-fourth to one-half of the normal annual freshwater flow in Florida Bay. Correlated with this are algal blooms, the collapse of the shrimp fishery, and hypersaline water pouring out between the keys onto the already-stressed coral reefs. The only way to restore the ecosystem of south Florida close to that approaching a natural condition would require that all the major agencies of state and federal governments are involved. All the stakeholders, such as business and environmental groups, need to be involved. If we can bring about a more integrated approach to living within our ecosystems, we are much more likely to save the fundamental structure of biodiversity.

As an understanding of sustainable development deepens, I believe a major portion will turn out to be biologically based. A prime example is tied to the 1993 Nobel Prize for chemistry. Kary Mullis shared the prize for the conception of the polymerase chain reaction (PCR). This reaction, which is just about the only true science in the popular work of fiction, *Jurassic Park*, has made it possible to do esoteric work such as the positive identification of the remains of Czar Nicholas II and his family. It is also a fundamental part of diagnostic medicine today. It is now no longer necessary to wait 3 days for a culture to be grown before receiving a diagnosis and prescription. The polymerase chain reaction is an important part of biotechnology and molecular biology.

The polymerase chain reaction only works because of an enzyme discovered in a bacterium found in the hot springs of Yellowstone. Science and society have benefited from these protected hot springs. In 1872, the area was set aside as the world's first national park—an act undertaken only on the basis of Yellowstone's scenic wonders. The fundamental goal of the National Biological Survey[1] is to organize our knowledge of biodiversity and obviate the need for that kind of luck. A survey indeed is a very old idea and a kind of activity that was going on even in the colonial period of this country. The concept of a biological survey received a great boost from Spencer Baird when he was Assistant Secretary of the Smithsonian and through the encouragement of C. Hart Merriam, the founder of the original U.S. Biological Survey. It is just basic biological good housekeeping to try to find out what we have and where it is. In turn, biodiversity can be enjoyed in a variety of ways and we can learn even more about it and thus help ourselves achieve sustainable development.

---

[1]The name of the National Biological Survey recently has been changed to the National Biological Service by the Secretary of the Interior, Bruce Babbitt.

REFERENCES

Bierregaard, R. O., Jr., T. E. Lovejoy, V. Kapos, A. A. dos Santos, and R. W. Hutchings. 1992. The biological dynamics of tropical rainforest fragments. BioScience 42:859-866.

Erwin, T. L. 1982. Tropical forests: Their richness in coleoptera and other arthropod species. Coleopt. Bull. 36(1):74-75.

Goulding, M. 1980. The Fishes and the Forest. University of California Press, Berkeley.

Lovejoy, T. E., R. O. Bierregaard, J. M. Rankin, and H. O. R. Schubart. 1983. Ecological dynamics of tropical forest fragments. Pp. 377-384 in S. L. Sutton, T. C. Whitmore, and A. C. Chadwick, eds., Tropical Rain Forest: Ecology and Management. Blackwell Scientific Publications, Oxford, England.

Lovejoy, T. E., J. M. Rankin, R. O. Bierregaard, K. S. Brown, L. H. Emmons, and M. E. Van der Voort. 1984. Ecosystem decay of Amazon forest fragments. Pp. 295-325 in M.H. Nitecki, ed., Extinctions. University of Chicago Press, Chicago.

Lovejoy, T. E. 1994. The quantification of biodiversity: An esoteric quest or a vital component of sustainable development? Phil. Trans. R. Soc. Lond. B 345:81-87.

Lovley, D. R., and J. C. Woodward. 1992. Consumption of freons CFC11 and CFC12 by anaerobic sediments and soils. Envir. Sci. Technol. 26:925-929.

MacArthur, R. H. 1972. Geographical Ecology: Patterns in the Distribution of Species. Harper and Row, N.Y.

May, R. M., J. H. Lawton, and N. E. Stork. 1995. Assessing extinction rates. Pp. 1-24 in J. H. Lawton and R. M. May, eds., Extinction Rates. Oxford University Press, Oxford, England.

Newell, R. I. E. 1988. Ecological changes in Chesapeake Bay: Are they the result of overharvesting the American oyster, *Crassostrea virginica*? Pp. 536-546 in M. P. Lynch and E. C. Krom, eds., Understanding the Estuary: Advances in Chesapeake Bay Research. Proceedings of a Conference. Chesapeake Research Consortium Publication CRC129, Baltimore, Md.

Nobre, C. A., P. J. Sellers, and J. Shukla. 1991. Amazonian deforestation and regional climate change. J. Climate 4:957-988.

Steadman, D. W., and S. L. Olson. 1985. Bird remains from an archaeological site on Henderson Island, South Pacfic: Man-caused extinctions on an "uninhabited" island. Proc. Natl. Acad. Sci. 82:6191-6195.

Wagner, W. L., D. R. Herbst, and S. H. Sohmer. 1990. Manual of the Flowering Plants of Hawaii. University of Hawaii Press and Bishop Museum Press, Honolulu.

Willis, E. O. 1974. Populations and local extinctions of birds on Barro Colorado Island, Panama. Ecol. Monogr. 44:153-219.

Wilson, E. O. 1985. The biological diversity crisis: A challenge to science. Issues Sci. Technol. 2(1):20-29.

Biodiversity: Why Is It Important?

RUTH PATRICK

Francis Boyer Chair of Limnology,
Academy of Natural Sciences of Philadelphia, Pennsylvania

The term biodiversity—the presence of a large number of species of animals and plants—is now generally recognized as being important by many people. To most people the existence of many species is important because their physiological differences furnish various sources of food, clothing, shelter, and medicine for man. Species also differ in their physical and chemical characteristics, which is particularly true of plants. For example we have cotton, flax, and rattan, which are all products of various species of plants and are used for different types of clothing. Likewise, the structure of plants, be they trees or palm fronds or bamboo poles, serve as a source of shelter for much of the human race. Therefore, from the human standpoint, the presence of a great number of species with different structures, different chemical compositions, and different lifespans form one of the most important bases of life for humans throughout our planet.

Although these aspects are very important to society, there are far more important aspects of biodiversity that many people do not realize. For example, it is the presence of different kinds of plants that make up the food of grazing animals, for example, grass, herbs, and fruits. It is this great diversity of plant life, with its varying food values, that enables the existence of many different kinds of animals.

Plants also, by their great diversity, form shelter and habitats for many different species. McArthur (1965) was the first to point out that trees, such as the conifers on Mount Desert Island, Maine, have many different habitats and that different species of birds occupy each of these habitats.

The diversity of plants is important for all animals, herbivores and omnivores included. The different nutritional values of fruits and seeds enable a species to get the variety of the chemicals that are necessary for its diet and hence to graze without exhausting the population of a given species. It has been noted by Janzen (1978) and others that certain species will feed on a given plant that is unpalatable to other species. This difference is due to secondary chemicals. The so-called primary chemicals are those that are necessary for the production of energy and nutrients for growth and reproduction of a species. The secondary chemicals are those that act as defense mechanisms or, in some cases, as attractants for reproduction. These chemicals vary greatly from species to species, as does the tolerance of the predators.

Often there is a specific species of insect that feeds on a given species of plant which contains certain secondary chemicals. In such cases, these secondary chemicals are not toxic at low concentrations to the predator. The predator ceases to prey when the toxic threshold for that species is reached. This condition allows the reduced population of the prey species to exist and reproduce. Many of these chemicals are known as allelochemics, allomones, and kairomones (Janzen, 1978).

Seeds that are high in nutrient value may have three or more classes of defensive compounds, such as one or more protease inhibitors or lecithins, alkaloids, or uncommon amino acids, glycosides, polyphenols, etc. As each group of compounds is known to be very diverse, it is unlikely that there would be many seeds with the same defenses by chance alone. These secondary compounds often are found in vacuoles or other structures so that they do not interfere with the ordinary metabolism of the plant.

Other secondary chemicals may be volatile and act as stimulators for pollination. For example, with the genus *Ophyrys* of orchids, several volatile chemicals that attract pollinators have been identified. The compounds so far have been grouped into three classes: terpenoids, fatty acid derivatives, and others. The terpenoids comprise mono-, sesqui-, and diterpenes (mainly hydrocarbons and derived alcohols). Most are mono- or bicyclic. The second group are fatty acid derivatives that are acyclic hydrocarbons, alcohols, keto compounds, and esters of short- and medium-chain length. The third group includes partly identified constituents that are either aromatic or contain nitrogen. It is these volatile chemicals that attract the pollinators (Bergström, 1978).

It is quite evident that the diversity of the biochemical composition of various types of plants has a great part to play in their survival and in keeping the predators from completely eliminating them.

Thus, we see that there is a great deal of chemical biodiversity in the natural world. Some species produce toxins to limit the predation of predators, whereas other species detoxify or destroy the toxicant so that the species can live in a given area. It is this great chemical diversity of the living world—in the production of toxicants that control predation, pheromones that stimulate reproduction, and the ability of some species to destroy chemicals that are toxic to other species—that restores the availability of the habitat for the functioning of the diverse natural fauna and flora.

Earth's climate is continually changing, and it is the diversity of species that allows the environment to be utilized throughout most of the year in temperate climates and continually in the tropics. Thus, energy and nutrients are transferred throughout the food web. The processes of decay and regeneration of chemicals that are a source of food for the organisms are active all year and are basic to the operation of terrestrial as well as aquatic ecosystems. Although most of the macroscopic species in temperate and cold areas seem to have very low metabolic rates during the cooler months, other small organisms and microorganisms that live in the soil may be very active. For example, there are certain bacteria that release nitrogen as N_2 to the atmosphere and thus prevent harmful concentrations of nitrates and ammonia from accumulating in the soil. In contrast, there are other bacteria and plants that are able to fix molecular nitrogen, or nitrogen in reduced forms, and convert it into a form that can be utilized easily by legumes. This ability to fix nitrogen and thus provide nutrients in the soil make the legumes and their associated bacteria one of the most valuable crops for soil enrichment. These bacteria would become too numerous if it were not for the protozoans, nematodes, small worms, and other micro- and macroscopic organisms in the soil that feed on them and control their populations.

From these examples, it is easy to understand that terrestrial ecosystems are dependent on a high diversity of macro- and microscopic organisms if the functioning of the ecosystem is to be efficient, so that the least amount of entropy accumulates in the system. Realizing the important function that these microorganisms play, considerable research is now being done, particularly with bacteria, to see if the genes involved in bacteria that fix nitrogen can be transferred to other crops so that agriculture would be less dependent on chemical fertilizers.

In the aquatic world, biodiversity is very important in maintaining the purity of the water for multiple uses by organisms as well as by man. The importance of biodiversity in streams and rivers was recognized first by my studies in the Conestoga River Basin in 1948 and the resulting paper (Patrick, 1949), in which I indicated that a large number of species with relatively small populations characterize natural streams that are unaffected by pollution. Furthermore, these species represented many different groups of organisms belonging to a great many different phyla. The importance of this diversity in the functioning of the ecosystem was emphasized by this and later studies.

These studies clearly indicated that not only were the numbers of species characteristic of natural ecosystems fairly similar in streams that were chemically and physically quite different, but the percentage of the total number of species in each stream that performed each of various functions—primary producers, detritivores, detritivore-herbivores, omnivores, and carnivores—were quite similar (Tables 3-1, 3-2a, 3-2b, and 3-2c). High diversity ensures the con-

TABLE 3-1 Numbers of Species

	Guadalupe Low Flow 1973		Potomac Low Flow 1968		Savannah Low Flow 1968	
	No.	%	No.	%	No.	%
Algae	53	29.1	87	38.3	44	25.3
Protozoa	66	36.3	49	21.6	40	23.1
Macro-invertebrates	17	9.3	18	8.0	21	12.1
Insects	29	16.0	41	18.1	41	24.0
Fish	17	9.3	32	14.0	27	15.5
Total	182		227		173	

tinuation of the functioning of the ecosystem, which typically consists of four stages of nutrient and energy transfer (Figure 3-1).

The bases of the food web are typically detritivores and primary producers—organisms that can fix the sunlight and CO_2 into carbohydrates. The detritus is made available by the action of a great many species (bacteria, fungi, protozoans, and insects). Typically the bacteria and fungi break down the more complex compounds present in the detritus into simpler forms which then are utilized by other organisms. The main function of the protozoans and invertebrates is to feed on the bacteria and the fungi and keep their populations in check so that overpopulation does not occur. It is well known that bacteria may produce autotoxins and thus greatly reduce the efficiency of their own populations if they become too numerous.

The main function of the insects, mainly stoneflies and cranefly larvae, is to break down detritus into various sized particles, so that it is more easily handled by other organisms in the ecosystem. The interaction of these organisms in the ecosystem is very important because it ensures that species do not overpopulate. For example, the bacterial population will become too large if protozoans are not present. Such overpopulation causes the system to function inefficiently and entropy increases. Of course, the physical condition of the water—its rate of flow and chemical constituents—affect these processes.

We do not understand all of the physical characteristics of the environment that are important in maintaining the species diversity of these ecosystems. As pointed out by Patrick and Hendrickson (1993), even though we tried to have diatom communities develop in conditions that were as similar as possible, there was a certain amount of variability that was unaccounted.

As in the terrestrial system, in riverine systems the transfer of energy and nutrients through the ecosystem and the destruction of wastes continues throughout the year at all seasons. An example is White Clay Creek in Delaware County, Pennsylvania, which has been studied extensively by scientists at the Stroud Water Research Laboratory, including myself.

TABLE 3-2a Guadalupe River—Food Types of Major Groups of Species

Guadalupe River—1973 Total Number of Species: 179	Algae	Protozoa	Macro-invertebrates	Insects	Fishes	Total Feeding Type	% of Distribution
Autotrophs	53	15				68	38.0
Auto- and Heterotroph (Microphagocytes)							
Detritus (Microphagocytes)		19	3			22	12.3
Detritus and Algae		32	4	7		43	24.0
Algae (Herbivore)				3	1	4	2.2
Detritus, algae, zooplankton, macroinvertebrates, fish (Omnivore)			7	4	8	19	10.6
Zooplankton, macroinvertebrates, fish (Carnivore)			3	12	8	23	12.9
Total Species in Each Category	53	66	17	26	17	179	100.0

TABLE 3-2b Potomac River—Food Types of Major Groups of Species

Potomac River—1968

Total Number of Species: 227	Algae	Protozoa	Macro-invertebrates	Insects	Fishes	Total Feeding Type	% of Distribution
Autotrophs	86					100	44.3
Auto- and Heterotroph (Microphagocytes)		14					
Detritus (Microphagocytes)		20	4			24	10.6
Detritus and Algae		15	9	3	3	30	13.3
Algae (Herbivore)				5	2	7	3.1
Detritus, algae, zooplankton, macroinvertebrates, fish (Omnivore)			4	20	12	36	15.9
Zooplankton, macroinvertebrates, fish (Carnivore)			1	13	15	29	12.8
Total Species in Each Category	86	49	18	41	32	226	100.1

TABLE 3-2c Savannah River—Food Types of Major Groups of Species

Savannah River—1968 Total Number of Species: 173	Algae	Protozoa	Macro-invertebrates	Insects	Fishes	Total Feeding Type	% of Distribution
Autotrophs	44					56	32.4
Auto- and Heterotroph (Microphagocytes)		1				1	0.6
Detritus (Microphagocytes)		12	3			15	8.7
Detritus and Algae		14	7	3		24	13.9
Algae (Herbivore)				10	1	11	6.3
Detritus, algae, zooplankton, macroinvertebrates, fish (Omnivore)		1	7	16	8	32	18.5
Zooplankton, macroinvertebrates, fish (Carnivore)			4	12	18	34	19.6
Total Species in Each Category	44	40	21	41	27	173	100.1

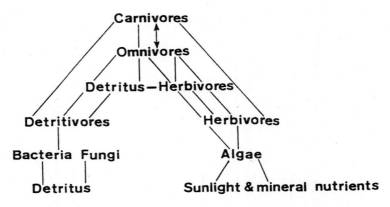

FIGURE 3-1   Pathways of nutrient and energy transfer (from Patrick, 1984).

Contrary to what many people have thought in the past, this stream is energetically active all year long. For example, Sweeney and Vannote (1981) found that there were roughly 50 species of mayflies living in the stream throughout the year. However, since they undergo developmental stages at different times of the year, some species of mayflies were rapidly growing, entering diapause, and emerging at different times throughout the year.

Figure 3-2 shows examples of the growth pattern of six mayflies. From this figure, it is evident that *Ephemerella subveria* had its most active period of growth from August through March, whereas *E. dorothea* had its most rapid growth from March through May. We find that *E. funeralis* had a fairly rapid period of growth in November and December and then again in April, whereas *E. verisimilis* was growing rapidly from April through May. *Ephemerella deficiens* had a fairly active period of growth in October and again in April and May and part of June, and *E. serrata* grew most actively in May, June, and July. Thus, throughout the year, one or another of the mayflies were grazing actively and converting food into energy and nutrients. Some of the nutrients, of course, were passed on to predators such as fish. The food of these mayflies is mainly detritus and algae.

Diatoms have a similar pattern of life history. There are a great many species of diatoms in White Clay Creek. Some of these species, such as *Cocconeis placentula* and *Surirella ovala*, reach their greatest densities during the spring; some form the largest populations during the summer and early fall months, such as *Melosira variens*; and others show the greatest population growth in the short day lengths of late fall and early winter, such as *Nitzschia linearis* and *Gomphonema olivaceum*.

Thus, we see that diatoms, which are the dominant source of food for these mayflies, change throughout the year, just as do the mayflies. It is these variabilities in population size that assure that the energy and nutrient transfer in the system will continue all year, and it is accomplished by different species. During each season of the year, there is a relatively high diversity of species present. This illustrates the importance of biodiversity in assuring a continued cycling of nutrients and energy.

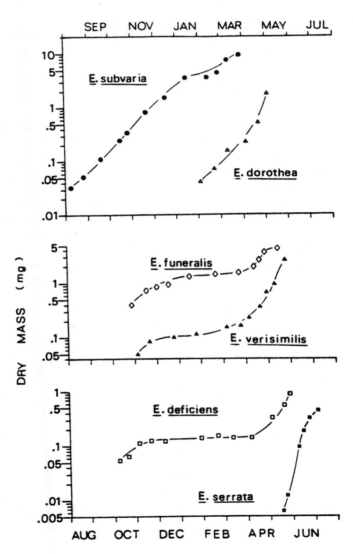

FIGURE 3-2 Average larval growth of six species of *Ephemerella* mayflies in White Clay Creek (from Sweeney and Vannote, 1981).

Furthermore, there seem to be similar patterns of species diversity in streams that are naturally very different. For this study, we selected rivers that, from a chemical and physical standpoint, had very different characteristics of the water. They were also in different geographic areas. The rivers were (1) the Guadalupe River in Texas (which is classed as a hard-water river because the calcium carbonate hardness is between 150-200 parts per million [ppm]); (2) the Potomac River, located between Maryland and Virginia (classified as a medium-hard river, with calcium carbonate at 60 to about 110 ppm); and (3) the Savannah River, located in the southeastern part of the United States (classified as a soft-water river, with a calcium hardness of less than 60 ppm).

It is interesting to note that the percentage of species for each of the major groups of organisms of the total number of species composing the fauna and flora are quite similar in these very different rivers (Tables 3-2a,b,c). Furthermore, the various groups of organisms that compose the major steps in the transfer of energy and nutrients through the system are similar (Tables 3-3a,b,c). Thus, the total number of species may vary in each of these respective rivers, but the percentage of the total number that is operative at a given functional level is similar. The kinds of species, as one would expect, vary greatly because of the many chemical and physical characteristics of the systems. From these aquatic studies, it is evident that there is not only a large number of species carrying out the functions of nutrient and energy transfer throughout the year, but a definite pattern is present.

In conclusion, it is evident that the many species composing the ecosystems of our planet have developed many unique chemical, physical, and structural characteristics and that they utilize many diverse strategies that ensure the functioning of the ecosystems of our planet.

REFERENCES

Bergström, G. 1978. Role of volatile chemicals in *Ophrys*-pollinator interactions. Pp 207-231 in J. B. Harborne, ed., Biochemical Aspects of Plant and Animal Coevolution. Academic Press, N.Y.

Janzen, D. H. 1978. The ecology and evolutionary biology of seed chemistry as related to seed predation. Pp. 163-206 in J. B. Harborne, ed., Biochemical Aspects of Plant and Animal Coevolution. Academic Press, N.Y.

McArthur, R. 1965. Patterns of species diversity. Biol. Rev. 40: 510-533.

Patrick, R. 1949. A proposed biological measure of stream conditions based on a survey of Conestoga Basin, Lancaster County, Pennsylvania. Proc. Acad. Nat. Sci. Phil. 101:277-341.

Patrick, R., and J. Hendrickson. 1993. Factors to consider in interpreting diatom change. Nova Hedwiggia 106:361-377.

Sweeney, B. W., and R. L. Vannote. 1981. *Ephemerella* mayflies of White Clay Creek: Bioenergetic and ecological relationships among coexisting species. Ecology 62(5):1353-1369.

PART

II

PATTERNS OF THE BIOSPHERE:
HOW MUCH BIODIVERSITY IS THERE?

The view from afar:
Planet Earth's changing mosaic.

4

Biodiversity at Its Utmost: Tropical Forest Beetles

TERRY L. ERWIN

Curator, Department of Entomology, National Museum of Natural History, Smithsonian Institution, Washington, D.C.

Life on Earth takes many forms and comes in all sizes, from microscopic one-celled plants to blue whales and human beings. Together these organisms and their interactions constitute our planet's biodiversity. Among this profusion of life are the beetles and their insect and arachnomorph relatives, which, taken together, constitute most of Earth's biodiversity (Erwin, 1982; Hammond, 1992; Robinson, 1986; Wilson, 1992). There are 1.4 million species of insects described in the scientific literature (Hammond, 1992), which is about 80% of all life currently recorded on Earth. Taxonomists, those who name and classify species, have been describing species of insects at about 4,400 per year for more than 235 years, and in the last 25 years, have described about 8,680 per year (±363). This written record is at best perhaps only 3.4% of the species actually living on the planet (Erwin, 1983a). Recent estimates of insect species, mostly in tropical forests, indicate that the descriptive process is woefully behind. These estimates indicate there may be as many as 30–50 million species of insects (Erwin, 1982, 1983b), making this pervasive terrestrial arthropod group 97% of global biodiversity. The familiar ants and grasshoppers, bees and beetles, houseflies and cockroaches, and spiders are but the tip of the iceberg of arthropod diversity; most species are small to very small tropical forest-dwelling forms that no one has seen or described on any adequate scale.

Insects and their relatives (spiders, ticks, centipedes, etc.) are the most dominant and important group of terrestrial organisms, besides humans, that affect life on Earth, often with an impact on human life. They affect human life in a multitude of ways—both for good and bad. Profound ignorance about insect life permeates most of human society, even among the highly educated. Insects

Weevils are a very diverse group of rainforest beetles.

and their relatives, in fact, are little credited for their beneficial environmental services and overblamed for their destructive activities. Despite lack of general human interest in insects, E. O. Wilson (1987:1) wrote that they are "the little things that run the world."

Insects and their relatives live on all continents and occupy microhabitats from deep in the soil and underground aquifers to the tops of trees and mountains, among the feathers of penguins on Antarctica, and even deep into caves and in our eyebrows. Many lineages have evolved adaptations for living on and under ice fields, others at the margins of hot springs, and still others on the open ocean. Land arthropods, by virtue of their pervasiveness, are incredibly important to the balance of life within ecosystems, e.g., pollination, nutrient recycling, and population control through vectoring diseases. Insects and their relatives eat virtually everything and compete even for the rocks under which they hide, mate, and rear their young. What would happen if all insects were removed from a habitat or natural community overnight? For one thing, most broadleaf trees and shrubs would not be pollinated, and there would be no fruits and seeds. For another, instead of penetrating dead matter, decomposers such as bacteria and fungi would live only on the surface, taking years or perhaps millennia to break the substrate down into recyclable nutrients for plants, and thus soils would be much less fertile. Many fish and birds, and even some mammals, would have no food and would cease to exist. In fact, insects seem to be one of nature's most important cornerstones on which most other types of life depend in one way or another.

Among the insects, the beetles are the most speciose, the most pervasive, and the most widespread across the face of the globe. During dry seasons in tropical forests, they are also the third most numerous individuals, after ants and termites, making up a full 12% of the total insect community (Erwin, 1989).

Beetles are found everywhere on our planet except in the deep sea. However, they do occur commonly in the sea's intertidal zone and estuarine salt flats (Erwin and Kavanaugh, 1980; Kavanaugh and Erwin, 1992; Lindroth, 1980). Beetles even occurred on Antarctica not long ago (Ashworth, personal communication, 1994). Most families of beetles, about 140 of them, are world-wide in distribution, and their species provide equivalent ecological services wherever they occur. The "play" is generally the same everywhere, only the "actors" themselves change from place to place.

We know of beetles from the Permian Period to the present (Arnol'di et al., 1992), a recorded history of some 250 million years. This history shows that two major faunal changes took place, the first in the mid-Jurassic Period when primitive lineages of beetles lost their dominance, and the second in the mid-Cretaceous Period, at which time modern forms acquired dominance over all other terrestrial arthropods. In terms of species and number of guilds (groups of species that fill similar ecological roles), they still have this dominance in nearly every biotope. By any broad measure, beetles are the most successful lineage of complex organisms ever to have evolved.

The *described* species of beetles, about 400,000+ (Hammond, 1992), comprise about 25% of all described species on Earth. This dominance of beetle taxa (any systematic category, such as species, genus, family, etc.) in the literature has resulted in Coleoptera being perceived as Earth's most speciose taxon. Thus, it has garnered further taxonomic attention from young taxonomists which in turn has resulted in more species of beetles being described than in other groups. Beetles are relatively easy to collect, prepare, and describe, significantly adding to their popularity. Such unevenness in taxonomic effort may or may not give us a false picture of true relative insect diversities. Nevertheless, the dominance of beetles has been used to arrive at an estimate of 30 million insects overall (Erwin, 1982), and even to designate the group most endeared to God (Gould, 1993). While this dominance may be arguable either scientifically or philosophically, it is certainly interesting. However, it does not address the real power that a knowledge of this extraordinary taxon might allow in evolutionary biology and conservation. What is neglected in the science of "coleopterology" is nearly everything except collecting, taxonomy, systematics, and a little autecology. Given that nearly everyone from naturalists, including Darwin and Bates to Edgar Allen Poe has or had "an inordinate fondness" (see Gould, 1993) for beetles, it seems strange that more attention is not given to them for use in interpreting environmental perturbations (Ashworth et al., 1991; Ashworth and Hoganson, 1993; Halffter and Favila, 1993), in understanding the rules (or nonrules) of assembly in tropical communities and biotopes (Erwin, 1985), and in environmental monitoring (Kremen, 1992; Kremen et al., 1993).

The reasons probably lie in the overwhelming numbers of species, individuals, and the ever-plodding course of traditional taxonomy. Potential users of data on beetles simply have to wait too long to get names; taxonomists have to wait too long to receive money to visit museums in which name-bearing type specimens are held; monographers take too long to produce documents with which users might identify their specimens by themselves; and specialists are reluctant to take on a large identification load for other scientists, such as ecologists and conservation biologists.

Given that millions of data points can be gathered in a very short time by sampling beetles (Table 4-1), far more than in any other group of diverse organisms (Adis et al., 1984; Allison et al., 1993; Basset, 1990, 1991; Erwin, 1982,

TABLE 4-1 Species Level Studies of Tropical/Subtropical Canopy/Subcanopy Beetles Using Insecticidal Fogging Techniques

	Est. vol. of foliage (m³)	No. of species	No. of specimens	Familes	Density	Species/m³/m	Specimen/species ratio	% Singletons
Allison/Miller (New Guinea)	2150	633	4840	54	2.25	0.29	7.65	50.7
Basset (Australia)[a]	4040	68	863	48	4.68	0.02	12.69	est. 19
Erwin (Panama)	1065	1250	8500	60	7.99	1.17	6.8	?
Erwin (Peru)[b]	2283	3429	15869	83	6.95	1.5	4.63	50.4
Stork (Brunei)	2690	859	4000	61	0.42	0.32	4.66	?
Stork (Sulawesi)	56550	1176	9158	?	0.16	0.02	7.79	?

[a]Restricted canopy fogging method.
[b]Includes six specific microhabitats, while others are predominately canopy rims with perhaps epiphytic growth.

1983a,b, 1988, 1989, 1991; Erwin and Scott, 1981; Farrell and Erwin, 1988; Kitching et al., 1993; Stork, 1991), how might we digest those data, turn them around to discern patterns that, once recognized and interpreted, can give us powers of prediction about the environment. With such an understanding, we could discern rich sites from slightly less rich sites for conservation (Rapid Assessment Program Team approach), or monitor life (environmental health) at those sites at a much finer resolution than is possible with vertebrates; and we could test much ecological theory also on a fine scale.

Neotropical beetles are second only to ants and flies (the latter in the wet season only) in numbers of free-ranging individuals of arthropods in the canopies and subcanopies of neotropical trees (termites are not usually free-ranging); Psocoptera are a distant third (Erwin, 1989). However, per species, beetles are not abundant (Figure 4–1). Beetles participate in virtually all aspects of ecosystem processes; they are predators, herbivores, folivores, detritivores, scavengers, fungivores, wood-eaters, and grazers, and they tunnel, mine, and chew nearly every substrate. Some are ectoparasites, others are nest parasites, some even live in the fur of vertebrates. Still others are subsocial, with adults participating in the raising of young. Knowledge of beetles, because they are *the* hyperdiverse group on the planet, offers direct insights into total biodiversity and the evolution of that biodiversity, as well as how this diversity is distributed in time and space across microenvironments, habitats, biomes, and seasons. A global perspective based on beetles could provide a much more fine-grained view of biodiversity than the coarse-grained one we get from less speciose groups such as jaguars, birds, and monkeys, which heretofore have garnered most of the attention.

The publication resulting from the National Forum on BioDiversity (Wilson

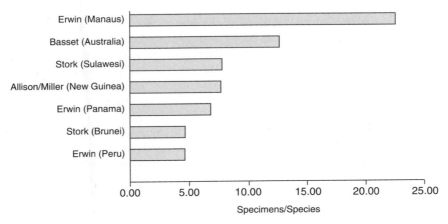

FIGURE 4–1 Relative abundance of species (large beetle-sampling programs).

and Peter, 1988), held in Washington, D.C. in 1986, included only 7 out of 521 pages devoted to insects (and only one speaker at the forum). At the most recent Biodiversity Forum (the Inaugural Symposium of the Consortium of Systematics and Biodiversity which formed the basis for this volume), there were six speakers on insects and three others whose contributions were at least partially based on insects (23%), a substantial realization in a mere 8 years within the biological community that biodiversity and the environment are insect dominated! If we are to understand the environment, which we must do if we are to successfully manage it, then we must have a better picture of the processes that brought about and maintained insect dominance since the Mesozoic Era.

Whether or not there are 30 million species (and, of that, 7.5 million species of beetles) or only a little more than the 1.4 million species that are already described, current human activity and that of the immediate future will exterminate a large percentage of these species (Erwin, 1988; Wilson, 1988). Attention must focus on the underlying evolutionary processes that have resulted in such diversity and evaluate these in terms of present human activities.

COLLECTION OF DATA

Because the interface between insects and their environment is at a small resolution, information they provide may well be critical for ecological restoration. Management will depend on what we really know rather than what we surmise. Conservation cannot now deal with insect information, but will be compelled to do so in the not-too-distant future. We will need a system for data gathering that is just now becoming available.

In Chapter 27 of this volume, Daniel Janzen describes his concept of an All Taxa Biodiversity Inventory (ATBI) for a 110,000 hectare site in Costa Rica. Such an undertaking, even in such a small area, will require methods other than those now employed for inventory, because the beetles alone are so pervasive and speciose anywhere in the tropics (along with all the other insects and their relatives) that completing an inventory would require generations of investigators.

One goal of Janzen's ATBI is to inventory *all* the taxa within a given area. A biotic inventory includes finding the area's species, classifying them, making voucher collections, and storing these data in a way that they are easily retrievable. Additional information about the species, either gathered during the process of inventorying or added later from literature or follow-up studies, can be piled on top of the four basic elements in a growing database.

The first ATBI area is destined to be at the Guanacaste Conservation Area (GCA), Costa Rica, a site with dry forest in lower elevations ascending through cloud forest and containing intermediaries between these levels. Based on my experience in (and data from) nearby Panama with a similar range of habitats, I estimate that GCA should have about 50,000 species of beetles. Since this estimate can be only a first approximation (but certainly within an order of magnitude), it is used for purposes of designing a sampling regime for the project; budget and time must be considered to be modifiable as the project narrows to better estimates.

Given GCA's latitude and altitudinal gradient, there are a minimum of 24 distinctive communities (forested and open habitats), each forest with a set of 15 or so microhabitats and each open area with 5 or so microhabitats (Erwin, 1991), all of which may contain different beetle faunules with perhaps as little as 20% species overlap, as was observed in my studies of 6 forest microhabitats at Pakitza, Peru. Each species of tree, shrub, and herb/grass may have its own host-specific species of beetles. Riparian strands in various watersheds will have different types of substrates, water quality, vegetation, etc., contributing to their distinctive biodiversities. In addition, the GCA is distinctly seasonal, hence both dry and wet seasons need to be sampled for each microhabitat (Erwin and Scott, 1981).

The sampling regime must consider the above in its attempt to record as many species as possible in the shortest amount of time. The guiding principles are as follows:

• Phase 1: mass co-occurrence sampling; rapid processing with bulk cold storage (dry and wet specimens, depending on Order); identification process using matching specimens; interim naming with alphanumerics; accumulation of data using linked spreadsheets, including curves showing sampling progress; and character filing with the Quick Taxonomic Assessment System (QTES).

• Phase 2: send target taxa and QTES data into the taxasphere (formal systematic literature) for formal species names;

• Phase 3: replace EXCEL 4.0 spreadsheet and QTES interim names with formal ones, transfer these data to the database at Instituto Nacional de Biodiversidad (INBio).

• Phase 4: generate illustrations and three-dimensional laser images; produce documents (lists, brochures, field guides, revisions, monographs, other analyses).

AN AGENDA FOR SAMPLING BEETLES IN AN ATBI

Sampling

The following criteria must be met for acquiring samples of beetles that can provide a reasonable inventory and serve both immediate and future needs of research, as well as determine to an order of magnitude the species present in the target area for use in subsequent sampling projects:

(1) Sampling assumes the use of a fogger and 3% Resmethrin (biodegradable with an LD50[1] better than aspirin, gone in 2 hours) for all microhabitats

---

[1] Dosage at which 50% of the organisms fail to survive.

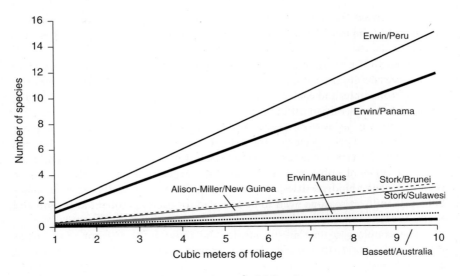

FIGURE 4–2 Accumulation of species (per m³ of foliage).

from 1 m above ground through the canopy rim (Erwin, 1982, 1983b, 1989, 1991); this should capture 1.7 species per m³ of foliage (see Figure 4–2) and much more in compacted microhabitats such as suspended dry leaves, vine tangles, and complex canopies. Leaf litter and soil layers are sampled by photoeclectors (Adis, 1984; Adis and Schubart, 1984) and sifting/Tolgren extractor techniques. Berlese banks can substitute for Tolgren, if electricity is available. The stratum of herbs and grasses is sweep-sampled by sweep-netting.

Methods of trapping by attraction and even passive traps that catch flying insects produce catch without biocontext, i.e., specimens that are not tied to any microhabitat, substrate, host plant, etc. Much time and effort goes into preparing, identifying, and storing such bulk lots, yet the quality of data is at the lowest level. These methods of collecting simply are not worth the effort, unless one is interested solely in recording presence of species in the general area or in building collections. However, these techniques can be used as a test of the methods that incorporate biocontext to determine if microhabitats exist that are not being sampled with the other techniques.

(2) A standard set of field data includes precise locality (latitude/longitude to seconds, and notes on permanent trail markers and topographic features if available; a Global Positioning System [GPS] device provides data on position and elevation); type of forest; type of microhabitat and its volume or surface area; information on species of plants (or other host); date; and collector(s). Lot numbers are assigned to each individual fogging collection, sifting series, photo-eclector sample, sweep series, etc. Thus, all specimens taken from the same

microhabitat or plant or trap at the same time get the same lot number so that the set of species, including nonbeetles, can be reassembled at a later date if desired (this faunule reassembly may be only a computer construct). Nonbeetle specimens will be directed to another Taxonomic Working Group (TWIG), along with appropriate sets of data.

(3) Specimens of beetles are preserved in 70% alcohol in the field. Alcohol must be changed the same day at the lab and subsequently each time the specimens undergo processing (see below). If specialists for nonbeetle groups are available at the time of fogging, dry specimens may be extracted by hand before the general sample goes into alcohol, so long as this does not delay the routine of the inventory for beetles and appropriate lot numbers are defined. Egg carton inserts are placed in funnels or on suspended sheets to catch specimens dry. After these are selected from the carton surface, the remaining specimens are dumped into the alcohol bottle.

(4) Sampling design involves taking replicate microhabitat samples in sets of 10 throughout each type of forest and open area. During preparation and data entry of the 10 samples, species accumulation curves and Chao's estimator (Colwell and Coddington, 1994) track the progress of the inventory. A complete inventory for smaller families will require fewer replicates, but the leaf-beetles and weevils will require many more than 10 replicates, based on data from over 5,000 species acquired at Pakitza, Peru, from 1988-1992. A decision needs to be made at the outset as to when to stop, because it will not be "humanly" possible, given today's resources, to get the "last" species on the list in the larger families. However, 100% likely will be reached in smaller families.

Preparation

All tropical forest samples are replete with beetles. The object of preparation should be to make the species and their whereabouts and abundance known in the shortest amount of time possible. Traditional preparation of all collected specimens, therefore, is not feasible. The following method leads to one prepared specimen per species per sample, with cold storage of the bulk lots (other specimens of the species) that easily can be accessed later by taxonomists and other workers who need series.

(1) Each sample lot is sorted to families using a 6.2 cm white ceramic dish with 70% alcohol. Families are gathered in small plastic lids set inside the bottom of a petri dish, the top of which is ringed with vaseline to create a seal when the specimens are sitting unworked. Parataxonomists and beginning graduate students can be trained to sort at this level quickly.

(2) Each family then is sorted in sequence to species, with one good specimen selected for pinning/pointing. The specimen is placed on damp filter paper inside another petri dish. On the filter paper, numbers 1 to 20 are written and the chosen specimen is placed next to a number according to its abundance in

A collection of rainforest insects.

the sample. If over 20 specimens of a species are in a sample (which is rare), a small label is written with the number, and this label is affixed to the pin to be removed later in the process (see below). The rest of the counted but unprepared specimens are returned to the lot vial of 70% alcohol, bar-coded, and sent to cold storage. Sorting to species across the Coleoptera can be done only by a highly trained taxonomist with long experience, and this person becomes the key to the entire project. Preparation and storage procedures can be handled by a technically trained person.

(3) Each specimen from the filter paper is pinned or pointed with Elmer's glue to pins or preprepared points in the traditional manner and placed in a unit tray with strips of numbers sequenced from 1 to 20. Each specimen is aligned next to one of these numbers according to its abundance in the sample.

(4) Preprinted labels are attached to each specimen as it is "identified-by-matching" using the synoptic collection. The name of the species is an alphanumeric in the form of "family coden + number" and lot number. All families of Coleoptera have a standard coden of four letters. The margin of the label is color-coded with pencil for instant recognition of microhabitat, although the lot number references this too. Once a family is represented by more than 100 species, "identifying-by-matching" becomes less and less efficient. For very large families, such as weevils, staphylinids, and chrysomelids, use of QTES is recommended.

Interim Identification

Each prepared specimen from a sample is compared with its corresponding family-level synoptic collection. Smaller families are easy to keep in one or a few units. Larger families may be subdivided by subfamily so that the amount of matching necessary for recognizing the specimen's status is kept to a minimum. As a specimen is identified or recognized as a species new to the synoptic collection, it is placed either in an interim unit tray awaiting entry of the data before going to the duplicate collection (those identified previously) or added to the synoptic collection (those determined as new), from where its data are entered. All species of a family that are sorted from the sample are labeled, then the data are entered in EXCEL before preparing the next family.

Data Storage

My EXCEL linked spreadsheet templates for families of beetles contain about 13 Kilobytes of forms that are based on microhabitats. Entry of data from a sample involves simply number of specimens per species per lot. The program automatically computes all basic information and accumulates the data on summary sheets for easy viewing. The program is exceedingly user-friendly.

Building Collections

The resulting synoptic-unit trays of families of beetles are ready for specialists at any time during the process if the specialist is on-site to make formal identifications. The duplicate collection—built from second through n occurrences of a species across samples (hence, it will not contain "uniques" [species known from single specimens] found only in the synoptic trays)—can be sent through the taxasphere regularly and results can be fed back into the EXCEL data system, making it easy to move the information to the INBio standard data files. As additional microhabitat replicates are sampled and specimens processed, those species represented by uniques in the synoptic collection will be duplicated and then can be sent through the taxasphere.

Serious taxonomists who must do a revisionary study immediately can read the database to find lots with series and arrange to extract those from cold storage themselves. Common species that are found in many or most lots will have that many more prepared specimens ready for study in the duplicate collection.

SUMMARY

The rate at which *all* the foregoing can be done is 58 specimens and 13 species per hour. Therefore, using the rate of accumulation for additional species found in Panama forest foliage, 1.7 per m^3 of microhabitat, we should be able to sample 70,000 m^3 of microhabitat and process 50,000 species of beetles in 2½ years. In other words, we know the actors and where they are standing on the stage, and each has a number hanging around its neck. The taxasphere is another creature, and getting formal names on the inventoried species is highly dependent on the group of beetles, its history of studies, and its current taxonomist(s).

The advantages of the TWIG protocol are that (1) it is far more rapid than any of its predecessors; (2) data byproducts allow diverse follow-up studies beyond the inventory process; (3) targeted taxa known to be important to users can be piped readily (and continuously) through the taxasphere; (4) space and storage facilities are minimized because samples mostly are stored cold in two-dram shell vials or petri dishes until needed by a dedicated specialist; and (5) dedicated specialists will "donate" their time to the collections as they select and prepare specimens from cold storage, hence building collections becomes a shared taxaspheric process.

Beyond the inventory itself, such questions as "do beetles form discreet assemblages in tropical forests, or in any biotope anywhere?" can be tested. If so, how that information might be used for answering scientific questions and for developing conservation strategies is of considerable interest. The objective of this kind of study would be to fill a large gap in our understanding of hyperdiversity. For example, (1) what percentage do beetles contribute to a sample? (2) What is the fidelity of beetle faunules to microhabitats? (3) What is the rate of species turnover across extensive geographic space in the tropics? (4) What is the rate of local species replacement among and between tropical microhabitats? (5) What proportion of the total beetle fauna inhabits arboreal versus forest floor habitats? (6) What is the rate of change in composition of faunules with respect to altitude?

This information does not now exist on any meaningful scale for any hyperdiverse group of organisms. Without this information, it is impossible to scale any kind of locally derived estimate of biodiversity to even a regional perspective. With this information, I believe we can get much closer to estimating the magnitude of life on the planet. And with these kinds of data from three or four ATBIs, much finer estimates can be made elsewhere of actual amounts of biodiversity that are based on fewer samples and made with quicker inventories.

REFERENCES

Adis, J. 1984. Seasonal igapo forests of Central Amazonian blackwater rivers and their terrestrial arthropod fauna. Pp 245-268 in H. Sioli, ed., The Amazon Limnology and Landscape Ecology of a Mighty Tropical River and Its Basin. W. Junk, Dordrecht, Netherlands.
Adis, J., Y. D. Lubin, and G. G. Montgomery. 1984. Arthropods from the canopy of inundated and terra firme forests near Manaus, Brazil, with critical consideration of the Pyrethrum-fogging technique. Stud. Neotrop. Fauna Envir. 19:223-236.
Adis, J., and H. O. R. Schubart. 1984. Ecological research on arthropods in Central Amazonian forest

ecosystems with recommendations for study procedures. Pp 111-114 in J. H. Cooley and F. B. Golley, eds., Trends in Ecological Research for the 1980's. NATO Conference Series I: Ecology. Plenum Press, N.Y.

Allison, A., G. A. Samuelson, and S. E. Miller. 1993. Patterns of beetle species diversity in New Guinea rain forest as revealed by canopy fogging: Preliminary findings. Selbyana 14:16-20.

Arnol'di, L. V., V. V. Zherikhin, L. M. Nikritin, and A. G. Ponomarenko. 1992. Mesozoic Coleoptera. Oxonian Press, New Delhi, India. 284 pp.

Ashworth, A. C., and J. W. Hoganson. 1993. The magnitude and rapidity of the climate change marking the end of the Pleistocene in the mid-latitudes of South America. Palaeogeogr. Palaeoclimatol. Palaeoecol. 101:263-270.

Ashworth, A. C., V. Markgraf, and C. Villagran. 1991. Late Quaternary climatic history of the Chilean Channels based on fossil pollen and beetle analyses, with an analysis of the modern vegetation and pollen rain. J. Quat. Sci. 6(4):279-291.

Basset, Y. 1990. The arboreal fauna of the rainforest tree *Argyrodendron actinophyllum* as sampled with restricted fogging: Composition of the fauna. Entomologist 109:173-183.

Basset, Y. 1991. The taxonomic composition of the arthropod fauna associated with an Australian rainforest tree. Aust. J. Zool. 39:171-190.

Colwell, R., and J. Coddington. 1994. Estimating the extent of terrestrial biodiversity through extrapolation. In D. L. Hawksworth, ed., The Quantification and Estimation of Organismal Biodiversity. Phil. Trans. R. Soc. Lond. (B) 345(1311):101-118.

Erwin, T. L. 1982. Tropical forests: Their richness in Coleoptera and other Arthropod species. Coleopt. Bull. 36(1):74-75.

Erwin, T. L. 1983a. Tropical forest canopies, the last biotic frontier. Bull. Entomol. Soc. Amer. 29(1):14-19.

Erwin, T. L. 1983b. Beetles and other arthropods of the tropical forest canopies at Manaus, Brasil, sampled with insecticidal fogging techniques. Pp. 59-75 in S. L. Sutton, T. C. Whitmore, and A. C. Chadwick, eds., Tropical Rain Forests: Ecology and Management. Blackwell Scientific Publications, Oxford, England.

Erwin, T. L. 1985. The taxon pulse: A general pattern of lineage radiation and extinction among Carabid beetles. Pp. 437-472 in G. E. Ball, ed., Taxonomy, Phylogeny, and Zoogeography of Beetles and Ants: A Volume Dedicated to the Memory of Philip Jackson Darlington, Jr., 1904-1983. W. Junk, The Hague, Netherlands.

Erwin, T. L. 1988. The tropical forest canopy: The heart of biotic diversity. Pp. 123-129 in E.O. Wilson and F. M. Peter, eds., BioDiversity. National Academy Press, Washington, D.C.

Erwin, T. L. 1989. Canopy arthropod biodiversity: A chronology of sampling techniques and results. Revista Peruana Entomologia 32:71-77.

Erwin, T. L. 1991. Natural history of the Carabid Beetles at the BIOLAT Rio Manu Biological Station, Pakitza, Peru. Revista Peruana Entomologia 33:1-85.

Erwin, T. L., and D. H. Kavanaugh. 1980. On the Identity of *Bembidion puritanum* Hayward (Coleoptera:Carabidae: Bembidiini). Coleopt. Bull. 34(2):241-242.

Erwin, T. L., and J. C. Scott. 1981. Seasonal and size patterns, trophic structure, and richness of Coleoptera in the tropical arboreal ecosystem: The fauna of the tree *Luehea seemannii* Triana and Planch in the Canal Zone of Panama. Coleopt. Bull. 34(3):305-322.

Farrell, B. D., and T. L. Erwin. 1988. Leaf–Beetles (Chrysomelidae) of a forest canopy in Amazonian Peru: Synoptic list of taxa, seasonality and host-affiliations. Pp. 73-90 in P. Jolivet, E. Petitpierre, and T. Hsiao, eds., The Biology of the Chrysomelidae. W. Junk, The Hague, Netherlands.

Gould, S. J. 1993. A special fondness for beetles. Nat. Hist. 102(1):4,6,8,10,12.

Halffter, G., and M. E. Favila. 1993. The Scarabaeidiae (Insecta:Coleoptera): An animal group for analyzing, inventorying and monitoring biodiversity in tropical rainforest and modified landscapes. Biol. Int. 27:15-21.

Hammond, P. 1992. Species inventory. Pp 17-39 in B. Groombridge, ed., Global Biodiversity: Status of the Earth's Living Resources. Chapman and Hall, London.

Kavanaugh, D. H., and T. L. Erwin. 1992. Extinct or extant? A new species of intertidal bembidiine (Coleoptera: Carabidae: Bembidiini) from the Palos Verdes Peninsula, California. Coleopt. Bull. 46(3):311-320.

Kitching, R. L., J. M. Bergelson, M. D. Lowman, S. McIntyre, and G. Carruthers. 1993. The biodiversity of arthropods from Australian rain forest canopies: General introduction, methods, sites, and ordinal results. Aust. J. Ecol. 18:81-191.

Kremen, C. 1992. Assessing the indicator properties of species assemblages for natural areas monitoring. Ecol. Appl. 2:203-217.

Kremen, C., R. K. Colwell, T. L. Erwin, D. D. Murphy, R. F. Noss, and M. A. Sanjayan. 1993. Terrestrial arthropod assemblages: Their use in conservation planning. Conserv. Biol. 7(4):796-808.

Lindroth, C. H. 1980. A revisionary study of the taxon Cillenus Samouelle, 1819, and related forms (Coleoptera: Carabidae, Bembidiini). Entomol. Scand. 11:179-205.

Robinson, M. H. 1986. The fate of the tropics and the fate of man. Zoogoer 5:4-10.

Stork, N. E. 1991. The composition of the arthropod fauna of Bornean lowland rain forest trees. J. Trop. Ecol. 7:161-180.

Wilson, E. O. 1987. The little things that run the world. Conserv. Biol. 1(4):344-346.

Wilson, E. O. 1988. The current state of biological diversity. Pp. 3-18 in E. O. Wilson and F. M. Peter, eds., BioDiversity. National Academy Press, Washington, D.C.

Wilson, E. O., and F. M. Peter. 1988. BioDiversity. National Academy Press, Washington, D.C. 521 pp.

Wilson, E. O. 1992. The Diversity of Life. Belknap Press, Cambridge, Mass. 424 pp.

Measuring Global Biodiversity and Its Decline

NIGEL E. STORK

Director, Cooperative Research Centre for Tropical Rainforest Ecology and Management, James Cook University, Cairns, Queensland, Australia

In recent years biologists have come to recognize just how little we know about the organisms with which we share the planet Earth. In particular, attempts to determine how many species there are in total have been surprisingly fruitless. In this chapter, I examine this and a number of related issues. I first consider the progress, or rather the apparent lack of progress, that we have made in describing organisms. Second, I consider how much we know about the biology, distribution, and threatened status of those species that have been described. Third, I examine some of the different methods that have been used to determine the global number of species. Finally, I examine the likelihood of extinction of species.

My remarks focus in large part (but not exclusively) on terrestrial arthropods, particularly insects. This is not simply because this happens to be my own special interest group, but rather because the issues discussed above have been pursued with greatest vigor for this taxon. In addition, on the basis of present evidence, insects appear to be the most speciose taxon on Earth and the one that is threatened with the greatest number of extinctions. In investigating the magnitude of biodiversity, I focus at the level of species, again because this is my own special interest. Those who might have greater interest in the genetic, landscape, or ecosystem level of biodiversity would find that our understanding of the magnitude of biodiversity is somewhat different. I do not examine local species richness per se, except in the context of the measurement of global species richness. For this subject, there are a number of important recent texts (e.g., Colwell and Coddington, 1994; Hammond, 1994; Soberon and Llorente, 1993).

THE LEGACY OF LINNAEUS

Some 237 years ago, Linnaeus published his *Systema Naturae*, a new binomial system for the naming of organisms in a classification that linked their morphological relationships (Linnaeus, 1758). He set in train a flurry of descriptions of species by a new breed of scientists—taxonomists. Since that time, some 1.4 to 1.6 million species have been named and described (Stork, 1988; Hammond, 1992) (see Figure 5-1). The precise figure is uncertain for several reasons, the main one being that there is no recognised central register of names for described species (although such registers do exist for a few groups), and therefore some species have been described many times. For example, someone who may be describing a species from India may be unaware that the species already has been described from Pakistan. In other cases, the natural variation of a species is unknown, and different forms of the same species are given different names. The common European "ten-spot ladybird," *Adalia decempunctata* L., for example, has at least 40 different synonyms, many of these having been used for the color morphs! Gaston and Mound (1993) noted that, although only some 4,000 species of mammals currently are recognised (Corbet and Hill, 1980), the collections at the Natural History Museum in London contain "types" (see "holotype," below) for 9,000 names. They also suggest that, for insects, the

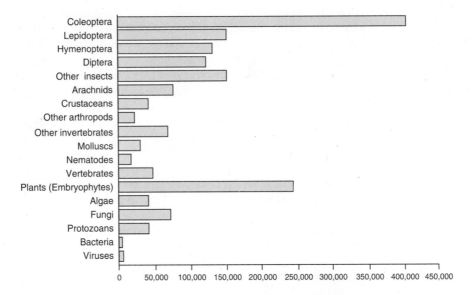

FIGURE 5-1 Number of species described for all organisms (after Hammond, 1992; Stork, 1993).

TABLE 5-1   Present Levels of Synonymy in Insects

| | Names | | | |
	Total Species	Currently Accepted	% Synonomy	Source
Odonata[a]	7,694	5,667	26	Bridges (1991)
Isoptera[b]	2,000	1,600	20	Snyder (1948)
Thysanoptera	6,479	5,062	22	L. A. Mound (unpublished data, 1993)
Homoptera				
Aleyrodidae	1,267	1,156	9	Mound and Halsey (1978)
Aphididae	5,900	3,825	35	Eastop and Hille Ris Lambers (1976)
Siphonaptera	2,692	2,516	7	D. J. Lewis (unpublished data)
Diptera				
Simuliidae	1,800	1,460	19	Crosskey (1987)
Lepidoptera				
Noctuidae	35,473	28,175	21	Poole (1989)
Papilionidae and Pieridae[a]	9,075	1,792	80	Bridges (1988a)
Lycaenidae and Riodinidae[a]	13,108	5,757	56	Bridges (1988c)
Hesperiidae[a]	8,445	3,589	58	Bridges (1988c)
Hymenoptera				
Chalcidoidea	22,533	18,601	18	Noyes (1990)

[a]Subspecific names are treated as synonyms.
[b]These figures are estimates, due to problems in data interpretation.

SOURCE: Gaston and Mound (1993).

level of synonymy is about 20% (Table 5-1). Therefore, although there may be some 1.8 to 2.0 million names used by taxonomists, in reality these probably represent some 1.4 to 1.6 million species.

Why is it that taxonomists apparently make so many "mistakes"? In part, this is because variation within species can make it difficult to assess whether a series of individuals represents one or more species and because often it is difficult to obtain representative samples of species from all parts of their range. However, a more practical problem for the high level of synonymies is that it can be very difficult for taxonomists to look at specimens of previously described species to check whether their specimens are of new species. When a species is described, one specimen usually is designated as the "holotype" (often called "type") for that species, and comparisons therefore need to be made against it. Probably more than 80% of all types are housed in the ancient collections of various European museums and herbaria—reflecting the former tradition of exploration by many of these countries during the nineteenth and early twentieth centuries. Someone working on the taxonomic revision of a group of beetles from West Africa, for example, might have to compare their specimens with types from many museums or herbaria in Europe or North America, and often this task is not feasible.

Not surprisingly, therefore, the current rate of description of about 15,000 new species a year (Stork, 1993; Hammond, 1992) is only about twice the average of 6,000 to 8,000 species described over the last 230 years (Figure 5-2). Even with the lowest estimates of global species diversity of 3 million, it would take 90-120 years to describe all species at this depressingly low rate. The solution to increasing the rate of description of the Earth's fauna and flora is not a simple one, and those who might prescribe setting up of a factory-like "conveyor belt" system of dedicated species describers should beware. Many entomologists are familiar with the name of Francis Walker, who worked at the British Museum of Natural History and described thousands of species of Coleoptera, Diptera, and Hymenoptera in this way. Many of his species have since been synonymized, involving an enormous amount of wasted time for later taxonomists. One obituary indicated that he died too late to save his reputation! Thus, the solution is not just to establish a cadre of species describers but rather to train many more taxonomists.

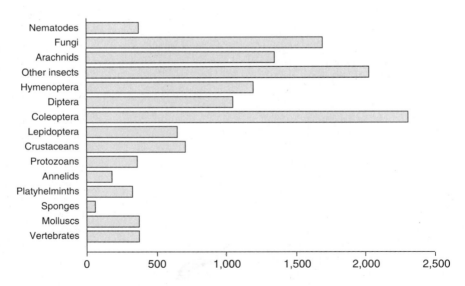

FIGURE 5-2 Average number of species described each year (from *Zoological Record*, 1978-1987) for selected animal groups (after Hammond, 1992).

WHAT DO WE KNOW OF SPECIES?

The question of "how many species are there?" has in many ways rather overshadowed other and perhaps more interesting and important questions. As Lawton (1993:4) wrote, "Intriguingly, I have never seen anybody discuss what we actually know about the 1.7 million (species) that do have names. Overwhelmingly the answer will be nothing, except where they were collected and what they look like." At modest (and, as I suggest later, probably quite reasonable) global estimates of around 5-15 million species, it is clear that we have named only about 10-30% of global species. And although many other unnamed species are housed in the accessions of museums and herbaria, these species or the information associated with them are effectively unavailable to the scientific community.

So what *do* we know of these species? For groups such as birds and large mammals, we have a very good understanding of their distribution, biology, and often their threatened or nonthreatened status. Even so, there are still massive gaps in our understanding of even the best-known groups, as Figure 5-3 shows. What then for other less well-known groups, such as plants, invertebrates, and fungi? Clearly, again there are some groups that have particular significance, whether for human health (e.g., mosquitos), agriculture (e.g., many crop plants), or other human interest (e.g., orchids), and the biology of these is sometimes much better known than for other related taxa. But for most groups, almost nothing is known of their distribution and biology.

Stork and Hine (unpublished, 1995) analyzed the distributions of beetles from revisions and descriptions extracted randomly from the 1987 *Zoological Record*. They found that 53% of the 186 species (whether newly described or previously described) included in the 46 taxonomic papers they examined were known only from a single locality. Furthermore, 13% were known from just single specimens. Beetles are about 25% of all described species, and if most other animals and plants are as little known as these, it would seem that more than half of described species are known from single localities. This will come as no surprise to those taxonomists working with some of the major collections in the world, since so many species are known just from type specimens.

Most estimates of present rates of extinction would suggest that many hundreds of thousands, if not millions, of species are threatened with extinction and that we are witnessing a mass extinction spasm as great or perhaps even greater than any in the geological record (May et al., 1995). This is discussed more fully at the end of the chapter. And yet only about 1,000 species are recorded as having become extinct in recent years (since 1600). If we know so little of the distribution and biology of most species, what then do we know of their "threatened" status?

Over the last 20 years the International Union for the Conservation of Nature has compiled Red Data Books for those species that are recognized as falling

FIGURE 5-3 Distribution of the localities known for 962 species of African passerine birds (from Hall and Moreau, 1970; Cotterill, personal communication, 1994). Birds are the best-known group for Africa, as for most parts of the world, and yet this map shows that there are still vast areas, such as in Angola, Mozambique, and some of the former French territories in West Africa for which we have almost no distributional data.

into one of their different threatened categories. Some of these Red Data Books, e.g., for groups such as primates and birds, are fairly comprehensive. For other groups, such as insects, only a few species are included as tokens of the probable but largely unknown threatened status of many other species. Some countries have produced their own Red Data Books, and those for the United Kingdom give an indication of just how high a proportion of species are threatened (see below).

World-wide, some 11% of mammals and birds are threatened with extinction. Equivalent values for British mammals and resident breeding birds are 29.1 and 55.7% respectively. In contrast, data on the threatened status of insects and other invertebrates are almost nonexistent. However, this does not mean that few are threatened, but rather that we have little information, since an average of about 13% of all British species of insects, spiders, and molluscs *are* threatened.

In summary, almost nothing is known of the distribution and threatened status of most organisms. A global estimate of, for example, 10 million species of all organisms would suggest that nothing is known of the distribution of 86% of species, 7% are known from just one locality, only 7% are known from more than one locality, and the threat of extinction is known for less then 0.5%.

HOW MANY SPECIES ARE THERE?

One of the central questions in the biodiversity debate has been "how many species are there?," for the answer has a bearing on other important and related issues such as loss of biodiversity and where on the Earth to focus attention for the conservation of biodiversity. As long ago as the 1800s, scientists were fascinated by the global number of species. Westwood (1883), for example, quoted another British entomologist, Ray, who suggested that there might be 20,000 species of insects in the world. We now know there are at least that many species in Britain alone! As more and more species have been described, the question of how many species there are has received less attention. Until the 1980s, most recent biology texts either ignored this issue or considered that the global total of species was only a little above the number of described species (i.e., 2-3 million species). Many thought that the total eventually would be found simply by describing all species. Since then, several methods of estimating the global number of species have been used, and these are considered below. Most of the methods used are based on simple or complex ecological or taxonomic patterns, and use empirical data as the basis for some kind of extrapolation or other.

Ratios of Known to Unknown Faunas

The simplest estimates of number of species are based on known ratios of groups of organisms. For example, Raven (1985) noted that, for mammals, birds, and other large and well-documented animals, there are roughly twice as many tropical as temperate species. If the same ratio is true for other organisms, he argued, then with 1.5 million species described and two-thirds of these being temperate, the global total would be 3 million. Although Gaston (1994) has shown that most species described in the late 1980s are from nontropical regions and from "nonmegadiverse" countries (Table 5-2), the distribution of most described species is not known. A survey of the countries represented by species in every one-hundredth drawer of the 11,500-drawer beetle collection at the Natural History Museum in London (Table 5-3) would suggest that for this collection, at least, there are more tropical than temperate species of described beetles (1,753 species sampled: 625 from temperate countries, 946 from tropical countries, 182 both tropical and temperate [e.g., Australia]). Since the estimated

TABLE 5-2   Analysis of the Geographical Distribution of the 24,000 Newly Described Species of Coleoptera (49% of species), Diptera (25%), and Hymenoptera (26%) That are Referenced in the Volumes of *Zoological Record* for the Years 1985-1989

Proportion of Species Descriptions from the Ten Countries from which the Most Species were Described in Each Case		Proportion of the Sum of Species Descriptions from the "Megadiversity" Countries		Proportion of the Total Number of Species Described over the 5-year Period from Different Faunal Regions	
Country	%	Country	%	Region	%
U.S.S.R.	8.7	China	6.2	Neartic[a]	8.3
China	6.2	Australia	5.8	Neotropical[b]	12.5
Australia	5.8	India	5.3	Palearctic[c]	19.2
India	5.3	Madagascar	3.6	Ethiopian[d]	12.1
United States	5.1	Brazil	2.8	Madagascar	3.7
Japan	4.6	Malaysia	2.0	Indian subcontinent	9.8
Papua New Guinea	4.1	Mexico	2.0	China, Japan, and Taiwan	13.7
Madagascar	3.6	Indonesia	1.7	Thailand to New Caledonia	12.7
South Africa	3.4	Peru	1.0	Australia	5.8
Brazil	2.8	Zaire	1.0	New Zealand	0.8
		Ecuador	0.7	Oceanic Islands	1.3
		Colombia	0.7		

[a]Includes the Arctic and temperate areas of North America and Greenland.
[b]Includes South America, the West Indies, Central America, and tropical Mexico.
[c]Includes Europe, Arabia, and Asia north of the Himalayas, but not China, Japan, and Taiwan.
[d]Includes the African subcontinent.

SOURCE: Gaston (1994).

175,300 species of beetles in this collection represent about half of all described species of beetles, Raven's suggestion that two-thirds of described species are tropical would seem to be correct. However, this does not take into account the fact that it is rare to find a new temperate species of insect, whereas most species in tropical samples are new to science.

Stork and Gaston (1990) used a slightly different approach to estimate the number of insects world-wide. They first noted that the ratio of the number of species of butterflies to all species of insects in the well-known British fauna is 67:22,000. They suggested that if this ratio were true world-wide, then (with an estimated 15,000-20,000 species of butterflies world-wide) this would indicate a total of 4.9 to 6.6 million species of insects.

These approaches are appealing in their simplicity, but how realistic are such extrapolations based on ratios for different groups or geographical areas? The evidence for this is conflicting. Geographical ranges for species are known to increase with increasing latitude and altitude (Stevens, 1990, 1992). How the relative numbers of species for different groups varies with latitude is not pursued here, but some evidence suggests that the relative proportions of different taxonomic groups or trophic groups, at least in terms of species, are the same in arthropod communities from temperate and tropical forest trees (see below and Stork, 1987, 1988, 1993). This is in spite of the fact that the number of species of arthropods in some tropical trees (up to 1,000 species per tree) is sometimes more than three to four times the number on temperate trees of equivalent size (in terms of canopy volume).

Hawksworth (1991) has used arguments based on the ratio of fungi to vascular plants in different regions of the world to predict that possibly there are more than 1.5 million species of fungi world-wide. He analyzed several well-known floras and recorded ratios of 1:1.4 to 1:6.0 for species of vascular plants to species of fungi and suggested that a higher ratio was more representative of floras world-wide. Given that there are an estimated 270,000 species of vascular plants world-wide, Hawksworth argued that this would give a conservative estimate of about 1.5 million species of fungi (including allowances for fungi in unstudied substrata). His data are for temperate floras, and the ratio of fungi to

TABLE 5-3 Number of Species of Beetles (and Predicted Total Number of Species of Beetles) for the Top 15 Countries Represented in the Natural History Museum in London

Number of Species Sampled	Predicted Number of Species in Collection	Country
143	14,300	*Australia*
142	14,200	South Africa
139	13,900	United States
134	13,900	*Brazil*
118	11,800	*India*
86	8,600	*Mexico*
65	6,500	Philippines
61	6,100	Japan
60	6,000	Malaysia
58	5,800	U.S.S.R.
49	4,900	*Indonesia*
44	4,400	*Colombia*
44	4,400	Guatemala
39	3,900	Papua New Guinea
37	3,700	*Madagascar*

NOTES: These data were produced by examining the species in every one-hundredth drawer of the 11,500-drawer collection of beetles. In total, 1,753 species were recorded, suggesting that the total holdings of identified species of beetles are 175,300. This does not include unidentified species in the accessions and several small separate collections of beetles. Data from "megadiversity" countries (see Table 5-2) are in italics.

plants may well be higher in tropical countries. If this is the case, then 1.5 million could be an underestimate of the numbers of species of fungi in the world.

Extrapolations from Samples

Erwin's 30 Million Species

The best known and most controversial estimates are those that have centered on the diversity of arthropods in the canopy of tropical forests (Figure 5-4). Erwin (1982:74) sketched out a novel method of estimating global species numbers for insects from samples he collected by applying knock-down insecticides to the tops of trees. His estimate has been quoted and misquoted so many times that I provide the full text here.

> "The tropical tree *Luehea seemannii* is a medium-sized seasonal forest evergreen tree with open canopy, large and wide-spaced leaves. The trees sampled (n=19) had few epiphytes or lianas generally, certainly not the epiphytic load normally thought of as being rich. These 19 trees over a three season sampling regime produced 955+ species of beetles, excluding weevils. In other samples now being processed from Brazil, there are as many weevils as leaf-beetles, usually more, so I added 206 (weevils) to the *Luehea* count and rounded to 1,200 for convenience. There can be as many as 245 species of trees in 1 hectare of rich forest in the tropics, often some of these in the same genus. Usually there are between 40 to 100 species and/or genera, *so I used 70 as an average*

FIGURE 5-4 (a) Insects can be sampled from the canopy of trees through the use of a fogging machine. This machine produces a warm rising cloud of a knock-down insecticide such as a synthetic nonresidual pyrethroid or natural pyrethrum. (b) The insects fall to the ground within a couple of hours and are collected on plastic sheets or on conical fabric trays, each 1 m^2 in area.

TABLE 5-4 Numbers of Host-Specific Species per Trophic Group on *Luehea seemannii*

Trophic Group	Number of Species (estimated)	% Host-Specific	Number of Host-Specifics (estimated)
Herbivores	682	20	136.4
Predators	296	5	14.8
Fungivores	69	10	6.9
Scavengers	96	5	4.8
Total	1,200		162.9

SOURCE: Erwin (1982).

number of genus-group trees where host-specificity might play a role with regard to arthropods. No data are available with which to judge the proportion of host-specific arthropods per trophic group anywhere, let alone the tropics. So conservatively, I allowed 20% of the *Luehea* herbivorous beetles to be host-specific (i.e., must use this tree species in some way for successful reproduction), 5% of the predators (i.e., are tied to one or more of the host-specific herbivores), 10% of the fungivores (i.e., are tied to fungus associated only with this tree), and 5% of the scavengers (i.e., are associated in some way with only the tree or with the other three trophic groups)" (see Table 5-4 above).

"Therefore, *Luehea* carries an estimated load of 163 species of host-specific beetles, a rather conservative estimate of 13.5%. I regard the other 86.5% as transient species, merely resting or flying through Luehea trees. *If 1 hectare has 70 such generic-group tree species, there are 11,410 host-specific species of beetles per hectare, plus the remaining 1,038 species of transient beetles, for a total of 12,448 species of beetles per hectare of tropical forest canopy.*

Beetles make up an estimated 40% of all arthropod species, therefore there are 31,120 species of arthropods in the canopy of 1 hectare of tropical forest. Based on my own observation, I believe the canopy fauna to be at least twice as rich as the forest floor and composed of a different set of species for the most part, so I added 1/3 more to the canopy figure to arrive at a grand total of 41,389 species per hectare of scrubby seasonal forest in Panama! What will there be in a rich forest? I would hope someone will challenge these figures with more data.

It should be noted that there are an estimated 50,000 species of tropical trees (R. Howard, via R. Eyde, pers. comm.). I suggested elsewhere (Erwin and Adis 1981) that tropical forest insect species, for the most part, are not highly vagile and have small distributions. If this is so, and using the same formula as above, starting with 162 host-specific beetles/tree species then there are perhaps as many as 30,000,000 species of tropical arthropods, not 1.5 million!"

In this way, Erwin raised previous estimates for the number of species in the world by almost a factor of 10, and because his estimate was based on real samples, it was seen by many to be quite credible. The timing was also important, for his figure of 30 million species is used by many to show, first, how little we know the Earth's fauna and flora, but also, when coupled with estimates of forest loss (Myers, 1989), to show that many species are threatened with extinction. Inevitably, the figure of 30 million species has become a political tool.

As May (1988, 1990) and Stork (1988, 1993) have suggested, the assumptions that Erwin made provide an agenda for research, and subsequently all of these have been tested. These are discussed below.

The first assumption relates to the host-specificity of insects to trees, and much evidence points to considerably lower numbers of species being specific to trees than Erwin suggested. Gaston (1992) for example, found that one indirect measure of host-specificity, the ratio of insect-to-plant numbers for different regions, was typically 10, with a maximum recorded of 25. Following Erwin's argument that 40% of insects are beetles, this would suggest that only 4 of the 10 host-specific insects per plant are beetles. This is far less than the 162 suggested by Erwin, but again, as with Hawksworth's estimate for the number of species of fungi (Hawksworth, 1991), the data for plant/insect ratios are from temperate countries only. In practice, although many species of insects may be specific to a single species of tree, many feed on several species or even whole genera or families. Canopy samples collected by Erwin and myself (Stork, 1991) (Figure 5-4) typically have many singletons (only one individual per species)

FIGURE 5-5 Preparation and sorting millions of individuals and thousands of species from canopy-fogging and other samples collected in Sulawesi.

TABLE 5-5 Diversity of Coleoptera (859 species; 3,919 individuals) and Chalcidoidea (739 species; 1,455 individuals)

Coleoptera				Chalcidoidea	
N (individuals)	N (species)	N (individuals) continued	N (species) continued	N (individuals)	N (species)
1	499	26	1	1	437
2	133	30	1	2	160
3	62	31	2	3	54
4	24	32	1	4	31
5	35	35	1	5	18
6	21	36	1	6	10
7	8	39	1	7	8
8	13	40	2	8	4
9	4	45	1	9	6
10	4	49	1	10	3
11	4	53	1	11	2
12	5	66	2	12	1
13	4	77	1	13	1
14	2	81	1	17	1
15	2	90	1	19	1
17	4	112	1		
18	2	129	1		
19	1	137	1		
22	3	140	1		
23	2	194	1		
24	3	235	1		

NOTES: Collected from 10 trees in Borneo using knock-down insecticide fogging (Stork, 1991, 1993). The number of individuals is shown for each species. For example, 499 species of Coleoptera each were represented by only one individual, 133 species each were represented by two individuals, and so on.

that probably do not feed on the tree in question. Table 5-5 shows the rank abundance curve for beetles and chalcid wasps fogged with knock-down insecticides from 10 trees in Borneo. For both groups, more than half of the species present were singletons. Botanists often record more than 100 species of trees in single hectares of tropical forest, and with such immense diversity of trees and other plants it should not be surprising to find that many species of insects are collected from trees with which they have no close association.

Thomas (1990) examined the number of species of *Heliconius* butterflies feeding on species of Passifloraceae. On 12 Central American sites, he found an average of 7.2 species of Passifloraceae and 9.7 species of Heliconiinae. With over 360 species of Passifloraceae being known for the neotropics, one might scale up (9.7/7.2 × 360) to produce an estimate of 485 species of Heliconiinae.

In practice, the total is only 66 species because these butterflies use different species of Passifloraceae in different parts of their range. There are other complications to this story as well, such as the fact that only 100-150 species of Passifloraceae are found below 1,500 m (the upper limit for Heliconiinae), but the principle is still the same. As May (1990) noted, the simple procedure of multiplying the average number of species per tree by the number of species of trees can be misleading, since the geographical ranges of plants and insects may differ.

May (1990) looked at the theoretical distribution function $pk(i)$, the fraction of canopy insects found on tree species k which utilize a total of i different species of trees. Further, he examined f, the proportion of species effectively specialized to each species of tree. Using reasonably accurate data on the known biologies of British beetles and their association with trees, he predicted that 10% were herbivores specific to the genus *Quercus*. A 5-year intensive study of the beetle fauna of Richmond Park, an area dominated by oak woodlands (Hammond and Owen, 1995), produced 983 species of beetles (the total is now over 1,095 species), which is about 25% of all British beetles. Part of this study included canopy-fogging of more than 40 oak trees plus several other species of trees at different seasons over 2 years, which produced 198 species of beetles (Hammond, 1994). Of the total (from all sampling methods), some 18% are species specifically associated with trees and some 3% are associated only with oaks (Hammond, personal communication, 1995). First estimates of the tree-specificity of tropical arthropods using May's function, $pk(i)$, and the beetle data from canopy-fogging 10 trees in Borneo (Table 5-5) also support a figure of less than 5% (Mawdsley and Stork, in press).

Erwin provided no data for his next assumption that beetles represent 40% of tropical canopy arthropods. I and my colleagues from the Natural History Museum in London sorted all arthropods to species in canopy samples from Brunei and found that, of the 3,000+ species, about 20-25% (859 species) were beetles (Stork, 1991). There has been no comparable tropical study to date with which to compare these results, but this figure is similar to the 18% of all British arthropods that is comprised by beetles (4,000 species). How the relative number of species for the four largest orders of insects (Coleoptera, Diptera, Hymenoptera, and Lepidoptera) vary in known (usually temperate) faunas was examined by Gaston (1991), who concluded that there may be strong latitudinal trends in the diversity of these taxa of which we are largely unaware. In spite of this, I found that there was a remarkable similarity in (1) the relative proportions of species from different guilds of insects from canopy samples from Brunei, the United Kingdom, and South Africa (Stork, 1987), and (2) the relative numbers of species for different families of beetles in canopy samples from Brunei and Panama (Stork, 1993). It is clear, then, that we are still some distance from determining the relative contributions of groups such as beetles, parasitic wasps, and flies to global diversity.

There are no published data that might provide an indication as to the relative richness of the arthropods of the canopy and of the ground, but two recent independent studies in Borneo and Sulawesi suggest that the 2:1 ratio in favor of the canopy in Erwin's estimate should be reversed. Hammond's (1992) preliminary analysis of large samples of beetles from canopy and ground samples from Sulawesi (see below), and subsequent more detailed analyses (Hammond et al., in press), suggest that only about 10% of the 4,000+ species of beetles collected in a hectare of lowland rain forest in Sulawesi are "canopy specialists" and that there are more than twice as many "ground specialists" as "canopy specialists." The conclusions of this study are largely supported by an analysis of more than 3,000 species of beetles collected in similar ways in Borneo (Mawdsley, 1995). Rather than exert too much effort in considering the ratio of species of arthropods from the canopy to the ground (for as Hammond et al., in press, have shown, a large part of the fauna is found in both ecotones), it might be better to examine the ratio of functional groups, such as herbivores to other trophic guilds.

In summary, Erwin probably has grossly overestimated the likely number of host-specific insects associated with a species of tree, but underestimated the contribution of insects other than beetles and those insects associated with the ground.

Estimates from Intensive Sampling in Sulawesi

One of the obvious problems with Erwin's method of estimating global species diversity is that his base data are from samples collected only from the canopy. In 1985, the Royal Entomological Society of London and the Indonesian Department of Science carried out a year-long study of the Dumoga-Bone area of north Sulawesi in Indonesia: Project Wallace. Some 200 entomologists carried out studies in this area, including an intensive sampling program for insects that was organized by the Natural History Museum in London. More than 6,000 species of beetles were collected by a wide variety of methods and then sorted to species (Hammond, 1990; Stork and Brendell, 1990) (Figure 5-5). Other groups of insects also were sampled intensively, and the data for one of these were used by Hodkinson and Casson (1991) to estimate the number of species in the world.

Hodkinson and Casson (1991) examined 1,690 species of collected Hemiptera and, after consultation with other specialists, estimated that 62.5% were new to science. They suggested that if the same proportion of new species was to be found world-wide (and they provided some statistical support for this) then there should be 184,000-193,000 species of Hemiptera world-wide. If 7.5 to 10% of all insects are Hemiptera, they argued, this would give a world total of 1.84 to 2.57 million species of all insects. They also suggested that if there are 500 species of trees in the Dumoga-Bone area and this produced 1,056 new species of Hemiptera, then 50,000 species of trees (see Erwin's method above) would

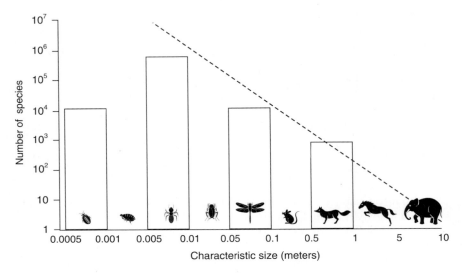

FIGURE 5-6 Using the relationship between the number of species and their body size for larger organisms, it is possible to back-predict for the less well-known smaller species to estimate the total number of species in the world (after May 1978, 1988).

give 105,600 new species. These plus the described species would give 187,300 species of Hemiptera and a total of 1.84 to 2.57 million for all insects world-wide.

There are several problems with their arguments (Stork, 1993). First, they assume that all species of Hemiptera in the area were collected and, since cumulative curves for the even more thoroughly sampled beetles (Figure 2 in Stork, 1993) show little leveling off, this seems very unlikely. Second, they fail to demonstrate that the ratio of described to undescribed species is representative of other areas in the world. Third, since many groups of Hemiptera include economically important species, it would seem probable that they have been subject to greater descriptive efforts than many other orders of insects. It would seem likely therefore, that the proportion of the world's species of insects that are Hemiptera is much less than 7.5-10%.

With some 6,000+ species of beetles having been sorted and a projected minimum total of 10,000 species estimated for the Dumoga-Bone area (Stork, 1993), it is possible to test their hypotheses. Using 15-20% as estimates of the proportion of the world's species of insects that are represented by beetles, there should be 3-4 million (based on 6,000 species of beetles sampled) or 5 to 6.7 million (based on the minimum projected total of 10,000 species of beetles in the Dumoga-Bone area) species of insects world-wide. What this clearly demonstrates is that, if the samples represent only 50% of the species in the area considered, then the estimates will be 50% under the true figure.

Inevitably, estimates based on such methods and on samples of insects will be open to criticism until much more intensive sampling or even complete inventories of some tropical areas are completed. Hopefully, much more accurate estimates of regional and global diversity will be some of the many benefits of the proposed All Taxa Biological Inventories (Janzen and Hallwachs, 1994).

Other Models for Estimation of Species

Many important ecological principles on the distribution and community structure of organisms have been determined over the last 30 or more years, and some of these have been used in the models discussed above. Several others also show some promise and are discussed here.

Body Size and Number of Species

May (1978, 1988) noted that for a wide range of organisms larger than 1 mm there is an inverse relationship between the number of species and body size. Fractal arguments would suggest that, in the relationship $S \sim L^{-x}$, where S is the number of species and L is body length, the factor x should be between 1.5 and 3.0. Using these figures and extrapolating down to 1 mm, May (1988) suggested global estimates of 10-50 million species (Figure 5-7). May's (1990) later ex-

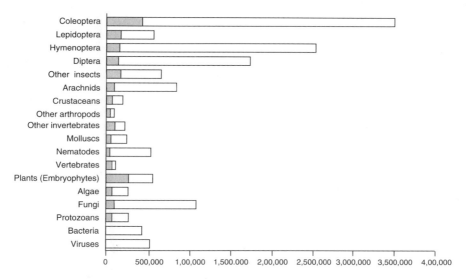

FIGURE 5-7   The projected number of species for different taxa based on an estimated global total of 12.5 million species (after Hammond, 1992; Stork, 1993); shaded area is the total number of described species from Figure 5-1.

trapolation to 0.2 mm gave an estimate of 10 million species, based on the premise that number of species increases 100-fold for a 10-fold reduction in length. Having just discovered a complex and species-rich fauna of mites and Collembola in sandy soils in France and Belgium, André et al. (1994) suggest that the extrapolation should be to at least 136-65 μm, which would add another 10 million species to May's (1990) total. Addition of even smaller organisms (Protozoa, Nematoda, Enchytraeidae, and Tardigrada), they argue, would further increase the total.

There are two major problems with this argument. First, the traditional species concept based on sexual reproduction does not appear to apply to very small organisms (or even to many larger organisms, such as some flowering plants). For these, it might be more practical to look at functional characteristics and genetic differences to separate them. Second, dispersal by air or by water is easier for very small organisms than larger ones, and they are therefore less likely to speciate, since the barriers necessary for the isolation of populations, and hence allopatric speciation, are less likely to occur. In practice then, below a body size of about 1.0 mm, number of species may decrease with decreasing body size. The implications from arguments of André et al. are that there are 10 million species or more of groups such as Collembola, Acarina, and Nematoda. So far, there is little evidence that this is the case. Intensive sampling of the nematode fauna of soil and leaf litter of lowland forest in Cameroon produced 483 morphospecies, with evidence that the number of new species was falling off with additional sampling (Lawton et al., 1995). It is therefore probable that, although the abundance of the soil invertebrate mesofauna is extremely high, the number of species involved may well be considerably less than that for insect groups such as beetles, flies, and Hymenoptera.

Species Turnover

One approach to measuring global species richness that has been employed in the marine environment is to measure how species accumulate with time and with distance. Theoretically, as additional samples are collected at increasing distances away from a central point, the number of species added will increase because of the added heterogeneity of the new habitats encountered. The difference between the species richness of samples from different sites is known as beta diversity or species turnover. It is theoretically possible, therefore, to estimate numbers of species based on estimates of beta diversity. Grassle and Maciolek (1992), for example, studied data for a range of species of macro-invertebrates from the deep sea that were collected over a 176 km transect of 14 stations at a depth of 2,100 m. They found 798 species of annelids, molluscs, and arthropods in the 90,677 individuals in the samples. Of these, 480 species had not been recorded before. They combined their samples over time at each site to examine patterns in the addition of new species as they moved along their transect (see Figure 11 in Grassle and Maciolek, 1992). The typical rarefaction curve they found showed a slowing down in the addition of species over the transect. They suggested that a straight line prediction from the upper end of this curve would indicate a linear rate of increase of one new species per km. They generalized this to one new species per km^2, which for an oceanic area of 3×10^8 km^2 deeper than 1,000 m would suggest several hundred million new species. Recognizing that the densities of organisms are much lower at the bottom of the deep sea than those on the continental shelf area they sampled, they scaled their estimate of the total number of benthic species to 10^7.

May (1992) criticized their estimate, suggesting that they falsely assumed a straight line for the upper part of their rarefaction curve, and that it would be more reasonable to assume a doubling of the number of benthic species on the basis of the number of new species relative to known species that they collected. May also criticized other points of their arguments.

A similar problem exists with other data that at first glance also would appear to support estimates of tens of millions of species. Erwin (1988, 1991) found that in beetle samples of 1,080 species produced by canopy-fogging in Amazonian forests, only 1% were shared between four different types of forest in the same area. He also cites the example of two samples of moths from sites in Bolivia and Peru that are 500 km apart which total 933 species (1,748 individuals) and 1,006 species (1,731 individuals) and which have only 60 species (3.2%) shared between sites. Since most species of insects are extremely rare in canopy (and other) samples (Table 5-5), the chance of catching the same species in two samples even at the same site is extremely low. In other words, although these are impressive figures for the number of species sampled, the data themselves are not sufficient to support high levels of endemism, nor do they provide any measure of how widespread these species are. Comparison of the percentage shared between subsamples at the same site and the percentage shared between different sites would provide more relevant information on local endemism.

Another way of looking at this problem is to consider the size of the regional pool of species. If we assume a modest global figure of 5.3 million for all species of insects (see Table 3 in Stork, 1993), that there are 1.2 million species in tropical South America, that 300,000 of these are beetles, and that half of these are widespread (but probably mostly not very abundant), then the regional beetle fauna where Erwin collected his canopy samples might be as high as 100,000. If this is the case, then his sample of 1,080 beetles from four types of forest in Manaus is very small and the 1% value for shared species hardly surprising. The assumptions may be completely wrong, but they need testing with the correct statistical tools and probably much larger samples.

Taxonomists' Views

The arguments presented above all rely on one important factor and that is the taxonomists' views of what constitutes a species. These views will differ depending on the taxa studied, the biology of the groups concerned, the techniques used to distinguish species, and how far individual taxonomists are prepared to go in order to clarify the number of species involved. Most of the above debate has concerned samples of beetles and other insects where a "morphospecies" concept has been adopted. Sometimes single species that appear to be valid on normal morphological characteristics later are found to comprise several species which can be distinguished only with molecular techniques (Adis, 1990). To this end, it is impossible at this stage to say how many such molecular species might exist.

EXTINCTIONS

In recent years, many eminent scientists have made ominous predictions about present and future rates of the extinction of species (Table 5-6). Why so few species actually have been recorded as extinct is evident from how little we know about the number of species on Earth and their distribution. One of the major problems with predicted extinction rates is that the groups most likely to be affected, in terms of numbers of species, are also those for which we have the least information. That the threat of extinction is real can be seen through the following analysis.

Mawdsley and Stork (1995) used British data on the threatened status of different groups to make predictions on the number of extinctions for groups such as insects (Table 5-7). They suggested that if the accuracy of the recorded Red Data Book status of groups of British animals such as birds, mammals, and invertebrates are comparable, and if these reflect their likelihood of extinction, then it is possible to use these data to estimate a "relative rate of extinction." The relative rate for threatened birds to insects is 4.3, implying that a British breeding bird is, on average, 4.3 times as likely to become extinct as an insect. Figures for endangered species give a higher relative rate of 9.8. Smith et al. (1993a) also used Red Data Book status in two ways to estimate extinction rates for well-known groups such as birds, mammals, and palms, but not for insects. They first examined the number of species added to the list of extinctions for animals during 1986-1990 and for plants during 1990-1992. Using these data and estimated numbers of species world-wide, they predicted that the time to extinction for 50% of species was 1,500 years for birds and 6,500 years for mammals. Second, they used data on the net changes of species in these groups towards extinction during these periods (i.e., changes in Red Data Book status of a species from "rare" through "probably extinct") to predict 50% extinction rates of 350 and 250 years for birds and mammals, respectively. Mawdsley and Stork argue that since about 1% of birds and mammals have become extinct since 1600, a mean relative extinction rate of 7.1 for birds to insects would suggest that 0.14% of species of insects have become extinct since 1600. Assuming a global total of 8 million species of insects, this would suggest that 11,200 have become extinct since that time. They also argue that, since the calculations of Smith et al. would suggest a 12- to 55-fold increase in extinctions of birds in the next 300 years, their relative extinction rate would predict an equivalent loss of 100,000-500,000 species of insects (i.e., 7-30 species per week). For an undescribed species of insect, it would appear that the chances of extinction may be greater than the chances of description!

Whether the assumptions made by Mawdsley and Stork (1995) match reality and whether it is possible to use models from the British fauna and flora to make global predictions is impossible to say, but their model at least can be back-tested within the British context. Of the 13,746 species included in the Red Data Book for insects (Shirt, 1987), 5% (99 species) have not been seen since 1900. Given a relative extinction rate of 7.1 for birds to insects, this would suggest that 11 species of birds should have disappeared in the same period. This prediction is confirmed by Sharrock (1974), who states that 11 species of birds have become extinct in Britain and Ireland, with 2 species later recolonizing during 1800-1949.

If the estimates of loss of 100,000-500,000 species of insects in the next 300 years, as predicted using the relative extinction rate, seem comparatively low, then perhaps it would be wise to look at the possible extinction threats for these and other groups. Groombridge (1992) shows that the probable cause of extinction for many species of birds and mammals was the introduction of other species and hunting. A comparable analysis of the extinct species of insects has not been made, but Mawdsley and Stork (1995) show that, for British beetles and butterflies, the overriding threats are land-use change and habitat destruction.

CONCLUDING REMARKS

What these arguments show is how little we actually know about some of the fundamental aspects of the biology and distribution of organisms. We cannot say how widespread species are, we do not know the size of the species pool, and we do not know how specific species are to a particular habitat, type of soil, type of forest, or, in some cases, a species of tree. They also indicate that for some of the most species-rich groups of organisms such as arthropods, annelids, nematodes, and fungi, the scale of sampling we have used so far is not sufficient to answer some of the most important questions in biology. If such questions are to be answered in the next 10-20 years, then we will need a program of action to inventory and assess the scale of biodiversity, such as *Systematics Agenda 2000* (Systematics Agenda 2000, 1994), which matches the Human Genome Project in scale and scope. Without such action, it is clear that many species will be doomed to extinction.

Estimating the number of species in the world has been bedeviled by the lack of evidence of biological patterns on a global scale that are supported by sound empirical data and anecdote. On present evidence, there seems little case

TABLE 5-6 Estimated Rates of Extinction

Estimate	% Global Loss Per Decade	Method of Estimation	Reference
One million species between 1975 and 2000	4	Extrapolation of past exponentially increasing trend	Myers (1979)
15-20% of species between 1980 and 2000	8-11	Estimated species-area curve; forest loss based on Global 2000 projections	Lovejoy (1980)
50% of species by 2000 or soon after; 100% by 2010-2025	20-30	Various assumptions	Ehrlich and Ehrlich (1981)
9% extinction by 2000	7-8	Estimates based on Lovejoy's calculations using Lanly's (1982) estimates of forest loss	Lugo (1988)
12% of plant species in neotropics, 15% of bird species in Amazon basin	—	Species-area curve ($z=0.25$)	Simberloff (1986)
2,000 plant species per year in tropics and subtropics	8	Loss of half the species in area likely to be deforested by 2015	Raven (1987)
25% of species between 1985 and 2015	9	As above	Raven (1988a,b)
At least 7% of plant species	7	Half of species lost over next next decade in 10 "hot-spots" covering 3.5% of forest area	Myers (1988)

0.2 to 0.3% per year	2-6	Half of rain forest species assumed lost in tropical rain forests are local endemics and become extinct with forest loss	Wilson (1988, 1989, 1993)
2-13% loss between 1990 and 2015	1-5	Species-area curve ($0.15 < z < 0.35$); range includes current rate of forest loss and 50% increase	Reid (1992)
Red Data Books for selected taxa: 50% extinct in 50-100 years (palms), 300-400 years (birds, mammals)	1-10	Extrapolating current recorded extinction rates and using the dynamics of threatened categories	Smith et al. (1993a,b)
Red Data Books for selected vertebrate taxa	0.6-5	Fitting of exponential extinction functions based on World Conservation Union categories of threat	Mace (1994)

NOTE: The estimated rates include species "committed" to extinction; see Heywood et al. (1994).

SOURCES: Mawdsley and Stork (1995), Reid (1992).

TABLE 5-7 "Relative Rates of Extinction" Estimated for Different Groups of British Animals

	1	2	3	4	5	6	7	8	9
	Number of British Species	Number of Threatened British Species	Number of Endangered British Species	% of British Species Threatened	% of World Species Threatened	% of British Species Endangered	Relative Extinction Rate: Threatened (from [4])	Relative Extinction Rate: Endangered (from [6])	Mean Relative Extinction Rate (from [7] and [8])
Mammals	55	16	No data	29.1	11	—	2.3	—	2.3
Birds	210	117	76	55.7	11	36.2	4.3	9.8	7.1
Insects	13,746[a]	1,786	506	12.9	0.07	3.7	1	1	1
Araneae	622	86	22	13.8	—	3.5	1.1	0.9	1
Mollusca	210[b]	30	10	14.2	0.4	4.8	1.1	0.8	1

[a]There are more than 22,000 species of British insects, but the Red Data Book (Shirt, 1987) only considers some of these.
[b]Only terrestrial, brackish, and freshwater molluscs are included.

NOTES: "Threatened" includes all categories and "endangered" is category 1 in the Red Data Book. Columns 7 and 8 are derived by comparing the relative percentages between taxa in columns 4 and 6, respectively. Note that the percentage of threatened British birds is an overestimate because of the data used to calculate this figure.

SOURCE: Mawdsley and Stork (1995).

to be made for estimates of 30 million or more global species; a more probable total is 5-15 million. If a single figure is to be selected, then that proposed by Hammond (1992) of 12.5 million species would seem reasonable at this stage (Figure 5-7). However, upward or downward revisions of this number could easily occur through the compilation of new and sound data on the species concept for some groups and on the distribution of major taxa such as insects, fungi, and other microorganisms in terrestrial systems, and for annelids, molluscs, and arthropods in marine systems.

ACKNOWLEDGMENTS

I thank F. Cotterill for drawing my attention to the distributional data for Passerine birds. I am most grateful to Harriet Eeley for assistance in preparing the figures and the manuscript.

REFERENCES

Adis, J. 1990. Thirty million arthropod species—too many or too few? J. Trop. Ecol. 6:115-118.

André, H. M., M. -I. Noti, and P. Lebrun. 1994. The soil fauna: The other last biotic frontier. Biodiv. Conserv. 3:45-46.

Bridges, C. A. 1988a. Catalogue of Papilionidae and Pieridae (Lepidoptera: Rhopalocera). Charles A. Bridges, Urbana, Ill.

Bridges, C. A. 1988b. Catalogue of Lycaenidae and Riodinidae (Lepidoptera: Rhopalocera). Charles A. Bridges, Urbana, Ill.

Bridges, C. A. 1988c. Catalogue of Hesperiidae (Lepidoptera: Rhopalocera). Charles A. Bridges, Urbana, Ill.

Bridges, C. A. 1991. Catalogue of the Family-Group, Genus-Group and Species-Group Names of the Odonata of the World. Charles A. Bridges, Urbana, Ill.

Colwell, R. K., and J. A. Coddington. 1994. Estimating terrestrial biodiversity through extrapolation. Phil. Trans. R. Soc. Lond. B. 344:1-25.

Corbet, G. B., and J. E. Hill. 1980. A World List of Mammalian Species. Cornell University Press, Ithaca, N.Y.

Crosskey, R. M. 1987. An annotated list of the world black flies (Diptera: Simuliidae). Pp. 425-520 in K. C. Kim and R. W. Merritt, eds., Black Flies. Ecology, Population Management and Annotated World List. Pennsylvania State University, University Park.

Eastop, V. F., and D. Hille Ris Lambers. 1976. Survey of the World's Aphids. W. Junk, The Hague, Netherlands.

Ehrlich, P., and A. Ehrlich. 1981. Extinction. The Causes of the Disappearance of Species. Random House, N.Y. 305 pp.

Erwin, T. L. 1982. Tropical forests: Their richness in Coleoptera and other arthropod species. Coleopt. Bull. 36:74-75.

Erwin, T. L. 1988. The tropical forest canopy: The heart of biotic diversity. Pp. 123-129 in E. O. Wilson and F. M. Peter, eds., BioDiversity. National Academy Press, Washington, D.C.

Erwin, T. L. 1991. How many species are there? Revisited. Conserv. Biol. 5:1-4.

Gaston, K. G. 1991. The magnitude of global insect species richness. Conserv. Biol. 5:183-196.

Gaston, K. G. 1992. Regional numbers of insects and plants. Func. Ecol. 6: 243-247.

Gaston, K. G. 1994. Spatial patterns of species description: How is our knowledge of the global insect fauna growing? Biol. Conserv. 67:37-40.

Gaston, K. G., and L. A. Mound. 1993. Taxonomy, hypothesis testing and the biodiversity crisis. Phil. Trans. R. Soc. Lond. B. 251:139-142.

Grassle, J. F., and N. J. Maciolek. 1992. Deep-sea species richness: Regional and local diversity estimates from quantitative bottom samples. Amer. Nat. 139:313-341.

Groombridge, B., ed. 1992. Global Biodiversity. Status of the Earth's Living Resources. Chapman and Hall, London.

Hall, B. P., and R. E. Moreau. 1970. An Atlas of Speciation in African Passerine Birds. British Museum Natural History, London.

Hammond, P. M. 1990. Insect abundance and diversity in the Dumoga-Bone National Park, N. Sulawesi, with special reference to the beetle fauna of lowland rain forest in the Toraut region. Pp. 197-254 in W. J. Knight and J. D. Holloway, eds., Insects and the Rain Forests of South East Asia (Wallacea). Royal Entomological Society of London, London.

Hammond, P. M. 1992. Species inventory. Pp. 17-39 in B. Groombridge, ed., Global Biodiversity. Status of the Earth's Living Resources. Chapman and Hall, London.

Hammond, P. M. 1994. Practical approaches to the estimation of the extent of biodiversity in speciose groups. Phil. Trans. R. Soc. Lond. B. 345:119-136.

Hammond, P. M., N. E. Stork, and M. J. D. Brendell. In press. Comparison of the composition of the beetle faunas of the canopy and other ecotones of lowland rainforest in Indonesia. In press in N. E. Stork and J. Adis, eds., Canopy Arthropods. Chapman and Hall, London.

Hammond, P. M., and J. Owen. 1995. The beetles of Richmond Park SSSI—a case study. English Nature Science 18:1-180.

Hawksworth, D. L. 1991. The fungal dimension of biodiversity: Magnitude, significance and conservation. Mycol. Res. 95:641-655.

Heywood, V. H., G. M. Mace, R. M. May, and S. N. Stuart. 1994. Uncertainties in extinction rates. Nature 368:105.

Hodkinson, I. D., and D. Casson. 1991. A lesser predilection for bugs: Hemiptera (Insecta) diversity in tropical forests. Biol. J. Linn. Soc. 43:101-109.

Janzen, D. H., and W. Hallwachs. 1994. All Taxa Biodiversity Inventory (ATBI) of Terrestrial Systems. A Generic Protocol for Preparing Wildland Biodiversity for Non-damaging Use. Report of National Science Foundation Workshop, 16-18 April, 1993, in Philadelphia. National Science Foundation, Washington, D.C. 132 pp.

Lawton, J. H. 1993. On the behaviour of autecologists and the crisis of extinction. Oikos 67:3-5.

Lawton, J. H., D. E. Bignell, G. F. Bloemers, P. Eggleton, and M. E. Hodda. 1996. Carbon flux and diversity of nematodes and termites in Cameroon forest soils. Biodiv. Conserv. 5:261-273.

Linnaeus, C. 1758. Systema Naturae, tenth ed. L. Salvii, Stockholm. 824 pp.

Lovejoy, T. E. 1980. A projection of species extinctions. Pp. 328-331 in G. O. Barney, Study Director, The Global 2000 Report to the President. Entering the Twenty-First Century, Vol. 2. Council on Environmental Quality, U.S. Government Printing Office, Washington, D.C.

Lugo, A. E. 1988. Estimating reductions in the diversity of tropical forest species. Pp. 58-70 in E. O. Wilson and F. M. Peter, eds., BioDiversity. National Academy Press, Washington, D. C.

Mace, G. M. 1994. Classifying threatened species: Means and ends. Phil. Trans. R. Soc. Lond. B. 344:91-97.

Mawdsley, N. M. 1995. Community structure of Coleoptera in a Bornean Lowland Forest. Ph.D. Thesis, Imperial College, University of London, London.

Mawdsley, N. M., and N. E. Stork. 1995. Species extinctions in insects: Ecological and biogeographical considerations. Pp. 322-369 in R. Harrington and N. E. Stork, eds., Insects in a Changing Environment. Academic Press, London.

Mawdsley, N. M., and N. E. Stork. In press. Host-specificity and the effective specialization of tropical canopy beetles. In press in N.E. Stork and J. Adis, eds., Canopy Arthropods. Chapman and Hall, London.

May, R. M. 1978. The dynamics and diversity of insect faunas. Pp. 188-204 in L. A. Mound and N. Waloff, eds., Diversity of Insect Faunas. Blackwell Scientific Publications, Oxford, England.

May, R. M. 1988. How many species are there on Earth? Science 241:1441-1449.

May, R. M. 1990. How many species? Phil. Trans. R. Soc. Lond. B. 330:293-304.

May, R. M. 1992. Bottoms up for the oceans. Nature 357:278-279.

May, R. M., J. H. Lawton, and N. E. Stork. 1995. Assessing extinction rates. Pp. 1-24 in J. H. Lawton and R. M. May, eds., Extinction Rates. Oxford University Press, Oxford, England.

Mound, L. A. M., and S. H. Halsey. 1978. Whitefly of the World. A Systematic Catalogue of the Aleyrodidae (Homoptera) with Host Plant and Natural Enemy Data. British Museum Natural History, London.

Myers, N. 1979. The Sinking Ark. A New Look at the Problem of Disappearing Species. Pergamon, N.Y. 307 pp.

Myers, N. 1988. Threatened biotas: "Hotspots" in tropical forests. Environmentalist 8:1-20.

Myers, N. 1989. Deforestation Rates in Tropical Forests and Their Climatic Implications. Friends of the Earth, London.

Noyes, J. S. 1990. The number of described Chalcidoid taxa in the world that are currently regarded as valid. Chalcid Forum 13:9-10.

Poole, R. W. 1989. Noctuidae. Lepid. Cat. 118:1-1314.

Raven, P. H. 1985. Disappearing species: A global tragedy. Futurist 19:8-14.

Raven, P. H. 1987. The scope of the plant conservation problem world-wide. Pp. 19-29 in D. Bramwell, O. Hamann, V. Heywood, and H. Synge, eds., Botanic Gardens and the World Conservation Strategy. Academic Press, London.

Raven, P. H. 1988a. Biological resources and global stability. Pp. 3-27 in S. Kawano, J. H. Connell, and H. Hidaka, eds., Evolution and Coadaptation in Biotic Communities. University of Tokyo Press, Tokyo.

Raven, P. H. 1988b. Our diminishing tropical forests. Pp. 119-122 in E. O. Wilson and F. M. Peter, eds., BioDiversity. National Academy Press, Washington, D.C.

Reid, W. V. 1992. How many species will there be? Pp. 55-73 in T. C. Whitmore and J. A. Sayer, eds., Tropical Deforestation and Species Extinction. Chapman and Hall, London.

Sharrock, J. T. R. 1974. The changing status of breeding birds in Britain and Ireland. Pp. 203-220 in D. L. Hawksworth, ed., The Changing Flora and Fauna of Britain. Academic Press for the Systematics Association, London.

Shirt, D. B., ed. 1987. British Red Data Books: 2. Insects. Nature Conservancy Council, Peterborough, England.

Simberloff, D. 1986. Are we on the verge of a mass extinction in tropical rain forests? Pp. 165-180 in D. K. Elliot, ed., Dynamics of Extinction. John Wiley and Sons, N.Y.

Smith, F. D. M., R. M. May, R. Pellew, T. H. Johnson, and K. S. Walter. 1993a. Estimating extinction rates. Nature 364: 494-496.

Smith, F. D. M., R. M. May, R. Pellew, T. H. Johnson, and K. R. Walter. 1993b. How much do we know about the current extinction rate? Trends Ecol. Evol. 8:375-378.

Snyder, T. E. 1948. Catalog of the Termites (Isoptera) of the World. Smithsonian Misc. Coll. 112:1-490.

Soberon, J. M., and J. B. Llorente. 1993. The use of species accumulation functions for the prediction of species richness. Conserv. Biol. 7:480-488.

Stevens, G. C. 1990. The latitudinal gradient in geographical range: How so many species coexist in the tropics. Amer. Nat. 133:240-256.

Stevens, G. C. 1992. The elevational gradient in altitudinal range: An extension of Rapoport's latititudinal rule to altitude. Amer. Nat. 140:893-911.

Stork, N. E. 1987. Guild structure of arthropods from Bornean rain forest trees. Ecol. Entomol. 12:69-80.

Stork, N. E. 1988. Insect diversity: Facts, fiction and speculation. Biol. J. Linn. Soc. 35:321-337.

Stork, N. E. 1991. The composition of the arthropod fauna of Bornean lowland rain forest trees. J. Trop. Ecol. 7:161-180.

Stork, N. E. 1993. How many species are there? Biodiv. Conserv. 2:215-232.

Stork, N. E., and M. J. D. Brendell, 1990. Variation in the insect fauna of Sulawesi trees with season, altitude and forest type. Pp. 173-190 in W. J. Knight and J. D. Holloway, eds., Insects and the Rain Forests of South East Asia (Wallacea). Royal Entomological Society of London, London.

Stork, N. E., and K. G. Gaston. 1990. Counting species one by one. New Scientist 1729:43-47.

Systematics Agenda 2000. 1994. Systematics Agenda 2000: Charting the Biosphere. Technical Report. Systematics Agenda 2000, a Consortium of the American Society of Plant Taxonomists, the Society of Systematic Biologists, and the Willi Hennig Society, in cooperation with the Association of Systematics Collections, N.Y. 34 pp.

Thomas, C. D. 1990. Fewer species. Nature 347:237.

Westwood, J. O. 1883. On the probable number of species in the Creation. Mag. Nat. Hist. VI:116-123.

Wilson, E. O. 1988. The current state of biological diversity. Pp. 3-18 in E. O. Wilson and F. M. Peter, eds., BioDiversity. National Academy Press, Washington, D.C.

Wilson, E. O. 1989. Threats to biodiversity. Sci. Amer. (September):108-116.

Wilson, E. O. 1993. The Diversity of Life. Belknap Press, Cambridge, Mass. 424 pp.

Butterfly Diversity and a Preliminary Comparison with Bird and Mammal Diversity

ROBERT K. ROBBINS

Research Entomologist, Department of Entomology, National Museum of Natural History, Smithsonian Institution, Washington, D.C.

PAUL A. OPLER

Managing Editor, National Status and Trends Report, Technology Transfer Center, Fort Collins, Colorado

Butterflies are among the best-known insects—an estimated 90% of the world's species have scientific names. As a consequence, their biology has been extensively investigated (Vane-Wright and Ackery, 1984), and they are perhaps the best group of insects for examining patterns of terrestrial biotic diversity and distribution. Butterflies also have a favorable image with the general public. Hence, they are an excellent group for communicating information on science and conservation issues such as diversity.

Perhaps the aspect of butterfly diversity that has received the most attention over the last century is the striking difference in species richness between tropical and temperate regions. For example, Bates (1875) wrote that it would convey some idea of the diversity of butterflies (in the neighborhood of Belém, a town near the mouth of the Amazon River) when he mentioned that about 700 species are found within an hour's walk of the town, whereas the total number found in the British Islands does not exceed 66, and the whole of Europe supports only 321. This early comparison of tropical and temperate butterfly richness has been well-confirmed (e.g., Owen, 1971; Scriber, 1973).

A general theory of diversity would have to predict not only this difference between temperate and tropical zones, but also patterns within each region, and how these patterns vary among different animal and plant groups. However, for butterflies, variation of species richness within temperate or tropical regions, rather than between them, is poorly understood. Indeed, comparisons of numbers of species among the Amazon basin, tropical Asia, and Africa are still mostly "personal communication" citations, even for vertebrates (Gentry, 1988a). In

*Butterflies are conspicuously
diverse in tropical forests.*

other words, unlike comparisons between temperate and tropical areas, these patterns are still in the documentation phase.

In documenting geographical variation in butterfly diversity, we make some arbitrary, but practical decisions. Diversity, number of species, and species richness are used synonymously; we know little about the evenness of butterfly relative abundances. The New World fauna makes up the preponderance of examples because we are most familiar with these species. By focusing on them, we hope to minimize the errors generated by imperfect and incomplete taxonomy. Although what is and is not a butterfly is technically controversial (e.g., Reuter, 1896; Kristensen, 1976; Scoble, 1986; Scoble and Aiello, 1990), we follow tradition (e.g., Bates, 1861; Kuznetsov, 1915, 1929; Ford, 1945) in which butterflies consist of skippers (Hesperioidea) and true butterflies (Papilionoidea).

The first three sections of this chapter summarize general patterns of butterfly diversity throughout the world, within the conterminous United States, and in the Neotropics, respectively. The fourth section points out, albeit preliminarily, how the distributions of butterflies—and presumably other insects—paint a different biogeographical picture of the world than the distributions of birds and mammals. Finally, we briefly discuss the significance of the observed patterns for conservation and for the study of diversity.

GLOBAL PATTERNS OF BUTTERFLY DIVERSITY

There are about 13,750 species of true butterflies in the world. Ehrlich and Raven (1965) estimated 12,000-15,000, and Robbins (1982) narrowed the range to 12,900-14,600 (including an estimate for undescribed species). Shields (1989) tabulated 13,688 described species. Since Robbins used sources published after 1965 and Shields used post-1982 papers for information, these estimates are somewhat independent, and their similarity indicates that 13,750 species of true butterflies is a reasonable "ballpark" figure, almost assuredly accurate within 10% and probably within 5%.

There are about 17,500 species of butterflies (true butterflies plus skippers) in the world. Ehrlich and Raven did not tabulate numbers of all butterflies, but Robbins estimated 15,900-18,225 species in the world, including estimates of undescribed species. Shields counted 17,280 species, including many synonyms but excluding undescribed species. The estimate of 17,500 species of butterflies in the world is again probably accurate within 10% (15,750-19,250 species).

The number of species of butterflies in each of the world's major biogeographical realms is presented in Figure 6-1, modified from Ackery et al. (1995) for the Ethiopian realm and from Robbins (1982) for the others. The estimates for the Nearctic, Palearctic, and Ethiopian realms are reasonably accurate because their faunas are fairly well-documented. The numbers for the more poorly known Oriental-Australian (innumerable islands) and Neotropical realms (literally hundreds of undescribed metalmark, skipper, and hairstreak butterflies) are less accurate, probably within 10% of actual species richnesses.

Species diversity is greater in tropical than temperate areas (Figure 6-1). Of the two northern temperate realms, the Palearctic has greater area and more species than the Nearctic. These temperate realms have fewer species than the primarily tropical Neotropical, Ethiopian, and Oriental-Australian realms. Among all realms, the Neotropics has the richest butterfly fauna, approximately equal to that of tropical Africa and Asia combined (Figure 6-1).

The area of lowland rain forest is greatest in the Neotropics (Raven, 1990), but is not responsible for the greater butterfly richness of Latin America. Liberia, the Malay Peninsula, and Panama have similar areas at about the same latitude (Figure 6-2), and their butterfly faunas are documented (Owen, 1971; Robbins, 1982; Eliot and D'Abrera, 1992). Panama is smaller than Liberia and the Malay

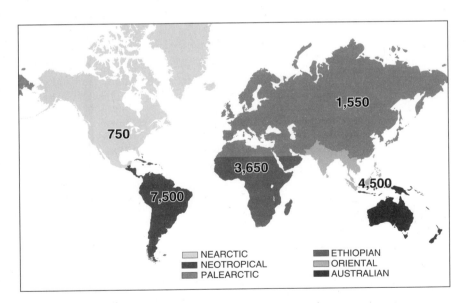

FIGURE 6-1 Number of species of butterflies by biogeographical realm. The 4,500 total is for the Oriental and Australian Realms combined. Ackery et al. (1995) listed 3,607 Ethiopian species. Robbins (1982) gave estimates for the other realms. There are an estimated 17,500 species of butterflies world-wide.

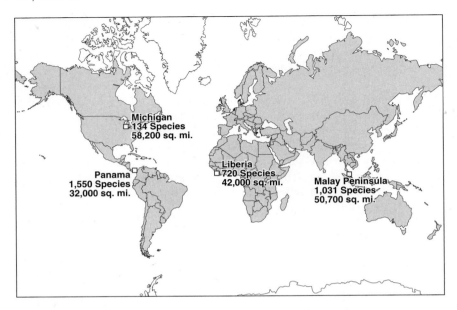

FIGURE 6-2 Area and number of species of butterflies for the Malay Peninsula (Eliot and D'Abrera, 1992), Liberia (Owen, 1971), and Panama (Robbins, 1982). Michigan (see Table 6-1) is included for comparison with an area in the temperate zone.

Peninsula, but has more than twice the number of species of butterflies than the latter areas (Figure 6-2). Since Central America is not an unusually rich part of the Neotropics (see below), its area of lowland rain forest alone is not likely to explain why the Neotropics have a greater butterfly diversity than the Old World tropics.

SPECIES RICHNESS OF U.S. BUTTERFLIES

Species of butterflies have been surveyed relatively well in most of the conterminous United States. For almost all states, it is reasonable to expect that more than 90% of the resident fauna has been discovered. Atlases of the county-by-county distribution of butterflies of the United States have been compiled over the past 20 years from published literature records, data from specimens in museums and private collections, and by conducting field surveys in poorly known geographic areas (Stanford and Opler, 1993; Opler, 1995, 1996).

By reviewing these atlases, we counted the number of true butterflies and skippers that are residents, regular colonists, and vagrants for each state (Table 6-1). Residents are species that reproduce yearly and can survive all seasons. Regular colonists are species that do not survive the winter, but which annually immigrate into the state and usually establish temporary breeding populations. Vagrants are species that have been reported in the state, but do not breed there

TABLE 6-1 The Number of Recorded Species of Butterflies That are Residents, Regular Colonists, and Vagrants for Each State of the Conterminous United States

State	Butterfly Residents and Colonists	Butterfly Vagrants	Breeding Birds[a]
Alabama	132	2	145
Arizona	246	80	246
Arkansas	127	25	130
California	225	25	286
Colorado	230	36	235
Connecticut	101	13	158
Delaware[b]	91	7	160
Florida	163	18	160
Georgia	151	8	160
Idaho[b]	154	6	?
Illinois	121	22	160
Indiana	123	19	151
Iowa	107	19	154
Kansas	133	50	175
Kentucky	116	17	153
Louisiana	117	15	158
Maine	88	13	176
Maryland	121	19	192
Massachusetts	93	19	177
Michigan	134	10	202
Minnesota	132	13	224
Mississippi	134	10	130
Missouri	125	31	175
Montana[b]	184	3	224
Nebraska	170	27	194
Nevada	181	26	224
New Hampshire	92	9	175
New Jersey	120	23	181
New Mexico	272	46	247
New York	119	19	220
North Carolina	140	11	178
North Dakota	132	11	171
Ohio	131	7	180
Oklahoma	146	16	180
Oregon	159	5	232
Pennsylvania	114	20	185
Rhode Island[b]	83	4	141
South Carolina	133	9	152
South Dakota	149	21	207
Tennessee[b]	112	12	160
Texas	290	133	300
Utah	197	18	220
Vermont[b]	66	6	175
Virginia	134	21	179
Washington	140	3	235
West Virginia	112	8	156
Wisconsin	133	12	203
Wyoming	197	13	222

[a]From Peterson (1963).
[b]Incomplete census for butterflies.

except on rare occasions. These categories were based on biology, caterpillar host plants, geographic range, and published reports of breeding. In some cases, category determinations were arbitrary decisions. For example, *Atalopedes campestris* is considered to be a resident in Maryland and Virginia and, although it may not survive every winter, it is judged to be a vagrant where recorded to the north.

The number of species of butterflies recorded per state ranges from 87 for Rhode Island, an incompletely sampled state, to 423 for Texas. Number of species increases from north to south. For example, along the Atlantic seaboard, species richness of butterflies increases steadily from the 101 recorded in Maine to the 181 in Florida. Among the Pacific coastal states, Washington has 143 species, Oregon 164, and California 250.

Texas has the richest butterfly fauna, influenced by the lower Rio Grande Valley. Even though only a few thousand acres of dry tropical forest habitat remain in a few parks, reserves, and refuges, many of Mexico's species of tropical lowland butterflies have been recorded in Cameron, Hidalgo, and Starr counties, largely as vagrants. Nevertheless, quite a few butterflies have their only breeding populations in the United States in the lower Rio Grande Valley.

A second trend is for high species richness to occur in states with greatest topographic diversity. The Rocky Mountain states have the greatest topographic diversity and have faunal connections through the mid-continental cordillera to both the Arctic and species-rich Mexico, through west Texas and southeastern New Mexico to its Sierra Madre Oriental, and through the very rich areas of southeastern Arizona and southwestern New Mexico to the Sierra Madre Occidental. These faunal connections are shown by species with southern biogeographic affinities that range northward at low to intermediate elevations from Mexico, and those that range southward at high elevations from the arctic and subarctic. All Rocky Mountain states have endemic western North American butterflies, but those bordering the Great Plains also include a significant number of eastern species in their faunas.

Another recognizable trend is for species richness to be greater in the west than in the east. While partly the result of relatively larger size of the states, greater topographic diversity, and their proximity to species-rich Mexico, it is nonetheless true that more butterflies per unit area are found in the richest areas of western states. Florida, the richest eastern state, has 163 species of residents and regular colonists, while the only western states having fewer species are Idaho (154), Oregon (154), and Washington (140).

The poorest region, relative to its latitude, is the alluvial Mississippi drainage, including the states of Arkansas (127), Illinois (121), Iowa (107), Louisiana (117), Mississippi (134), and Missouri (125). The somewhat higher regional species richness in Arkansas, Mississippi, and Missouri is no doubt due to their modest topographic relief. Historical factors such as the relatively recent flooding of the Mississippi embayment also may have had a role.

NEOTROPICAL DIVERSITY

To document how diversity varies within the Neotropics, we tabulated species richness at single localities (Figure 6-3). We did not distinguish tropical breeding residents from strays because, with the exception of some migrants (e.g., Beebe, 1949, 1950a, 1950b, 1951), there is little evidence that tropical butterflies stray to areas where they do not breed.

La Selva field station is situated in lowland rain forest on the Atlantic side of Costa Rica and is relatively well-collected for the larger butterflies. The number of recorded Papilionidae, Pieridae, and Nymphalidae is 204 species (DeVries, 1994). These three families comprise about one-third of the fauna in Panama (Robbins, 1982), the Amazon Basin (Robbins et al., 1996), and at Itatiaia, a national park primarily above 500 m in Rio de Janeiro state (Zikán and Zikán, 1968). Consequently, 600-650 species is probably a reasonable estimate of La Selva's species richness. Since Belém (Brazil) is not nearly so well-documented, the 700 species recorded by Bates (1875) is a minimal value. For Madre de Dios, Peru, 1,231 species have been recorded since 1979 at the Tambopata Reserve (5,500 ha), and 1,300 species were recorded on five field trips averaging less than 3 weeks each to Pakitza (<4,000 ha), Manu National Park (Robbins et al.,

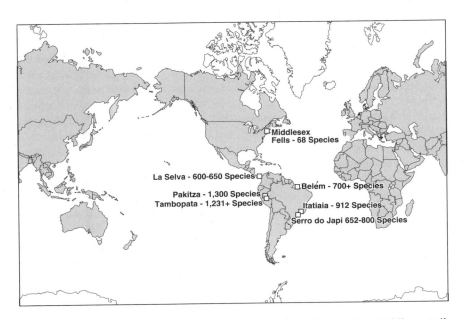

FIGURE 6-3 Number of species of butterflies at neotropical sites. The Middlesex Fells Reservation in Massachusetts is included for comparison with a site in the temperate zone (from Robbins, 1993b).

1996). For southern Brazil, 912 species of butterflies were found over 5 decades at Itatiaia (Zikán and Zikán, 1968; 17 species of Acraeini were omitted in this paper), and 652 species were collected in 8 years at Serra do Japi (750-1,286 m), a reserve near Jundiai, São Paulo (Brown, 1992). Documentation for the Peruvian and Brazilian sites includes museum vouchers.

Even though the number of species at Pakitza and Tambopata continue to increase with each field trip, each of these sites already has more species of butterflies than most, if not all, countries in tropical Africa and Asia (Robbins, 1993a). Clearly, explanations for the greater butterfly diversity of the Neotropics need to account for the extraordinary richness that may occur at single neotropical sites (the within-habitat, alpha, and point diversity concepts of MacArthur, 1969; Whittaker, 1972; Pielou, 1975).

The high butterfly diversity at Pakitza and the Tambopata Reserve is not unique. From what we know about the distribution of neotropical butterflies, there appears to be a band of high butterfly diversity from southern Colombia to the Peru-Bolivia border, ranging eastward from the base of the Andes to the Brazilian states of Acre and Rondônia. This band also appears to extend, with slightly decreased diversity, along the eastern base of the Andes in Venezuela and Bolivia, but documentation is poor.

The upper Amazonian band of high butterfly diversity very roughly consists of two faunal zones and is not uniformly high in diversity. The Rio Madeira drainage in the south has a distinct dry season from about May to September, approximately 2,000 mm annual precipitation (Erwin, 1983, 1991; Terborgh, 1983), and a well-documented high butterfly diversity. Besides Pakitza and the Tambopata Reserve, Jaru and Cacaulandia, Rondônia, Brazil, appear to have similarly high diversities (Brown, 1984; Emmel and Austin, 1990). Alternately, much of the drainage of the Rio Solimões (upper Amazon River) in the north lacks a distinct dry season, has more than 3,000 mm annual precipitation (Gentry, 1988b), and supports a poorly documented butterfly fauna. From museum collections, we infer that parts of eastern Ecuador and the Iquitos, Loreto, Peru, areas are very rich, although sites in the vicinity of Pantoja, Loreto, Peru, on the Rio Napo are relatively poor in species. The faunas of the Rio Solimões and Rio Madeira mix in parts of Acre, Brazil, and Ucayali, Peru, which consequently may be the richest areas in the world for butterflies (Brown and Lamas, personal communications, 1993).

BUTTERFLIES, BIRDS, AND MAMMALS

If patterns of species richness and endemism were similar for different groups of organisms, then knowing these patterns for any group, such as mammals, would be sufficient to determine "biological" priorities among potential sites for conservation. However, at the scale of 100 km^2, butterfly diversity is not well correlated with the diversity of other groups of organisms in temperate

England (Prendergast et al., 1993). For the Neotropics, although lakes and caves do not affect butterfly diversity at a site, they can affect the diversity of bats and freshwater birds. Further, species richness of lowland plants is correlated with precipitation (Gentry, 1988b), and the same is probably true for butterflies. (The data are scanty, but areas with less than 1,500 mm annual precipitation have fewer species of butterflies than wetter areas.) However, the diversity of neotropical mammals does not appear to be correlated with precipitation (Emmons, 1984). Consequently, patterns of butterfly diversity in the Neotropics are not expected to be strongly correlated with the patterns of mammals.

We recorded the number of breeding birds for each state in the United States (Table 6-1). Because diversity is a function of area, we expected the numbers of breeding birds and nonvagrant butterflies for each state to be positively correlated. After omitting incompletely documented states, a Spearman rank correlation coefficient was indeed highly significant ($r=0.606$, $n=42$, $p<0.001$). However, the increase in butterfly diversity from north to south (discussed above) is less pronounced in birds than butterflies. For example, Florida, Georgia, and South Carolina each have fewer breeding birds (152-160) than Maine or Massachusetts (176-177). Whereas the bird fauna of New Mexico is about 10% greater than that of Wyoming, the butterfly fauna is nearly 40% greater.

We graphed the percentage of the world's species of butterflies that occur in each major biogeographical realm (Figure 6-4) with similar percentages for

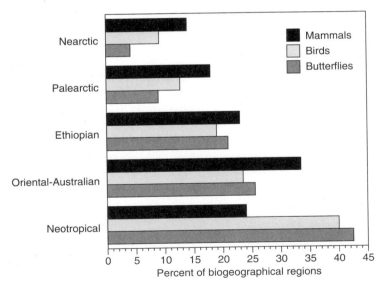

FIGURE 6-4 Proportions of species of butterflies, terrestrial mammals (Cole et al., 1994), and nonmarine birds (Welty and Baptista, 1988) that occur in each of the world's major biogeographical realms.

breeding nonmarine birds (Welty and Baptista, 1988) and terrestrial mammals (Cole et al., 1994). Although slightly different boundaries were used by each author, species richness of butterflies is strongly correlated with diversity of birds, but not with diversity of mammals. For example, the neotropical realm is the richest region for butterflies and birds (40-43%), but fewer than 25% of the world's mammals are neotropical.

Perhaps the major difference between the diversities of butterflies, birds, and mammals is that butterflies are more "tropical" than birds, which, in turn, are more "tropical" than mammals. The percentage of the world's species that occur in the northern temperate Holarctic (including the Nearctic and Palearctic) is 32% for mammals, 21% for birds, and 13% for butterflies. Although there are approximately 2 species of butterflies for every species of bird worldwide, birds greatly outnumber butterflies in the Arctic, have about equal numbers of species as butterflies in temperate North America, and are outnumbered by butterflies in the Neotropics (Table 6-2). Very roughly, an upper Amazonian site will have 3-4 times more species of mammals (including bats)(Emmons, personal communication, 1993), 5 times more species of birds, and 15 times more species of butterflies than a temperate North American site.

DISCUSSION

Among the well-known taxonomic groups of terrestrial animals, butterflies have the greatest number of species. With 17,500 species, they are three to five times more numerous than mammals (Wilson and Reeder, 1993), amphibians (Zug, 1993), mosquitos (Wilkerson, personal communication, 1993), termites (Nickle, personal communication, 1993), or dragonflies (Louton, personal communication, 1993). There are approximately two species of butterflies for every species of nonmarine bird (Welty and Baptista, 1988), and a bit less than three species of butterflies for every one of reptiles (Zug, 1993).

TABLE 6-2 Number of Breeding Birds and Butterflies for Greenland, Georgia, Panama, and Colombia

Locality	Butterflies	Birds
Greenland	5	57
Georgia	151	160
Panama	1,550	710
Colombia	3,100[a]	1,556

[a]Estimate is probably low.

SOURCES: Birds: Panama (Ridgely, 1976); all others (Welty and Baptista, 1988); Butterflies: Greenland (Wolff, 1964); Georgia (this paper); Panama (Robbins, 1982); Colombia (Brown, 1991).

Approximately 90% of the world's species of butterflies have been described taxonomically (Robbins et al., 1996). The great majority of the more than 1,500 undescribed species occur in the Neotropics, which means that diversity studies in this realm, even if restricted to the better-known families (Papilionidae, Pieridae, Nymphalidae), need to include systematists specializing on the neotropical fauna. Since the Papilionidae, Pieridae, and Nymphalidae make up about one-third of this fauna, at least in the lowlands (Robbins et al., 1996), it is possible to focus on these families and still estimate the number of Hesperiidae, Lycaenidae, and Riodinidae.

Although hypotheses can be suggested to explain why butterflies and mammals have different patterns of species richness, we have focused on the implications for conservationists. If butterfly distributions are representative of other insects, then strategies for conserving insects should not be based on the diversity patterns of terrestrial vertebrates, particularly mammals. Further, butterflies are far more "tropical" than mammals, which emphasizes the importance of protecting tropical areas for conserving terrestrial arthropods. From a biological viewpoint, it might be more reasonable to expect a high correlation between butterfly and plant diversity, but this hypothesis appears to be largely unexamined.

The attention given to the magnitude of differences between tropical and temperate areas appears to have had the unfortunate effect of obscuring difficulties in defining, measuring, and comparing diversity. Given that the abundance of tropical insects fluctuates markedly, changing the species composition at a site from year to year (e.g., Wolda, 1992), is the species richness of butterflies in the vicinity of Belém the number of species occurring there in 1 day, 2 months, or 5 years? If the comparison is with England's depauperate fauna, it does not matter which "definition" is used for Belém's species richness. But, if the comparison is with Manaus and Iquitos, then it may first be necessary to understand the dynamics of tropical butterfly communities.

CONCLUSIONS

There is much to be done in developing a general theory that will predict the patterns of diversity that we are beginning to document. We do not know of any current theory that predicted that the upper Amazon basin at the eastern base of the Andes would have the highest species richness in the world or that single sites there might have more species of butterflies than tropical countries on other continents. The Rio Madeira drainage is the richest area for birds and the Rio Solimões has the greatest diversity for canopy trees (Gentry, 1988a); however, neither theory nor correlations of diversity with other groups allows us to predict which of these areas will be richer for butterflies. There is yet a long way to go before we understand tropical butterfly diversity.

ACKNOWLEDGMENTS

For discussion, advice, and unpublished information, we thank P. Ackery, K. Brown, J. Burns, M. Casagrande, G. Ceballos, R. Cole, L. Emmons, D. Harvey, G. Lamas, J. Louton, O. Mielke, A. Navarro, D. Nickle, K. Philip, R. Stanford, R. Thorington, G. Tudor, R. Wilkerson, and D. Wilson. For reading the manuscript and making constructive suggestions, we are most appreciative to P. Ackery, G. Austin, J. Brown, A. Gardner, J. Glassberg, G. Lamas, S. Miller, and O. Shields.

REFERENCES

Ackery, P. R., C. R. Smith, and R. I. Vane-Wright, eds. 1995. Carcasson's African Butterflies. An Annotated Catalogue of the Papilionoidea and Hesperioidea of the Afrotropical Region. CSIRO, Canberra, Australia. 803 pp.

Bates, H. W. 1861. Contributions to an insect fauna of the Amazon Valley—Lepidoptera—Papilionidae. J. Entomol. 1:218-245.

Bates, H. W. 1875. The Naturalist on the River Amazons, fourth ed. John Murray, London. 394 pp.

Beebe, W. 1949. Migration of Papilionidae at Rancho Grande, north-central Venezuela. Zoologica (New York) 34:119-126.

Beebe, W. 1950a. Migration of Danaidae, Ithomiidae, Acraeidae and Heliconidae (butterflies) at Rancho Grande, north-central Venezuela. Zoologica (New York) 35:57-68.

Beebe, W. 1950b. Migration of Pieridae (butterflies) through Portachuelo Pass, Rancho Grande, north-central Venezuela. Zoologica (New York) 35:189-196.

Beebe, W. 1951. Migration of Nymphalidae (Nymphalinae), Brassolidae, Morphidae, Libytheidae, Satyridae, Riodinidae, Lycaenidae and Hesperiidae (Butterflies) through Portachuelo Pass, Rancho Grande, north-central Venezuela. Zoologica (New York) 36:1-16.

Brown, K. S., Jr. 1984. Species diversity and abundance in Jaru, Rondônia (Brazil). News Lepid. Soc. 1984:45-47.

Brown, K. S., Jr. 1991. Conservation of neotropical environments: Insects as indicators. Pp. 349-404 in N. M. Collins and J. A. Thomas, eds., Conservation of Insects and Their Habitats. Academic Press, London.

Brown, K. S., Jr. 1992. Borboletas da Serra do Japi: Diversidade, hábitats, recursos alimentares e variação temporal. Pp. 142-187 in L. P. C. Morellato, organizer, História Natural da Serra do Japi: Ecologia e Preservação de um Área Florestal no Sudeste do Brasil. Unicamp/Fapesp, Campinas, Brazil.

Cole, F. R., D. M. Reeder, and D. E. Wilson. 1994. A synopsis of distribution patterns and the conservation of mammal species. J. Mammology 75:266-276.

DeVries, P. J. 1994. Patterns of butterfly diversity and promising topics in natural history and ecology. Pp. 187-194 in L. A. McDade, K. S. Bawa, H. S. Hespenheide, and G. S. Hartshorn, eds., La Selva, Ecology and Natural History of a Neotropical Rain Forest. University of Chicago Press, Chicago.

Ehrlich, P. R., and P. H. Raven. 1965. Butterflies and plants: A study in coevolution. Evolution 18:586-608.

Eliot, J. N., and B. D'Abrera. 1992. The Butterflies of the Malay Peninsula. Malayan Nature Society, Kuala Lumpar, Malaysia. 595 pp.+69 plates.

Emmel, T. C., and G. T. Austin. 1990. The tropical rain forest butterfly fauna of Rondônia, Brazil: Species diversity and conservation. Trop. Lepid. 1:1-12.

Emmons, L. H. 1984. Geographic variation in densities and diversities of non-flying mammals in Amazonia. Biotropica 16:210-222.

Erwin, T. L. 1983. Tropical forest canopies: The last biotic frontier. Bull. Entomol. Soc. Amer. 29:14-19.

Erwin, T. L. 1991. Natural history of the carabid beetles at the BIOLAT Biological Station, Rio Manu, Pakitza, Peru. Revista Peruana de Entomologia 33:1-85.

Ford, E. B. 1945. Butterflies. The New Naturalist. Collins, London. 368 pp.

Gentry, A. H. 1988a. Tree species richness of upper Amazonian forests. Proc. Natl. Acad. Sci. 85:156-159.

Gentry, A. H. 1988b. Changes in plant community diversity and floristic composition on environmental and geographical gradients. Ann. Missouri Bot. Gard. 75:1-34.

Kristensen, N. P. 1976. Remarks on the family-level phylogeny of butterflies (Insecta, Lepidoptera, Rhopalocera). Zeit. Zool. Syst. Evol.-forschung. 14:25-33.

Kuznetsov, N. Y. 1915, 1929. Fauna of Russia and Adjacent Countries, Lepidoptera. Vol. 1, Introduction. Translated (1967) by Israel Program for Scientific Translations (published for the U.S. Department of Agriculture and the National Science Foundation), Jerusalem. 305 pp.

MacArthur, R. H. 1969. Patterns of communities in the tropics. Pp. 19-30 in R. H. Lowe-McConnell, ed., Speciation in Tropical Environments. Academic Press, London.

Opler, P. A. 1995. Conservation and management of butterfly diversity in North America. Pp. 316-324 in A. S. Pullin, ed., Ecology and Conservation of Butterflies. Chapman and Hall, London.

Opler, P. A. 1996. Lepidoptera of North America. 2. Distribution of the butterflies (Papilionoidea and Hesperioidea) of the Eastern United States. Contributions of the C. P. Gillette Museum of Insect Biodiversity, Colorado State University, Fort Collins.

Owen, D. F. 1971. Tropical Butterflies. Clarendon Press, Oxford, England. 214 pp.

Peterson, R. T. 1963. The Birds (Life Nature Library). Time-Life Books, N.Y. 192 pp.

Pielou, E. C. 1975. Ecological Diversity. John Wiley and Sons, N.Y. 165 pp.

Prendergast, J. R., R. M. Quinn, J. H. Lawton, B. C. Eversham, and D. W. Gibbons. 1993. Rare species, the coincidence of diversity hotspots and conservation strategies. Nature 365:335-337.

Raven, P. H. 1990. Endangered realm. Pp. 8-31 in M. C. Christian, managing ed., The Emerald Realm: Earth's Precious Rain Forests. National Geographic Society, Washington, D.C.

Reuter, E. 1896. Ueber die Palpen der Rhopaloceren. Ein Beitrag zur Erkenntnis der verwandtschaftlichen Beziehungen unter den Tagfaltern. Acta Soc. Sci. Fennica. 22:xvi+578 pp.

Ridgely, R. S. 1976. A Guide to the Birds of Panama. Princeton University Press, Princeton, N.J. 394 pp.

Robbins, R. K. 1982. How many butterfly species? News Lepid. Soc. 1982:40-41.

Robbins, R. K. 1993a. Comparison of butterfly diversity in the neotropical and Oriental regions. J. Lepid. Soc. 46:298-300.

Robbins, R. K. 1993b. Middlesex Fells Reservation, Middlesex County, Massachusetts. Pp. 97-99 in J. S. Glassberg, Butterflies Through Binoculars: A Field Guide to Butterflies in the Boston, New York, and Washington Region. Oxford University Press, N.Y.

Robbins, R. K., G. Lamas, O. H. H. Mielke, D. J. Harvey, and M. M. Casagrande. 1996. Taxonomic composition and ecological structure of the species-rich butterfly community at Pakitza, Parque Nacional del Manu, Peru. In D. Wilson and A. Sandoval, eds., La Biodiversidad del Sureste del Peru: Manu, Biodiversity of Southeastern Peru. Editorial Horizonte, Lima, Peru.

Scoble, M. J. 1986. The structure and affinities of the Hedyloidea: A new concept of the butterflies. Bull. Brit. Mus. (Nat. Hist.) Entomol. 53:251-286.

Scoble, M. J., and A. Aiello. 1990. Moth-like butterflies (Hedylidae: Lepidoptera): A summary, with comments on the egg. J. Nat. Hist. 24:159-164.

Scriber, J. M. 1973. Latitudinal gradients in larval feeding specialization of the world Papilionidae. Psyche 80:355-373.

Shields, O. 1989. World numbers of butterflies. J. Lepid. Soc. 43:178-183.

Stanford, R. E., and P. A. Opler. 1993. Atlas of Western USA Butterflies, Including Adjacent Parts of Canada and Mexico. Published by the authors, Denver and Fort Collins, Colo. 275 pp.

Terborgh, J. 1983. Five New World Primates. Princeton University Press, Princeton, N.J. 260 pp.

Vane-Wright, R. I., and P. R. Ackery, eds. 1984. The Biology of Butterflies. Symposia of the Royal Entomological Society of London. Academic Press, London. 429 pp.

Welty, J. C., and L. Baptista. 1988. The Life of Birds, fourth ed. Saunders College Publishing, N.Y. 698 pp.

Whittaker, R. H. 1972. Evolution and measurement of species diversity. Taxon 21:213-251.

Wilson, D. E., and D. M. Reeder. 1993. Mammal Species of the World, second ed. Smithsonian Institution Press, Washington, D.C. 1,312 pp.

Wolda, H. 1992. Trends in abundance of tropical forest insects. Oecologia 89:47-52.

Wolff, N. L. 1964. The Lepidoptera of Greenland. Medd. om Gron. 159(11):1-74+21 pl.

Zikán, J. F. and W. Zikán. 1968. Inseto-fauna do Itatiaia e da Mantiqueira. III. Lepidoptera, Pesquisas Agropecuarias Brasil 3:45-109.

Zug, G. R. 1993. Herpetology: An Introductory Biology of Amphibians and Reptiles. Academic Press, San Diego, Calif. 527 pp.

The Global Biodiversity of Coral Reefs: A Comparison with Rain Forests

MARJORIE L. REAKA-KUDLA

Professor, Department of Zoology, University of Maryland, College Park

There has been increasing concern over declining global biodiversity due to overexploitation and habitat destruction by humans, who now consume 20-40% of the global net terrestrial primary productivity (Ehrlich and Ehrlich, 1992; Ehrlich and Wilson, 1991; Wilson, 1992). Tropical communities are particularly important in the global economics of biodiversity, because it is here that human populations are increasing most rapidly, monetary resources will be most strained, and problems of food production, pollution, and environmental change will be most acute during the twenty-first century.

Two of the most diverse natural communities on Earth, coral reefs and rain forests, both occur in the tropics. Coral reefs resemble rain forests in their biologically generated physical complexity, high species diversity, elaborate specializations of component species, and coevolved associations between species. Rain forests and coral reefs usually are considered to represent the two pinnacles of biodiversity on Earth, yet no detailed attempts to quantify the total species diversity on coral reefs have been made. This chapter describes why coral reefs are important for all societies to conserve and manage for the future, addresses the need for training specialists in the systematics of marine organisms (particularly those who would study the rich and poorly known tropical regions), and provides the first quantified estimate of the total biodiversity of global coral reefs as compared to rain forests.

THE VALUE AND CURRENT STATUS OF CORAL REEFS

Although they generally inhabit nutrient-poor waters, coral reefs are one of the most productive ecosystems on Earth (Grigg et al., 1984). Their fishes and invertebrates traditionally have been and continue to be a critical source of protein for the world's tropical coastal countries. This fact will become increasingly important over the next century and beyond, because it is precisely these countries that are experiencing explosive population growth and some of the most severe coastal degradation. Failure to ameliorate the deterioration of marine waters and to provide management plans for the sustainable use of reef fisheries will remove an essential source of protein for these human populations (Norse, 1993).

In addition, coral reefs form ramparts that enclose lagoons, which are the primary nurseries and feeding grounds for fishes and other organisms. Historically, these protected embayments have facilitated the development of human transportation and commerce systems along tropical coastal areas. Today, they protect human populations from hurricane and wave damage, making coastlines secure for navigation, fishing, and tourism. Bioerosion of the carbonate reef framework and the calcareous shells of reef organisms provide almost all of the sand that comprises most tropical beaches. Tourism on coral reefs is an increasingly important economic resource for developing countries, but such activity must be managed sustainably if tourism is to remain a viable source of income. Geologically, reefs are associated with oil repositories. The diversity on reefs represents a largely unknown and untapped source of genetic material that has potentially great medical, pharmaceutical, and aquacultural use.

However, the coral reefs of the world are endangered by overexploitation, chemical and oil pollution, sedimentation, and eutrophication (resulting from deforestation, construction, and agricultural runoff), as well as large-scale environmental hazards such as increased ultraviolet light exposure and temperature anomalies (Allen, 1992; Hallock et al., 1993; Hughes, 1994; Kuhlmann, 1988; Sebens, 1994). World-wide episodes of coral "bleaching" (loss of symbiotic algae that provide nutrients and increase rates of calcification; Brown and Ogden, 1993; Glynn, 1993; Ogden and Wicklund, 1988; Williams and Bunkley-Williams, 1990), mass mortalities of reef-dwelling organisms (sometimes encompassing entire geographic regions, such as the Caribbean-wide mortality of an ecologically important sea urchin; Hughes et al., 1985; Lessios et al., 1984), and declining abundances of coral and other reef species (Porter and Meier, 1992) have been documented over the last decade.

Some of the above trends have received attention from the media and have been recognized by congressional hearings or other institutional initiatives (e.g., D'Elia et al., 1991). However, the value, risk of loss, and even the amount of biodiversity on coral reefs (and in the marine environment in general) has received relatively little attention by the scientific community, which is based largely in northern temperate zones and is focused primarily on the terrestrial environment (Gaston and May, 1992). Also, although ecotourism and recreation in tropical environments, including coral reefs, has become increasingly popular in recent years, the public generally may be unaware of the urgency and importance of environmental degradation in these tropical marine habitats. In particular, the public has not been adequately informed of the declining numbers of trained biologists who are capable of estimating the amount of biodiversity on coral reefs, documenting how reef biodiversity is biogeographically distributed, and targeting where the biodiversity on global coral reefs is most at risk.

There is a crisis of declining numbers of trained systematists (scientists who analyze the relationships between lineages, the evolutionary trajectories, and the biogeographic distributions of organisms, and signify these patterns in hierarchical taxonomic classifications so that field-caught organisms can be identified and their ecological roles studied). This crisis is especially acute for systematists of marine invertebrates and algae (Feldmann and Manning, 1992; Gaston and May,

Along with rainforests, coral reefs represent the second pinnacle of biodiversity on Earth.

1992; John, 1994). There is an urgent need for major national and international initiatives in the systematics and biodiversity of the global marine biota, particularly of the richest and least known components, the tropical marine biota.

The next section provides an estimate of the global biodiversity of coral reefs and suggests that the lack of prior availability of such information is due to the fact that marine environments in general and tropical marine habitats in particular are understudied and relatively poorly known. This prevents us from assessing which marine habitats of the world are in the most serious jeopardy and, most importantly, prevents us from being able to recognize extinctions if they occur or even from assessing the potential for extinctions. I argue that extinctions in marine macrofauna (particularly on coral reefs) are likely to have been more frequent than we have thought, and that the potential for loss of coral reef species—with all of their inherent genetic and ecological value—over the next 100 years is great.

A QUANTIFIED ESTIMATE OF BIODIVERSITY ON CORAL REEFS

There currently are about 1.8 million described species in all environments on Earth, although various authors have estimated that 5-120 million species exist and that <10-50% of the Earth's species may be known (Ehrlich and Wilson, 1991; Erwin, 1982, 1988; Gaston, 1991; Grassle and Maciolek, 1992; May, 1988, 1990, 1992; National Science Board Task Force on Biodiversity, 1989; Stork, 1988; Wilson, 1988, 1992). Most international concern over declining biodiversity has been focused on terrestrial environments, particularly the rapidly vanishing rain forests. This concern is justified, given the discoveries over the last decade of how many species are present in these tropical wonderlands, the potential uses of such genetic diversity, the potential effect of burning rain forests on global climate, and the shocking rates—graphically relayed by satellite—at which these habitats are being eclipsed by the activities of humans. Until recently, however, the amount of biodiversity and its possible decline in marine environments has received little attention.

It has been recognized that marine environments have more higher-level taxonomic diversity than terrestrial environments. Among all macroscopic organisms, there are 43 marine phyla and 28 terrestrial phyla (of the 33 animal phyla, 32 live in the sea and only 12 inhabit terrestrial environments), and 90% of all known classes are marine (Angel, 1992; May, 1994; Pearse, 1987; Ray, 1985, 1988, 1991). Many of today's marine animal phyla originated or diversified during the Cambrian evolutionary radiation more than 500 million years ago, whereas plants and then animals invaded land later in the Paleozoic Era (approximately 200-400 million years ago; Signor, 1994). This long separation of evolutionary pathways among marine lineages has resulted in a greater variety of body plans, greater functional and biochemical diversity, and greater "endemism" in major groups of marine compared to terrestrial animals (64% of animal phyla inhabit only the sea, while only 5% of animal phyla live exclusively on land; May, 1994).

In addition to containing great higher-level taxonomic diversity, some marine environments also contain high species diversity. For example, the deep seas are repositories of biodiversity. Based on samples of 1,597 species of marine macrofauna from soft bottoms off the east coast of North America (255 to 3,494 m depths), Grassle and Maciolek (1992) calculated that the global deep sea fauna, because of the huge area it occupies, may include 10 million species (mostly polychaete worms, crustaceans, and molluscs; but see lower estimates in May, 1992, 1994; Poore and Wilson, 1993). Most of these species from the deep sea are rare (90% of the species sampled by Grassle and Maciolek comprised <1% of the individuals, and 28% of the species in the entire fauna were collected only once).

Local marine habitats also contain high numbers of species. Grassle and Maciolek (1992) reported 55-135 species in individual 30 × 30 cm cores of ocean floor sediment at 2,100 m depth. In a shallower soft-sediment environment in south Australia, more than 800 species were found within a 10 m^2 area (Poore and Wilson, 1993). Hughes and Gamble (1977) obtained 350 species from intertidal soft substrates around a reef flat on Aldabra Atoll, and 6 liters of sediment on Oahu yielded 158 species of polychaete worms (Butman and Carlton, 1993).

Coral reef communities also contain high local diversities of species. These communities can be divided into three main components: (1) the suprabenthic fishes; (2) sessile epibenthic organisms that provide the complex structure of the reef (hard and soft corals, sponges, coralline and fleshy algae); and (3) the cryptofauna, which includes organisms that bore into the substrate (primarily sponges, polychaete and sipunculan worms, and bivalves), sessile encrusters living within bioeroded holes and crevices (e.g., bryozoans, sponges, tunicates, polychaete worms), and motile nestlers inhabiting bioeroded holes and crevices (e.g., polychaete and sipunculan worms, echinoderms, molluscs, and especially crustaceans). Although we usually think of coral reefs in terms of the first two components, in fact most of the diversity as well as biomass of coral reef communities is included within the cryptofauna, which is the functional equivalent of insects in the rain forest.

On the Great Barrier Reef of Australia, 350 species of hermatypic (reef-building) corals (33 of them endemic) are recognized, and 242 species are known from one island (Ishigaki) in the Indo-West Pacific (Veron, 1985). Fifteen hundred species of reef fishes have been reported from the Great Barrier Reef (Sale, 1977), 496 species of fishes are known from the Bahamas and adjacent waters (Bohlke and Chapin, 1968), 442 species of fishes have been recorded in the Dry Tortugas (Florida)(Longley and Hildebrand, 1941), and 517 species of fishes occur on Alligator Reef (Florida) alone (Starck, 1968). Using rotenone to sample the fishes on small areas of reef, single collections have yielded 67-200 species in the Bahamas and Palau, respectively (Goldman and Talbot, 1976; Smith, 1978a). Bohnsack (1979) reported 10-23 species of reef fishes living around single coral heads on Big Pine Key, Florida.

The reef-associated cryptofauna also is diverse at several scales. Taylor (1968) found 320 molluscan species in a 31,000 km² area of the Seychelles. Peyrot-Clausade (1983) documented 776 species of motile cryptofauna (four phyla) in dead coral from one reef flat in Moorea. One species of coral, *Oculina arbuscula*, in Florida provided habitat for 309 species of organisms larger than 0.2 mm (McClosky, 1970). Fifty-five species of decapod crustaceans have been reported to live in the coral *Pocillopora damicornis* in the Pearl Islands, Panama, and up to 101 species of decapod crustaceans were found in *P. damicornis* in the Indo-West Pacific (Abele, 1976; Abele and Patton, 1976; Austin et al., 1980). Bruce (1976) documented 620 species of shrimps and prawns that are commensal on different species of corals. Jackson (1984) documented 46 species of encrusting cheilostome bryozoans between 0-21 m depth in Jamaica. On a smaller scale, Gibbs (1971) found up to 220 species (8,265 individuals) of boring cryptofauna in single colonies of dead coral, and Grassle (1973) reported 103 species of polychaete worms in one colony of living coral. Thus, although no comprehensive all taxa biodiversity inventories (Yoon, 1993) of coral reef habitats have been made, it seems likely that diversity within as well as between coral reef habitats is extraordinarily high.

The total species diversity on global coral reefs has been difficult to quantify precisely, however. The most current information on the number of species contained within a group of organisms is found in monographic treatments of the systematics and evolutionary biology of individual taxa, but different monographs often target different taxonomic levels (groups of species within a genus, or groups of genera, families, or superfamilies) and usually include species from all of the habitats occupied by the taxon (freshwater, estuarine, terrestrial, marine), making it difficult to tally the numbers of species on global coral reefs among all taxa.

Using the concepts of island biogeography, known and calculated areas of the major marine and terrestrial regions of the globe, and several testable assumptions about the biogeographical distribution and abundance of marine species, one can calculate the described and expected species diversity of coastal marine organisms, tropical coastal marine organisms, and coral reef organisms in comparison to that of rain forests. My laboratory at the University of Maryland is in the process of testing with empirical data the generalities about relationships between species richness in different habitats and biogeographical realms that are present in or inferred from the literature and used in the present calculations. Further additions to the database may alter the numerical results slightly, but are not expected to substantially change the conclusions. Also, the results presented here can be modified and updated easily as more data become available, since the assumptions and mathematical relationships are identified.

Available data and calculations (Table 7-1) reveal that the terrestrial realm includes about 33%, global rain forests 2%, coastal zones 8%, tropical seas 24%, and tropical coastal zones 2% of the global surface area. Global coral reefs

TABLE 7-1 Area Relationships of Coastal Marine Zones and Terrestrial Regions of the Globe (all areas are $n \times 10^6$ km^2)

Zones and Regions	km^2	% of Earth
Global surface area	511	100
Global land area	170.3	33.3
Global rain forests	11.9	2.3
Global oceans	340.1	66.7
Tropical seas	123.0	24.0
Global coastal zones	40.9	8.0
Tropical coastal zones	9.8	1.9
Coral reefs	0.6	0.1

SOURCES: Data were taken or calculated from information provided in Kuhlmann (1988), the Rand McNally Atlas (1980), Ray (1988), Smith (1978b), and Wilson (1988).

comprise about 0.1% of the Earth's surface, 6% of tropical coastal zones, and 5% of the area of global rain forests.

Table 7-2 shows that there are approximately 1,450,000 currently described species of terrestrial organisms (about 78% of the global biota), about 100,000 currently described species of symbiotic organisms (about 5% of the total), and approximately 318,000 described species of aquatic organisms (17% of the global total). Of aquatic organisms, my calculations from data in Pennak (1989) and other sources revealed that about 26,000 species, or about 13% of the species of macroscopic invertebrates overall, inhabit freshwater. Although about 40% of the world's species of fishes occur in freshwater (Ray, 1988), only 5-10% of macroalgal species live in freshwater environments (John, 1994, and personal communication, 1995). The above independently derived figure of about 13% freshwater invertebrates is in good agreement with May's (1994) data (drawn from a tabulation of species in all benthic and pelagic marine and freshwater animal phyla), which show that about 12% of all aquatic species live in freshwater. Consequently, to assess the number of marine species within the relatively little-known microscopic algae, viruses and bacteria, and protistans (all of whose affinities for freshwater or marine habitats might be expected to be closer to that of macroalgae and invertebrates than to fishes), the proportion of marine versus freshwater species was estimated to be about 90% and 10%, respectively.

Thus, of the 318,000 described species of aquatic organisms, a total of about 274,000 species was estimated to be marine (including approximately 180,000 species of macroscopic marine invertebrates; 36,000 species of micro- and macroscopic marine algae; and 58,000 species of other marine groups such as vertebrates, protistans, viruses and bacteria; Table 7-2). Therefore, about 15% of global described species are marine (a figure independently obtained by May, 1994). If only macroscopic marine species are included due to uncertainties in the taxonomy of microorganisms, there would be about 200,000 described species of marine macroalgae, macroinvertebrates, and chordates, or about 11% of the total described global species.

From the total numbers of described species of marine animals and plants (above), one can calculate the number of species in *global coastal zones* by estimating that about 80% of all marine species occur in the coastal zones (National Science Board Task Force on Biodiversity, 1989; Ray, 1988, 1991). This figure probably is conservative. Over 90% of all marine species are benthic (bottom-living) rather than pelagic (May, 1988, 1994). Almost all marine macroalgae live in benthic (John, 1994, and personal communication, 1995) sunlit environments, and oceanic phytoplankton comprise only 9-11% of all algal species (Sournia and Ricard, 1991).

One then can calculate the number of described marine species that should occur in *tropical coastal* and *coral reef* environments based on the global area of these regions and current knowledge of biogeographic patterns. These calculations employ known theoretical and empirical relationships between the rate at

TABLE 7-2 Species Diversity of Major Groups of Living Organisms

Organisms	Number of Described Species (to nearest 1,000)	% of Total Described Species (@ 1.87 million)
Terrestrial Organisms		
Terrestrial chordates	23,000	1.2
Insects	950,000	50.8
Noninsect and noncrustacean arthropods	80,000	4.3
Other terrestrial invertebrates (molluscs, nematodes, annelids, platyhelminths, etc.)	57,000	3.0
Fungi	70,000	3.7
Terrestrial plants	270,000	14.4
Total Terrestrial Species	1,450,000	77.5
Aquatic Organisms		
Algae	40,000	2.1
All marine algae[a]	36,000	1.9
All freshwater algae[a]	4,000	0.2
Marine macroalgae	4,000-8,000	0.2-0.4
Freshwater macroalgae	450	<0.1
Viruses and procaryotes	10,000	0.5
Marine viruses and procaryotes[a]	9,000	0.5
Freshwater viruses and procaryotes[a]	1,000	<0.1
Protozoa	40,000	2.2
Marine protozoans[a]	36,000	1.9
Freshwater protozoans[a]	4,000	0.2
Macroinvertebrates		
Marine macroinvertebrates	180,000	9.6
Freshwater macroinvertebrates	26,000	1.4
Chordates		
Marine chordates	13,000	0.7
Freshwater chordates	9,000	0.5
Total Marine Species		
All taxa	274,000	14.7
Macrobiota	197,000-201,000	10.5-10.7
Total Freshwater Species		
All taxa	44,000	2.4
Macrobiota	35,000	1.9
Total Aquatic Species	318,000	17.0
Symbiotic Organisms		
Total Symbiotic Species	100,000	5.3
Total Global Described Biodiversity	1,868,000	—

[a]Assumes that the proportions of marine and freshwater species are 90% and 10%, respectively (see text).

SOURCES: Data were taken, calculated, or updated from Barnes (1984), Brusca and Brusca (1990), Ehrlich and Wilson (1991), Hammond (1992), Hawksworth (1991), John (1994, and personal communication, 1995), Margulis and Schwartz (1988), May (1988, 1991, 1992, 1994), Parker (1982), Pearse (1987), Pennak (1989), Raven and Wilson (1992), Ray (1988, 1991), Systematics Agenda 2000 (1994), and Wilson (1988).

which numbers of species change with area ($S=cA^z$, where S is number of species, c is a constant, A is area, and z is a scaling factor that usually falls between 0.2 and 0.3; MacArthur and Wilson, 1967; May, 1975, 1994; Wilson, 1989, 1992). Where $z=0.25$, a reduction of 90% in area coincides with a reduction of about half of the species present, which approximates natural situations for faunas on islands of different sizes or where habitat destruction has reduced the amount of area available to species.

Using the above biogeographical equations and the assumptions that tropical coastal zones are approximately twice as rich in species (or, as modeled here, that z, the exponent in the above equation, is twice as high in tropical as in temperate faunas) and are as well studied as those at higher latitudes, tropical coastal zones should include about 195,000 total described species and 143,000 described species of macrobiota, given their area (Table 7-3). A review of data and inferences in the literature suggests that the assumption of double area-specific diversity in the tropics is realistic but may be conservative. Although data often are not available on an area-specific scale, there are two to at least three times more species in tropical than temperate environments for most (though not all) groups of organisms (Angel, 1992; May, 1986a, 1988; Raven and Wilson, 1992; Rex et al., 1993; Stevens, 1989; Stork, 1988). Also, because of the assumption that the tropical coastal zone is as well studied as the global coastal zone (which likely is not met), the values presented likely underestimate true biodiversity in tropical coastal zones.

Similarly, using the above area relationships and assuming that the complex coral reef substrate contains approximately twice as many species per unit area (or, that z is twice as large) and is as well studied as level-bottom (sand, mud) habitats in the same biogeographical region, there are about 93,000 described species of all coral reef taxa and 68,000 species of described coral reef macrobiota on Earth. Although Abele (1976) reports that 53 species of crustaceans occupy coral habitat (*P. damicornis*) compared to 16 species in sandy beach habitats on the Pacific coast of Panama, biodiversity in coarse- versus level-bottom marine habitats probably needs to be more extensively quantified to document this assumption. Thusly calculated, though, the total described species on coral reefs represents only about 5% of the described global biota.

In contrast, rain forests may account for more than 70% of the described global biota (Table 7-3). If 90% of currently described terrestrial species occurred in rain forests (as do primates; Mittermeier, 1988) and if all groups were as well known as primates, then rain forests would include about 1,305,000 described species. This yields an underestimate of the true number of species in rain forests, however, since about two-thirds of currently described species (mostly insects) are thought to occur in temperate regions (due to more intensive study there), and there probably are two undescribed species of tropical insects for every described species of temperate insect (May, 1986a, 1988). Other estimates, incorporating the high probability that large numbers of undescribed

TABLE 7-3 Calculated and Expected Species Diversity on Global Coral Reefs for all Taxa and Macrobiota

Organisms	Number of Described Species (to nearest 1,000)	% of Total Described Species (@ 1.87 million)
Total Described Marine Species[a]	274,000	14.7
Macroscopic Described Marine Species[a]	200,000	10.7
Animals	193,000	10.3
Algae	4,000-8,000	0.2-0.4
Total Described Coastal Species (if 80% of all marine species are coastal)	219,000	11.7
Macroscopic Described Coastal Species (if 80% of macroscopic marine animals and most marine macroalgae are coastal)	160,000	8.6
Animals	154,000	8.2
Algae	4,000-8,000	0.2-0.4
Total Described Tropical Coastal Species (if communities in the tropical coastal zone are as well studied and twice as diverse as those at higher latitudes; tropical coastal zone=24% of global coastal zone[b])	195,000	10.4
Macroscopic Described Tropical Coastal Species (same assumptions)	143,000	7.6
Animals	138,000	7.4
Algae	3,000-7,000	0.2-0.4
Total Described Coral Reef Species (if reef communities are as well studied and twice as diverse as those on nonreef level bottoms; coral reefs=6% of tropical coastal zone[c])	93,000	5.0
Macroscopic Described Coral Reef Species (same assumptions)	68,000	3.6
Animals	66,000	3.5
Algae	2,000-3,000	0.1-0.2
Global Rain Forest Species (1) if 90% of all currently described terrestrial species[a] live in rain forests	1,305,000	72.5
(2) independent conservative estimate of described and undescribed species in rain forests (see text)	2,000,000	—
(3) potential number of described and undescribed species in rain forests (see text)	20,000,000	—

TABLE 7-3 Continued

Organisms	Number of Described Species (to nearest 1,000)	% of Total Described Species (@ 1.87 million)
Expected Global Coral Reef Species (if coral reefs are as diverse and as well studied as rain forests; global coral reefs=5% of the area of global rain forests[d]):		
From (1) above	618,000	34.3
From (2) above	948,000	—
From (3) above	9,477,000	—

[a]From Table 7-2.

[b]$S=cA^z$; S=number of species obtained empirically from the number of described marine and coastal species above; $z=0.265$ and 0.133 for species in tropical and high latitudes, respectively; A=area known from Table 7-1; and c is provided by solution of the equation.

[c]$S=cA^z$; A is known from Table 7-1; $z=0.265$ and 0.133 for reef and level-bottom tropical coastal habitats, respectively; c is determined above [b], and S is provided by solution of the equation.

[d]$S=cA^z$; A is known from Table 7-1 for rain forests and coral reefs; S is known and c is calculated for rain forests; $z=0.25$ for rain forests and coral reefs; c is the same as for rain forests and S is calculated for coral reefs.

species occur in tropical rain forests, indicate that rain forests likely contain 2 to >20 million species (Ehrlich and Wilson, 1991; Wilson, 1989). Two million species will be used as a conservative estimate of species in rain forests hereafter. Although rain forests cover 20 times more global surface area than coral reefs (Table 7-1), and thus one would expect fewer species on global coral reefs than rain forests, the calculated number of described species on coral reefs (93,000) still is extraordinarily low.

Based on the area of the globe that they occupy compared to that of rain forests, coral reefs should be comprised of about 600,000-950,000 total species (34-53% of currently described global species; Table 7-3), assuming that rain forests have 1-2 million species, that the two environments are equally studied, and that similar ecological and evolutionary processes operate on coral reefs as in rain forests (i.e., coral reefs would have the same biodiversity as rain forests if they occupied equal global area). If rain forests included 10 million species and coral reefs had equivalent area-specific diversity, coral reefs would contain 4,739,000 species; and if 20 million species existed in rain forests, coral reefs would contain more than 9 million species (Table 7-3). The true number of species on global coral reefs probably is at least 950,000, because 2 million species in rain forests is likely to be a conservative figure.

EVALUATION OF THE RESULTS AND THEIR IMPLICATIONS

The difference between the figures for *described* species on global coral reefs (93,000 for all species and 68,000 for macrobiota) versus *expected* total species on global coral reefs (at least 950,000) suggests two hypotheses: (1) only about 10% of all species on reefs have been studied and described, or conversely, that ≥90% of the species on the world's coral reefs remain undiscovered (note that these calculations are based on conservative figures for the number of species in rain forests, so that the expected number of species on coral reefs may be larger and the proportion of described species lower on coral reefs [especially for microorganisms] than is represented here); (2) alternatively, the assumption that similar ecological and evolutionary processes generate and maintain diversity in coral reef and rain forest communities may be incorrect, and coral reefs indeed may have lower area-specific diversity than rain forests due to biological or historical constraints that affect diversification or extinction.

Several lines of evidence suggest that the first rather than the second hypothesis is correct. High numbers of undocumented species are likely in coral reef environments because, being far from the location of most systematists and biologists, tropical environments are less studied than those in temperate latitudes (Diamond, 1989; Erwin, 1988; Gaston and May, 1992; May, 1994; Wilson, 1985, 1988). For example, 80% of ecological researchers and 80% of insect taxonomists are based in North America and Europe, in contrast to 7% in Latin America and tropical Africa, and about 78% of borrowed botanical specimens go to North American and European institutions compared to those in the Neotropics or African tropics (Gaston and May, 1992). Tropical marine environments provide even further barriers to study because they require scuba diving and fairly extensive logistic support for investigation.

In addition, as Gaston and May (1992) have shown, both the number of taxonomists and the level of scientific effort devoted to the systematics of "other" invertebrates (crustaceans, molluscs, echinoderms, cnidarians, sponges, and helminths, most of which are marine) is 2 orders of magnitude less than that devoted to the systematics of tetrapod vertebrates and 1 order of magnitude less than that devoted to plants. These authors also document the aging of the work force (see also Feldmann and Manning, 1992; Manning, 1991). The small numbers within this aging but dedicated and productive cadre of marine systematists are particularly alarming at a time when technological advances and steady progression of knowledge about marine diversity is bearing spectacular fruit.

For example, two new phyla of marine invertebrates, the Loricifera (small worm-like organisms that live between sand grains) and the Vestimentifera (large tube-dwelling worms without a mouth or intestine but a large proboscis-like structure that contains millions of symbiotic bacteria) were described from soft sediments and hydrothermal vents, respectively, as recently as 1983 and 1985 (Grassle, 1986; Raven and Wilson, 1992). Since the communities associated with hydrothermal vents were first discovered in 1977, more than 20 new families or subfamilies, 50 new genera, and over 100 species have been described (Grassle, 1989). One of the largest species of sharks, the megamouth, was described within a new family as recently as 1976 (Raven and Wilson, 1992).

Recent systematic studies indicate that concealed sibling species (morphologically similar and previously classified within one species) are more common in marine taxa than previously thought. Even in large commercially important decapods, 18 distinct new species were recognized within 2 previous species of deep-dwelling crabs (Feldmann and Manning, 1992). One of the commonest and most important species of reef-building corals in the relatively well-known Caribbean region (*Montastrea annularis*) recently was found to consist of 4 species (Knowlton et al., 1992). Despite only slight morphological differences, non-overlapping biochemical characteristics (coincident with differences in life history) clearly distinguished 6 sibling species in a worm that is a well-known indicator of pollution, *Capitella capitata* (Grassle and Grassle, 1976). Recently established specific differences between the endangered Kemp's ridley sea turtle and the widespread olive ridley (Bowen et al., 1991) demonstrate the lack of systematic effort that has been devoted to marine organisms as well as the importance of systematics in conservation and management issues.

As recently as 1992, researchers discovered in the marine plankton a major new archaebacterial group in which genetic relationships to their nearest rela-

The Smithsonian Tropical Research Institute marine station in the San Blas Islands, Panama.

tives (microorganisms in hot springs) are as distant as those between plants and animals (Fuhrman et al., 1992). Similarly, scientists discovered only in the 1980s that photosynthetic marine picoplankton, too small to have been detected previously, are extremely abundant and account for a significant proportion of global primary productivity. In 1988, Chisholm et al. described a new group of these picoplankton that are free-living relatives of *Prochloron*, the hypothesized ancestor of chloroplasts in higher plants. As recently as 1989, newly discovered marine viruses (bacteriophages) were found to be so abundant that one-third of the marine bacterial population could experience a phage attack each day (Bergh et al., 1989).

Further evidence that large numbers of species on coral reefs remain undiscovered comes from the fact that the percentages of described species (about 10%) generated in the above calculations (Table 7-3) are in the same general range as those found for other relatively little studied or tropical groups in which the number of known versus unknown species were counted or estimated. For example, calculation of the overall proportion of described versus unknown species in the *Systematics Agenda 2000 Technical Report* (1994) shows that only 1-12% are thought to be described (this range represents the percentages obtained when the minimum and maximum estimates of species remaining to be discovered are summed for all groups, divided by the total known plus unknown species, multiplied by 100, and this percentage subtracted from 100%). Tabulation of data for individual groups in the *Systematics Agenda 2000 Techical Report* reveals that only 21% of global crustacean species and 26% of global molluscan species have been described, although these taxa represent some of the best-studied groups of marine invertebrates and are commercially important. In other studies, Grassle and Maciolek (1992) found that 31% of the peracarid crustaceans (mostly isopods and amphipods) had been described in soft sediments from the deep sea off eastern North America. From shallower sediments in southern Australia, however, Poore and Wilson (1993) reported that only 10% of the relatively well-studied isopods were known; they suggest that, due to great regional differences in the extent to which the oceans have been studied, probably only 5% of marine invertebrates are known from the oceans overall. About 36% of the species of polychaete worms and about 63% of the relatively well-known molluscan species in Grassle and Maciolek's (1992) samples from the deep sea off North America had been described. About 17% of the total species of algae have been described (John, 1994; Systematics Agenda 2000, 1994). Only 1%, 1-10%, 4-7%, and 2-3%, respectively, of the estimated total species of the poorly studied viruses, bacteria, fungi, and nematodes are described (Systematics Agenda 2000, 1994).

Among terrestrial organisms, only about 7-9% of the global species of spiders and mites, and only about 9-11% of global species of insects have been described (Systematics Agenda 2000, 1994), despite the fact that entomologists who work on insects and spiders represent about 30% of taxonomists and these large groups receive considerable taxonomic attention because of their economic importance (Gaston and May, 1992). Estimates within taxa of insects (Gaston, 1991) indicate that 11-33% of beetles are described; only 10-13% of some beetle families (e.g., the speciose staphilinids) have been described, but 25-50% of others (pselaphids, curculionids, carabids, chrysomelids) are known. Up to half of the flies and the large, relatively well-known butterflies and moths probably have been described (but a much lower proportion of the small, more cryptic microlepidopterans are known). Except for the bees (where about half of the species are described), the proportions of described species in the major hymenopteran superfamilies (e.g., ichneumonoid and chalcidoid wasps, ants) are lower, ranging from 17-25%. Similarly, in the hemipteran and some homopteran (cicadellid) bugs, only 20-33% of the total species are described (although about 50% of the economically important coccoid Homoptera are known).

Thus, an estimate of 10% described species on global coral reefs is not unreasonable in terms of what is known of other (especially predominantly tropical) groups. This estimate is further considered realistic because all of the less well-known marine groups are included in this total estimate for coral reefs, the estimate encompasses some very poorly known regions (e.g., areas in the Indo-West Pacific), and marine realms are still relatively little studied.

Additionally, although we tend to think of coral reef communities in terms of their flamboyant fishes, large sessile organisms such as corals, and large colorful benthic invertebrates such as lobsters, most species on coral reefs are small in body size, as shown in Figure 7-1 for reef-dwelling mantis shrimps. Indeed most cryptic species on coral reefs are *constrained* to small body sizes by the sizes of bioeroded holes in the reef, whose refuge they must obtain in order to survive intense fish predation (Moran and Reaka, 1988; Reaka, 1985, 1986, 1987; Reaka-Kudla, 1991; Wolf et al., 1983). Several authors (Hutchinson and MacArthur, 1959; May, 1986b, 1988; Morse et al., 1985) have documented this skewed size distribution, with vastly more small than large species, for almost all groups of animals (but see Fenchel, 1993; May, 1994; and Stork, Chapter 5, this volume; for the microbiota; i.e., those smaller than 1-5 mm).

Small organisms almost always are poorly observed and known (Gaston, 1991; May, 1978; Mayr, 1969) because they often live in cryptic or interstitial environments. This is true in coral reefs as well, where my laboratory has recorded several hundred thousand small macroscopic (>5 mm) motile reef organisms of 12 or more phyla living within holes and crevices in the upper 10 cm of 1 m^2 of reef substrate (Moran and Reaka, 1988; Moran and Reaka-Kudla, 1990; Reaka, 1985, 1987; Reaka-Kudla, 1991; also see other references on cryptofauna given above). In addition to their cryptic habits, these motile organisms often are crepuscular or nocturnal (and thus often are unobserved even by field biologists; Dominguez and Reaka, 1988). Collection of organisms from these three-dimensional calcareous crypts is difficult and labor intensive, leading to their strong underrepresentation in many ecological and systematic studies. Com-

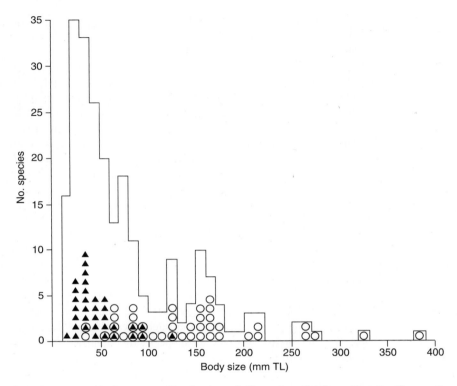

FIGURE 7-1 Size frequency distribution of all species of Atlanto-East Pacific mantis shrimps (Stomatopoda, Crustacea). Closed triangles represent species with known abbreviated larval development, and open circles signify species with known long-distance larval dispersal.

prising the greatest proportion of species on reefs, these cryptic invertebrates are the ecological equivalents of insects in the rain forest, and they usually are overlooked when the diversity of a coral reef is considered.

In addition to inadequate study of these species on local scales, the number of small species is likely to be underestimated on a global scale because of their restricted geographic distributions. One of the strongest correlations in marine biology is the association between small body size of macroscopic marine animals and the production of only a few relatively large larvae that have abbreviated developmental times (where the brooded young either emerge from the parent's protection as relatively large juveniles or the parents produce large larvae with short planktonic stages; both are characterized by short dispersal and restricted geographic distributions; Hansen, 1978; Jablonski, 1986a; Jablonski and Lutz, 1983; Reaka, 1980; Reaka and Manning, 1981, 1987; Strathmann and Strathmann, 1982). In contrast, species that attain large body sizes within their lineage commonly produce large numbers of small swimming larvae that feed in the plankton for extended periods, resulting in broad geographic distributions.

Figure 7-2 shows that the body sizes of species of reef-dwelling mantis shrimps are significantly correlated with the sizes of their geographic ranges (Reaka, 1980). Fenchel (1993) has suggested that there may be fewer rather than more species at the smallest end of the range of body sizes (1-5 mm), and that these microscopic species have larger geographic ranges, larger population sizes, and may be less vulnerable to extinction than the species just above this size (see also May, 1986b, 1988, 1994; Stork, Chapter 5, this volume). The lack of ecological and detailed systematic knowledge for most of these minute taxa, however, may obscure the number of species and the sizes of the geographic ranges, and this argument does not affect the macrobiota (generally larger than 5-10 mm) discussed in the present study. Therefore, because most macroscopic spe-

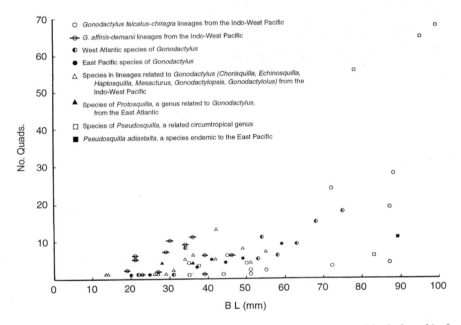

FIGURE 7-2 The relationship between body size (maximum mm total body length) of species and extent of the geographic range (number of 5 × 5° latitudinal × longitudinal quadrants in which each species has been recorded) in world-wide lineages of coral-dwelling mantis shrimps (Stomatopoda, Crustacea). Each species is signified by one datum, and different symbols represent different taxonomic lineages. See Reaka (1980) for different analyses in which body size of species correlated significantly with other measures of the size of the geographic range for species within and between lineages.

Coral with black band disease.

cies are small (e.g., 10-50 mm) on coral reefs and most small species have relatively small geographic ranges, poorly sampled areas (e.g., some areas of the Indo-West Pacific) are very likely to contain undocumented species with endemic distributions that do not extend into better sampled areas.

It seems highly likely, therefore, that global coral reefs contain a very large number of species that are undescribed. Rapid environmental degradation in coastal environments thus is likely to place at risk great amounts of genetic and species diversity that we have not yet even discovered.

On the other hand, if the ecological and evolutionary processes that govern the diversity of coral reef communities differ from those operating in rain forests, and if coral reefs truly are less diverse per unit area than rain forests (the second hypothesis given in the introduction to this section), the estimate of the number of species expected on global coral reefs (≥950,000), and thus the percentage of unknown species on coral reefs, would be too large.

Although much remains to be learned, considerable evidence suggests, in contrast to the second hypothesis, that many ecological processes governing local diversity in coral reef and rain forest communities are similar (Connell, 1978, 1979; Jackson, 1991; Ray and Grassle, 1991). Both reefs and rain forests are continually exposed to disturbance (including predation), and species diversity is maximized at intermediate intensities of these disruptions (Connell, 1978). On larger scales, similar patterns of recruitment, life history and resource specializations, physical gradients, and other historical factors generate the extraordinary adaptations and species richness seen in both environments (although the specific combinations of these factors affecting diversity of reefs and rain forests at any one time may vary; see Jackson, 1991; May, 1986b, 1988; Ray and Grassle, 1991; Reaka, 1985, 1987; Reaka-Kudla, 1991; and Sebens, 1994, for additional discussions of ecological processes affecting diversity in these environments). Further support for the idea that similar evolutionary processes govern diversity in coral reefs and rain forests is provided by graphs that trace the history of biodiversity in marine invertebrates, vascular plants, vertebrates, and insects from the Paleozoic Era onward, showing parallel patterns over time (Signor, 1994).

On the other hand, the overwhelming abundance of species of insects (Erwin, 1982, 1988; May, 1978, 1988, 1994; Stork, 1988) and their Paleozoic radiation (apparently even before the rise of the flowering plants)—accompanied by relatively low extinction rates throughout their history and continuing increases in diversification (Labandeira and Sepkoski, 1993)—suggests that something (perhaps related to their small body size and its relationship to their habitat; May, 1978, 1988) fosters exceptionally high diversity in this lineage of terrestrial arthropods. It is probably still too early to know whether or not similar high diversity might be found in comparably sized small marine invertebrates in structurally complex habitats if they were adequately studied.

To accommodate the perceived relatively low diversity of described marine species versus the terrestrial biota, it has been suggested that the geographic ranges of marine organisms are large, extinctions are unlikely, and diversification (at least after the Cambrian radiation of major body plans) is relatively slow because of the presence of a fluid medium. The aqueous medium is considered to buffer local variations and promote long-distance dispersal, which in turn allows recolonization of locally disturbed sites and connects distant populations genetically (see Angel, 1992; Hutchinson, 1959; May, 1994; Norse, 1993; Pielou, 1979). The fact that relatively few marine extinctions have been observed (Carlton et al., 1991) bolsters the opinion that marine organisms are at a relatively low risk of extinction in the modern world compared to terrestrial species. Indeed only 195 molluscs and crustaceans (0.2% of present described species) compared to 229 vertebrates (0.5% of present described species) have been certified as extinct since 1600 (Smith et al., 1993). The latter authors caution, however, that figures of endangered and extinct species strongly reflect the intensity of scientific study devoted to the group and thus should be applied only to well-known groups such as vertebrates and palms.

It is true that dispersing larvae can swamp genetic differences among adjacent populations, retard rates of diversification, and confer resistance to extinction (Hansen, 1978; Jablonski 1986a,b, 1991; Jablonski and Lutz, 1983; Reaka and Manning, 1981, 1987). However, the common inference that most marine species have dispersing larvae and are at low risk of extinction relies on the most conspicuous species in marine environments (e.g., some starfish, crustaceans, molluscs, and fish), which are large in body size and hence produce large numbers of dispersing larvae and have large geographic ranges. This perception ignores the fact that the greatest proportion of species within marine macroscopic lineages are small in body size and thus are likely to have relatively abbreviated larval development and small geographic ranges (Figures 7-1, 7-2). These small, numerous species are relatively vulnerable to extinction.

Paleontological data show that, at normal background levels of extinction, species with restricted geographic ranges are more susceptible to extinction than those with broader ranges (Hansen, 1978, 1980, 1982; Jablonski, 1980, 1982, 1986a,b, 1991; Scheltema, 1978; Valentine and Jablonski, 1983; Vermeij, 1987; see also Reaka, 1980; Reaka and Manning, 1981, 1987). Vermeij (1993) and Stanley (1986) did not find a correlation between small geographic range and high rates of extinction in certain molluscs, but Stanley points out that species with large geographic ranges can be fragmented by heavy predation into smaller populations which then suffer high extinction, reducing the strength of the correlation in some cases. Also, tropical species are particularly susceptible to extinction, as evidenced by the striking demise of reef communities at each of the major mass extinctions (Jablonski, 1991). The background rates of extinction for marine invertebrates (1-10% of species per million years, Jablonski, 1991) are vastly lower than the extinctions that potentially could result from present-day environmental alterations (Diamond, 1989; Ehrlich and Wilson, 1991; Smith et al., 1993), and the long narrow coastlines of coral reefs are especially vulnerable to habitat degradation and fragmentation.

Therefore, this study proposes that undocumented diversity and—of particular importance at the present time—undocumented contemporary extinctions are likely to be higher than we realize in marine environments because there are many more relatively small, cryptic, and unstudied macroscopic species in coral reef environments than generally recognized. Not only is it likely that undocumented extinctions already have taken place, but the *potential* for future extinction in macroscopic species on coral reefs is higher than generally realized because of the preponderance of diminutive species with small geographic ranges in these environments.

CONCLUSIONS

These analyses suggest that about 93,000 total described species of all taxa occur on coral reefs, which represents about 5% of the described global biota. These numbers are considerably lower than the number of species that are estimated to occur in rain forests. However, coral reefs occupy only 5% of the global area of rain forests. If coral reefs were equivalently studied and contained as much biodiversity as rain forests per km², and if rain forests contained 2 million species, then coral reefs should include approximately 950,000 species. The difference between the numbers of described (93,000) versus expected (950,000) species suggests that coral reefs are repositories of very high undocumented species diversity. Most species on coral reefs are relatively small and cryptic, and difficult to observe and collect. This, in combination with the fact that tropical environments and particularly tropical marine habitats receive less study than those at higher latitudes or terrestrial sites, suggests that many coral reef taxa are indeed very poorly known.

Furthermore, associated with their relatively small size and abbreviated larval dispersal, *most* species on coral reefs are likely to have small geographic ranges, rendering them vulnerable to extinction. Although coral reef communities do not achieve the phenomenal global diversities found in rain forests because of their smaller area, and although it remains unclear whether or not the cryptic reef biota may rival the extraordinary diversity exhibited by insects per unit area, this study suggests that coral reefs may contain far more species than previously supposed (which is congruent with the elaborate specializations and biological interactions found in reef communities), and that very large amounts of this biodiversity may be lost due to human activities before they are even discovered if appropriate scientific study and conservation measures are not taken.

ACKNOWLEDGMENTS

I extend my greatest appreciation to Edward O. Wilson for his encouragement and support throughout my professional career. His scholarship; his intuitive sense of the crucial, central issues of the field; and his personal integrity and warmth have inspired a generation of naturalists. In addition, I thank Hector Severeyn for his participation in gathering data during the early stages of this work, David John and Rita Colwell for information on biodiversity in algae and microorganisms, Joel Cracraft for the provision of a new reference, Mitchell Tartt for another important reference and assistance during preparation of the manuscript, and Don Wilson for helpful comments on the manuscript. Dr. Paul Mazzocchi, Dean of Life Sciences at the University of Maryland, devoted considerable time, logistic, and financial support during development of the Inaugural Symposium of the Consortium for Systematics and Biodiversity and the present volume, for all of which I am grateful.

REFERENCES

Abele, L. 1976. Comparative species composition and relative abundance of decapod crustaceans in marine habitats of Panama. Mar. Biol. 38:263-278.

Abele, L., and W. K. Patton. 1976. The size of coral heads and the community biology of associated decapod crustaceans. J. Biogeogr. 3:35-47.

Allen, W. H. 1992. Increased dangers to Caribbean marine ecosystems. BioScience 42:330-335.

Angel, M. 1992. Managing biodiversity in the oceans. Pp. 23-59 in M. A. Petersen, ed., Diversity of Ocean Life: An Evaluative Review. Center of Strategic International Studies, Washington, D.C.

Austin, A. D., S. A. Austin, and P. F. Sale. 1980. Community structure of the fauna associated with the coral *Pocillopora damicornis* (L.) on the Great Barrier Reef. Aust. J. Mar. Freshwater Res. 31:163-174.

Barnes, R. D. 1984. Invertebrate Zoology. W. B. Saunders Company, Philadelphia.

Bergh, O., K. Y. Borsheim, G. Brabtak, and M. Heldal. 1989. High abundance of viruses found in aquatic environments. Nature 340:467-468.

Bohlke, J., and C. Chapin. 1968. Fishes of the Bahamas and Adjacent Tropical Waters. Livingston Publishing Company, Wynnewood, Pa.

Bohnsack J. A. 1979. The Ecology of Reef Fishes on Isolated Coral Heads: An Experimental Approach with an Emphasis on Island Biogeographic Theory. Ph.D. Dissertation, University of Miami, Coral Gables, Fla.

Bowen, B. W., A. B. Meylan, and J. C. Avise. 1991. Evolutionary distinctiveness of the endangered Kemp's ridley sea turtle. Nature 352:709-711.

Brown, B. E., and J. C. Ogden. 1993. Coral bleaching. Sci. Amer. 268:64-70.

Bruce, A. J. 1976. Shrimps and prawns of coral reefs, with special reference to commensalism. Pp. 37-94 in O. A. Jones and R. Endean, eds., Biology and Geology of Coral Reefs, Vol. 3. Academic Press, N.Y.

Brusca, R. C., and G. J. Brusca. 1990. Invertebrates. Sinauer Associates, Sunderland, Mass.

Butman, C. A., and J. T. Carlton. 1993. Biological Diversity in Marine Systems. National Science Foundation, Washington, D.C.

Carlton, J. T., G. J. Vermeij, D. R. Lindberg, D. A. Carlton, and E. C. Dudley. 1991. The first historical extinction of a marine invertebrate in an ocean basin: The demise of the eelgrass limpet *Lottia alveus*. Biol. Bull. 180:72-80.

Chisholm, S. W., R. J. Olson, E. R. Zettler, R. Goericke, J. B. Waterbury, and N. A. Welschmeyer. 1988. A novel free-living prochlorophyte abundant in the ocean euphotic zone. Nature 344:340-343.

Connell, J. H. 1978. Diversity in tropical rain forests and coral reefs. Science 199:1302-1310.

Connell, J. H. 1979. Tropical rain forests and coral reefs as open non-equilibrium systems. Pp. 141-162 in R. M. Anderson, B. D. Turner, and L. R. Taylor, eds., Population Dynamics. Twentieth Symposium of the British Ecological Society, London.

D'Elia, C. F., R. W. Buddemeier, and S. V. Smith. 1991. Workshop on Coral Bleaching, Coral Reef Ecosystems and Global Change: Report Proceedings. Maryland Sea Grant College Pub. No. UM-SG-TS-91-03. University of Maryland, College Park.

Diamond, J. M. 1989. The present, past and future of human-caused extinctions. Phil. Trans. R. Soc. Lond. B 325:469-477.

Dominguez, J. H., and M. L. Reaka. 1988. Temporal activity patterns in reef-dwelling stomatopods: A test of alternative hypotheses. J. Exp. Mar. Biol. Ecol. 117:47-69.

Ehrlich, P. R., and A. H. Ehrlich. 1992. The value of biodiversity. Ambio 21:219-226.

Ehrlich, P. R., and E. O. Wilson. 1991. Biodiversity studies: Science and policy. Science 253:758-762.

Erwin, T. 1982. Tropical forests: Their richness in Coleoptera and other arthropod species. Coleopt. Bull. 36:74-75.

Erwin, T. 1988. The tropical forest canopy: The heart of biotic diversity. Pp. 123-129 in E. O. Wilson and F. M. Peter, eds., BioDiversity. National Academy Press, Washington, D.C.

Feldmann, R. M., and R. B. Manning. 1992. Crisis in systematic biology in the "Age of Biodiversity." J. Paleontol. 66:157-158.

Fenchel, T. 1993. There are more small than large species? Oikos 68:375-378.

Fuhrman, J. A., K. McCallum, and A. A. Davis. 1992. Novel major archaebacterial group from marine plankton. Nature 356:148-149.

Gaston, K. 1991. The magnitude of global insect species richness. Conserv. Biol. 5:283-296.

Gaston, K. J., and R. M. May. 1992. Taxonomy of taxonomists. Nature 356:281-282.

Gibbs, P. E. 1971. The polychaete fauna of the Solomon Islands. Bull. Brit. Mus. (Nat. Hist.) Zool. 21:99-211.

Glynn, P. W. 1993. Coral reef bleaching: Ecological perspectives. Coral Reefs 12:1-17.

Goldman, B., and F. H. Talbot. 1976. Aspects of the ecology of coral reef fishes. Pp. 125-154 in O. A. Jones and R. Endean, eds., Biology and Geology of Coral Reefs, Vol. 3, Biology 2. Academic Press, N.Y.

Grassle, J. F. 1973. Variety in coral reef communities. Pp. 247-270 in O. A. Jones and R. Endean, eds., Biology and Geology of Coral Reefs, Vol. 2, Biology 1. Academic Press, N.Y.

Grassle, J. F. 1986. The ecology of deep-sea hydrothermal vent communities. Adv. Mar. Biol. 23:301-362.

Grassle, J. F. 1989. Species diversity in deep-sea communities. Trends Ecol. Evol. 4:12-15.

Grassle, J. P., and J. F. Grassle. 1976. Sibling species in the marine pollution indicator, *Capitella* (Polychaeta). Science 192:567-569.

Grassle, J. F., and N. J. Maciolek. 1992. Deep-sea species richness: Regional and local diversity estimates from quantitative bottom samples. Amer. Nat. 139:313-341.

Grigg, R. W., J. J. Polovina, and M. J. Atkinson. 1984. Model of a coral reef ecosystem. III. Resource limitation, community regulation, fisheries yield and resource management. Coral Reefs 3:23-27.

Hallock, P., F. E. Mueller-Karger, and J. C. Halas. 1993. Coral reef decline. Natl. Geogr. Res. Expl. 9:358-378.

Hammond, P. 1992. Species inventory. Pp. 17-39 in B. Groombridge, ed., Global Biodiversity: Status of the Earth's Living Resources. Chapman and Hall, London.

Hansen, T. A. 1978. Larval dispersal and species longevity in Lower Tertiary gastropods. Science 199:885-887.

Hansen, T. A. 1980. Influence of larval dispersal and geographic distribution on species longevity in neogastropods. Paleobiology 6:193-207.

Hansen, T. A. 1982. Modes of larval development in Early Tertiary neogastropods. Paleobiology 8:367-377.

Hawksworth, D. L. 1991. The fungal dimension of biodiversity: Magnitude, significance, and conservation. Mycol. Res. 95:441-456.

Hughes, R., and J. Gamble. 1977. A quantitative survey of the biota of intertidal soft substrata on Aldabra Atoll, Indian Ocean. Phil. Trans. R. Soc. Lond. B 279:324-355.

Hughes, T. P. 1994. Castrophes, phase shifts, and large-scale degradation of a Caribbean coral reef. Science 265:1547-1551.

Hughes, T. P., B. D. Keller, J. B. C. Jackson, and M. J. Boyle. 1985. Mass mortality of the echinoid *Diadema antillarum* Philippi in Jamaica. Bull. Mar. Sci. 36:377-384.

Hutchinson, G. E. 1959. Homage to Santa Rosalia, or why are there so many kinds of animals? Amer. Nat. 93:145-159.

Hutchinson, G. E., and R. H. MacArthur. 1959. A theoretical ecological model of size distributions among species of birds. Amer. Nat. 93:117-125.

Jablonski, D. 1980. Apparent versus real biotic effects of transgression and regression. Paleobiology 6:397-407.

Jablonski, D. 1982. Evolutionary rates and modes in Late Cretaceous gastropods: Role of larval ecology. Proc. Third N. Amer. Paleontol. Conv. 1:257-262.

Jablonski, D. 1986a. Larval ecology and macroevolution in marine invertebrates. Bull. Mar. Sci. 29:565-587.

Jablonski, D. 1986b. Background and mass extinctions: The alternation of macroevolutionary regimes. Science 231:129-133.

Jablonski, D. 1991. Extinctions: A paleontological perspective. Science 253:754-757.

Jablonski, D., and R. A. Lutz. 1983. Larval ecology of marine benthic invertebrates: Paleobiological implications. Biol. Rev. 58:21-89.

Jackson, J. B. C. 1984. Ecology of cryptic coral reef communities. III. Abundance and aggregation of encrusting organisms with particular reference to cheilostome Bryozoa. J. Exp. Mar. Biol. Ecol. 75:37-57.

Jackson J. B. C. 1991. Adaptions and diversity of reef corals. BioScience 41:475-482.

John, D. M. 1994. Biodiversity and conservation: An algal perspective. Phycologist 38:3-15.

Knowlton, N., E. Weil, L. A. Weight, and H. M. Guzman. 1992. Sibling species in *Montastraea annularis*, coral bleaching, and the coral climate record. Science 255:330-333.

Kuhlmann, D. H. H. 1988. The sensitivity of coral reefs to environmental pollution. Ambio 17:13-21.

Labandeira, C. C, and J. J. Sepkoski, Jr. 1993. Insect diversity in the fossil record. Science 261:310-315.

Lessios, H. A., D. R. Robertson, and J. D. Cubit. 1984. Spread of *Diadema* mass mortality through the Caribbean. Science 226:335-337.

Longley, W., and S. Hildebrand. 1941. Systematic catalogue of the fishes of the Totugas, Fla., with observations on color, habits and local distribution. Papers Tortugas Lab. 34:1-331.

MacArthur, R. H., and E. O. Wilson. 1967. The Theory of Island Biogeography. Princeton University Press, N.J.

Manning, R. B. 1991. The importance of taxonomy and museums in the 1990's. Mem. Queensland Mus. 31:205-207.

Margulis, L., and K. V. Schwartz. 1988. Five Kingdoms: An Illustrated Guide to the Phyla of Life on Earth. W. H. Freeman, San Francisco.

May, R. M. 1975. Patterns of species abundance and diversity. Pp. 81-120 in M. Cody and J. M. Diamond, eds., Ecology of Species and Communities. Harvard University Press, Cambridge, Mass.

May, R. M. 1978. The dynamics and diversity of insect faunas. Pp. 188-204 in L. A. Mound and N. Waloff, eds., Diversity of Insect Faunas. Blackwell Scientific Publications, Oxford, England.

May, R. M. 1986a. How many species are there? Nature 326:514-515.

May, R. M. 1986b. The search for patterns in the balance of nature: Advances and retreats. Ecology 67:1115-1126.

May, R. M. 1988. How many species are there on Earth? Science 241:1441-1449.

May, R. M. 1990. How many species? Phil. Trans. R. Soc. Lond. B 330:292-304.

May, R. M. 1991. A fondness for fungi. Nature 352:475-476.

May, R. M. 1992. Bottoms up for oceans. Nature 357:278-279.

May, R. M. 1994. Biological diversity: Differences between land and sea. Phil. Trans. R. Soc. Lond. B 343:105-111.

Mayr, E. 1969. Principles of Systematic Zoology. McGraw-Hill, N.Y.

McCloskey, L. 1970. The dynamics of the community associated with a marine scleractinian coral. Int. Revue Ges. Hydrobiol. 55:13-81.

Mittermeier, R. A. 1988. Primate diversity and the tropical forest: Case studies from Brazil and Madagascar and the importance of the megadiversity countries. Pp. 145-154 in E. O. Wilson and F. M. Peter, eds., BioDiversity. National Academy Press, Washington, D.C.

Moran, D. P., and M. L. Reaka. 1988. Bioerosion and the availability of shelter for benthic reef organisms. Mar. Ecol. Progr. Ser. 44:249-263.

Moran, D. P., and M. L. Reaka-Kudla. 1990. Effects of disturbance: Disruption and enhancement of coral reef cryptofaunal populations by hurricanes. Coral Reefs 9:215-224.

Morse, D. R., J. H. Lawton, M. M. Dodson, and M. H. Williamson. 1985. Fractal dimension of vegetation and the distribution of arthropod body lengths. Nature 314:731-733.

National Science Board Task Force on Biodiversity. 1989. Loss of Biological Diversity: A Global Crisis Requiring International Solutions. National Science Board, Committee on International Science, National Science Foundation, Washington, D.C.

Norse, E. A., ed. 1993. Global Marine Biological Diversity. Island Press, Washington, D.C.

Ogden, J., and R. Wicklund, eds. 1988. Mass Bleaching of Coral Reefs in the Caribbean: A Research Strategy. National Oceanic and Atmospheric Administration National Undersea Research Program, Washington, D.C.

Parker, S. P. 1982. Synopsis and Classification of Living Organisms. McGraw-Hill, N.Y.

Pearse, V. B. 1987. Living Invertebrates. Blackwell Scientific Publications, Oxford, England.

Pennak, R. W. 1989. Freshwater Invertebrates of the United States, third ed., Protozoa to Mollusca. John Wiley and Sons, N.Y.

Peyrot-Clausade, M. 1983. Transplanting experiments of motile cryptofauna on a coral reef flat of Tulear (Madagascar). Thalassagraphica 6:27-48.

Pielou, E. C. 1979. Biogeography. Wiley Interscience, N.Y.

Poore, G. C. B., and G. D. F. Wilson. 1993. Marine species richness. Nature 361:579.

Porter, J. W., and O. W. Meier. 1992. Quantification of loss and change in Floridian reef coral populations. Amer. Zool. 32:625-640.

Rand McNally. 1980. Rand McNally World Atlas. Rand McNally and Company, N.Y.

Raven, P. H., and E. O. Wilson. 1992. A fifty-year plan for biodiversity surveys. Science 258:1099-1100.

Ray, G. C. 1985. Man and the sea: The ecological challenge. Amer. Zool. 25:451-468.

Ray, G. C. 1988. Ecological diversity in coastal zones and oceans. Pp. 36-50 in E. O. Wilson and F. M. Peter, eds., BioDiversity. National Academy Press, Washington, D.C.

Ray, G. C. 1991. Coastal zone biodiversity patterns. BioScience 41:490-498.

Ray, G. C., and J. F. Grassle. 1991. Marine biological diversity. BioScience 41:453-457.

Reaka, M. L. 1980. Geographic range, life history patterns, and body size in a guild of coral-dwelling mantis shrimps. Evolution 34:1019-1030.

Reaka, M. L. 1985. Interactions between fishes and motile benthic invertebrates on reefs: The significance of motility vs. defensive adaptations. Proc. Fifth Int. Coral Reef Congr. 5:439-444.

Reaka, M. L. 1986. Biogeographic patterns of body size in stomatopod Crustacea: Ecological and evolutionary consequences. Pp. 209-235 in R. H. Gore and K. L. Heck, eds., Biogeography of the Crustacea. Balkema Press, Rotterdam, Netherlands.

Reaka, M. L. 1987. Adult-juvenile interactions in benthic reef crustaceans. Bull. Mar. Sci. 41:108-134.

Reaka-Kudla, M. L. 1991. Processes regulating biodiversity in coral reef communities on ecological vs. evolutionary time scales. Pp. 61-70 in E. C. Dudley, ed., The Unity of Evolutionary Biology. Dioscorides Press, Portland, Oreg.

Reaka, M. L., and R. B. Manning. 1981. The behavior of stomatopod Crustacea, and its relationship to rates of evolution. J. Crustacean Biol. 1:309-327.

Reaka, M. L., and R. B. Manning. 1987. The significance of body size, dispersal potential, and habitat for rates of morphological evolution in stomatopod Crustacea. Smithsonian Contr. Zool. 448:1-46.

Rex, M. A., C. T. Stuart, R. R. Hessler, J. A. Allen, H. L. Sanders, and G. D. F. Wilson. 1993. Global-scale latitudinal patterns of species diversity in the deep-sea benthos. Nature 365:636-639.

Sale, P. F. 1977. Maintenance of high diversity in coral reef fish communities. Amer. Nat. 111:337-359.

Scheltema, R. S. 1978. On the relationship between dispersal of pelagic veliger larvae and the evolution of marine prosobranch gastropods. Pp. 303-322 in B. Battaglia and J. A. Beardmore, eds., Marine Organisms. Plenum Press, N.Y.

Sebens, K. P. 1994. Biodiversity of coral reefs: What are we losing and why? Amer. Zool. 34:115-133.

Signor, P. W. 1994. Biodiversity in geological time. Amer. Zool. 34:23-32.

Smith, C. L. 1978a. Coral reef fish communities: A compromise view. Envir. Biol. Fishes 3:109-128.

Smith, S. V. 1978b. Coral-reef area and the contribution of reefs to processes and resources of the world's oceans. Nature 273:225-226.

Smith, F. D. M., R. M. May, R. Pellew, T. H. Johnson, and K. S. Walter. 1993. Estimating extinction rates. Nature 364:494-496.

Sournia, A. C. D., and M. Ricard. 1991. Marine phytoplankton: How many species in the world ocean? J. Plankton Res. 12:1039-1099.

Stanley, S. M. 1986. Population size, extinction, and speciation: The fission effect in Neogene Bivalvia. Paleobiology 12:89-110.

Starck, W. A. 1968. List of fishes of Alligator Reef, Florida, with comments on the nature of the Florida Reef fauna. Undersea Biol. 1:1-40.

Stevens, G. C. 1989. The latitudinal gradient in geographical range: How many species coexist in the tropics? Amer. Nat. 133:240-256.

Stork, N. E. 1988. Insect diversity: Facts, fiction and speculation. Biol. J. Linn. Soc. 35:321-337.

Strathmann, R. R., and M. F. Strathmann. 1982. The relationship between adult size and brooding in marine invertebrates. Amer. Nat. 119:91-101.

Systematics Agenda 2000. 1994. Systematics Agenda 2000: Charting the Biosphere. Technical Report. Systematics Agenda 2000, a Consortium of the American Society of Plant Taxonomists, the Society of Systematic Biologists, and the Willi Hennig Society, in cooperation with the Association of Systematics Collections, N.Y. 34 pp.

Taylor, J. 1968. Coral reef-associated invertebrate communities (mainly molluscan) around Mahe, Seychelles. Phil. Trans. R. Soc. Lond. B 254:129-206.

Valentine, J. W., and D. Jablonski. 1983. Speciation in the shallow sea: General patterns and biogeographic controls. Pp. 201-226 in R. W. Sims, J. H. Price, and P. E. S. Whalley, eds., Evolution, Time, and Space: The Emergence of the Biosphere. Academic Press, N.Y.

Veron, J. E. 1985. Aspects of the biogeography of hermatypic corals. Proc. Fifth Int. Coral Reef Congr. 4:83-88.

Vermeij, G. J. 1987. Evolution and Escalation: An Ecological History of Life. Princeton University Press, N.J.

Vermeij, G. J. 1993. Biogeography of recently extinct marine species: Implications for conservation. Conserv. Biol. 7:391-397.

Williams, E. H., Jr., and L. Bunkley-Williams. 1990. The world-wide coral reef bleaching cycle and related sources of coral mortality. Atoll Res. Bull. 335:1-71.

Wilson, E. O. 1985. Time to revive systematics. Science 230:1227.

Wilson, E. O. 1988. The current state of biological diversity. Pp. 3-18 in E. O. Wilson and F. M. Peter, eds., BioDiversity. National Academy Press, Washington, D.C.

Wilson, E. O. 1989. Threats to biodiversity. Sci. Amer. (September):108-116.

Wilson, E. O. 1992. The Diversity of Life. Belknapp Press, Cambridge, Mass. 424 pp.

Wolf, N. G., E. B. Bermingham, and M. L. Reaka. 1983. Relationships between fishes and mobile benthic invertebrates on coral reefs. Pp. 89-96 in M. L. Reaka, ed., The Ecology of Deep and Shallow Coral Reefs, Vol. 1. National Oceanic and Atmospheric National Undersea Research Program, Washington, D.C.

Yoon, C. K. 1993. Counting creatures great and small. Science 260:620-622.

CHAPTER
8

Common Measures for Studies of Biodiversity: Molecular Phylogeny in the Eukaryotic Microbial World

MITCHELL L. SOGIN
Director, Program in Molecular Evolution

GREGORY HINKLE
Post-Doctoral Researcher

Marine Biological Laboratory, Woods Hole, Massachusetts

Biodiversity is surveyed at many different levels using a broad array of criteria. For many naturalists, richness in biodiversity is measured in terms of the number of species of a given genus per m^2. Others are more concerned with the number of genera, classes, kingdoms, etc., within an ecosystem. Although the fundamental concepts in systematic biology that underlie these studies are well established, different properties are used to infer evolutionary relationships for different kinds of organisms. Consequently, above the level of species, equivalence among taxonomic ranks for fungi, plants, and animals is rather poor. Within the microbial world, the assignment of taxa to a particular species is generally meaningless. Accordingly, systematic descriptions of biodiversity can be contentious, and sometimes are more appropriately described as forms of political ecology. If systematic biologists and ecologists are to complete an inventory of the taxosphere, its full importance will be appreciated only if a "common currency" can be established, one that will allow genetic diversity in one group to be calibrated against that observed in any other evolutionary assemblage.

Systematic biology is the oldest of biological disciplines, yet it is a dynamic field that periodically undergoes major change in response to novel ideas or new technology. More than 100 years ago, Haeckel (1894) revolutionized taxonomy by establishing the concept of phylogeny, while improvements in the microscope paved the way to discovery of the microbiota. Today biologists employ new methodologies to gain insights into evolutionary events that gave rise to today's biosphere. The use of "macromolecular sequences as documents of evolutionary history" (Zuckerkandl and Pauling, 1965), coupled to advances in com-

Fungi: a highly diverse, but poorly known group.

putational biology, rapidly have transformed systematics and hence studies of biodiversity into experimental sciences.

Molecular techniques and databases offer a means to survey and quantify biodiversity through the acquisition, storage, and comparison of linear sequences of amino acids and nucleotides, the very fundamentals of life. The comparison of genetic elements that have been transmitted from generation to generation makes possible the measurement of genetic differences between members of populations, species, and even between kingdoms of organisms. Molecular data provide a practical metric for assessing biodiversity and registering organisms in the same evolutionary framework in which morphological and biochemical differences among organisms arose. Phylogenetic inferences generated from comparisons of homologous sequences of nucleotides or amino acids provide more than simple dichotomous branching topologies; they offer quantitative measures of phylogenetic depth and hence information about antiquity of genetically distinct lineages. Although it seems unlikely that field biologists ever will collect complete genome information, tools for acquiring sequences for a limited number of genes that meet the criteria of reliable evolutionary markers may one day go hand-in-hand with monoculars and butterfly nets.

Nowhere has the impact of molecular approaches on systematics been more significant than in the microbial world (Woese, 1987). In response to the failure of more traditional taxonomic approaches (those based on comparisons of phenotypic characters), the microbiological research community has seized upon molecular tools not only to reevaluate ideas about the composition of major groups of organisms but also to probe for novel biodiversity (Pace et al., 1986). In some cases, previously unknown microbial taxa may constitute significant components of the community biomass (Fuhrman et al., 1993).

Sequence comparisons of the larger ribosomal RNAs (rRNAs) or their coding regions have gained widespread acceptance among microbial systematists for inferring phylogenetic frameworks because these molecules are evolutionarily homologous and functionally equivalent in all organisms, their sequence changes sufficiently slowly to allow measurements of even the largest genealogical distances, and they do not undergo transfer between species (Sogin, 1989). Unlike small molecules such as the 5S rRNAs, the 16S-like rRNAs contain a large number of variable sites (>1,400-2,000 positions) that provide large sets of characters for phylogenetic inferences. Of equal importance to broad studies of biodiversity is the mosaic-like arrangement of genetic elements that display different rates of evolutionary change within 16S-like rRNA (Sogin and Gunderson, 1987). Regions with high rates of change are interspersed among moderately conserved or nearly invariant domains. The genetic difference between even the most distantly related organisms can be estimated from comparisons of well-conserved regions of small subunit rRNAs, and yet rapidly evolving domains permit use of the very same genes to infer relationships between species within the same genus.

Several thousand complete sequences of small subunit rRNAs that are deposited in public databases are curated by the Ribosomal RNA Database Project (RDP) (Larsen et al., 1993). Unlike archival databases such as EMBL and GenBank, the RDP collection of sequences is a dynamic data structure that maintains the aligned sequences required for phylogenetic inference. The explosive growth in the number of sequences in the RDP mirrors the remarkable track record of rRNA sequence comparisons as a molecular tool for inferring ancient evolutionary history. More than any other molecule, small subunit rRNA offers phylogenetic frameworks that are consistent with the biology of the representative taxa.

Comparisons of rRNA genes have exerted a profound influence on systematics and our understanding of early evolution. Unlike the standard "five kingdoms" (plants, animals, fungi, protists, and bacteria) (Whittaker, 1969) that are presented in today's textbooks, molecular studies define three primary lines of descent (Eukarya, Bacteria, and Archaea) (Woese et al., 1990). The evolutionary tree shown in Figure 8-1 is based on similarities in small subunit rRNAs and hence can be regarded as a phylogenetic framework describing early events in the history of life in this biosphere. Just within the eukaryotic subtree, there are numerous revelations. For example, instead of being relatively recent innovations, eukaryotes represent a lineage that may be as old as the archaebacterial and eubacterial kingdoms (Sogin et al., 1989). The earliest diverging lineages are represented by diplomonads, microsporidians, and trichomonads (Leipe et al., 1993). These organisms lack many features characteristic of eukaryotic cells, e.g., mitochondria, typical golgi, and complex cytoskeletons. The early lineages are followed by a series of independent branches of protists, and then by the "higher" kingdoms of fungi, plants, animals, and at least two additional complex evolutionary assemblages, the stramenopiles and the alveolates.

The stramenopile assemblage includes diatoms, oomycetes, bicosoecids, labyrinthulids, brown algae, and many chromophytes (exclusive of dinoflagellates, cryptomonads, and haptophytes). Many of these species are photosynthetic autotrophs containing plastids with chlorophylls a and c. At one time, they were classified as eukaryotic algae along with rhodophytes (red algae), chlorophytes (green algae most closely related to plants), and chromophytes (recognized today as a polyphyletic group containing similar plastid pigments). Other stramenopiles are nonpigmented heterotrophs that feed on microorgan-

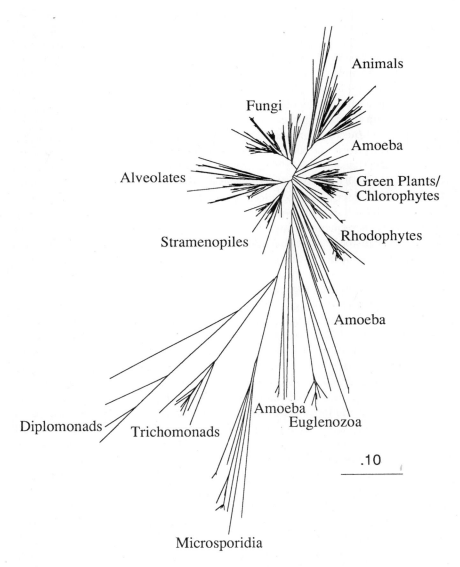

FIGURE 8-1 An unrooted molecular phylogeny of eukaryotes. A computer-assisted method was used to align the 16S-like rRNA sequences from 600 diverse eukaryotes. Pairwise comparisons of all positions that could be aligned unambiguously were used to compute values of structural similarity for all pairs of sequences. After conversion to evolutionary distances using a Kimura correction, a Neighbor Joining tree was inferred using the computer program PHYLIP. The length of segments indicates extent of molecular change. The scale bar represents 10 nucleotide changes per 100 positions.

isms. They can be important inter-
mediates in the food chain, and bicosoecids may be responsible for
consuming as much as 10% of the
marine prokaryotic biomass or picoplankton.

The alveolates include ciliates,
dinoflagellates, and apicomplexans
(Leipe et al., 1994; Patterson, 1989;
Patterson and Sogin, 1992). Ciliates
are defined by the presence of both
macro- and micronuclei and frequently have many flagella. The

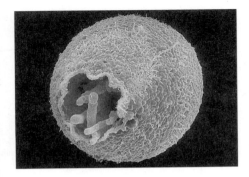

Testate amoeba with pseudopods.

micronucleus harbors germline genetic information, while the macronucleus
contains "processed" DNA used by the cell's transcription machinery. At one
time, ciliates were regarded as protists most closely related to animals. In contrast, dinoflagellates (including photosynthetic and nonphotosynthetic forms)
were thought to represent an eukaryotic lineage that diverged very early because of their unusual chromatin structure. Apicomplexans are united by an
ultrastructural feature described as the apical complex. Generally parasites of
mammalian species, apicomplexa cause such devastating diseases as malaria
or histoplasmosis. Molecular studies of rRNA and actin genes demonstrate
the phylogenetic affinity of apicomplexans, ciliates, and dinoflagellates
(Bhattacharya et al., 1991; Gajadhar et al., 1991). They are described as alveolates
because ciliates, dinoflagellates, and apicomplexans have alveoli (membranous,
spherical structures found at the flagellar base).

The nearly simultaneous separation of plants, fungi, animals, alveolates,
and stramenopiles (described as the eukaryotic "crown") occurred approximately 1 billion years ago[1]. Although we are unable to resolve the exact branching sequence among the crown groups, it is now apparent that animals and
fungi shared a more recent common evolutionary history that excluded all other
eukaryotic groups (Wainright et al., 1993).

Glimpses of other important events in the evolution of eukaryotes are offered by the detailed branching patterns within the crown groups. For example, stramenopiles (see Figure 8-2A) include morphologically diverse organisms with very different lifestyles. Yet all members of the group have flagella
with tripartite flagellar hairs that are not found in any other eukaryotes. The
flagellar hairs may have conferred a major ecological advantage for these organ-

---

[1]This calibration is based on knowledge of the time of divergence for organisms that are well
preserved in the fossil record and by calculating the rate of change of nucleotides in their rRNAs.
Assuming a relatively constant incorporation of mutations—a "molecular clock"—we can estimate
when the crown groups first arose.

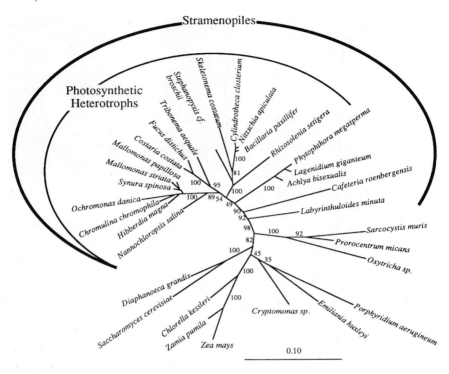

FIGURE 8-2a Maximum likelihood phylogenetic inference for 16S-like rRNAs from stramenopiles and outlying eukaryotes. The percentage of 110 bootstrap resamplings that support topological elements in maximum likelihood inferences are shown. Branch lengths represent relative evolutionary distances. The scale bar corresponds to 10 changes per 100 positions. A bootstrap value of 82% lends some support to the notion of a close relationship between alveolates and stramenopiles. Given the low bootstrap value for the common branch that unites haptophytes, cryptomonads, and red algae, the relationship among these outgroups is completely unresolved.

isms by allowing thrust reversal during swimming and thus enhanced ability to entrap prey (Patterson, 1989).

Even more important to our understanding of eukaryotic history is the early divergence of nonphotosynthetic taxa within the stramenopiles. Heterotrophs including the slime net *Labyrinthuloides minuta* and the heterotrophic marine flagellate *Cafeteria roenbergensis* diverged prior to the relatively late separation of oomycetes from photosynthetic groups (Leipe et al., 1994). This branching pattern affects how we interpret the evolution of plastids. The simplified phylogeny in Figure 8-2B summarizes our current understanding of the evolution of stramenopiles. A heterotrophic protist with two flagella and tubular mitochondrial cristae developed tripartite tubular hairs. This adaptive morphology gave rise to the labyrinthulids, bicosoecids, and the oomycetes. The last common ancestor of the stramenopile assemblage was probably a heterotrophic flagellate. If true, autotrophy within the stramenopiles was independent of the other major autotrophic assemblages: the green plants, cryptomonads, dinoflagellates, hapto-phytes, and rhodophytes. Endosymbiotic origins of plastids in eukaryotes there-fore must have occurred several times. At this time, we cannot distinguish between multiple primary events involving cyanobacterial ancestors or a single primary event followed by multiple secondary endosymbioses of heterotrophic and phototrophic eukaryotes. In light of the emerging molecular data, the alter-native hypothesis (independent loss of chloroplasts in bicosoecids, labyrinth-ulids, oomycetes, and other heterotrophic stramenopiles) is no longer the most parsimonious explanation for the distribution of photosynthetic phenotypes in eukaryotic phylogenies.

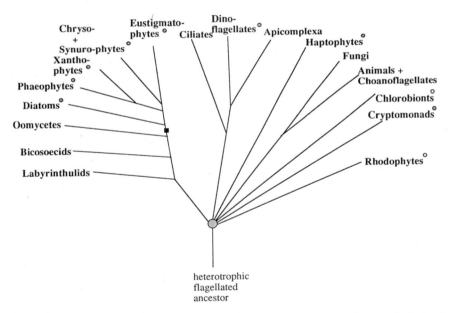

FIGURE 8-2b Interpretive drawing of stramenopile evolution. The diagram is derived from the rRNA framework. The last common ancestor to the crown groups was a bi-flagellated heterotroph with tubular mitochondrial cristae. Areas of uncertainty (the unresolved polytomy at the base of the eukaryotic crown and the relative branching pattern of oomycetes, diatoms, and the remaining stramenochromes) are indicated by a shaded square. The occurrence of chloroplasts with chlorophyll a and c is indicated by a shaded circle at the taxon's name. Other autotrophic groups are marked with an open circle.

In addition to identifying assemblages at the level of kingdoms, comparisons of rRNA genes have resolved relationships between nearly identical taxa. Phylogenetic studies of the rRNAs from fungal symbionts of neotropical leaf-cutting (tribe Attini) ants reveal a remarkable evolutionary story (Chapela et al., 1994; Hinkle et al., 1994). The symbiotic relationship of attine ants with fungi has allowed exploitation by the ants of a range of food resources that are not otherwise available (Wilson, 1971). Because of a rich fossil record and numerous morphological characters for cladistic analyses, relationships among attine ants are well understood. In contrast, there has been much disagreement concerning the phylogenetic affinities of the symbiotic attine fungi due to the general absence of useful morphological characters. Some investigators perceive attine fungi as a polyphyletic group made up of both Ascomycotines and Basidiomycotines (Kermarrec et al., 1986; Weber 1966, 1972). Others argue that the attine fungi are all closely related species (Cherrett et al., 1989). Because of these uncertainties, many authors refer to all strains of attine fungi collectively as *Attamyces bromatificus* (Cherrett et al., 1989) within the polyphyletic Fungi Imperfecti. The rRNA phylogeny in Figure 8-3 clearly demonstrates that four of the five taxa sampled represent distinct basidiomycete genera. Only the fungal symbionts of *Sericomyrmex* and *Trachymyrmex* are sufficiently similar (two nucleotide changes) to permit assignment within the same genus. More important, the evolutionary relationships of the "higher" attine ants (*Atta, Cyphomyrmex, Sericomyrmex,* and *Trachymyrmex*) and their fungal symbionts are entirely congruent. Although we are not able to resolve the branching order of free-living *Agaricus, Lepiota,* and the fungal symbiont of *Apterostigma,* the congruency of the trees for ants and fungi is evidence of stable coevolution over millions of years. This observation prompts one to consider what the fate of one of these ant lineages would be if an environmental insult were to cause the fungal partner to become extinct; it is quite likely that such an event would lead to loss of the ant lineage as well. In nature, there are many examples of symbioses that involve animals or plants and physically distinct, yet metabolically integrated, fungal or microbial partners. In those cases where coevolution has occurred over extended periods of time, preservation of biodiversity may require maintenance of both members of the partnership. Molecular systematics offers a tool for assessing coevolution of symbiotically linked, but evolutionarily distant taxa.

The value of the molecular databases extends beyond the inference of phylogenetic trees. Given the large collection of rRNA gene sequences, molecular systematists can pinpoint taxonomic positions or phylotypes for organisms of uncertain phylogenetic affinity. The gut symbionts of surgeonfish are just one of many examples where traditional morphological criteria lead to the incorrect taxonomic description of a microbe. Because of its enormous size (typically $80 \times 600\mu$), the surgeonfish symbiont *Epulopiscium fishelsoni* was originally described as an unusual protist (Fishelson et al., 1985). Contrary evidence indicative of an

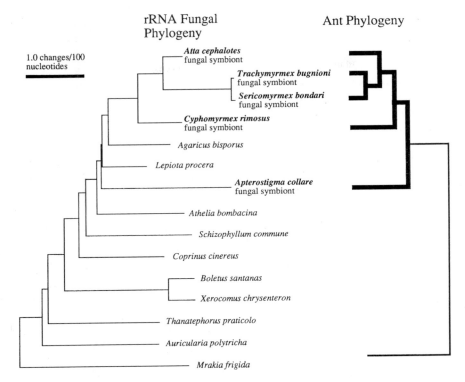

rRNA Fungal
Phylogeny

Ant Phylogeny

FIGURE 8-3 Distance matrix tree for homobasidiomycete fungi based on sequences of 16S-like rRNA genes and ant phylogeny based on parsimony analysis of morphological data. The tree for fungal phylogeny was inferred from the comparison of positions that can be aligned unambiguously in all sequences of full-length 16S-like rRNA genes that have been reported for homobasidiomycetes. In this tree, distance matrices were used to infer relationships for the fungal symbionts of attine ants and representatives of diverse lineages of basidiomycetes. *Mrakia frigida* was used as the outgroup. The topology of the phylogeny for attine ants is derived from a cladistic analysis of 44 morphological characters in prepupal worker larvae from 51 species of attines (Schultz and Meier, 1995). The topology of the phylogeny for the corresponding symbionts of the "higher" attine ants and fungi is delineated with a thick line.

affinity to prokaryotes came from electron microscopical studies (Clements and Bullivant, 1991). Instead of nuclei, the giant symbionts contain nucleoids with no hint of surrounding membranes, similar to those of bacteria. Their flagella resemble those of bacteria rather than the classic 9+2 microtubular structures found in eukaryotic cilia. Gross morphology and ultrastructure presented two contradictory phylogenetic hypotheses. When portions of the 16S rRNA genes of the *Euplopiscium* symbiont were sequenced, the molecular data demonstrated incontrovertibly their phylogenetic affinity with the low g+c Gram positive eubacteria (Angert et al., 1993). Carefully controlled in situ hybridizations with fluorescent rRNA probes confirmed that the symbionts were the source of the sequenced rRNA genes.

The discovery of prokaryotic giants forces biologists to reconsider one of the dominant paradigms of microbial evolution and diversity. Size is a frequently cited but incorrect criterion for differentiating eukaryotes and prokaryotes. Although cell volumes in eukaryotes are typically 100- to 1,000-fold greater than in prokaryotes, their size range overlaps. The chlorophyte *Nanochlorum eukaryotum* contains a mitochondrion, chloroplast, and a nucleus within a 1-2μ cell. Very long, skinny bacteria (>200μ × 0.75 to 8.0μ) have been described that contain nominal amounts of cytoplasm due to their spiral morphology or presence of large liquid vacuoles. Since the 80 × 600μ cell dimensions of prokaryotic symbionts in surgeonfish are far greater than the examples cited above, the theoretical constraints on cell size imposed by prokaryotic cell architecture and apparent absence of intracellular vesicular transport are no longer credible.

Perceived size constraints for prokaryotes, and hence interpretations of microfossils according to size, have dominated traditional hypotheses about the evolutionary history of eukaryotes (Schopf and Oehler, 1976). The evolution of large size is considered to be a consequence of the emergence of the most fundamental eukaryotic characters, the endomembrane system and cytoskeleton. The absence of large cells from the fossil record predating 1.5 to 2.0 billion years ago is taken as evidence for the more recent appearance of cells with nuclei. It now becomes clear that the assignment of eukaryotic status to microfossils is strictly an operational definition; the existence of small eukaryotes such as *Nanochlorum* and the discovery of giant prokaryotes lacking cytoskeletons and a known vesicular transport system raises suspicions about the phylogenetic significance of size differences between eukaryotes and prokaryotes. This forces reexamination of the fossil record as it pertains to the evolutionary origins of eukaryotes. Molecular data in the form of unanticipated diversity of rRNA sequences is consistent with the interpretation that eukaryotic lineages diverged before 3.0 billion years ago (Sogin, 1991).

The radical departure of the bacterial symbionts in surgeonfish from normal prokaryotic cell architecture underscores the likelihood of unknown diversity among microorganisms that cannot be cultured in the laboratory. Molecular methods now offer a window for viewing diversity in natural microbial populations. Polymerase chain reaction (PCR) methods permit the isolation of genes that represent evolutionary homologues from natural populations (Pace et al., 1986). As in the case of rRNA genes from the symbionts of surgeonfish, rRNA genes isolated from natural populations of organisms can authenticate phylogenetic affinity and assess organismal diversity in natural populations. This can be achieved without the requirement that the organism is isolated and cultured. This molecular strategy for assessing microbial diversity has been used recently to characterize organisms from extreme environments as well as more mundane ecosystems, including blue water open ocean. Both have dramatically extended our knowledge and changed our perspective of microbial biodiversity.

Among the three primary domains of life, the Archaea (or archaebacteria) are the least well-understood in terms of phylogenetic diversity and physiology. Such "extremophiles" are difficult to isolate and propagate in the laboratory. Barnes et al. (1994) used a combination of polymerase chain reaction (PCR) techniques and rRNA sequence analysis to describe the enormous phylogenetic depth of the Archaea present in the sediment from "Jim's Black Pool," a hot spring in Yellowstone National Park. The difficulty in culturing the organisms that thrive in boiling or near-boiling water with traditional techniques made suspect any assessment of the microbial diversity in these extreme environments. Using PCR techniques that forestall the need to culture organisms, Barnes et al. discovered that the majority of the 98 16S-like rRNA clones they recovered were from previously undescribed Archaeal species; the sequences of a significant fraction of clones bore very little similarity to those from previously described Archaea. The genetic distances estimated from comparison of the rRNA sequences are indicative of distant phylogenetic relationships. Thus, the physiological variation within a group heretofore thought to be well circumscribed in all likelihood is much greater than commonly believed. Since many of these gene sequences were represented by single clones, it is highly unlikely that the samples have exhaustively measured the diversity of even a simple 27 m^2 pool. The potential usefulness of temperature-stable, biologically active products from Archaea such as those sampled by Barnes and colleagues should be an incentive for the identification and cataloging of the species comprising these unusual communities.

While assessment of microbial diversity in extreme environments is a relatively recent phenomenon, scientists have been studying the microbes living in the open oceans for many years. Nevertheless, the general inability to culture the majority (by some estimates 99%; Fuhrman et al., 1992) of the bacteria found in the open ocean has left enormous gaps in our knowledge of marine biological processes. Certain groups with easily identifiable characteristics, such as cyanobacteria and their photosynthetic pigments, are well known and have been the subject of extensive study. More recently, however, the use of molecular techniques has established the presence of entire groups of microorganisms that had escaped detection by marine scientists who employed traditional methods.

Analyses of 16S-like rRNA genes that have been amplified randomly by PCR from marine water samples by a number of investigators (DeLong, 1992; Fuhrman et al., 1992, 1993; Schmidt et al., 1991) have demonstrated the presence of great numbers of novel organisms. Furthermore, the general absence of duplicates in analyses of surface and deep water from the Pacific Ocean demonstrates the absence of any dominant group of marine microorganisms, and hence greater than anticipated microbial diversity. Since the evolutionary distance between these undescribed lineages and those of other well-studied groups is so large, we must presume that they have evolved substantially different physiologies. Ecosystem models of marine processes based on preconceived notions of the composition of marine microbial communities may be seriously in error (Fuhrman et al., 1993).

Another surprise has been the discovery of marine archaebacteria in coastal surface waters (DeLong, 1992) and at depths of up to 500 m (Fuhrman et al., 1992) in environments that long have been thought to contain only bacteria (eubacteria) and eukaryotes. Since archaebacteria generally thrive in extreme environments (e.g., high temperature, low pH, high salt), there is now great interest in learning how these organisms survive in the sea alongside eukaryotes and bacteria. The discovery of archaebacteria throughout the upper layers of the ocean is another example of how molecular techniques can perceive organisms that somehow have escaped notice despite their general ubiquity and great numbers. Taken together, the few studies to date that have quantified the diversity of marine microorganisms show that we know surprisingly little about the majority of the organisms living in by far the largest environment on the Earth. Comparisons of universally conserved homologous DNA sequences such as 16S-like rRNA offer the opportunity to both quantify this diversity as well as identify organisms with physiologies that are potentially of use to man.

ACKNOWLEDGMENTS

We are grateful for support from the G. Unger Vetleson Foundation and the National Institutes of Health Grant GM32964.

REFERENCES

Angert, E. R., K. D. Clements, and N. R. Pace. 1993. The largest bacterium. Nature 361:239-241.

Barnes, S. M., R. E. Fundyga, M. W. Jeffries, and N. R. Pace. 1994. Remarkable archaeal diversity detected in a Yellowstone National Park hot spring environment. Proc. Natl. Acad. Sci. 91:1609-1613.

Bhattacharya, D., S. K. Stickel, and M. L. Sogin. 1991. Molecular phylogenetic analysis of actin genic regions from *Achlya bisexualis* (Oomycota) and *Costaria costata* (Chromophyta). J. Mol. Evol. 33:4275-4286.

Chapela, I., S. A. Rehner, T. R. Schultz, and U. G. Mueller. 1994. Evolutionary history of the symbiosis between fungus-growing ants and their fungi. Science 266:1691-1694.

Cherrett, J. M., R. J. Powell, and D. J. Stradling. 1989. The mutualism between leaf-cutting ants and their fungi. Pp. 93-120 in N. Wilding, N. M. Collins, P. M. Hammond, and J. F. Webber, eds., Insect Fungus Interactions. Academic Press, N.Y.

Clements, K. D., and S. Bullivant. 1991. An unusual symbiont from the gut of surgeonfishes may be the largest known prokaryote. J. Bacteriol. 173:5359-5362.

DeLong, E .F. 1992. Archaea in coastal marine environments. Proc. Natl. Acad. Sci. 89:5685-5689.

Fishelson, L., W. L. Montgomery, and A. A. Myrberg. 1985. A unique symbiosis in the gut of tropical herbivorous surgeonfish (Acanthuridae: Teleostei) from the Red Sea. Science 229:49-51.

Fuhrman, J. A., K. McCallum, and A. A. Davis. 1992. Novel major archaebacterial group from marine plankton. Nature 356:148-149.

Fuhrman, J. A., K. McCallum, and A. A. Davis. 1993. Phylogenetic diversity of subsurface marine microbial communities from the Atlantic and Pacific oceans. Appl. Envir. Microbiol. 59:1294-1302.

Gajadhar, A. A., W. C. Marquardt, R. Hall, J. Gunderson, E. V. A. Carmona, and M. L. Sogin. 1991. Ribosomal RNA sequences of *Sarcocystis muris*, *Theileria annulata*, and *Crypthecodinium cohnii* reveal evolutionary relationships among apicomplexans, dinoflagellates, and ciliates. Mol. Biochem. Parasitol. 45:147-154.

Haeckel, E. 1894. The Gastraea-Theory, the phylogenetic classification of the animal kingdom and the homology of the germ-lamellae. Quart. J. Microscop. Sci. 14:142.

Hinkle, G., J. K. Wetterer, T. R. Schultz, and M. L. Sogin. 1994. Phylogeny of the attine ant fungi based on analysis of small subunit ribosomal RNA gene sequences. Science 266:1695-1697.

Kermarrec, A., M. Decharme, and G. Febvay. 1986. Leaf-cutting ant symbiotic fungi: a synthesis of recent research. Pp. 231-246 in C. S. Lofgren and R. K. Vander Meer, eds., Fire Ants and Leaf-cutting Ants. Westview, Boulder, Colo.

Larsen, N., G. J. Olsen, B. L. Maidak, M. J. McCaughey, R. Overbeek, T. J. Macke, T. L. Marsh, and C. R. Woese. 1993. The ribosomal database project. Nucleic Acids Res. 21(Suppl.):3021-3023.

Leipe, D. D., J. H. Gunderson, T. A. Nerad, and M. L. Sogin. 1993. Small subunit ribosomal RNA of *Hexamita inflata* and the quest for the first branch in the eukaryotic tree. Mol. Biochem. Parasitol. 59:41-48.

Leipe, D. D., P. O. Wainright, J. H. Gunderson, D. Porter, D. J. Patterson, F. Valois, S. Himmerich, and M. L. Sogin. 1994. The stramenopiles from a molecular perspective: 16S-like rRNA sequences from *Labyrinthuloides minuta* and *Cafeteria roenbergensis*. Phycologia 33:369-377.

Pace, N. R., D. A. Stahl, D. J. Lane, and G. J. Olsen. 1986. The analysis of natural microbial populations by ribosomal RNA sequences. Adv. Microbial Ecol. 9:1-55.

Patterson, D. J. 1989. Stramenopiles: Chromophytes from a protistan perspective. Pp. 357-379 in J. C. Green, B. S. C. Leadbeater, and W. L. Diver, eds., The Chromophyte Algae: Problems and Perspectives. Clarendon Press, Oxford, England.

Patterson, D. J., and M. L. Sogin. 1992. Eukaryote origins and protistan diversity. Pp. 13-46 in H. Hartman and K. Matsuno, eds., The Origin and Evolution of Prokaryotic and Eukaryotic Cells. World Scientific Publishing Company, Singapore.

Schmidt, T. M., E. F. DeLong, and N. R. Pace. 1991. Analysis of a marine picoplankton community by 16S rRNA gene cloning and sequencing. J. Bacteriol. 173:4371-4378.

Schopf, J. W., and D. Z. Oehler. 1976. How old are the eukaryotes? Science 193:47-49.

Schultz, T. R., and R. Meier. 1995. A phylogenetic analysis of the fungus-growing ants (Hymenoptera: Formicidae: Attini) based on morphological characters of the larvae. Syst. Entomol. 20:337-370.

Sogin, M. L. 1989. Evolution of eukaryotic microorganisms and their small subunit ribosomal RNAs. Amer. Zool. 29:487-499.

Sogin, M.L. 1991. Early evolution and the origin of eukaryotes. Current Opinion Gen. Dev. 1:457-463.

Sogin, M. L., and J. H. Gunderson. 1987. Structural diversity of eukaryotic small subunit ribosomal RNAs: Evolutionary implications. Ann. N.Y. Acad. Sci. 503:125-139.

Sogin, M. L., J. H. Gunderson, H. J. Elwood, R. A. Alonso, and D. A. Peattie. 1989. Phylogenetic significance of the Kingdom concept: An unusual eukaryotic 16S-like ribosomal RNA from *Giardia lamblia*. Science 243:75-77.

Wainright, P. O., G. Hinkle, M. L. Sogin, and S. K. Stickel. 1993. The monophyletic origins of the metazoa: An unexpected evolutionary link with fungi. Science 260:340-343.

Weber, N. A. 1966. Fungus-growing ants. Science 153:587.

Weber, N. A. 1972. Gardening Ants, The Attines. The American Philosophical Society, Philadelphia. 146 pp.

Whittaker, R. H. 1969. New concepts of kingdoms of organisms. Science 163:150-160.

Wilson, E. O. 1971. The Insect Societies. Harvard University Press, Cambridge, Mass.

Woese, C. R. 1987. Bacterial evolution. Microbiol. Rev. 51:221-271.

Woese, C. R., O. Kandler, and M. L. Wheelis. 1990. Towards a natural system of organisms: Proposal for the domains Aarchaea, Bacteria, and Eucarya. Proc. Natl. Acad. Sci. 87: 4576-4579.

Zuckerkandl, E., and L. Pauling. 1965. Molecules as documents of evolutionary history. J. Theor. Biol. 8:357-366.

PART III

THREATS TO BIODIVERSITY: WHAT HAVE WE LOST AND WHAT MIGHT WE LOSE?

Deforestation is a stark reminder
of lost biological wealth.

CHAPTER
9

The Rich Diversity of Biodiversity Issues

NORMAN MYERS

*Visiting Fellow, Green College, Oxford University, Oxford, United Kingdom,
and Senior Fellow, World Wildlife Fund-US, Washington, D.C.*

As the biodiversity question becomes more prominent among scientists and the public, it becomes more challenging. It is turning out to be a rich complex of interacting factors (did you ever hear that phrase applied to an ecosystem?). In this chapter, we shall look at a selection of these issues.

While biodiversity, and indeed life itself, is the key characteristic of our planet, we know more about the total numbers of atoms in the universe than about Earth's complement of species. We spend more on exploring distant planets than on documenting the abundance and variety of life here on Earth. Scientific absurdity as this may be, taxonomists and systematists themselves are becoming an endangered entity. In fact, the one thing we know for certain about biodiversity is that it is declining at ever faster rates.

HOTSPOTS REVISITED

Fortunately, we are learning that much biodiversity is located in small areas of the planet. As much as 20% of plant species and a still higher proportion of animal species are confined to 0.5% of Earth's land surface. These species are endemic to their areas, so if the local habitats are eliminated, these species will suffer extinction. The areas in question are indeed threatened with imminent habitat destruction. It is the two attributes together that cause the areas to be designated "hotspots" (Myers, 1988, 1990a). The concept has been much advanced in recent years. In some localities, it applies well, in others less so (Bibby et al., 1992; Curnutt et al., 1994; Dinerstein and Wikramanayake, 1993). From the start, the hotspots concept was explicitly not intended to apply much outside the tropics and subtropics, since species in temperate and boreal zones tend to feature far less localized endemism than those from lower latitudes (cf., Prendergast et al., 1993). The hotspots thesis has underpinned the MacArthur Foundation's investment of $148 million in its biodiversity program during 1985-1995, and it has featured prominently in the Global Environment Facility's $335 million funding of biodiversity during the program's first 3 years.

The hotspots approach has been limited largely so far to tropical forests and Mediterranean-type zones. This is not to say that other biomes do not qualify; they simply have not been subjected to much analysis for hotspots. For example, coral reefs are likely to feature hotspots (see Reaka-Kudla in Chapter 7, Thomas in Chapter 24). Something similar applies to certain tropical lakes and wetlands, and we shall briefly review them here.

Tropical Lakes

A number of lake and river ecosystems are unusually rich in biodiversity. World-wide lakes and rivers contain at least 8,400 fish species, or roughly 40% of Earth's species of fish that have been identified to date (and almost 20% of all vertebrates). In turn, this means that these freshwater ecosystems support almost one-quarter of the planet's known biodiversity in less than 0.01% of the planet's water (Groombridge, 1992; Nelson, 1984).

At the same time, freshwater areas overall probably are being degraded and eliminated globally at a rate faster than that of tropical forests, i.e., the fastest rate in the world for extensive biomes, even though their expanse is but a small part of that of tropical forests. In the United States, for instance, half of the 5.8 million km of rivers and streams are polluted to a significant degree, and 360,000 km have been channelized in the name of flood control, while 75,000 sizeable dams block nearly every river outside Alaska, leaving only 2% of rivers free-flowing (Carr, 1993; Palmer, 1994). As a result, 20% of fish species, 36% of crayfish, and 55% of mussels are endangered or have become extinct, by contrast with only 7% of the mammals and birds in the United States (Master, 1990; Williams et al., 1993).

Much the same applies in other regions of the developed northern hemisphere (Dynesius and Nilsson, 1994). Despite this downside assessment, freshwater ecosystems—wetlands too—receive little conservation attention compared with tropical forests, coral reefs, and Mediterranean-type regions (Greenwood, 1992; Stiassny and Raminosoa, 1994).

By virtue of their isolation, lakes often form "ecological islands." In turn, this situation can lead to a high degree of speciation and endemism, producing exceptional biodiversity in terms of fish faunas, especially in the tropics (Lowe-McConnell, 1993; Payne, 1986). Yet lakes generally do not feature large floras—not, at least, so far as has been determined. But they may well feature sizeable invertebrate faunas alongside the fish communities, presumably with parallel levels of endemism—though, again, all too little is known about this aspect of their species richness. Note, however, that in Lake Baikal in Russia—with its 2,000+ species, of which 1,500 are endemic—there are 800 plant species and 1,100 invertebrate species to go with 50 fish species (Brooks, 1950; Kinystautas, 1987).

If we assume that the fish concentrations count as an "indicator" of unusual biodiversity in other species categories as well, then three East African lakes may well merit hotspot status. Moreover, many observers would urge conservation for them on grounds of their remarkably speciose fish faunas alone, with all that it implies for evolutionary radiation. That is to say, and as we shall see below, there is a qualitative rationale for enhanced conservation beyond sheer numbers of species.

Lakes Victoria, Tanganyika, and Malawi feature at least 1,167 identified fish species, with a true total of perhaps 1,450 species. Around 885 (76%) of the known species are endemic. These "fish swarms" occur in approximately 121,500 km^2, which means that 11% of Earth's species of freshwater fish are confined to 0.08% of Earth's land surface (Barlow and Lisle, 1987; Keenleyside, 1991; Lowe-McConnell, 1993). Another two smaller lakes, Edward and George, contain at least 60 endemic species. The evolution of these Rift Valley lakes has produced a phenomenon of explosive speciation, generating exceedingly rich fish communities with far greater differentiation than in any other tropical lakes. Indeed the chain of lakes, encompassing some 775 endemic cichlid fish species (plus almost 100 other endemic fish species), must be regarded as more significant for the study of evolution than the Galapagos Islands (Echelle and Kornfield, 1984; Greenwood, 1981). It even is considered that certain of the cichlids can speciate in as little as 200 years (Greenwood, personal communication, 1995). However, the basic biology of the leading lake in the chain, Lake Malawi, has yet to be elucidated, and there is next to no scientific program for long-term research of a substantive and systematized sort.

Lake Malawi's expanse—28,231 km^2—contains at least 500 cichlid species, 495 of them (99%) endemic, plus roughly 40 other fish species, approximately 20 of them (50%) endemic (Tweddle, 1991). Each year a good number of new species is discovered, which means that the true total could be rather higher than the accepted figure (the same applies to the other two lakes). Indeed the latest estimate (Greenwood, personal communication, 1995) suggests a total of 800 cichlid species. Yet Lake Malawi is only one-eighth the size of North America's Great Lakes, which feature only 173 fish species, fewer than 10 of them endemic. Lake Malawi is suffering from sediment deposition due to soil erosion in its watersheds, pollution from industrial installations and hinterland agriculture, extensive overfishing, and proposed introduction of alien species.

In Lake Victoria's 65,000 km^2, there are—or rather there used to be—more than 300 cichlid species, around 130 (43%) of them endemic, plus almost 40 other species, some 20 of them (ca. 50%) endemic (Witte et al., 1992). On top of the speciocity of the lake's fish flock, the evolutionary youth of the biota (the lake is less than 750,000 years old) and its ecological diversity make the fish community an unrivalled subject for comparative biology and evolutionary study of vertebrates. Yet introduced predators, among other problems, already have reduced the cichlid stock by 200 species (some 67%), and are likely to reduce the endemics by 90% within another decade at most (Kaufman, 1992; Lowe-McConnell, 1993; Witte et al, 1992). This must rank as the greatest extinction spasm of vertebrates in recent times.

As for Lake Tanganyika with its 28,399 km², there are 172 cichlid species and 115 other species of fish, 220 of them (76%) endemic (Lowe-McConnell, 1993). The least rich lake in the chain, Lake Tanganyika nonetheless features almost one-quarter more fish species than Europe's total of 192 species. (However, when one considers the invertebrates as well, the lake may turn out to be the most species rich of the three; Coulter, 1991.) Fortunately, Lake Tanganyika is subject to little pervasive degradation as yet, though there is much sediment deposition from rivers draining eroded catchments, and there are plans for various forms of disruptive development in the lake and its environs, notably oil extraction.

Indeed, both Lake Malawi and Lake Tanganyika could rapidly follow the impoverishing experience of Lake Victoria, unless there are precautionary conservation measures put in place ahead of time. To this extent, and on the basis of their spectacular cichlid faunas alone, there is much scope for anticipatory (as opposed to salvaging) conservation. It is likely that conservation of these lake ecosystems would safeguard large numbers of species of invertebrates as well, though the scale of this spinoff benefit is impossible to determine at this stage.

In sum, these three lakes probably feature some 1,450 fish species, or 17% of the world's 8,400 species of freshwater fish (and 48% of continental Africa's ca. 3,000 species of fish) in 0.08% of Earth's land surface. Almost 900 known species (76% of the known species) are endemic. Apart from these three, few other lakes could qualify even as subsidiary hotspot areas. The next two candidates, Lake Atitlan in Guatemala and Lake Lanao in the Philippines, possess only a few dozen endemic fish species each.

Tropical Wetlands

Wetlands are areas that contain, for part of the year at least, enough water to foster the development of specialized communities of plants and animals adapted to waterlogged conditions. They include marshes, fens, bogs, swales, wet heaths and moorlands, peatlands, floodplains, deltas, and estuaries. Some of these types of ecosystems are as different from one another as are forests and savannahs (Dugan, 1993; Finlayson and Moser, 1991; Kusler et al., 1994; Mitsch and Gosselink, 1993; Whigham et al., 1993; Williams, 1991). Many dense human communities live in or near wetlands, and they traditionally have sought to drain or otherwise modify wetlands in order to make them better suited to human needs. As a result, we have lost half the world's wetlands during the course of this century. In the year 1800, there were 80,000 km² of wetlands in the "lower 48" United States, but by 1975 there remained less than half that much, and today an additional 1,800 km² are drained and filled each year (Groombridge, 1992; see also Tiner, 1984). In Asia, more than 5,000 km² of wetlands are lost each year to agriculture, irrigation, dam construction, and the like.

With 8.5 million km² or almost 6% of Earth's land surface (roughly the same as the expanse of tropical forests or the United States), wetlands include such areas as the Pantanal Swamp in Brazil, the Sudd Swamp in Sudan, the Usumacinta Delta in Mexico, the Okavango Swamp in Botswana, the inner Niger

Disappearing wetlands threaten wading bird populations.

Delta in Mali, the Kafue Flats in Zambia, the Indus Delta in Pakistan, and the Sundarbans in Bangladesh. They are unusually rich in both plant and animal life—though, while there are lots of data on mammals and birds, there is all too little information on other vertebrates, let alone invertebrates. Of the United States and Canada's 17,000 plant species, almost 4,700 are aquatic species and another 2,000 are associated with wetlands. North American wetlands, comprising 8% of the land area, feature 39% of the region's plant species.

Regrettably, we do not know much about endemism—a key factor in hotspots—in wetlands. Because wetland areas within one country or region often are interconnected by virtue of their aquatic continuum, there may not be much endemism. The global situation in this respect remains almost entirely undocumented.

An area with prime potential to rank as a wetlands hotspot is the Pantanal Swamp. It is the largest wetland in South America and perhaps in the world, with almost 130,000 km² (an area similar to that of Illinois). Four-fifths of its expanse are in Brazil, other sectors being in Paraguay and Bolivia (Alho et al., 1988; Mittermeier et al., 1987). Consisting mostly of the low-lying (100 m altitude) floodplain of the River Paraguai and several tributary rivers, and interspersed with various types of mesic to dry forest, savannah and grassland, the area is the most important in South America for waterfowl. Well over 500 plant species are reported from the Pantanal Swamp. Little is documented concerning vertebrates, let alone invertebrates, and there is no such indication of endemism. One reasonably can assume that the species complement is sizeable, and that endemism is moderate.

The area has been disturbed since the early 1970s by agricultural encroachment (some ranches exceed 1,000 km^2), plus a good deal of deforestation, dam construction, and mining and industrial installations. The water stocks have been contaminated by chemicals from agriculture, mining, and industry. There is also a growing problem with fires, commercial overfishing, and poaching for wildlife products (e.g., caiman skins) and the live-animal trade. In short, a largescale biota is being progressively depleted. The Brazilian government has placed less than 1,500 km^2, hardly 1%, under protected status, and this seems to be theoretical protection for the most part.

Another notable wetland area is the Sudd Swamp on the River Nile in southern Sudan. In fact, it comprises a series of large swamps enclosing several substantial lakes, making up a present expanse of 16,500 km^2 of permanently flooded lands and an additional 15,000 km^2 of seasonally flooded lands. There are at least 500 plant species, of which only 100 or so occur in the central sector that is permanently flooded and features 7,000 km^2 of a monoculture of papyrus, but only one such species is known to be endemic. Over 500 bird species have been recorded, and just under 100 fish species. The area also supports 400,000 people and 800,000 cattle (Carp, 1988; Cobb, 1983; Dryver and Marchand, 1985; Howell et al., 1988; Moghraby and Sammani, 1985).

Much of the biota is threatened by the Jonglei Canal, a 360 km conduit that would cause much of the Nile's water to bypass the Sudd, reducing the permanent swamps by 21-25% and the floodplains by 15-17% compared to the late 1980s. A fully operational canal could well cause the Nile to decline to the level characteristic of the early 1950s, whereupon there would be an 80% reduction in the expanse of the swamp and a 58% loss of floodplains, hence entraining all manner of adverse repercussions for the biota. As it happens, the construction of the canal has been suspended since 1983, due to political, technical, and financial factors—mainly the civil war that persists in southern Sudan.

This summary account of two prime wetland areas indicates that there is substantial biodiversity at stake there. At the same time, this review, like that of the three East African lakes, shows that, while certain areas may contain unusual concentrations of biodiversity facing unusual threat, they cannot qualify as other than second-order hotspots because their stocks of species hardly compare with those of hotspot areas in tropical forests, coral reefs, and Mediterranean-type zones, especially as concerns endemism. The poorest tropical forest hotspots are the southwestern Ivory Coast, with 2,770 plant species, 200 of them (7%) endemic, in an area of 4,000 km^2 of primary forest; and southwestern Sri Lanka, with some 1,000 plant species, 500 of them (50%) endemic, in 700 km² of primary forest. Nonetheless, tropical wetlands and certain lakes are generally richer, at least in particular categories of species, than any other terrestrial sectors except the three biomes cited.

OTHER CATEGORIES OF BIODIVERSITY

Biodiversity occurs not only at the level of species, but embraces both species subunits and ecosystems. Consider species populations. If, as is likely, an average species comprises hundreds of genetically distinct populations, there are billions of such populations world-wide (Ehrlich and Daily, 1993). Because of their ecological differentiation, some populations will be better equipped than others to adapt to the swiftly changing environmental conditions of the foreseeable future. We may well find that, if the depletion of biodiversity continues unabated, we shall lose, say, 50% of all species, and many if not most of the surviving species will have lost, say, 90% of their populations and perhaps 50% of their genetic variability. This latter outcome is all the more probable insofar as human-induced attrition of populations' habitats tends to occur around the fringes of populations' ranges—precisely the sectors that often feature the greatest genetic variability insofar as it is here that populations encounter the greatest variety of environmental conditions. Given this prognosis, we might well ask a key question: which form of biodepletion (species, populations, genetic variability) will eventually induce the more adverse biospheric repercussions?

To illustrate the point of population depletion, consider the case of wheat. In 1994, the species flourished across an expanse of 232 million hectares, featuring a rough average of 2 million stalks per hectare. This means that the total number of individuals in 1994 exceeded 450 trillion, probably a record (Myers, 1995). As a species, then, wheat is the opposite of endangered. But because of a protracted breeding trend toward genetic uniformity, wheat has lost the great bulk of its populations and most of its genetic variability. In extensive sectors of the original range of wheat, wild strains have all but disappeared, there is a virtual "wipeout" of endemic genetic diversity. Of Greece's native wheats, 95% have become extinct; and, across Turkey, wild progenitors of wheat find sanctuary from grazing animals only in graveyards and castle ruins. Collections of wheat germplasm are viewed as quite inadequate (Damania, 1993; Knutson and Stoner, 1989).

ARE EARTH'S BIOTAS STRESSED?

Let us now consider a further dimension of the biotic crisis that receives next to no attention and that could markedly affect the eventual outcome. Certain biotic communities may be "stressed" following events of the late Pleistocene, leaving them unduly prone to extinction. For want of a better term, they may lack "ecological resilience" or "survival capacity" of a long-term sort.

The evidence for this, such as it is, lies in the fact that a mini-mass extinction started to overtake Earth's biotas during the late Pleistocene. After very long periods of steadily growing biodiversity (Signor, 1990), there was a marked decline starting some 30,000 years ago and continuing until about 1,000 years ago. Whether through human overhunting or climatic change or both, the large mammalian fauna of several entire regions—notably North and South America, Oceania, and Madagascar—lost more than 100 genera (including 70% of large North American mammals; Martin and Klein, 1984). Following this extinction spasm, there has been continuing elimination of vertebrate species (plus some plant species), albeit not on such a spectacular scale as the late-Pleistocene episode, until the onset of the present mass extinction, which started roughly in the middle of this century. These recent eliminations were due almost entirely to human activities. We have no idea how many associated species—especially invertebrate species—likewise have been lost, but they must have been quite numerous.

So today's biodiversity already was somewhat depauperate before the arrival of the unprecedentedly severe human impact from around 1950 onwards. Many surviving biotas have surely been affected adversely along the way, at least through depletion of their subspecies and populations (Kauffman and Walliser, 1990). Large numbers of species must have lost much of their genetic variability, hence their ecological adaptability—leaving them the more vulnerable to summary extinction.

In short, the present mass extinction may have had some of its origins in events long past. It would be difficult indeed to determine how far today's biotas are not only impoverished but have been subject to stress in the manner postulated. How should we define and document the stressing factors involved? What criteria could we invoke to evaluate the processes at work, let alone their present-day upshot? Thus far, the overall question has received scant research attention, despite its possibly potent signficance for the capacity of today's biotas to resist extinction pressures.

CONSEQUENCES FOR EVOLUTION

The loss of large numbers of species, let alone species subunits, will be far from the only outcome of the present biotic debacle, supposing it proceeds unchecked. There is likely to be a significant disruption of certain basic processes of evolution (Myers, 1985). The forces of natural selection and speciation can work only with the "resource base" of species and subunits available (Eldredge, 1991; Raup, 1991)—and, as we have seen, this crucial base is being grossly reduced. To cite the graphic phrasing of Michael Soule and Bruce Wilcox (1980:8) "Death is one thing; an end to birth is something else." Given what we can discern from the geologic record, the recovery period, i.e., the interval until speciation capacities generate a stock of species to match today's in abundance and variety, will be protracted. After the late Cretaceous crash, between 5 and 10 million years elapsed before there were bats in the skies and whales in the seas. Following the mass extinction of the late Permian, when marine invertebrates lost roughly half their families, as many as 20 million years were needed before the survivors could establish even half as many families as they had lost (Jablonski, 1991).

The evolutionary outcome this time around could prove yet more drastic. The critical factor lies with the likely loss of key environments. Not only do we appear set to lose most if not virtually all tropical forests, but there is progressive depletion of tropical coral reefs, wetlands, estuaries, and other biotopes with exceptional abundance and diversity of species and with unusual complexity of ecological workings. These environments have served in the past as preeminent "powerhouses" of evolution, meaning they have thrown up more species than other environments. Virtually every major group of vertebrates and many other large categories of animals have originated in spacious zones with warm, equable climates, notably the Old World tropics and especially their forests (Darlington, 1957; Mayr, 1982). The rate of evolutionary diversification— whether through proliferation of species or through emergence of major new adaptations—has been greatest in the tropics, especially in tropical forests. Tropical species, notably those in tropical forests, appear to persist for only brief periods of geologic time, which implies a high rate of evolution (Jablonski, 1993; Stanley, 1991).

As extensive environments are eliminated wholesale, moreover, the current mass extinction applies across most if not all major categories of species. The outcome will contrast sharply with the end of the Cretaceous, when not only placental mammals survived (leading to the adaptive radiation of mammals, eventually including man), but also birds, amphibians, and crocodiles, among many other nondinosaurian reptiles, persisted. In addition, the present extinction spasm is eliminating a large portion of terrestrial plant species world-wide, by contrast with episodes of mass extinction in the prehistoric past, when terrestrial plants survived with relatively few losses in many parts of the world (Knoll, 1984)—and thus supplied a resource base on which evolutionary processes could start to generate replacement species of animals forthwith. If this biotic substrate is markedly depleted within the foreseeable future, the restorative capacities of evolution will be the more diminished.

All this will carry severe implications for human societies throughout the recovery period, which is estimated to be a minimum of 5 million years, possibly several times longer. Just 5 million years would be 20 times longer than humankind itself has been a species. The present generation is effectively imposing a decision on the unconsulted behalf of at least 200,000 follow-on generations. This must rank as the most far-reaching decision ever taken on behalf of such a large number of people during the whole course of human history. Suppose that Earth's population maintains an average of 2.5 billion people during the next 5 million years (rather than the present 5.7 billion), and that the generation time remains 25 years. The total number affected will amount to 500 trillion people.

WHAT SHALL WE DO ABOUT IT ALL?

We are reaching a stage when there is less and less scope for protection of biodiversity through the traditional strategy of parks and reserves. These protected areas total more than 8,000 units covering 7 million km², or 5% of Earth's land surface (Western et al., 1994). True, there is great need for many more such areas. Ecologists and biogeographers consider that, in tropical forests alone, we should at least triple the expanse protected, while recognizing that one-third of existing parks and reserves already are subject to agricultural encroachment and other forms of human disruption (Groombridge, 1992). Additional forms of degradation are likely to stem in the future from pollution: as much as 1 million km² or almost 15% of remaining tropical forests soon could become subject to acid precipitation (Rodhe and Herrera, 1988). There eventually could be further and more widespread degradation from enhanced UV-B radiation and global warming (Gates, 1993; Peters and Lovejoy, 1992; Woodwell, 1990). We could set aside the whole of Amazonia, declaring it one huge park, and then build a fence around it 50 m high. That still would not prevent it from being depleted through atmospheric pollution and climatic change.

This all means that protected areas are becoming far less of the sufficient conservation response they once were considered to be. The time has arrived when, as a bottom-line conclusion, we must recognize that we ultimately can safeguard biodiversity only by safeguarding the biosphere as well—with all that this entails for agriculture, industry, energy, and a host of other sectors, especially the growth of both population and consumption (Myers, 1994). Increasingly too, it is apparent that we ultimately may find that we inhabit a world where there are no more protected areas: either because they have been overrun by landless peasants and grandscale pollution, or because we have learned to manage all our landscapes in such a manner that there is automatic provision for biodiversity (McNeely, 1990).

Let us conclude with a reflection on the Global Environment Facility, an initiative that has provided $335 million over the program's first 3 years to assist biodiversity. This is quite the largest such dispensation ever made. But compare it with what is at stake. Every year, just the commercial value to just the rich nations of just the present array of plant-derived pharmaceuticals is 150 times greater (Principe, 1996). More significant still, the world-wide amount spent annually on "perverse" subsidies, i.e., subsidies that inadvertently foster overloading of croplands, overgrazing of rangelands, profligate burning of fossil fuels, wasteful use of water, overcutting of forests and overharvesting of fisheries (to cite but a few examples of activities that also reduce biodiversity) is

12 times greater again. Note too that the shortfall in spending to support those 120 million couples of the developing world who possess the motivation to reduce their fertility but lack facilities for family planning, is $2.4 billion. If we were to take care of these unmet needs—which should be catered on humanitarian grounds even if there were no population problem at all—we would reduce the ultimate total of the world's population by at least 2 billion people (Bongaarts, 1994) and massively reduce pressures on species' habitats.

HOW LONG DO WE HAVE?

While formulating our responses to the mass extinction crisis, we need to bear in mind the length of time still available to us. The critical criterion for our efforts is not whether we are doing far more than before, but whether it will be enough—and that in turn raises the question of "enough by when?" How soon might we cross a threshold after which our best efforts could prove to be of little avail?

Of course, not all habitats are going to be destroyed outright within the immediate future. But that is hardly the point. What looks set to eliminate many if not most species in the long run will be the "fragmentation effect," i.e., the break up of extensive habitats into small isolated patches that are too small to maintain their stocks of species into the indefinite future. This phenomenon has been widely analyzed through the theory of island biogeography, and appears to be strongly supported through abundant empirical evidence, albeit with a good number of variations on the general theme. True, the process of ecological equilibration, with its delayed fall-out effects, will take an extended period to exert its full depletive impact; in some instances, it will be decades and even centuries before species eventually disappear. But the ultimate upshot, which is what we should be primarily concerned with, will be the same.

Consider the environmental degradation that already has occurred. Through dynamic inertia, it will continue to exert an increasingly adverse effect for a good way into the future, no matter how vigorously we try to resist the process: much potential damage is already "in the pipeline." An obvious example is acid rain, which will keep on inflicting injury on biotas by reason of pollutants already deposited though not yet causing apparent harm. Similarly, tropical forests will suffer desiccation through climatic changes induced by deforestation that already has taken place. Desertification will keep on expanding its impact through built-in momentum. Ozone-destroying CFCs now in the atmosphere will continue their work for a whole century even if we were to cease releasing them forthwith. There is enough global warming in store through past emissions of greenhouse gases to cause significant climatic change no matter how much we seek to slow it, let alone halt it.

In light of this on-going degradation of the biosphere, let us suppose, for the sake of argument, that in the year 2000 the whole of humankind were to be removed from the face of the Earth in one fell swoop. Because of the many environmental perturbations already imposed, with their impacts persisting for many subsequent decades, gross biospheric impoverishment would continue and thus serve to eliminate further large numbers of species in the long term (Myers, 1990b).

To consider a specific calculation, note Simberloff's (1986) calculations as concerns Amazonia. If deforestation continues at recent rates until the year 2000 (it is likely to accelerate in much of the region), but then halts completely, we should anticipate an eventual loss of about 15% of the species of plants in Amazonia. Were the forest cover to be ultimately reduced to those areas now set aside as parks and reserves, we should anticipate that 66% of the species of plants would disappear, together with almost 69% of the species of birds and similar proportions of other major categories of animals.

As a result of the potential biodiversity depletion that humankind already has engendered, it is realistic to prognose that there will be large numbers of extinctions in a post-2000 world, even if it were relieved of humankind's continuing disruptions. For sure, this is a highly pessimistic prognosis. The writer is anxious to avoid undue doom and gloom: we must do all we can, while we can, to limit the biotic debacle ahead. There is certainly a great deal that we still can do to contain the ultimate catastrophe. But let us not delude ourselves into supposing that there is plenty of time to make leisurely plans—"leisurely," that is, in light of the corner into which we already have painted ourselves, together with millions of fellow species. Time is of the essence, and we should take a cool look at how much maneuvering room is left to us for response. Incisive and urgent action is at a premium—which seems an appropriate point on which to end this chapter.

REFERENCES

Alho, C. J. R., T. E. Lacher, and H. C. Goncalves. 1988. Environmental degradation in the Pantanal ecosystem. BioScience 38:164-171.

Barlow, C. G., and A. Lisle. 1987. The biology of the Nile perch. Biol. Conserv. 39:269-289.

Bibby, C. J., N. J. Collar, M. J. Crosby, M. F. Heath, C. Imboden, T. H. Johnson, A. J. Long, A. J. Stattersfield, and S. J. Thirgood. 1992. Putting Biodiversity on the Map: Global Priorities for Conservation. International Council for Bird Preservation, Cambridge, England.

Bongaarts, J. 1994. Population policy options in the developing world. Science 263:771-776.

Brooks, J. R. 1950. Speciation in ancient lakes. Quart. Rev. Biol. 5:30-60.

Carp, E. 1988. Directory of Afro-Tropical Wetlands of International Importance. Conservation Monitoring Centre, Cambridge, England.

Carr, J. R. 1993. Protecting ecological integrity: An urgent societal goal. Yale J. Int. Law 18:297-306.

Cobb, S. 1983. Jonglei's fragile ecosystem. New Sci. (Oct. 27th):287.

Coulter, G. C., ed. 1991. Lake Tanganyika and Its Life. Oxford University Press, Oxford, England.

Curnutt, J., J. Lockwood, L. Hang-Kwan, P. Nott, and G. Russell. 1994. Hotspots and species diversity. Nature 367:326-327.

Damania, A. B., ed. 1993. Biodiversity and Wheat Improvement. Wiley, Chichester, England.

Darlington, P. J. 1957. ZooGeography: The Geographical Distribution of Animals. Wiley, N.Y.

Dinerstein, E., and E. D. Wikramanayake. 1993. Beyond "hot spots": how to prioritize investments to conserve biodiversity in the Indo-Pacific region. Conserv. Biol. 7:53-65.

Dryver, C. A., and M. Marchand. 1985. Taming the Floods: Environmental Aspects of Floodplain Development in Africa. Environmental Database on Wetland Interventions, Leiden, Netherlands.

Dugan, P., ed. 1993. Wetlands in Danger: A World Conservation Atlas. Oxford University Press, Oxford, England.

Dynesius, M., and C. Nilsson. 1994. Fragmentation and flow regulation of river systems in the northern third of the world. Science 266:753-759.

Echelle, A. A., and I. Kornfield. 1984. Evolution of Fish Species Flocks. University of Maine Press, Orono, Maine.

Ehrlich, P. R., and G. C. Daily. 1993. Population extinction and saving biodiversity. Ambio 22:64-68.

Eldredge, N. 1991. The Miner's Canary. Prentice-Hall, N.Y.

Finlayson, M., and M. Moser, eds. 1991. Wetlands. Facts on File, N.Y.

Gates, D. M. 1993. Climate Change and Its Biological Consequences. Sinauer Associates, Sunderland, Mass.

Greenwood, P. H. 1981. The Haplochromine Fishes of the East African Lakes. Kraus International Publications, Munich, Germany.

Greenwood, H. 1992. Are the major fish faunas well known? Netherlands J. Zool. 42(2-3):131-138.

Groombridge, B., ed. 1992. Global Biodiversity: Status of the Earth's Living Resources. Chapman and Hall, London.

Howell, P., M. Lock, and S. Cobb, eds. 1988. The Jonglei Canal: Impact and Opportunity. Cambridge University Press, Cambridge, England.

Jablonski, D. 1991. Extinctions: A palaeontological perspective. Science 253:754-757.

Jablonski, D. 1993. The tropics as a source of evolutionary novelty through geological time. Nature 364:142-144.

Kauffman, E. G., and O. H. Walliser, eds. 1990. Extinction Events in Earth History. Springer-Verlag, N.Y.

Kaufman, L. 1992. Catastrophic change in species-rich freshwater ecosystems: The lessons of Lake Victoria. BioScience 42:846-858.

Keenleyside, M. H. A., ed. 1991. Cichlid Fishes: Behavior, Ecology and Evolution. Chapman and Hall, London.

Kinystautas, A. 1987. The Natural History of the U.S.S.R. McGraw-Hill, N.Y.

Knoll, A. H. 1984. Patterns of extinction in the fossil record of vascular plants. Pp. 21-68 in M. H. Nitecki, ed., Extinctions. University of Chicago Press, Chicago.

Knutson, L., and A. K. Stoner, eds. 1989. Biotic Diversity and Germplasm Preservation: Global Imperatives. Kluwer Academic Publishers, Dordrecht, Netherlands.

Kusler, J. A., W. J. Mitsch. and J. S. Larson. 1994. Wetlands. Sci. Amer. 270:64-70.

Lowe-McConnell, R. H. 1993. Fish faunas of the African Great Lakes: origins, diversity and vulnerability. Conserv. Biol. 7:634-643.

Master, L. 1990. The imperilled status of North American Aquatic Animals. Biodiv. Network News 3(3):1-2, 7-8.

McNeely, J. A. 1990. The future of national parks. Environment 32(1):16-20, 36-41.

Martin, P. S., and R. G. Klein, eds. 1984. Quaternary Extinctions: A Prehistoric Revolution. University of Arizona Press, Tucson.

Mayr, E. 1982. The Growth of Biological Thought: Diversity, Evolution and Inheritance. Harvard University Press, Cambridge, Mass.

Mitsch, W. J., and J. G. Gosselink. 1993. Wetlands. Van Nostrand and Reinhold, N.Y.

Mittermeier, R. A., I. de Gusmao Camara, M. T. J. Padua, and J. S. Blank. 1987. The Pantanal Region of Brazil: Conservation Problems and Action Priorities. World Wildlife Fund-US, Washington, D.C.

Moghraby, A. I., and M. Sammani. 1985. On the environmental and socio-economic impact of the Jonglei Canal project, Southern Sudan. Envir. Conserv. 12:41-48.

Myers, N. 1985. The end of the lines. Nat. Hist. 94:2,6,12.

Myers, N. 1988. Threatened biotas: "hot spots" in tropical forests. Environmentalist 8:187-208.

Myers, N. 1990a. The biodiversity challenge: expanded hot-spots analysis. Environmentalist 10:243-256.

Myers, N. 1990b. Mass extinctions: What can the past tell us about the present and the future? Global Planet. Change 82:175-185.

Myers, N. 1994. Population and biodiversity. Pp. 117-136 in F. Graham-Smith, ed., Population: The Complex Reality. North American Press/Fulcrum, Golden, Colo.

Myers, N. 1995. Population and biodiversity. Ambio 24(1):56-57.

Nelson, J. S. 1984. Fishes of the World, second ed. Wiley-International, N.Y.

Palmer, T. 1994. Lifelines: The Case for River Conservation. Island Press, Washington, D.C.

Payne, A. I. 1986. The Ecology of Tropical Lakes and Rivers. John Wiley, N.Y.

Peters, R. L., and T. E. Lovejoy, eds. 1992. Consequences of the Greenhouse Warming to Biodiversity. Yale University Press, New Haven, Conn.

Prendergast, J. R., R. M. Quinn, J. H. Lawton, B. C. Eversham, and D. W. Gibbons. 1993. Rare species: the coincidence of diversity hotspots, and conservation strategies. Nature 365:335-337.

Principe, P. P. 1996. Monetizing the pharmocological benefits of plants. Pp. 191-218 in M. J. Balick, E. Elisabetsky, and S. Laird, eds., Medicinal Resources of the Tropical Forest: Biodiversity and Its Importance to Human Health. Columbia University Press, N.Y.

Raup, D. M. 1991. Extinction: Bad Genes or Bad Luck? Norton, N.Y.

Rodhe, H., and R. Herrera, eds. 1988. Acidification in Tropical Countries. Wiley, Chichester, England.

Signor, P. W. 1990. The geologic history of diversity. Ann. Rev. Ecol. Syst. 21:509-539.

Simberloff, D. 1986. Are we on the verge of a mass extinction in tropical rain forests? Pp. 165-180 in D. K. Elliott, ed., Dynamics of Extinction. Wiley, N.Y.

Soule, M. E., and B. A. Wilcox. 1980. Conservation biology: its scope and its challenge. Pp. 1-8 in M. E. Soule and B. A. Wilcox, eds., Conservation Biology: An Evolutionary-Ecological Perspective. Sinauer Associates, Sunderland, Mass.

Stanley, S. M. 1991. The New Evolutionarey Timetable. Basic Books, N.Y.

Stiassny, M. L. J., and M. Raminosoa. 1994. The fishes of the inland waters of Madagascar. Ann. Madagascar Mus. Zool. 275:133-149.

Tiner, R. W. 1984. Wetlands of the United States: Current Status and Recent Trends. U.S. Fish and Wildlife Service, Washington, D.C.

Tweddle, D. 1991. Twenty years of fisheries research in Malawi. Malawi Fish. Bull. 7:1-43.

Western, D., R. M. Wright, and S. C. Strum, eds. 1994. Natural Connections: Perspectives in Community-Based Conservation. Island Press, Washington, D.C.

Whigham, T., D. Bykyjova, and S. Hejny, eds. 1993. Wetlands of the World: Inventory, Ecology and Management. Kluwer Academic Publishers, Dordrecht, Netherlands.

Williams, J. D., M. L. Warren, K. S. Cummings, J. L. Harris, and R. J. Neves. 1993. Conservation status of freshwater mussels of the United States and Canada. Fisheries 18:6-22.

Williams, M., ed. 1991. Wetlands: A Threatened Landscape. Blackwell Scientific Publications, Oxford, England.

Witte, F., T. Goldschmidt, P. C. Goudswaard, W. Ligtvoet, M. J. P. von Oijen, and J. H. Wanink. 1992. Species extinction and concommitant ecological changes in Lake Victoria. Netherlands J. Zool. 42:214-232.

Woodwell, G. M., ed. 1990. The Earth in Transition: Patterns and Processes of Biotic Impoverishment. Cambridge University Press, N.Y.

CHAPTER 10

Human-Caused Extinction of Birds

DAVID W. STEADMAN

Curator of Birds, Florida Museum of Natural History,
University of Florida, Gainesville

Because they are so conspicuous and appealing to the human senses of sight and sound, birds always have attracted more than their fair share of our zoological attention. Almost by necessity, therefore, birds have played a prominent role in our understanding of the processes by which species become rare, endangered, and finally extinct. The resulting literature on avian conservation biology has proliferated for decades and now is part of the information explosion, with all of its benefits and frustrations.

Each year we read of additional species of birds whose existence no longer can be demonstrated. One of the latest is the Colombian grebe, *Podiceps andinus*, whose demise in highland Colombia is attributed to the loss of wetlands, the introduction of exotic fish, and hunting (Fjeldsa, 1993). Declaring a species extinct can be a tricky business (Diamond, 1987); the discovery of even one living individual, regardless of the long-term viability of the species, refutes the claim. In other words, negative evidence (such as finding no grebes) can be refuted by even one bit of positive evidence (finding a grebe). To discover a few living Colombian grebes, however, is unlikely to prevent extinction of the species, given the vulnerability of very small populations to demographic stochasticity (Caughley, 1994; Gabriel and Burger, 1992), catastrophes (Lande, 1993), genetic viability (Lynch et al., 1995), and disease (Wilson et al., 1994).

Most species declared extinct by ornithologists never have been rediscovered. One exception is the Cebu flowerpecker, *Dicaeum quadricolor*, endemic to the Philippine island of Cebu (5,088 km^2), where 8 of the 12 endemic subspecies of birds already are gone (Dutson et al., 1993). Considered extinct since 1906, up to four individuals of the Cebu flowerpecker were observed in 1992 and 1993 in the last remaining patch (<2 km²) of closed canopy forest on the entire island. The prospect of long-term survival for the Cebu flowerpecker is remote, given the scarcity of its habitat and its small population. Only 2 years after its redis- covery, the Cebu flowerpecker may be gone for good.

This chapter discusses extinction, the final stage of endangerment. Extinc- tion really is forever, in spite of what we are led to believe in dinosaur movies. Extinction is occurring today at unprecedented rates across a broad range of terrestrial and aquatic habitats (McNeely, 1992). Because species richness of birds is so high in tropical forests, the single most prolific cause of endanger- ment and extinction in birds (and many other groups of organisms) is the de- struction of tropical forests (Phillips et al., 1994; Whitmore and Sayer, 1992; Wilson, 1992; Remsen, 1995). Deforestation and other types of habitat loss also deplete avian communities in temperate (Willson et al., 1994; see Figure 10-1) and high latitude areas, which typically have fewer species of birds to lose than tropical areas.

We have no evidence that any of the species of birds now endangered or that have gone extinct in recent millennia would be in their predicament if not for human activity. While extinction does occur naturally, human impact has increased rates of extinction by orders of magnitudes over background rates (Steadman et al., 1991; Wilson, 1992), and therefore is the only significant cause of our current "biodiversity crisis."

FIGURE 10-1 An adult and three downy chicks of the Least Bittern (*Ixo- brychus exilis*) at Presque Isle State Park, Pennsylvania. Like so many spe- cies of birds, the Least Bittern is strictly dependent upon wetlands, and there- fore becomes rarer and more localized as wetlands are lost to human activities.

BACKGROUND

A comprehensive attempt to set global conservation priorities for birds (i.e., to avoid further extinction) has been compiled by Bibby et al. (1992), who identify 2,609 species of birds (27% of all living species) with breeding ranges of less than 50,000 km², designating these as "restricted-range species" (RRS). Sets of these species tend to occur together on islands or in well-defined areas of a particular continental habitat, especially tropical or montane forests. An "endemic bird area" (EBA) is any place where two or more RRS are sympatric. The 221 EBAs each contain 2-67 RRS. Many of the EBAs correspond roughly or rather exactly with centers of endemism in other organisms, such as vascular plants, butterflies, amphibians, reptiles, or mammals. The total numbers of both EBAs and RRS are about evenly divided between islands and continents. The tropics have 76% of all EBAs and more than 90% of all RRS.

The data on EBAs and RRS provide a rough but informative idea of the potential for extinction of birds in upcoming decades. For example, the country with the most RRS is Indonesia (Table 10-1), where most islands are unprotected. The EBA with the most RRS is the Solomon Islands, where none of the land is officially protected (Table 10-2). Of the 10 EBAs with the most RRS, only one has more than 15% of its land under protection. We are rapidly approaching the point of no return for hundreds of species.

Humans cause the extinction of birds in four major ways: (1) direct predation; (2) the introduction of nonnative species; (3) the spread of disease; and (4) habitat degradation or loss. Direct predation includes hunting (killing living birds), gathering eggs, or removing nestlings for captive rearing. The introduc-

TABLE 10-1 Ten Countries With the Most Restricted-Range Species (RRS) of Birds

Country	Occurring RRS	Confined RRS	Threatened RRS	Number of EBAs[a]
Indonesia	411	339	95	24
Peru	216	106	51	18
Brazil	201	122	67	19
Colombia	189	61	51	14
Papua New Guinea	172	82	18	12
Ecuador	159	32	38	11
Venezuela	120	40	17	8
Philippines	111	106	36	9
Mexico	102	59	23	14
Solomon Islands	96	43	16	4

[a]Endemic bird areas.

SOURCE: Modified from Bibby et al. (1992:Appendix 2).

TABLE 10-2   Ten Endemic Bird Areas (EBA) With the Most Restricted-Range Species (RRS) of Birds

EBA[a]	Occurring RRS	Confined RRS	Threatened RRS	Land Area (10^3 km^2)	% Land Area Protected
Pacific Islands, Indonesia					
Solomon Islands	67	42	10	32	0
New Britain and New Ireland	57	34	4	46	0-5
Tanimbar and associated islands	46	23	10	5.6	0-5
Northern Moluccas	44	27	6	29	0
Central New Guinean midmountains	41	27	2	98	5-10
Central America					
Costa Rican and Panamanian highlands	54	51	2	27	10-15
South America					
Ecuadorian dry forests	51	44	14	57	5-10
Southeast Brazil	48	41	23	50	10-15
Tepuis	42	36	0	35	60-70
Western Andes of Colombia and Ecuador	41	35	12	27	10-15

[a]Under natural conditions, each of these EBAs is primarily tropical forest of some sort.

SOURCE: Modified from Bibby et al. (1992:Appendix 1).

tion of nonnative animals exposes indigenous birds to new predators, competitors, parasites, or pathogens, and thus is related to categories (1) and (3). Habitat degradation or loss can be direct (deforestation, draining wetlands, plowing prairies, toxic pollution, etc.) or due to encroachment by nonnative plants (category 2).

Much of our biodiversity crisis is due to human impact of recent centuries, especially the past few decades. Cooperative research by archaeologists, geologists, and biologists has shown, however, that most plant and animal communities were not pristine in preindustrial times. Our skills with tools and fire have set us apart from other animals for tens of millennia. All human societies, even nonagricultural hunter-gatherers, have had various effects on their environment (e.g., Betancourt and Van Devender, 1981; Burney, 1993; Diamond, 1992; Klein, 1992; Martin, 1990; Martin and Klein, 1984; Steadman, 1991). Hunting, fishing, and gathering often focus on certain species or groups of species (Redford, 1992). Agriculture affects natural ecosystems in more diverse ways, including modifications of landscape, soils, and water supply through deforestation, erosion, channeling, flooding, draining, etc., as well as the elimination or propagation of selected species of plants and animals.

An understanding of past plant and animal communities is important for long-term estimates of community stability and change (Betancourt et al., 1990; Webb et al., 1993), such as how thoroughly and rapidly ecosystems might recover from disturbances. The responses of plants and animals to future climatic changes will be compromised by human-caused habitat fragmentation (Peters, 1988). Some species may be unable to disperse across tracts of disturbed habitat. When feasible, past distributions can aid in planning translocation programs, with an overall goal of preserving species assemblages that at least approach those of a less disturbed state.

CONTINENTS

The distinction between continents and islands is useful, although there is a continuum in land area and isolation from small, remote oceanic islands, such

Barranca del Cobre, Chihuahua, Mexico. On level ground, corn is cultivated here by the Tarahumara. The largest pine trees have been removed, resulting in the extinction of the Imperial Woodpecker (Campephilus imperialis), *a close relative of the Ivory-billed Woodpecker* (C. principalis), *which is extinct in the United States and nearly so in its only other locale, Cuba.*

as Ascension or St. Helena, to large "continental" islands, such as Madagascar, Borneo, or New Guinea, to a relatively small and isolated continent such as Australia. Regardless, we now may be at a threshold where human-caused extinctions of birds—heavily biased thus far toward species on islands (compare Tables 10-3 and 10-4)—will occur at comparable rates on continents. For example, the number of threatened restricted-range species on continents compared to islands is 247 versus 165 (Table 10-1) and 51 versus 32 (Table 10-2). The most influential factor in this new trend is habitat loss, fueled by advances in human technology and population pressure.

Another important factor in endangerment and extinction, much more species-specific than habitat loss, is direct exploitation. This may take the form of hunting (often for food; "gamebirds" such as waterfowl, currasows, guans, chachalacas, pheasants, quail, grouse, pigeons, and doves are most affected) or taking young as pets (raptors, parrots, and songbirds are the most exploited, although many other types of birds are also taken; see chapters in Beissinger and Snyder, 1992). The combination of habitat loss and direct exploitation has had verifiable impacts on one or more species from virtually every family of birds.

Unlike on some islands (see next section), we know little about the prehistoric impact of people on birds on most continents. A partial exception is North America, where people first arrived about 11,000 years ago (Haynes, 1992). Between 20 and 40 species of birds became extinct at this time, probably because of ecological dependencies on the more than 40 species of ground sloths, mammoths, mastodons, horses, tapirs, camels, and other large mammals that died out across the continent (Steadman and Martin, 1984). If humans were involved in the mammal extinctions (Martin, 1990), then they indirectly caused the extinctions of the birds as well. Before the large mammal communities collapsed, the carrion-feeding California condor lived as far away from California as Texas, Florida, and New York (Emslie, 1987; Steadman and Miller, 1987). Most

TABLE 10-3 Minimum Estimates of Human-Caused Extinction of Continental Birds since A.D. 1600

Continent	Number of Species
North America	5-7
Central America	3
South America	3
Africa	1
Europe and temperate Asia	3-4
Tropical Asia	1
Australia	1
Total	17-20

SOURCES: Mountfort (1988) and herein.

TABLE 10-4 Minimum Estimates of Extinction of Island Birds

Region	Number of Species Prehistoric[a]	A.D. 1600-1899	A.D. 1900-1994
Pacific Ocean	90	28	23
Indonesia[b]	0	0	2
Indian Ocean	11	30	1
Philippines[b]	0	0	1
Caribbean Sea	34	2	1
New Guinea and Melanesia	10	2	3
Atlantic Ocean	3	3	1
Mediterranean Sea	10	0	0
Total	158	65	32

[a]The prehistoric category consists of species from prehistoric cultural contexts, and includes only species already described. Many other prehistoric extinct species have been found but remain undescribed. For each region, an even greater number of extinct species are undiscovered because of incomplete archeological sampling.

[b]There is no interpretable prehistoric record of birds from Indonesia or the Philippines.

SOURCES: Modified from Johnson and Stattersfield (1990) and Milberg and Tyrberg (1993).

other scavenging birds perished completely. In the next 10,000 years before the arrival of Europeans, however, only two North American birds are known to have become extinct: a flightless duck, *Chendytes lawi*, of the Pacific coast (Morejohn, 1976; Guthrie, 1992) and a small turkey, *Meleagris crassipes*, from the Southwest (Rea, 1980).

Bones from late prehistoric archaeological sites in North America document birds such as the trumpeter swan, Mississippi kite, swallow-tailed kite, whooping crane, sandhill crane, long-billed curlew, great auk, Carolina parakeet, ivory-billed woodpecker, common raven, and fish crow in localities well outside of their modern ranges. While intertribal exchange of birds might account for some of these range extensions, most seem to reflect former indigenous populations. Prehistoric hunting, trapping, and habitat modification, as well as natural climatic change, may have caused these range contractions. It probably is no coincidence that three of these same species (the great auk, Carolina parakeet, ivory-billed woodpecker) are now extinct and the whooping crane is endangered. Some birds, such as the California condor, bald eagle, and golden eagle, were hunted for feathers, bones, or ceremonial purposes more than for food (Bates et al., 1993; Emslie, 1981; Rea, 1983; Simons, 1983).

In the past 200 years, at least five species of North American birds have been lost (great auk, Labrador duck, passenger pigeon, Carolina parakeet, ivory-billed woodpecker). Each of these, except the woodpecker, represented a monotypic genus. Two other species, the Eskimo curlew and Bachman's warbler, are either extinct or virtually so. The California condor, whooping crane, red-cockaded woodpecker, black-headed vireo, golden-cheeked warbler, and Kirtland's warbler persist today only in dangerously small, localized populations. Without management, three of these species already would be extinct. Even if no more habitat is lost, some of these last eight species are likely to become extinct in the next 200 years, as may others that are known to be endangered, in decline, or localized today (such as greater and lesser prairie chickens, sage grouse, piping plover, bristle-thighed curlew, common nighthawk, red-headed woodpecker, sedge wren, loggerhead shrike, California gnatcatcher, cerulean warbler, and Henslow's sparrow; reviewed in Ehrlich et al., 1992). Losing 20-25 species per millennium may seem slow to politicians and economists, but is a devastatingly high rate of extinction from an evolutionary standpoint. We have no evidence that speciation will offset any of these losses.

As serious as this situation seems, North American birds may be facing much less extinction in the next century or two than avifaunas from many other parts of the world, especially tropical regions. Each country or EBA listed in Tables 10-1 and 10-2 is tropical. With its many social, economic, political, legal, and ethical complexities (Rudel, 1993; Rush, 1991), tropical deforestation continues at rates far beyond sustained yield. Many tropical regions, especially those that are mountainous or covered with swamp forests, remain poorly surveyed for birds. In fact, it is often the case that localities are not surveyed until after human disturbances (clearing for roads, airstrips, settlements, etc.) allow access.

Long-term (decadal or more) ecological data are lacking for most tropical localities. Where such data have been gathered, decreases in species richness have been detected. At the San Antonio cloud forest in the western Andes of Colombia, for example, the approximate number of species of forest birds declined from 128 in 1911, to 104 in 1959, to 92 in 1989-1991 (Kattan et al., 1994). Caused primarily by forest fragmentation, these "local extinctions" are steps leading toward full extinction of species. Although all feeding guilds are involved, large canopy frugivores and understory insectivores have lost the most species at San Antonio.

Throughout the tropics, but perhaps especially in South America and Melanesia, the extent of local, regional, and full extinction of birds is undocumented for many species. This clearly justifies the call for more surveys (Kattan et al., 1994). Many such surveys will be fascinating biologically but ultimately futile for preventing extinction as long as so much habitat continues to be lost. How personally frustrating it must be for scientists such as Gretton et al. (1993), who determined that only 20-30 pairs of Gurney's pitta (*Pitta gurneyi*) still exist, yet territories of this colorful but secretive songbird are being lost each year to deforestation in Thailand. The gap between scientific knowledge and conservation policy often is large (Remsen, 1995). It may be safe to say that no single country on any continent has preserved enough habitat to secure the long-term survival of its current avifauna.

ISLANDS

The relatively small land areas of islands result in small populations of organisms that tend to be more vulnerable to extinction than those on continents (Diamond, 1985). Of 108 species of birds known to have become extinct since A.D. 1600, 97 (90%) lived on islands (Johnson and Stattersfield, 1990). By their calculations (97 extinct versus 1,750 extant species on islands, 11 extinct versus 7,500 extant species on continents), the probability of extinction during the past 4 centuries has been about 40 times greater on islands than continents. Although both 97 and 11 are underestimates (see Tables 10-3 and 10-4), this ratio probably is more or less valid.

The plight of island birds did not begin, however, in A.D. 1600. As summarized in Table 10-4, even more human-caused extinctions already had occurred on islands in prehistoric times (Milberg and Tyrberg, 1993; Steadman, 1995). Because of how incompletely we have sampled the zooarchaeological record of island birds thus far, the known prehistoric extinctions are a small fraction of those that actually occurred. These losses seem to have been due to the same processes that still exterminate species on islands today: predation by humans and introduced species, nonnative pathogens, and habitat destruction (Collar and Andrew, 1988; Kirch, 1983; Olson, 1977; Olson and James, 1982; Savidge, 1987; Steadman et al., 1984, 1990).

The losses of birds on oceanic islands consist of: (1) extinction (loss of all populations of a species); (2) extirpation (loss of a species on an entire island, with other population[s] surviving elsewhere); and (3) reduced population (loss of individuals from a surviving population). The last two categories are steps leading toward genuine extinction, which represents irreversible losses rather than the short-term fluctuations in populations near continental source areas that biogeographers often call "extinctions" when studying faunal turnover. Research in the Galapagos Islands (Steadman et al., 1991), Hawaiian Islands (James, 1987), and Tonga (Steadman, 1993) has shown that natural (prehuman) rates of genuine extinction in island birds may be at least 2 orders of magnitude less than posthuman rates.

Nearly all islands in Oceania (Melanesia, Micronesia, and Polynesia; Figure 10-2) were inhabited prehistorically by humans (Irwin, 1992). Birds provided fat, protein, bones, and feathers for the human colonists, who also cleared forests, cultivated crops, and raised domesticated animals (Steadman, 1989). The prehistoric extinction of Pacific island birds is known from studying bones from archaeological sites. Micronesia is not as well studied as Polynesia, but seems to differ only in details of taxonomy and chronology (Steadman and Intoh, 1994). Although Melanesian islands also have lost a variety of seabirds and landbirds (Balouet and Olson, 1989), a larger percentage of the indigenous avifauna survives on large Melanesian islands than on Polynesian or Micronesian islands, or on small Melanesian islands. This may be due to the buffering effects that steep

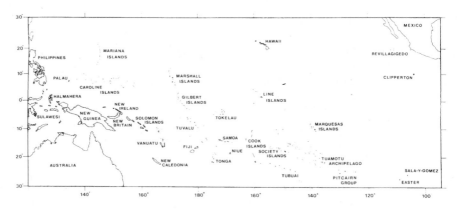

FIGURE 10-2   The Pacific Ocean, showing major island groups.

terrain, cold and wet montane climates, and human diseases have had on human impact. As predicted from biogeographic theory (MacArthur and Wilson, 1967), birds tend to be easier to extinguish on low, flat islands (which often are small) than on high, steep islands (which often are large).

While Pacific islands have a well-earned reputation as the source of much modern extinction of birds, most species of landbirds already were extinct when Captain Cook opened the region to European influence 220 years ago. Lost were as many as 2,000 species of birds, dominated by flightless rails, but also including moas, petrels, prions, pelicans, ibises, herons, swans, geese, ducks, hawks, eagles, megapodes, kagus, aptornithids, sandpipers, gulls, pigeons, doves, parrots, barn-owls, strigid owls, owlet-nightjars, and many types of passerines (Milberg and Tyrberg, 1993; Steadman, 1995). Assuming that about 9,600-9,700 species of birds exist today (Monroe and Sibley, 1993; Sibley and Monroe, 1990), the world avifauna would be about 20% richer in species had islands of the Pacific remained unoccupied by humans.

In the Hawaiian Islands alone, at least 62 endemic species of birds have become extinct since the arrival of Polynesians nearly 2,000 years ago (James and Olson, 1991; Olson and James, 1991). As is true throughout Oceania, the number of extinct species known from the Hawaiian Islands increases with each new season of archaeological or paleontological field work. Most of the extinct Hawaiian species were gone before the arrival of Europeans, including most species within the spectacular endemic radiations of cardueline finches (the Hawaiian "honeycreepers") and flightless geese, ducks, ibises, and rails (James et al., 1987; Olson and James, 1982, 1984). In New Zealand, at least 44 endemic species of landbirds have become extinct in the past millennium, featuring many endemic species of moas, waterfowl, hawks, rails, and passerines (Anderson, 1989; Holdaway, 1989; Worthy and Holdaway, 1993).

Outside of the Hawaiian Islands and New Zealand, prehistoric Polynesian birds are known from the Marquesas Islands, Society Islands, Cook Islands, Henderson Island, Easter Island, Samoa, Tonga, and Polynesian outliers in Melanesia (Steadman, 1995). In these island groups, rails, pigeons, doves, and parrots have undergone the most extinction. Rails have lost more species than any other family. Most island rails, in the Pacific and elsewhere, were flightless forest species rather than volant wetland or grassland species. Any Pacific island with a relatively thorough prehistoric record of birds has yielded bones of one or more endemic species of flightless or nearly flightless rails. The three islands with the best fossil records (Ua Huka, Mangaia, and 'Eua) each have produced two to four endemic species of flightless rails; the records from even these islands are incomplete. Two species of rails occur among only seven bones of landbirds known from remote Easter Island (see below). If not for human activities, at least 800 islands in Oceania probably would be inhabited today by flightless rails. Assuming one to four endemic species per island, rails alone may account for as many as 2,000 species of birds that would be alive today had people not colonized Oceania. The only surviving flightless rails in the tropical Pacific (east of New Guinea) are three species of *Gallirallus* (from Okinawa, Guam, and Solomon Islands; Diamond, 1991), *Nesoclopeus woodfordi* of the Solomon Islands (Hadden, 1981), and *Porzana atra* of Henderson Island (Graves, 1992). All except the last species is endangered. Just think of the possibilities for studying comparative systematics, evolution, biogeography, ecology, and behavior if hundreds or thousands of species of flightless rails still were alive. Sadly, we have wiped out nearly everything, leaving only some bones from one of nature's most dramatic examples of adaptive radiation.

Seabird losses in Polynesia have been greatest for petrels and shearwaters, although the distribution and population size of various albatrosses, storm-petrels, tropicbirds, frigatebirds, boobies, gulls, and terns have been reduced as well. Species nesting on or within the ground have been lost to predation from nonnative mammals and erosion of topsoil associated with deforestation. Abbott's booby, now confined to Christmas Island in the eastern Indian Ocean, once was widespread in the South Pacific (Figure 10-3). One island away from extinction, Abbott's booby is one of many examples of seabirds whose modern breeding range is but a tiny fraction of what it once was.

The loss of native birds is more complete on remote Easter Island (160 km², elevation 507 m) than on any other island of its size in Oceania. Although Easter Island was forested at first human contact about 1500 years ago, deforestation was virtually complete by about 550 years ago (Flenley et al., 1991). Small samples of bird bones, associated with Polynesian artifacts 600-900 years old (Steadman et al., 1994), show that Easter Island once sustained at least 22 species of seabirds, including 12 tubenoses (albatrosses, fulmars, petrels, prions, shearwaters, storm-petrels). Only 1 of the 22 species of seabirds still nests on Easter Island itself, while 7 of them occur on one or two of its offshore islets. When

FIGURE 10-3 Because so many species and populations already are lost, determining the natural distribution of island birds depends on studying bones. The tibiotarsi (A-D) and tarsometatarsi (E-H) of Abbott's booby *Papasula abbotti* are shown. A, C, E, and G are modern specimens of *P. abbotti abbotti* from Christmas Island, Indian Ocean. B, D, F, and H are archaeological specimens of an extinct subspecies from the Marquesas Islands, *P. abbotti costelloi*. The Marquesas Islands are 11,200 km east of Christmas Island, the only remaining breeding site for the species. Scale bars=20 cm. Drawing by Virginia Carter Steadman.

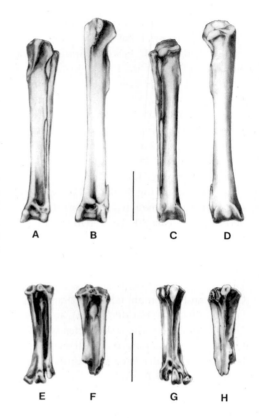

fully known, the prehistoric seabird fauna of Easter Island probably will exceed 30 species, more than any other Polynesian island. Bones also provide the only evidence that indigenous landbirds once lived on Easter Island, these being 6 extinct endemic species (a heron, 2 rails, 2 parrots, and an owl).

On Ua Huka (78 km², elevation 855 m), Marquesas Islands (Table 10-5), the Hane archaeological site has yielded thousands of bird bones (Steadman, 1989, 1991). Eight of the 20 species of seabirds from the Hane site no longer nest on Ua Huka or its tiny offshore islets. Most of the 12 other seabirds nest only on the islets, not on Ua Huka itself. The most common landbirds from Hane are 6 species of pigeons and doves, only 1 of which survives on Ua Huka. One subspecies of seabird and 8 species of landbirds from Hane are extinct.

The prehistoric exploitation of birds can be evaluated precisely at Tangatatau Rockshelter (site MAN-44), Mangaia (52 km², elevation 169 m), Cook Islands (Kirch et al., 1992; Steadman and Kirch, 1990). The age of cultural deposits at MAN-44 ranges from about 1,000 years old (zones 1A-1B) to 200 years old (zone 17). The eight extinct and five extirpated species of landbirds from MAN-

TABLE 10-5 Resident Native Birds from Ua Huka, Marquesas Islands

Native Birds	Bones from Hane Site	Modern Record[a]
Seabirds		
Shearwaters, Petrels		
Puffinus pacificus	x	x
Puffinus nativitatis	x	—
Puffinus lherminieri	x	—
Bulweria cf. *bulwerii*	x	x
Pterodroma rostrata	x	—
Pterodroma cf. *alba*	x	—
Pterodroma small sp.	x	—
Storm-Petrels		
Nesofregetta fuliginosa	x	x
Fregetta grallaria	x	—
Tropicbirds		
Phaethon lepturus	x	x
Boobies		
Sula sula	x	x
Sula leucogaster	x	x
Sula dactylatra	x	—
*Papasula abbottii costelloi	x	—
Frigatebirds		
Fregata minor	x	x
Fregata ariel	x	x
Terns		
Sterna lunata	—	x
Sterna fuscata	x	x
Anous stolidus	x	x
Anous minutus	x	x
Procelsterna cerulea	—	x
Gygis microrhyncha	x	x
Landbirds		
Herons		
Egretta sacra	x	x
Sandpipers		
Prosobonia cf. *cancellata*	x	—
Rails		
*Porzana new sp.	x	—
*Gallirallus new sp.	x	—
Pigeons, Doves		
Gallicolumba rubescens	x	—
*Gallicolumba nui	x	—
*Ptilinopus mercierii	x	—
Ptilinopus dupetithouarsii	x	x
Ducula galeata	x	—
*Macropygia heana	x	—

continued

TABLE 10-5 *Continued*

Native Birds	Bones from Hane Site	Modern Record[a]
Landbirds—*continued*		
Parrots		
Vini ultramarina	x	—
Vini vidivici	x	—
Vini sinotoi	x	—
Swifts		
Collocalia ocista	—	x
Kingfishers		
Halcyon godeffroyi	x	—
Monarch Flycatchers		
cf. Myiagra new sp.	x	—
Pomarea iphis	—	x
Warblers		
Acrocephalus mendanae	x	x
Seabirds		
Total species	20	14
Combined total species	22	
Landbirds		
Total species	15	5
Combined total species	18	

*=extinct taxon.

[a]Nineteenth or twentieth century specimen or sight record from Ua Huka or its offshore islets.

NOTE: Dashes indicate no record.

SOURCE: Modified from Steadman (1991).

44 are rails, sandpipers, pigeons, doves, and parrots whose bones dominate zones 1-4 and decline sharply by zone 5 (Table 10-6). Most species of indigenous landbirds are not recorded above zones 5-7, which are 700 to 500 years old. The most common flightless rail, *Porzana rua*, is last recorded in zone 8. Two species of doves are the only extinct/extirpated landbirds recorded above zone 8. Each of the four surviving landbirds from MAN-44 (a duck, rail, kingfisher, and warbler) tolerates forest clearance. The bristle-thighed curlew (*Numenius tahitiensis*), a rare shorebird that nests in Alaska, has been killed and eaten for centuries on its Pacific island wintering areas (Table 10-6), where about 50% of the adults become flightless during wing molt (Marks, 1993).

Because of its depauperate modern avifauna, the Kingdom of Tonga did not qualify as an EBA in Bibby et al. (1992). Bones from caves on 'Eua (85 km^2, elevation 300 m) indicate, however, that at least 27 species of landbirds lived on this Tongan island before the arrival of people about 3,000 years ago (Steadman,

TABLE 10-6   Bones of Landbirds from Tangatatau Rockshelter (site MAN-44), Mangaia, Cook Islands, Ranging from Zone 1A (Oldest) to 17 (Youngest)

Native Birds	Zone									Total
	1A	1B	2-3	4	5-7	8	9-14	15	17	
Migratory Shorebirds										
Plovers										
Pluvialis dominica	—	—	2	—	—	—	—	—	—	2
Curlews										
Numenius tahitiensis	5	—	2	2	—	—	—	1	—	10
Native Landbirds										
Ducks										
Anas superciliosa	—	—	3	1	2	2	8	11	3	30
Rails										
**Gallirallus ripleyi	10	5	12	16	1	—	—	—	—	44
Porzana tabuensis	—	—	2	1	—	1	—	1	1	6
**Porzana rua	11	44	41	21	2	1	—	—	—	120
**Porzana new sp.	—	2	1	1	—	—	—	—	—	4
**Porphyrio? new sp.	—	1	—	—	—	—	—	—	—	1
Sandpipers										
**Prosobonia new sp.	1	1	1	1	—	—	—	—	—	4
Pigeons, Doves										
*Gallicolumba erythroptera	3	5	3	5	—	—	—	—	—	16
**Gallicolumba new sp.	—	—	—	—	—	—	7	1	3	11
**Gallicolumba nui	3	3	—	3	—	—	—	—	—	9
*Ptilinopus rarotongensis	1	4	4	2	—	—	—	1	—	12
*Ducula aurorae	1	2	1	1	—	—	—	—	—	5
*Ducula galeata	3	1	5	2	—	1	—	—	—	12
Parrots										
*Vini kuhlii	14	41	15	23	1	—	—	—	—	94
**Vini vidivici	42	14	3	14	2	—	—	—	—	75
[*/**Vini kuhlii/vidivici]	2	2	—	5	—	—	—	—	—	9
Kingfishers										
Halcyon mangaia	1	2	2	2	—	—	—	—	1	8
Warblers										
Acrocephalus kerearako	6	6	1	2	—	—	—	—	—	15
Total NISP										
All Species	103	133	98	102	8	5	15	15	8	487
Migratory Shorebirds	5	0	4	2	0	0	0	1	0	12
Native Landbirds	98	133	94	100	8	5	15	14	8	475
*/** Native Landbirds	91	125	86	94	6	2	7	2	3	416
% NISP */** Native Landbirds										
of All Nonfish Bones	72	57	5	21	4	0.4	2	0.5	2	11
Total Species										
All	13	14	16	16	5	4	2	5	4	19
Migratory Shorebirds	1	0	2	1	0	0	0	1	0	2
Native Landbirds	12	14	14	15	5	4	2	4	4	17
*/** Native Landbirds	10	12	10	11	4	2	1	2	1	13

*=extant species, extirpated on Mangaia.
**=extinct species.

NOTES: Dashes indicate no record. Taxa in brackets are not different from others listed more specifically. NISP represents number of identified specimens.

SOURCE: Updated from Steadman and Kirch (1990).

1993, 1995; Table 10-7). Only 6 of these same species still survive on 'Eua. If they were still alive, most species of 'Eua's extinct forest birds probably would qualify as RRS under the criteria of Bibby et al. (1992), thereby changing considerably the international conservation priority of Tongan birds.

The 'Euan avifauna has been depleted irreversibly. As sampled thus far, the prehuman landbirds consisted of 27 forest and no nonforest species, compared to 9 forest and 4 nonforest species today (Table 10-7). Forest frugivores/granivores have declined from 12 to 4, nectarivores from 4 to 1, omnivores from 3 to 0, insectivores from 6 to 3, and predators from 2 to 1. These losses are more complete for ground-dwelling than midlevel/understory or canopy species, which might be expected with predation from humans, rats, dogs, and pigs. The means of pollination and/or seed dispersal for various Polynesian forest trees undoubtedly has been restricted or eliminated by the loss of so many nectarivorous, frugivorous, and granivorous birds (Franklin and Steadman, 1991).

DISCUSSION

Humans cause the extinction of birds by overhunting, by introducing non-native species, by spreading disease, and through habitat destruction. For many, perhaps most, species of birds in danger of extinction today, habitat destruction is the most serious threat. Loss of habitat is why tropical regions now have more potential for extinction than temperate or polar regions. Even in the United States, however, with our many and often admirable environmental laws, much natural habitat has been lost or seriously degraded. We are fortunate that more species are not already gone. Given current trends in human population growth and habitat loss, this good fortune is unlikely to persist in the United States or anywhere else. That 20% of all living species and 70% of all threatened species of birds are confined to only 2% of the Earth's land surface (Bibby et al., 1992) is both a blessing and a curse. On the one hand, it means that we might be able to save many species by protecting relatively small areas. On the other hand, we can lose many species by allowing just these same places to be degraded.

Across the world, extinction rates of birds increase whenever humans enter a previously uninhabited region. Human colonization of the Pacific islands alone led to the extinction of as many as 2,000 species of birds, dominated by flightless rails, but also including species in many other families. A crude but conservative estimate of overall losses of birds in Oceania is that an average of at least 10 species (including 2-3 that are endemic to the island and 2-3 that are endemic to the island group) has been lost already on each of the approximately 800 major islands, yielding a minimum total loss of perhaps 2,000 species. Each of the 16 Polynesian islands that has yielded 300 or more prehistoric bird bones approaches or exceeds 20 extirpated species, and none of these records is complete.

TABLE 10-7 Chronology and Community Ecology of Indigenous Resident Landbirds from 'Eua, Tonga

Resident Landbirds	Pre-human Record[a]	Archaeological Record[b]	Extant in 1988	Guild[c]
Herons				
Egretta sacra	—	X	X	NF
**Nycticorax* new sp.	X	—	—	GP
Ducks				
Anas superciliosa	—	—	X	NF
Hawks				
Accipiter cf. *rufitorques*	X	—	—	DR
Megapodes				
**Megapodius alimentum*	X	X	—	GF
Megapodius pritchardi	X	—	—	GF
**Megapodius* new sp.	X	X	—	GF
Rails				
**Gallirallus* new sp.	X	—	—	GO
Gallirallus philippensis	—	X	X	NF
Porzana tabuensis	—	X	—	NF
**Gallinula* new sp.	X	—	—	GO
Porphyrio porphyrio	—	X	X	NF
Pigeons, Doves				
Gallicolumba stairi	X	X	—	GF
**Didunculus* new sp.	X	X	—	MF
Ptilinopus porphyraceus	X	X	X	CF
Ptilinopus perousii	X	X	X	CF
**Ducula david*	X	X	—	CF
**Ducula* new sp.	X	X	—	CF
Ducula pacifica	—	X	X	CF
Parrots				
Vini solitarius	X	—	—	CN
Vini australis	X	X	—	CN
**Eclectus* new sp.	X	X	—	CF
Barn-Owls				
Tyto alba	—	X	X	NR
Swifts				
Collocalia spodiopygia	—	X	X	AI
Kingfishers				
Halcyon chloris	X	X	X	MI
Trillers				
Lalage maculosa	X	X	X	MI
*/**cf.*Lalage* sp.	—	X	—	MI
Whistlers, Robins				
*/***Eopsaltria* sp.	X	—	—	MI
Monarchs				
Clytorhyncus vitiensis	X	X	—	MI
Myiagra sp.	X	X	—	MI
Warblers				
*/***Cettia* sp.	X	—	—	MI
Thrushes				
Turdus poliocephalus	X	X	—	MF

continued

TABLE 10-7 Continued

Resident Landbirds	Pre-human Record[a]	Archaeological Record[b]	Extant in 1988	Guild[c]
Starlings				
Aplonis tabuensis	X	X	X	MF
Honeyeaters				
*Myzomela cardinalis	X	X	—	CN
Foulehaio carunculata	X	X	X	CN
White-Eyes				
**Zosteropidae new sp.	X	—	—	CO
Total Species	27	26	13	
Total */** Species	21	14	3	
# of Sites/# of Bird Bones	1/401	14/888	—	
Guild Totals[c]				
AI	0	1	1	
CF	5	6	3	
CN	4	3	1	
CO	1	0	0	
DR	1	0	0	
GF	4	3	0	
GO	2	0	0	
GP	1	0	0	
MF	3	3	1	
MI	6	5	2	
NF	0	4	4	
NR	0	1	1	
Total Forest Species (All Categories of Guilds except NF)	27	22	9	
Food Categories				
Frugivores	12	12	4	
Nectarivores	4	3	1	
Omnivores	3	0	0	
Insectivores	6	6	3	
Predators	2	1	1	
Height Categories				
Aerial	0	1	1	
Canopy	10	9	4	
Midlevel/understory[d]	10	9	4	
Ground	7	3	0	

*=extirpated species.

**=extinct species.

[a]>3000 years before the present (BP)(1 site).

[b]3000-200 years BP (15 different sites).

[c]AI=aerial insectivore; CF=canopy frugivore/granivore; CN=canopy nectarivore; CO=canopy omnivore; DR=diurnal raptor; GF=ground frugivore/granivore; GO=ground omnivore; GP=ground predator; MF=midlevel/understory frugivore/granivore; MI=midlevel/understory insectivore; NF=nonforest species; NR=nocturnal raptor. Certain distinctions between C and M are arbitrary.

[d]Includes the hawk and owl.

NOTE: Dashes indicate no record.

SOURCE: From Steadman (1993, 1995).

The rapidity of extinction on oceanic islands has been influenced by the ruggedness of terrain and the size and permanence of human populations. After the arrival of humans, extinction of birds is what we have come to expect; survival is the exception. While it is too late to maintain or restore the entire natural avifaunas of most oceanic islands, rigorous habitat protection, predator control, and translocation can save much of what is left (Franklin and Steadman, 1991). Nevertheless, because most island biotas are so degraded, programs for endangered species on islands face difficult challenges to ensure the long-term survival of the species that remain. We know what needs to happen to save most species, but without improved conservation funding and coordinated changes in human activities, many of the challenges will not be met.

On tropical islands in all oceans, the prehistoric and ongoing extinction of birds has consequences far beyond losing the birds themselves. For example, the loss of hundreds of populations and a few entire species of Pacific seabirds probably has influenced marine food webs, in which seabirds are top consumers. Extinct Pacific island landbirds undoubtedly were involved in the pollination and seed dispersal of indigenous plants, many of which may lack natural means of intra- or interisland dispersal today.

CONCLUSIONS

Rather than despair at what already is lost, I would argue that the extinct species of birds should inspire us to save those that remain. I also would hope to elevate the scientific status of recently extinct species so that their study is regarded as an important component of modern biology. Hundreds of species that should be living today exist now only as skins or bones in museums. Who will study these "relics of a lost world" (Graves, 1993) to learn more about their phylogeny, biogeography, and ecology? Very few young ornithologists are being trained in systematics, especially in areas other than molecular systematics. This is a serious situation, given that even such a widespread and locally common species as the yellow warbler (*Dendroica petechia*), when studied more comprehensively than ever before (2,500 skins examined by Browning, 1994), has been found to include two previously unnamed subspecies in Alaska and northern Canada. As pointed out so clearly by Trombulak (1994:590), training in conservation biology should include much more than "applied ecology and field population genetics."

Whether their last gasp was thousands, hundreds, or only tens of years ago, virtually all birds lost since the last ice age would still be alive if not for humans; they would be feeding, preening, singing, nesting, molting, and doing anything else that living birds do. You could use them to test behavioral hypotheses. You could see them during field work, vacations, or maybe on the way to work. They would be illustrated in field guides and be part of your overall biodiversity consciousness. (A field guide to South Pacific birds would depict as many species as a field guide to South American birds!) Pacific island rails would compete with Darwin's finches and Hawaiian finches as "textbook" examples of adaptive radiation.

We need to squeeze out as much knowledge as we can about recently extinct species of birds. This chapter began by noting that, relative to other animals, birds are conspicuous and therefore more studied. This is supported by the low rate at which new living species are being discovered. The number of "good" new species of birds described per year has varied from 6.0 in 1938-1941 to 2.6 in 1941-1955, 3.5 in 1956-1965, 3.1 in 1966-1975, 2.4 in 1976-1980, and 2.4 in 1981-1990 (Vuilleumier et al., 1992). While estimates of valid descriptions of species, just like estimates of extinction, have margins of error, sometime within the past decade we probably reached the point where more living species of birds are going extinct per year than are being newly described. This is a debt that cannot be repaid.

REFERENCES

Anderson, A. 1989. Prodigious birds: Moas and moa-hunting in prehistoric New Zealand. Cambridge University Press, Cambridge, England.

Balouet, J. C., and S. L. Olson. 1989. Fossil birds from late Quaternary deposits in New Caledonia. Smithsonian Contr. Zool. 469:1-38.

Bates, C. D., J. A. Hamber, and M. J. Lee. 1993. The California condor and the California Indians. Amer. Indian Art 19:40-47.

Beissinger, S. R., and N. F. R. Snyder, eds. 1992. New World Parrots in Crisis: Solutions from Conservation Biology. Smithsonian Institution Press, Washington, D.C.

Betancourt, J. L., and T. R. Van Devender. 1981. Holocene vegetation in Chaco Canyon, New Mexico. Science 214:656-658.

Betancourt, J. L., T. R. Van Devender, and P. S. Martin, eds. 1990. Packrat Middens: The Last 40,000 Years of Biotic Change. University of Arizona Press, Tucson.

Bibby, C. J., N. J. Collar, M. J. Crosby, M. F. Heath, C. Imboden, T. H. Johnson, A. J. Long, A. J. Stattersfield, and S. J. Thirgood. 1992. Putting Biodiversity on the Map: Priority Areas for Global Conservation. International Council for Bird Preservation, Cambridge, England.

Browning, M. R. 1994. A taxonomic review of *Dendroica petechia* (Yellow Warbler) (Aves: Parulinae). Proc. Biol. Soc. Wash. 107:27-51.

Burney, D. A. 1993. Recent animal extinctions: Recipes for disaster. Amer. Sci. 81:530-541.

Caughley, G. 1994. Directions in conservation biology. J. Anim. Ecol. 63:215-244.

Collar, N. J., and P. Andrew. 1988. Birds to Watch: the ICBP World Checklist of Threatened Birds. International Council for Bird Preservation Tech. Pub. 8:1-303.

Diamond, J. M. 1985. Population processes in island birds: Immigration, extinction and fluctuation. International Council for Bird Preservation Tech. Pub. 3:17-21.

Diamond, J. M. 1987. Extant unless proved extinct? Or, extinct unless proved extant? Conserv. Biol. 1:77-79.

Diamond, J. M. 1991. A new species of rail from the Solomon Islands and convergent evolution of insular flightlessness. Auk 108:461-470.

Diamond, J. 1992. The Third Chimpanzee. Harper Collins, N.Y.

Dutson, G. C. L., P. M. Magsalay, and R. J. Timmins. 1993. The rediscovery of the Cebu Flowerpecker *Dicaeum quadricolor*, with notes on other forest birds of Cebu, Philippines. Bird Conserv. Int. 3:235-243.

Ehrlich, P. R., D. S. Dobkin, and D. Wheye. 1992. Birds in Jeopardy. Stanford University Press, Stanford, Calif.

Emslie, S. D. 1981. Prehistoric agricultural ecosystems: Avifauna from Pottery Mound, New Mexico. Amer. Antiq. 46:853-861.

Emslie, S. D. 1987. Age and diet of fossil California Condors in Grand Canyon, Arizona. Science 237:768-770.

Fjeldsa, J. 1993. The decline and probable extinction of the Colombian Grebe *Podiceps andinus*. Bird Conserv. Int. 3:221-234.

Flenley, J. R., S. M. King, J. Jackson, C. Chew, J. T. Teller, and M. E. Prentice. 1991. The late Quaternary vegetational and climatic history of Easter Island. J. Quat. Sci. 6:85-115.

Franklin, J., and D. W. Steadman. 1991. The potential for conservation of Polynesian birds through habitat mapping and species translocation. Conserv. Biol. 5:506-521.

Gabriel, W., and R. Burger. 1992. Survival of small populations under demographic stochasticity. Theor. Pop. Biol. 41:44-71.

Graves, G. R. 1992. The endemic land birds of Henderson Island, Southeastern Polynesia. Wilson Bull. 104:32-43.

Graves, G. R. 1993. Relic of a lost world: A new species of Sunangel (Trochilidae: *Heliangelus*) from "Bogota." Auk 110:1-8.

Gretton, A., M. Kohler, R. V. Lansdown, T. J. Pankhurst, J. Parr, and C. Robson. 1993. The status of Gurney's Pitta *Pitta gurneyi*, 1987-1989. Bird Conserv. Int. 3:351-367.

Guthrie, D. A. 1992. A late Pleistocene avifauna from San Miguel Island, California. Los Angeles County Nat. Hist. Mus. Sci. Ser. 36:319-327.

Hadden, D. 1981. Birds of the North Solomons. Wau Ecology Institute Handbook No. 8, Wau, Papua New Guinea.

Haynes, C. V., Jr. 1992. Contributions of radiocarbon dating to the geochronology of the peopling of the New World. In A. Long and R. S. Kra, eds., Radiocarbon Dating After Four Decades. Springer-Verlag, N.Y.

Holdaway, R. N. 1989. New Zealand's pre-human avifauna and its vulnerability. New Zealand J. Ecol. 12(Suppl.):115-129.

Irwin, G. 1992. The Prehistoric Exploration and Colonization of the Pacific. Cambridge University Press, Cambridge, England.

James, H. F. 1987. A late Pleistocene avifauna from the island of Oahu, Hawaiian Islands. Documents des Laboratoires de Geologie de la Faculte des Sciences de Lyon 99:221-230.

James, H. F., and S. L. Olson. 1991. Descriptions of thirty-two species of birds from the Hawaiian Islands: part II. Passeriformes. Ornithol. Monogr. 46:1-88.

James, H. F., T. W. Stafford, Jr., D. W. Steadman, S. L. Olson, P. S. Martin, A. J. T. Jull, and P. C. McCoy. 1987. Radiocarbon dates on bones of extinct birds from Hawaii. Proc. Natl. Acad. Sci. 84:2350-2354.

Johnson, T. H., and A. J. Stattersfield. 1990. A global review of island endemic birds. Ibis 132:167-180.

Kattan, G. H., H. Alvarez-Lopez, and M. Giraldo. 1994. Forest fragmentation and bird extinctions: San Antonio eighty years later. Conserv. Biol. 8:138-146.

Kirch, P. V. 1983. Man's role in modifying tropical and subtropical Polynesian ecosystems. Archaeol. Oceania 18:26-31.

Kirch, P. V., J. R. Flenley, D. W. Steadman, F. Lamont, and S. Dawson. 1992. Prehistoric humans impacts on an island ecosystem: Mangaia, Central Polynesia. Natl. Geogr. Res. Expl. 8:166-179.

Klein, R. G. 1992. The impact of early people on the environment: The case of large mammal extinctions. Pp. 13-34 in J. E. Jacobsen and J. Firor, eds., Human Impact on the Environment: Ancient Roots, Current Challenges. Westview Press, Boulder, Colo.

Lande, R. 1993. Risks of population extinction from demographic and environmental stochasticity, and random catastrophes. Amer. Nat. 142:911-927.

Lynch, M., J. Conery, and R. Bürger. 1995. Mutation accumulation and the extinction of small populations. Amer. Nat. 146:489-518.

MacArthur, R. H., and E. O. Wilson. 1967. The Theory of Island Biogeography. Princeton University Press, N.J.

Marks, J. S. 1993. Molt of Bristle-thighed Curlews in the northwestern Hawaiian Islands. Auk 110:573-587.

Martin, P. S. 1990. 40,000 years of extinctions of the "planet of doom." Palaeogeogr. Palaeoclimatol. Palaeoecol. 82:187-201.

Martin, P. S., and R. G. Klein, eds. 1984. Quaternary Extinctions. University of Arizona Press, Tucson.

McNeely, J. A. 1992. The sinking ark: Pollution and the world-wide loss of biodiversity. Biodiv. Conserv. 1:2-18.

Milberg, P., and T. Tyrberg. 1993. Native birds and noble savages—a review of man-caused prehistoric extinctions of island birds. Ecography 16:229-250.

Monroe, B. L., Jr., and C. G. Sibley. 1993. A World Checklist of Birds. Yale University Press, New Haven, Conn.

Morejohn, G. V. 1976. Evidence of the survival to recent times of the extinct flightless duck *Chendytes lawi* Miller. Smithsonian Contr. Paleobiol. 27:207-211.

Mountfort, G. 1988. Rare Birds of the World. Stephen Greene Press, Lexington, Mass.

Olson, S. L. 1977. A synopsis of the fossil Rallidae. Pp. 509-525 in S. D. Ripley, Rails of the World. David R. Godine, Boston.

Olson, S. L., and H. F. James. 1982. Fossil birds from the Hawaiian Islands: Evidence for wholesale extinction by man before Western contact. Science 217:633-635.

Olson, S. L., and H. F. James. 1984. The role of Polynesians in the extinction of the avifauna of the Hawaiian Islands. Pp. 768-780 in P. S. Martin and R. G. Klein, eds., Quaternary Extinctions. University of Arizona Press, Tucson.

Olson, S. L., and H. F. James. 1991. Descriptions of thirty-two species of birds from the Hawaiian Islands: Part I. Non-Passeriformes. Ornithol. Monogr. 45:1-88.

Peters, R. L. 1988. The effect of global climatic change on natural communities. Pp. 450-461 in E. O. Wilson and F. M. Peter, eds., BioDiversity. National Academy Press, Washington, D.C.

Phillips, O. L., P. Hall, A. H. Gentry, S. A. Sawyer, and R. Vasquez. 1994. Dynamics and species richness of tropical forests. Proc. Natl. Acad. Sci. 91:2805-2809.

Rea, A. M. 1980. Late Pleistocene and Holocene turkeys in the Southwest. Los Angeles County Mus. Nat. Hist. Contr. Sci. 330:209-224.

Rea, A. M. 1983. Once a River. University of Arizona Press, Tucson.

Redford, K. H. 1992. The empty forest. BioScience 42:412-422.

Remsen, J.V., Jr. 1995. The importance of continued collecting of bird specimens to ornithology and bird conservation. Bird Conserv. Int. 5:145-180.

Rudel, T. K. 1993. Tropical Deforestation. Columbia University Press, N.Y.

Rush, J. 1991. The Last Tree: Reclaiming the Environment in Tropical Asia. The Asia Society, N.Y.

Savidge, J. A. 1987. Extinction of an island forest avifauna. Ecology 68:660-668.

Sibley, C. G., and B. L. Monroe, Jr. 1990. Distribution and Taxonomy of Birds of the World. Yale University Press, New Haven, Conn.

Simons, D. D. 1983. Interactions between California Condors and humans in prehistoric far western North America. Pp. 470-494 in S. R. Wilbur and J. A. Jackson, eds., Vulture Biology and Management. University of California Press, Berkeley.

Steadman, D. W. 1989. Extinction of birds in Eastern Polynesia: A review of the record, and comparisons with other Pacific island groups. J. Archaeol. Sci. 16:177-205.

Steadman, D. W. 1991. Extinction of species: Past, present, and future. Pp. 156-169 in R. L. Wyman, ed., Global Climate Change and Life on Earth. Routledge, Chapman and Hall, N.Y.

Steadman, D. W. 1993. Biogeography of Tongan birds before and after human impact. Proc. Natl. Acad. Sci. 90:818-822.

Steadman, D. W. 1995. Prehistoric extinctions of Pacific island birds: Biodiversity meets zoo-archeology. Science 267:1123-1131.

Steadman, D. W., E. C. Greiner, and C. S. Wood. 1990. Absence of blood parasites in indigenous birds and introduced birds from the Cook Islands, South Pacific. Conserv. Biol. 4:398-404.

Steadman, D. W., and M. Intoh. 1994. Biogeography and prehistoric exploitation of birds from Fais Island, Yap, Federated States of Micronesia. Pac. Sci. 48:116-135.

Steadman, D. W., and P. V. Kirch. 1990. Prehistoric extinction of birds on Mangaia, Cook Islands, Polynesia. Proc. Natl. Acad. Sci. 87:9605-9609.

Steadman, D. W., and P. S. Martin. 1984. Extinction of birds in the Late Pleistocene of North America. Pp. 466-477 in P. S. Martin and R. G. Klein, eds., Quaternary Extinctions. University of Arizona Press, Tucson.

Steadman, D. W., and N. G. Miller. 1987. California condor associated with spruce-jack pine woodland in the Late Pleistocene of New York. Quat. Res. 28:415-426.

Steadman, D. W., G. K. Pregill, and S. L. Olson. 1984. Fossil vertebrates from Antigua, Lesser Antilles: Evidence for late Holocene human-caused extinctions in the West Indies. Proc. Natl. Acad. Sci. 81:4448-4451.

Steadman, D. W., T. W. Stafford, Jr., D. J. Donahue, and A. J. T. Jull. 1991. Chronology of Holocene vertebrate extinction in the Galapagos Islands. Quat. Res. 35:126-133.

Steadman, D. W., P. Vargas, and C. Cristino. 1994. Stratigraphy, chronology, and cultural context of an early faunal assemblage from Easter Island. Asian Perspect. 16:59-77.

Trombulak, S. C. 1994. Undergraduate education and the next generation of conservation biologists. Conserv. Biol. 8:589-591.

Vuilleumier, F., M. LeCroy, and E. Mayr. 1992. New species of birds described from 1981 to 1990. Bull. Brit. Ornithol. Club, Cent. Suppl. 112A:267-309.

Webb, T., III, P. J. Bartlein, S. Harrison, and K. H. Anderson. 1993. Vegetation, lake levels, and climate in eastern United States since 18,000 yr B.P. Pp. 415-467 in H. E. Wright, T. Webb, III, and J. E. Kutzbach, eds., Global Climates Since the Last Glacial Maximum. University of Minnesota Press, Minneapolis.

Whitmore, T. C., and J. A. Sayer, eds. 1992. Tropical Deforestation and Species Extinction. Chapman and Hall, London.

Willson, M. F., T. L. De Santo, C. Sabag, and J. J. Armesto. 1994. Avian communities of fragmented south-temperate rainforests in Chile. Conserv. Biol. 8:508-520.

Wilson, E. O. 1992. The Diversity of Life. Belknap Press, Cambridge, Mass.

Wilson, M. H., C. B. Kepler, N. F. R. Snyder, S. R. Derrickson, F. J. Dein, J. W. Wiley, J. M. Wunderle, Jr., A. E. Lugo, D. L. Graham, and W. D. Toons. 1994. Puerto Rican Parrots and potential limitations of the metapopulation approach to species conservation. Conserv. Biol. 8:114-123.

Worthy, T. H., and R. N. Holdaway. 1993. Quaternary fossil faunas from caves in the Punakaiki area, West Coast, South Island, New Zealand. J. R. Soc. New Zealand 23:147-254.

Global Warming and Plant Species Richness: A Case Study of the Paleocene/Eocene Boundary

SCOTT L. WING

Associate Curator, Department of Paleobiology, National Museum of
Natural History, Smithsonian Institution, Washington, D.C.

There is widespread concern over the possibility that greenhouse gases generated by human activity may cause global warming. Much effort is being devoted to monitoring changes in climate and organisms, and to modeling the possible effects of greenhouse gases on climatic and biological systems. Programs of monitoring and modeling clearly are necessary, but the geological history of the Earth and its biota have not been mined thoroughly enough for information about global climate and ecosystem response to climatic change.

Fossils and sedimentary rocks form a record of changes in the Earth's climate and of biotic responses over geological time. This record can be used in two main ways to develop a more complete understanding of global climate and of the long-term effects of climatic change on ecological systems. First, comparing climatic conditions indicated by fossils and sedimentary rocks with computer simulations of global climate for the same period is the only way we have of testing the ability of climate models to simulate conditions other than those that exist today. If models can successfully simulate climatic patterns known to have existed in the past, then we can have greater confidence in their predictions about the future. Second, the fossil record provides our sole opportunity to examine the biological consequences of climatic change without waiting for them to happen in "real time." If particular kinds of climatic change are associated with specific patterns of faunal and floral turnover, then we have some basis for anticipating future biotic responses to climatic change.

This chapter focuses on climatic changes and biotic events that took place during the Paleocene-Eocene transition, about 55 million years ago. With the exception of the Holocene deglaciation, the Paleocene-Eocene transition is probably the best documented example of a geologically rapid warming of global climate, although just how rapidly it occurred is still a matter of investigation. Our distance in time from these events places severe limitations on our ability to determine the rates of environmental and biological changes, and especially the synchroneity of events over large geographic areas. Typically, events that are less than 10,000 years apart appear synchronous, and records from different parts of the world cannot be correlated with greater than 100,000 year resolution. Still, the Paleocene-Eocene warming event is a close match for extreme predictions of human-induced global warming in terms of the absolute magnitude of warming and final climatic conditions. This makes the Paleocene-Eocene interval a valuable example for improving our understanding of the effect of rapid global warming on terrestrial ecosystems.

RECONSTRUCTING CLIMATES OF THE PAST

Climatic conditions for the distant past can be inferred in a variety of ways, but the two techniques most widely applied to continental paleoclimates are based on fossil plants. One method, called the nearest living relative (NLR) method, relies on the assumption that fossil species grew in climates similar to those preferred by their extant relatives. The validity of the NLR method depends on how closely related the fossil and extant species are, how much evolutionary change there has been in climatic preferences of the lineage, and the degree to which the range of the extant species is controlled by climate. Living taxa that have relictual distributions or few species are especially unreliable paleoclimatic indicators because they are far from occupying the full range of climatic conditions they can tolerate physiologically (e.g., *Ginkgo, Metasequoia*). NLR inferences about early Cenozoic climate are more reliable if they are based on diverse, nonrelictual taxa that have strong climatic limitations on their present-day distribution.

Palms are a good example of a speciose group with clear climatic limits on their distribution. Palms do not naturally occur in regions where means of cold months are less than about 6°C, or where frosts persist for more than two days (Wing and Greenwood, 1993; Figure 11-1). Experimental work on a wide variety of living species of palms shows that sensitivity to cold temperatures is typical of the whole family and results from absent or poorly developed physiological mechanisms of frost hardening (Sakai and Larcher, 1987). The combination of geographic and physiological information on palms makes a convincing case that extinct species in the family are unlikely to have had significantly more frost tolerance than their living relatives, so palms can be used with confidence to fix lower limits on paleotemperature estimates for early Cenozoic floras.

The second major method for inferring paleoclimate from fossil plants, leaf physiognomy, relies on correlations between the shape and size of leaves and climate that are observed in living floras. Generally, floras growing under

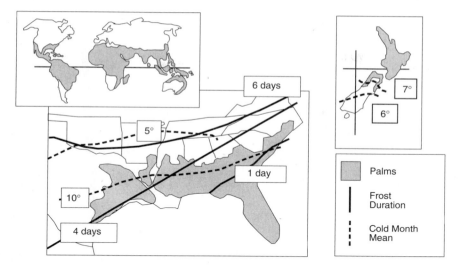

FIGURE 11-1   Natural distribution of living palms in relation to cold-month mean temperature (CMM) and average number of days of frost per year.   Note close correspondence between the 6°C CMM and the limit of palms in both North America and New Zealand.   Data were compiled by Wing and Greenwood (1993).

warmer climates will have a higher proportion of species that have leaves with entire margins (nontoothed leaves). The correlation between mean annual temperature (MAT) and the proportion of species with entire margins was first observed over 80 years ago, but has been documented and quantified most extensively by Wolfe (1979) using floras from East Asia (Figure 11-2). However the high correlation exhibited in Wolfe's data set is due in part to the exclusion of living floras from seasonally dry climates. Although the relationship between the percentage of species with entire margins in a local flora and MAT has been used widely to obtain numerical estimates of MAT for Cenozoic fossil floras, it may give misleading results for fossil floras that grew under seasonally dry conditions.

More recent attempts to make leaf physiognomic analysis more general and robust have used additional descriptors of leaves and climate (Wolfe, 1993). Many characteristics of leaf shape and size are correlated with temperature and rainfall conditions and can be used to infer paleoclimatic conditions. Wolfe (1993) scored a large number of Central and North American floras based on characteristics of leaf size and shape (Figure 11-3a). These leaf characteristics then were compared with parameters of temperature and rainfall for the sites using ordination analysis (Wolfe, 1993). Relationships between physiognomic variables of leaves and climatic variables (Figure 11-3b) then can be used to infer

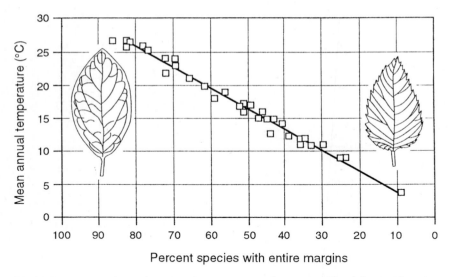

FIGURE 11-2   Correlation between the percentage of species in a local flora with entire-margined leaves (leaf on left) and the mean annual temperature (MAT) of the site.   Data are from mesic East Asian forests (Wolfe, 1979).   Linear regression of this data set yields the equation: MAT=1.14 + (0.31 × percentage of species with entire leaf margins). R^2=0.98, $p<0.001$, standard error of the estimate is 0.79°C.

climatic conditions for fossil leaf assemblages based on their physiognomy. Wolfe's Central and North American data set also has been analyzed using multiple regression techniques (Wing and Greenwood, 1993).   Mean annual temperature estimates based on the multiple regression approach have errors of 2-4°C; precipitation estimates generally are no more than ballpark figures.

In spite of the limitations on the precision of paleoclimate inferences, there is strong agreement between estimates using the different paleobotanical methods and between estimates derived from plant and vertebrate fossils (Hutchison, 1982; Markwick, 1994; Wing, 1991; Wing and Greenwood, 1993).   There is also general congruence between marine temperature curves derived from oxygen isotope studies and terrestrial temperature curves derived from fossil plant and animal evidence (Wing et al., 1991; Wolfe and Poore, 1982).   The consistency of paleoclimate estimates based on different fossil data sets and methodologies suggests that the estimates are robust.   This is important, because paleoclimate estimates based on paleontological data indicate a world sharply different from the modern one with climatic conditions that are difficult to explain in terms of the climate systems observed today.

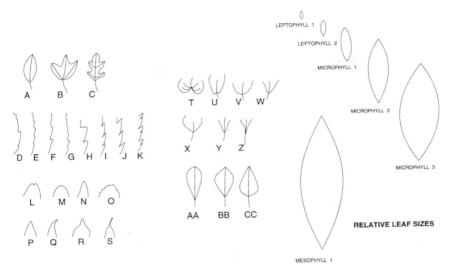

FIGURE 11-3a   Leaf character states used in multivariate analysis of relationships between physiognomy and climate (Wolfe, 1993). Character states: A-C=general margins, A=untoothed leaf, B=palmately lobed untoothed leaf, C=pinnately lobed untoothed leaf; D-K=toothed margins, D=irregularly spaced, E=regularly spaced, F=distantly spaced, G=closely spaced, H=rounded, I=appressed, J=acute, K=compound; L-S=shapes of the apex, L=apex emarginate, M-O=apex rounded, P-Q=apex acute, R-S=apex attenuate; T-Z= shapes of the base, T-U=base cordate, V-W=base rounded, X-Z=base acute; AA-CC=shapes of the leaf, AA=obovate leaf, BB=elliptic leaf, CC=ovate leaf.   Leaf-size categories are approximately 0.23×.

THE EQUABLE CLIMATE PARADOX

Global climate through most of Earth's history has been much warmer than at present. The early Eocene has long been recognized as the warmest part of the Cenozoic and one of the warmest periods of global climate in the last 100 million years (Miller et al., 1987; Savin et al., 1975; Wolfe, 1978). During the early Eocene, polar icecaps were absent in both hemispheres (Crowley and North, 1991), and midlatitude continental interiors had much warmer winters than they do now (Hickey, 1977; MacGinitie, 1974; Wing, 1991; Wing and Greenwood, 1993).

The causes for such globally warm climate are not well understood. Higher sea levels, lack of polar icecaps, and more dispersed land masses all have been thought to play a role in maintaining a warmer world. In recent years, general circulation models (GCMs) have been applied to paleoclimates (e.g., Crowley and North, 1991). To produce "predictions" for temperature and precipitation for a

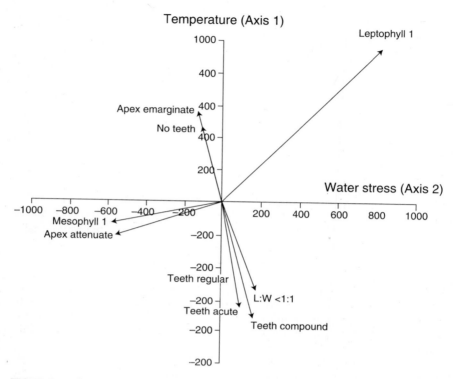

FIGURE 11-3b Reciprocal averaging plot showing relationship of climatic and physiognomic variables (Wolfe, 1993). The largest leaf-size category and attenuate apices are more abundant in wetter climates. All tooth features are more abundant in colder climates, as are narrow leaves with a length to width ratio of less than 1:1. Entire margins and emarginate apices are correlated with warmer climates.

time in the past, GCMs use starting conditions such as ancient coastline positions, paleotopography, continental positions, sea-surface temperature gradients, and global ice distribution. Weather patterns are generated using the same equations that describe the dynamics of the atmosphere today, then the weather patterns are averaged to yield a paleoclimate.

Application of these models to past periods of equable climate like the early Eocene consistently yields simulations that are more like the modern world than paleontological data indicate. The early Eocene is arguably the most interesting case of conflict between proxy data and model output, because it is close enough to the present that the proxy data are extensive, and uniformitarian assumptions about the climatic tolerances of animal and plant lineages are probably valid. Additionally, there is little evidence for very high CO_2 levels in the early

Eocene atmosphere, so this factor cannot be invoked in an unconstrained fashion to explain differences between proxy data and model output.

The sharpest discrepancy between model output and proxy data occurs in mid- to high latitude continental interiors because the low thermal capacity of land and the isolation of continental interiors from the moderating influence of the oceans result in model predictions of high seasonal variation in temperature in these areas (Sloan, 1994; Sloan and Barron, 1990, 1992). There are literally hundreds of early Eocene plant and animal fossil localities scattered across much of the continental interior of North America. The distribution of frost-intolerant forms (e.g., palms, cycads, crocodilians), as well as weak development of seasonal growth rings in some fossil wood, high diversity of small arboreal frugivores and insectivores requiring year-round food resources, and the physiognomy of fossil dicot foliage, all yield similar conclusions. In the Early Eocene, the interior of North America as far as 50°N experienced no significant winter freezing (e.g., annual minimum temperatures were $>-10°C$, frost durations were less than 1-2 days). Mean temperatures in cold months were certainly above freezing, and probably higher than 5°C in most areas. Mean annual temperature was 15-20°C (Wing and Greenwood, 1993). There are fewer proxy data available for the interiors of other continents, but southern Australia, which was 50-60°S in the early Eocene, has fossil faunas and floras indicating even warmer winter temperatures than similarly high latitudes in the interior of North America (Greenwood and Wing, 1995). The proxy data consistently show equable climates with low seasonality of temperature and warm winters. GCM simulation results consistently yield highly seasonal climates in continental interiors, with freezing temperatures far south of fossil sites containing frost intolerant forms (Sloan and Barron, 1990, 1992; Figure 11-4).

There are three basic kinds of explanation for warmth in the mid- to high latitudes during the early Eocene: (1) more heat was transported from equatorial to polar regions by ocean currents or winds, (2) more solar radiation was absorbed at high latitudes because albedo was lower, and (3) more heat was retained at high latitudes, and everywhere else, because of higher concentrations of "greenhouse" gases such as water vapor, methane, or CO_2. The persistent discrepancy between GCM output and proxy data arises because mechanisms in the first two categories—such as increased heat transport by ocean currents or no polar ice—are not strong enough to explain the levels of warmth indicated by proxy data, especially in continental interiors (Walker and Sloan, 1992). The last mechanism, a CO_2, methane, or water-vapor greenhouse, implies that tropical as well as high-latitude regions would have been substantially warmer than at present. The best proxy data for the Eocene tropics, however, indicate little difference from present temperatures (Adams et al., 1990; Graham, 1994). Furthermore, neither proxy data nor geochemical modeling are consistent with Eocene CO_2 levels more than about 900 parts per million (ppm), and it is possible

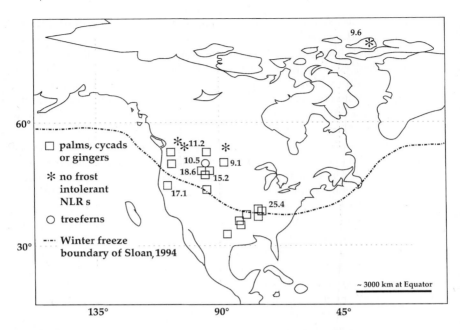

FIGURE 11-4 Eocene paleogeographic reconstruction of North America, showing frost-sensitive plants and high MAT estimates far north of the line where GCM simulation results indicate freezing temperatures. Numbers next to sites are MAT estimates based on multiple regression analysis of physiognomic data as described by Wing and Greenwood (1993). Proxy data are from Wing and Greenwood (1993). Model results are from Sloan (1994); simulation is based on present CO_2 level without large lakes in Wyoming.

that Eocene CO_2 levels were no different from those at present (Berner, 1991; Cerling, 1991).

The resolution of the equable climate paradox apparently will come from better GCMs, not from reinterpretation of the fossil record. The significance of the "equable climate paradox" is that it reveals the strong tendency for climate models to yield results that are more like the present than they should be. If similar problems plague the prediction of future climates under higher levels of atmospheric CO_2, our predictions about the magnitude and rate of global climatic change may be far less accurate than we think.

THE TERMINAL PALEOCENE EVENT

The global climate warmed considerably from the late Paleocene to the early Eocene, roughly 57-52 million years ago. This warming trend has been quantified in the marine realm by measuring changes in oxygen isotope ratios in the tests of benthic and planktonic foraminifera (one-celled, amoeba-like organisms that secrete calcareous shells) recovered from deep sea cores (Miller et al., 1987; Pak and Miller, 1992; Figure 11-5) and in North America by physiognomic and floristic analyses of fossil plant assemblages (Hickey, 1977, 1980; Wing et al., 1991; Wolfe, 1978). In the mid-latitudes of interior North America, MAT increased from approximately 10°C to nearly 20°C between the late Paleocene and the mid- to early Eocene, a period of about 3 million years (Wing et al., 1991). Although this rate of change averages to only a few thousandths of a degree per

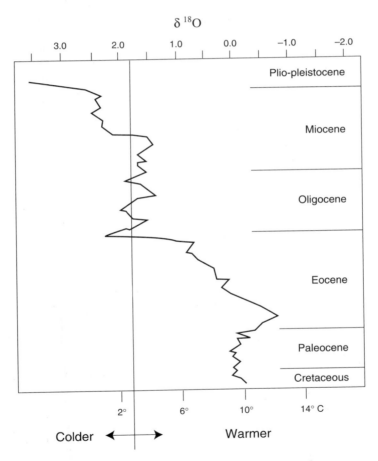

FIGURE 11-5 Cenozoic $\delta^{18}O$ curve showing major fluctuations in deep ocean temperature over the last 65 million years (redrawn from Miller et al., 1987). Temperature equivalents for $\delta^{18}O$ scale assume no polar icecaps.

millennium over the whole period, there is growing evidence that the rate of climatic warming was not constant.

In cores that recover laminated ocean bottom sediments, it is possible to resolve time in tens of thousands of years, even in the early Cenozoic (Kennett and Stott, 1991). Recent detailed stratigraphic studies of the Paleocene/Eocene boundary interval recovered in cores from the southern Pacific and southern Indian Oceans have shown an excursion in isotope values that occurred approximately 100,000 years before the Paleocene/Eocene boundary (approximately 55.1 million years ago; Kennett and Stott, 1991; Pak and Miller, 1992; Figure 11-6). The sudden increase in the light isotope of oxygen (decrease in $\delta^{18}O$ values) took place over less than 10,000 years and is thought to represent an interval when the temperature of bottom waters and surface waters in mid- to high latitude oceans increased by 5-8°C. This geologically short period of time, marked by isotopic shifts, is referred to here as the "Terminal Paleocene Event."

Oxygen isotope analyses of planktonic foraminifera tests indicate that surface waters in the equatorial Pacific maintained a temperature of about 20°C during the Terminal Paleocene Event, but analyses of the tests of tropical benthic species imply warming similar to that seen in the benthic and planktonic foraminifera of higher latitudes (Zachos et al., 1993). The isotopic shift coincides

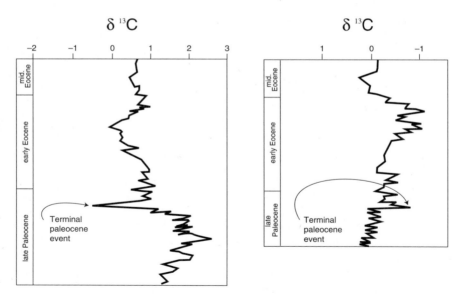

FIGURE 11-6 Terminal Paleocene excursion in carbon and oxygen isotope values (redrawn from Pak and Miller, 1992). The Terminal Paleocene Event occurred about 100,000 years before the Paleocene/Eocene boundary and was associated with a major extinction of benthic foraminifera (Kennett and Stott, 1991).

with the most severe extinction (about 50% of species) of benthic foraminifera from the deep ocean during the Cenozoic; the extinction is believed to have been caused by rapid warming of deep ocean waters and a concomitant decrease in their dissolved oxygen levels (Kennett and Stott, 1991; Zachos et al., 1993). Following the negative isotopic excursion (warming) there was a rapid rebound in the positive direction (cooling), followed by a renewed decrease in $\delta^{18}O$ values that continued to a Cenozoic minimum in the mid-early Eocene about 52-53 million years ago. This minimum indicates the time of maximum global warmth during the Cenozoic.

The terminal Paleocene oxygen isotopic excursion is paralleled by a similar excursion in carbon isotope values (Kennett and Stott, 1991). The shift from ^{13}C to ^{12}C is probably related to the source and rate of delivery of organic carbon to the deep ocean, and possibly to decreased rates of oxidation of organic material in bottom waters (Kennett and Stott, 1991). The rapid isotopic and biotic changes during the Terminal Paleocene Event all may relate to a change in ocean circulation in which the source for bottom waters shifted from cool high latitudes to warm low latitudes (Kennett and Stott, 1989, 1991; Rea et al., 1990).

In this hypothesis, global warming during the later Paleocene increased the warmth, and therefore decreased the density, of high-latitude surface waters. At some point, warm, oxygen-poor saline waters generated by evaporation in the low latitudes exceeded the density of cooler, more oxygenated waters from the polar regions, and deep ocean circulation began to move in a poleward direction. This is the opposite of present-day conditions in which dense bottom water is created in cold, moderately saline high-latitude oceans. The hypothesized "reversed" bottom-water flow at the end of the Paleocene transported heat to mid- and high latitudes very effectively, resulting in a rapid increase of bottom water temperatures, but also a sudden warming of mid- to high latitude surface waters and continental surfaces as the warm bottom water upwelled (Brass et al., 1982; Pak and Miller, 1992; Rea et al., 1990; Zachos et al., 1993). Although the precise triggering mechanism that led to the sudden reversal of bottom water circulation is not understood, the effects of the Terminal Paleocene Event appear to have been global, because rapid changes are seen at this time in sediments deposited in continental and shallow marine environments as well as in the deep sea.

One effect of increased poleward heat transport by deep ocean currents would have been a decrease in latitudinal temperature gradients. There is direct evidence for decreased temperature gradients both in ocean surface waters and on the continents (Greenwood and Wing, 1995; Zachos et al., 1992). Decreased surface temperature gradients would be expected to lead to a reduced intensity of zonal atmospheric circulation, which is largely driven by equator-to-pole temperature contrasts. Reduced surface wind velocities were responsible for the sharp decline in the size of wind-blown dust grains in Paleocene/Eocene boundary sediments of the central Pacific; sluggish winds are not capable of carrying larger dust particles far out to sea (Janecek and Rea, 1983; Rea et al., 1985). A secondary effect of reduced wind velocities may have been a decrease in coastal upwelling zones, which are powered by wind shear (Stott, 1992).

Greater warmth at mid- to high latitudes also may have resulted in increased precipitation and chemical weathering on land surfaces. Evidence for increased chemical weathering is seen in the sudden, widespread increase during the Terminal Paleocene Event of kaolinitic clays that are characteristic of leached soils formed under "tropical" weathering regimes (Robert and Chamley, 1991; Hovan and Rea, 1992; Gibson et al., 1993).

THE PALEOCENE-EOCENE TRANSITION ON LAND

The carbon isotope excursion associated with the Terminal Paleocene Event is a geochemical marker that can be detected in continental carbon sources such as hydroxyapatite preserved in tooth enamel of fossil mammals and calcium carbonate preserved in fossil soil nodules. Because the largest short-term reservoir in the global carbon budget is the ocean, shifts in the $^{13}C/^{12}C$ ratio of the oceans, such as those observed during the Terminal Paleocene Event, should force similar isotopic shifts in the carbon reservoirs represented by the atmosphere and continental ecosystems. Investigations of Paleocene/Eocene boundary sections in the Bighorn Basin of northern Wyoming have detected a carbon isotope excursion that corresponds to the Terminal Paleocene Event in direction, magnitude, and timing, thus establishing a datum that permits correlation of terrestrial and marine events (Koch et al., 1992).

The terrestrial carbon isotope excursion coincides with the Clarkforkian/Wasatchian boundary, a major and rapid change in the composition of mammalian faunas which has long been thought to be approximately correlated with the Paleocene/Eocene boundary (Wood et al., 1941). It is at the beginning of the Wasatchian that North American faunas first included representatives of the perissodactyla (odd-toed ungulates), artiodactyla (even-toed ungulates), and euprimates (Gingerich, 1989; Rose, 1981a). Similar faunal changes are known to have taken place at about the same time in Europe and Asia (Rose, 1981a; Russell and Zhai, 1987), and in Europe the faunal change also is associated with the carbon isotope excursion (Sinha and Stott, 1993). The nearly simultaneous appearance of these forms on all three northern continents indicates that high latitude land corridors in Greenland and Beringia were open to mammalian migrants (Maas et al., 1995; McKenna, 1983; Rose, 1981a), but the place of origin for these groups is still not known. Although rates of mammalian taxonomic turnover were high near the Paleocene/Eocene boundary, there is no evidence for a substantial decline in the number of mammalian genera (Wing et al., 1995). In spite of maintaining high numbers of genera, there is evidence that, at the level of species, latest Paleocene mammalian faunas from North America were characterized by high dominance of a few species, possibly an indication of stressful conditions (Rose, 1981b).

The effect of the Terminal Paleocene Event on terrestrial floras is less well understood, although it has been known for some time that floral similarity between North America and Europe reached a maximum in the early Eocene (Tiffney, 1985). Two studies have shown a substantial (about 30%) reduction in the number of species of plants near the end of the Paleocene in North America, approximately coincident with the mammalian faunal exchange and the Terminal Paleocene Event (Frederiksen, 1994; Wing et al., 1995; Figure 11-7). This rapid decrease in the number of species of plants at the end of the Paleocene was geologically short-lived (about 1 million years); by the mid- to early Eocene (about 53 million years ago), as global climate reached its maximum Cenozoic warmth, floral richness had recovered to levels considerably higher than those

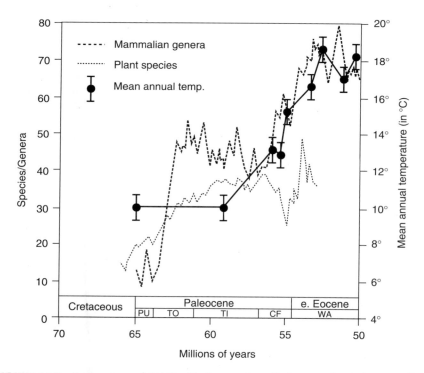

FIGURE 11-7 Comparison of MAT with the number of plant species and mammalian genera through the Paleocene and earliest Eocene. MAT estimates are based on physiognomic analysis of floras from western Wyoming. Plant and mammal taxonomic richness are based on data compiled from Wyoming and southern Montana. Abbreviations for provincial ages: PU=Puercan, TO=Torrejonian, TI=Tiffanian, CF=Clarkforkian, WA=Wasatchian. Figure was redrawn from Wing et al. (1995).

preceding the Terminal Paleocene Event. Stratigraphic resolution of the terminal Paleocene drop in floral richness is not yet sufficient to determine whether it coincides with increasing temperatures at the start of the Terminal Paleocene Event, decreasing temperatures at the end of the Terminal Paleocene Event, or is spread throughout the entire excursion.

The fluctuations in numbers of species of plants near the Paleocene/Eocene boundary are part of a long-recognized change in the composition of North American floras (Leopold and MacGinitie, 1972). Paleocene assemblages are dominated by species belonging to largely deciduous groups with modern north temperate distributions (e.g., *Metasequoia*, Betulaceae, Cercidiphyllaceae, Hamamelidaceae, Juglandaceae; Figure 11-8), whereas Eocene floras typically have many species belonging to largely subtropical evergreen families (e.g., Annonaceae, Lauraceae, Leguminosae, Myristicaceae, Palmae, Zamiaceae; Figure 11-9).

In the present, warmer climates are correlated positively with larger numbers of species of plants per unit area, particularly if regions of strong aridity are excluded (Gentry, 1988). In view of this correlation, a simplistic expectation for the response of plant species richness during the Paleocene/Eocene transition would be an increase in number of species as global temperatures warmed. Instead, the number of species of plants declined in two widely separated areas of North America: the northern Rocky Mountains (Wing et al., 1995), and the Gulf Coastal Plain (Frederiksen, 1994). The geographic scope of the decline argues against it being an artifact of local sampling or preservational effects, and implies that the cause is of continental or greater geographic scope.

A SCENARIO FOR PALEOCENE/EOCENE PLANT EXTINCTIONS

What causal connection is there, if any, between the climatic changes of the Terminal Paleocene Event, the decrease in plant diversity, and the shift in mammalian faunal composition? The stratigraphic resolution and geographic scope of our knowledge of faunal and floral change across this time period is increasing rapidly, but it is not yet sufficient to provide a clear answer. What follows is a preliminary hypothesis that is amenable to testing as the paleontological database improves.

A key to understanding the cause of the loss of species among plants is the difference between the curves of species richness for mammals and plants. In the present, the diversity of both groups of organisms generally increases toward the equator, but the migrational abilities are obviously quite different. The difference in rates of migration provides an explanation for greater loss of species among plants than mammals at the Paleocene/Eocene boundary.

The Paleocene flora of the northern continents long has been noted for its relatively low diversity and high homogeneity, with many lineages of deciduous plants distributed throughout the middle and high latitudes of North

FIGURE 11-8 Typical Paleocene plant fossils of interior western North America. Based on thickness of the fossil organic compression and living relatives, most of these were probably deciduous trees. Toothed leaves are typical of the Paleocene midlatitude floras. A. *Cercidiphyllum* sp., related to katsura tree, a native of East Asia; B. *Ulmus* (*Chaetoptelea*) *microphylla*, related to living elms in Central America; C. Betulaceous leaf (birch family), probably an extinct genus; D. *Alnus* (alder) sp. leaf.

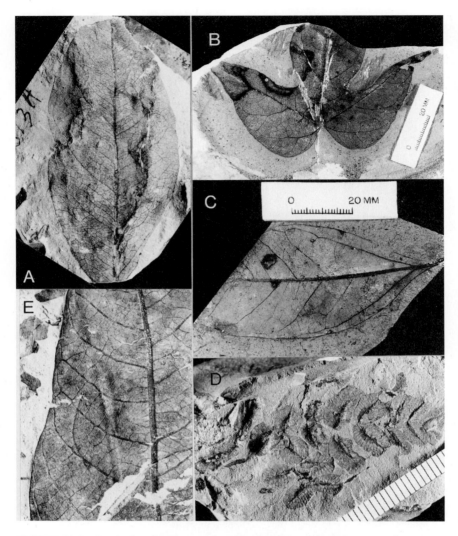

FIGURE 11-9 Typical early Eocene plant fossils of interior western North America. Based on thickness of the fossil organic compression and living relatives, most of these were probably broad-leaved evergreens. Large, entire-margined leaves are typical of Eocene midlatitude floras. A. Leaflet of a legume similar to living species of *Machaerium* B. Menispermaceous leaf (moonseed family), probably a vine or liana; C. Lauraceous leaf (avocado family); D. *Salvinia preauriculata*, a floating aquatic fern characteristic of Eocene floras in midlatitude North America. Living *Salvinia* species occur in subtropical to tropical climates; E. Apocynaceous leaf (oleander family).

America, Europe, and Asia (Boulter and Kvacek, 1989; Brown, 1962; Guo et al., 1984; Koch, 1963). Warming at middle and high latitudes during the Terminal Paleocene Event and the early Eocene is likely to have caused local extinction at the southern ends of the ranges of cool-adapted lineages with broad circumpolar distributions. Warming at middle and high latitudes also would have permitted the poleward migration of subtropical and tropical taxa in both mammals and plants. For floras in the northern Rocky Mountains and the Gulf Coastal Plain of North America, there was a significant delay, as much as several hundred thousand years, between the loss of species at the Terminal Paleocene Event and the arrival of subtropical elements in the early Eocene. It was this lag that created the rapid drop and rapid recovery of plant richness at middle latitudes. What in mammals shows up as a rapid intercontinental mixing of faunas manifests itself among plants as an extinction event at mid- to high latitudes, followed perhaps 100,000 years later by a major immigration of species from the south.

The migrational lag explanation seems at first glance unlikely because plant "migration" can be very rapid geologically; the movement of forest taxa in the wake of the retreating Holocene glaciers totally modified the vegetation of vast areas of North America in 5,000-10,000 years (Overpeck et al., 1992). The Holocene example is not wholly applicable to the Eocene situation, however, because of the different continental configuration in the Eocene, and because Holocene plants were migrating into a recently denuded landscape that probably was not occupied by established forest vegetation.

In the early Eocene, the Tethys Sea still formed a wide barrier between the northern and southern continents; it separated Africa from Europe and South America from North America, although there were probably islands in both the proto-Mediterranean and the proto-Caribbean (Figure 11-10). This ocean crossing may have been one factor that slowed the migration of subtropical and tropical plants into North America. Tropical taxa may have had limited access to Europe along the northern shore of the Tethys, but the straits of Turgai divided Europe from eastern Asia at about the present-day position of the Caspian Sea, so the appearance of subtropical forms in Europe also might have been delayed. By contrast, Asia had a broad land connection with tropical continental areas during the Eocene, so that the ranges of terrestrial plants could have shifted northward without confronting an oceanic barrier.

The paleogeographic differences between Asia and North America can be used to help resolve whether a migrational lag was important in creating the pattern of change in number of species across the interval of the Paleocene-Eocene boundary. Because of eastern Asia's connection to tropical land masses, the migration of lineages from warm climates into middle latitudes should have been much easier than in North America or Europe. If this hypothesis is correct, Paleocene-Eocene floral immigration into mid-latitude Asia should have been more rapid, and the sharp drop in plant species richness should have had a duration of thousands or tens of thousands of years—almost certainly too short

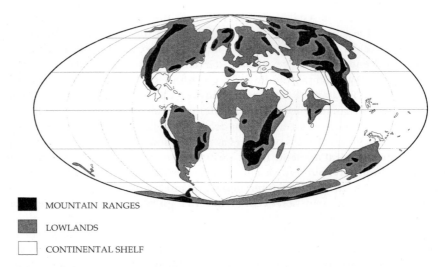

■ MOUNTAIN RANGES

■ LOWLANDS

□ CONTINENTAL SHELF

FIGURE 11-10 Global paleogeographic reconstruction of the early Eocene. Note ocean barriers that separated the European and North American land masses from Africa and South America. Only Asia had a significant land connection to tropical latitudes. Figure was redrawn from Smith et al. (1994).

to detect in the fossil record. Finding the same pattern of diversity change in Asia and North America would remove migrational delay as a reasonable hypothesis for the decline in species numbers observed in North America.

The cause of the loss of species during the Paleocene-Eocene transition also can be evaluated by comparing the pattern of floral immigration within North America. If the delayed immigration of subtropical forms to the northern Rocky Mountain region was largely a consequence of the water gap between North America and South America rather than slow migration across the North American continent, then forms from warm climates should appear at essentially the same time in the southern and northern part of the continent. If migrants adapted to warm climates show up detectably earlier in the south, this would indicate a slow rate of migration across the continent, suggesting that the rate of northward shift in the ranges of plants was limited by the pace of warming, or by the ability of the immigrant species to gain footholds in the established vegetation of North America.

Two possible general patterns emerge from a consideration of events at the Paleocene/Eocene boundary. First, species richness of plants is much more dramatically affected by the climatic change than is taxonomic richness of mammals. This should not be surprising in light of the sensitivity of plants to climate, but it has been proposed that plants are generally more "extinction resistant" than other groups of organisms (Knoll, 1984). This conclusion was based on the observation that most major lineages of plants survived mass extinctions at the Permian/Triassic and Cretaceous/Tertiary boundaries, whereas higher taxonomic groups of animals were decimated. Analysis of the Paleocene/Eocene extinctions of plants supports the idea that the cause of an extinction (e.g., climate versus bolide impact) may be more important in understanding its effects than its "size," as measured by the percentage of taxa that are lost.

The Paleocene/Eocene extinction of plants also illustrates the principle that the rate of climatic change may be as or more important than its direction in causing extinction of plants. Rapid climatic shifts initially may decrease species richness of plants through accelerated local extinction regardless of whether climate is warming or cooling.

Through the last 65 million years of Earth's history, the middle and high latitudes have been more affected by both warming and cooling events than the tropics because their temperature is more easily influenced by changes in the efficiency and direction of heat transport by oceanic and atmospheric circulation. Although the image of the tropics as climatically invariant has been thoroughly debunked by increased understanding of the Miocene through Pleistocene history of Amazonia (Hoorn, 1994; Van der Hammen and Absy, 1994), over the long term it is the middle and high latitudes that have been most dramatically affected by fluctuations in global climate and most frequently afflicted by large-scale extinction of plants. During the Pleistocene, the comings and goings of glaciers in the north temperate regions resulted in the intermittent sterilization of large areas, certainly a more severe form of disruption than the rainfall and temperature fluctuations experienced at lower latitudes. Relative climatic stability on geological time scales is probably one important factor leading to larger numbers of species in the tropics.

CONCLUSIONS

The fossil record can be used in two different and very productive ways to understand global climatic change and its effects on biotic systems. First, it is a testing ground for climate models—the only way we can find out if they are overtuned to reproduce present conditions. The example of the enigma of equable climate suggests that there is a great deal of room for improving general circulation models of climate, and that predictions generated through GCMs have a bias toward reproducing conditions similar to those of the present. This is a very serious problem for understanding the probable effects of anthropogenic addition of CO_2 or other greenhouse gases to the atmosphere.

The fossil record also provides our sole opportunity to examine the biological consequences of climatic change without waiting for them to happen in "real time." Although causation is difficult to prove in an historical record (or, often enough, even in the laboratory), the pattern of biotic response to environmental change can indicate probable processes. In particular, we should be on the alert for patterns that are consistent with threshold responses to climatic forcing factors (DiMichele and Phillips, 1995).

Paleontologists have invoked climatic change as a major factor in extinction and biotic turnover for as long as there have been paleontologists. But generally this has been an afterthought. A concerted study of the biotic effects of climatic change involves choosing time intervals for the climatic events that are known to occur, rather than searching for climatic change once a dramatic biotic change has been noted. This is the only way to find out if some climatic change occurs without catastrophic biological effects. Such studies require combining research on paleoecology, the history of biotic diversity, and extinction, with paleoclimatic reconstructions. Mixing these approaches will provide results of practical importance, but also will raise fascinating biological issues relating to ecosystem response, threshold effects, and biogeography.

Some view the globally warm periods of the past as a prologue to the future. Are they? This is not a question we can answer yet. We do know that climatic models "tuned" to the present do a poor job of reproducing the climate of past greenhouse worlds. We also know that the present is an unusual time in Earth's history, and not a simple key to the past. Clearly this raises doubts about our ability to predict climatic change or biotic responses under elevated CO_2 conditions.

If the present is not the key to the past, is the past likely to reveal much about the future? Here I think the answer is a cautious "yes." The process of interpretation is not simple, but by enlarging the set of "worlds" we have to explain, study of the past will increase the generality of our understanding of the hypercomplex climatic and biological systems that we are trying to predict. The fossil record is difficult to use, but it provides the only examples we have of events and states that occurred but can no longer be observed directly. Figuring out how to interpret the fossil record is now more important than ever. So is the advice given by Herodotus: "Study the past."

REFERENCES

Adams, C. G., D. E. Lee, and B. R. Rosen. 1990. Conflicting biotic and isotopic evidence for tropical sea-surface temperatures during the Tertiary. Palaeogeogr. Palaeoclimatol. Palaeoecol. 77:289-313.

Berner, R. A. 1991. A model for atmospheric CO_2 over Phanerozoic time. Amer. J. Sci. 291:339-376.

Boulter, M. C., and Z. Kvacek. 1989. The Palaeocene flora of the Isle of Mull. Special Papers Palaeontol. 42:3-149.

Brass, G. W., J. R. Southam, and W. H. Peterson. 1982. Warm saline bottom water in the ancient ocean. Nature 296:620-623.

Brown, R. W. 1962. Paleocene flora of the Rocky Mountains and Great Plains. U.S. Geol. Surv. Prof. Paper 375:1-119.

Cerling, T. E. 1991. Carbon dioxide in the atmosphere: Evidence from Cenozoic and Mesozoic paleosols. Amer. J. Sci. 291:377-400.

Crowley, T. J., and G. R. North. 1991. Paleoclimatology. Oxford University Press, Oxford, England.

DiMichele, W. A., and T. L. Phillips. 1995. The response of hierarchically-structured ecosystems to long-term climatic change: A case study using tropical peat swamps of Pennsylvanian age. Pp. 174-184 in A. H. Knoll and S. Stanley, eds., Effects of Past Global Change on Life. National Academy Press, Washington, D.C.

Frederiksen, N. O. 1994. Paleocene floral diversities and turnover events in eastern North America and their relation to diversity models. Rev. Palaeobot. Palynology 82:225-238.

Gentry, A. H. 1988. Changes in plant community diversity and floristic composition on environmental and geographical gradients. Ann. Missouri Bot. Garden 75:1-34.

Gibson, T. G., L. M. Bybell, and J. P. Owens. 1993. Latest Paleocene lithologic and biotic events in neritic deposits of southwestern New Jersey. Paleoceanography 8:495-514.

Gingerich, P. D. 1989. New earliest Wasatchian mammalian fauna from the Eocene of northwestern Wyoming: Composition and diversity in a rarely sampled high-floodplain assemblage. Univ. Mich. Papers Paleontol. 28:1-97.

Graham, A. 1994. Neotropical Eocene coastal floras and $^{18}O/^{16}O$-estimate warmer vs. cooler equatorial waters. Amer. J. Bot. 81:301-306.

Greenwood, D. E., and S. L. Wing. 1995. Eocene climate and latitudinal temperature gradients on land. Geology 23:1044-1048.

Guo, S.-X., Z.-H. Sun, H.-M. Li, and Y.-W. Dou. 1984. Paleocene megafossil flora from Altai of Xinjiang. Bull. Nanjing Inst. Geol. Palaeontol. Academia Sinica 8:119-146.

Hickey, L. J. 1977. Stratigraphy and paleobotany of the Golden Valley Formation (early Tertiary) of western North Dakota. Mem. Geol. Soc. Amer. 150:1-181.

Hickey, L. J. 1980. Paleocene stratigraphy and flora of the Clark's Fork Basin. In P. D. Gingerich, ed., Early Cenozoic Paleontology and Stratigraphy of the Bighorn Basin, Wyoming. Univ. Mich. Papers Paleontol. 24:33-49.

Hoorn, C. 1994. Fluvial palaeoenvironments in the intracratonic Amazonas Basin (early Miocene-early middle Miocene, Colombia). Palaeogeogr. Palaeoclimatol. Palaeoecol. 109:1-54.

Hovan, S. A., and D. K. Rea. 1992. Paleocene/Eocene boundary changes in atmospheric and oceanic circulation: A southern hemisphere record. Geology 20:15-18.

Hutchison, J. H. 1982. Turtle, crocodilian, and champsosaur diversity changes in the Cenozoic of the north-central region of western United States. Palaeogeogr. Palaeoclimatol. Palaeoecol. 37:149-164.

Janecek, T. R., and D. K. Rea. 1983. Eolian deposition in the northeast Pacific Ocean: Cenozoic history of atmospheric circulation. Bull. Geol. Soc. Amer. 94:730-738.

Kennett, J. P., and L. D. Stott. 1989. Warm saline bottom water in early-middle Eocene: Isotopic evidence from Antarctica. Trans. Amer. Geophys. Union (EOS) 70:363.

Kennett, J. P., and L. D. Stott. 1991. Abrupt deep sea warming, palaeoceanographic changes and benthic extinctions at the end of the Palaeocene. Nature 353:225-229.

Knoll, A. H. 1984. Patterns of extinction in the fossil record of vascular plants. Pp. 21-68 in M. H. Nitecki, ed., Extinctions. University of Chicago Press, Chicago.

Koch, B. E. 1963. Fossil plants from the lower Paleocene of the Agatdalen (Angmartussut) area, Central Nugssuaq peninsula, northwest Greenland. Undersea Gronlands Geol. 38:1-120.

Koch, P. L., J. C. Zachos, and P. D. Gingerich. 1992. Correlation between isotope records in marine and continental carbon reservoirs near the Palaeocene/Eocene boundary. Nature 358:319-322.

Leopold, E. B., and H. D. MacGinitie. 1972. Development and affinities of Tertiary floras in the Rocky Mountains. Pp. 147-200 in A. Graham, ed., Floristics and Paleofloristics of Asia and Eastern North America. Elsevier Publishing Company, Amsterdam.

Maas, M. C., M. R. L. Anthony, P. D. Gingerich, G. F. Gunnell, and D. W. Krause. 1995. Mammalian generic diversity and turnover in the late Paleocene and early Eocene of the Bighorn and Crazy Mountains Basins, Wyoming and Montana (U.S.A.). Palaeogeogr. Palaeoclimatol. Palaeoecol. 115:181-208.

MacGinitie, H. D. 1974. An early middle Eocene flora from the Yellowstone-Absaroka volcanic province, northwestern Wind River Basin, Wyoming. Univ. Calif. Pub. Geol. Sci. 108:1-103.

Markwick, P. 1994. "Equability," continentality, and Tertiary "climate": The crocodilian perspective. Geology 22:613-616.

McKenna, M. C. 1983. Cenozoic paleogeography of North Atlantic land bridges. Pp. 351-399 in M. H. P. Bott, S. Saxov, M. Talwani, and J. Thiede, eds., Structure and Development of the Greenland-Scotland Ridge, Plenum Press, N.Y.

Miller, K. G., R. G. Fairbanks, and G. S. Mountain. 1987. Tertiary oxygen isotope synthesis, sea level history, and continental margin erosion. Paleoceanography 2:1-19.

Overpeck, J. T., R. S. Webb, and T. Webb. 1992. Mapping eastern North American vegetation change of the past 18 ka: No-analogs and the future. Geology 20:1071-1074.

Pak, D. K., and K. G. Miller. 1992. Paleocene to Eocene benthic foraminiferal isotopes and assemblages: Implications for deepwater circulation. Paleoceanography 7:405-422.

Rea, D. K., M. Leinen, and T. R. Janecek. 1985. Geologic approach to the long term history of atmospheric circulation. Science 227:721-725.

Rea, D. K., J. C. Zachos, R. M. Owen, and P. D. Gingerich. 1990. Global change at the Paleocene-Eocene boundary: Climatic and evolutionary consequences of tectonic events. Palaeogeogr. Palaeoclimatol. Palaeoecol. 79:117-128.

Robert, C., and H. Chamley. 1991. Development of early Eocene warm climates as inferred from clay mineral variations in ocean sediments. Palaeogeogr. Palaeoclimatol. Palaeoecol. 89:315-331.

Rose, K. D. 1981a. The Clarkforkian Land-Mammal Age and mammalian faunal composition across the Paleocene-Eocene boundary. Univ. Mich. Papers Paleontol. 26:1-197.

Rose, K. D. 1981b. Composition and species diversity in Paleocene and Eocene mammal assemblages: An empirical study. J. Vert. Paleontol. 1:367-388.

Russell, D. E., and R.-J. Zhai. 1987. The Paleogene of Asia: Mammals and stratigraphy. Mémoires du Muséum National d'Histoire Naturelle Sciences de la Terre. Tome 52.

Sakai, A., and W. Larcher. 1987. Frost Survival of Plants: Responses and Adaptations to Freezing Stress. Springer-Verlag, Berlin. 321 pp.

Savin, S. M., R. G. Douglas, and F. G. Stehli. 1975. Tertiary marine paleotemperatures. Bull. Geol. Soc. Amer. 86:1499-1510.

Sinha, A., and L. D. Stott. 1993. Estimation of atmospheric CO_2 change across the Paleocene/Eocene boundary. 1993 Meeting Abstracts with Program, Society for Sedimentary Geology, p. 34.

Sloan, L. C. 1994. Equable climates during the Eocene: Significance of regional paleogeography for North American climate. Geology 22:881-884.

Sloan, L. C., and E. J. Barron. 1990. Equable climates in Earth history? Geology 19:489-492.

Sloan, L. C., and E. J. Barron. 1992. Eocene climate model results: Quantitative comparison to paleoclimatic evidence. Palaeogeogr. Palaeoclimatol. Palaeoecol. 93:183-202

Smith, A. G., D. G. Smith, and B. M. Funnell. 1994. Atlas of Mesozoic and Cenozoic Coastlines. Cambridge University Press, Cambridge, England. 99 pp.

Stott, L. D. 1992. Higher temperatures and lower atmospheric pCO_2: A climate enigma at the end of the Paleocene epoch. Paleoceanography 7:395-404.

Tiffney, B. H. 1985. Perspectives on the origin of the floristic similarity between eastern Asia and eastern North America. J. Arnold Arbor. 66:73-94.

Van der Hammen, T., and M. L. Absy. 1994. Amazonia during the last glacial. Palaeogeogr. Palaeoclimatol. Palaeoecol. 109:247-261.

Walker, J. C. G., and L. C. Sloan. 1992. Something is wrong with climate theory. Geotimes 37:16-18.

Wing, S. L. 1991. Comment on: "Equable climates in Earth history?" an article by L. C. Sloan and E. Barron. Geology 19:539-540.

Wing, S. L., and D. R. Greenwood. 1993. Fossils and fossil climate: The case for equable continental interiors in the Eocene. In J. R. L. Allen, B. J. Hoskins, B. W. Sellwood, and R. A. Spicer, eds.,

Palaeoclimates and their Modeling with Special Reference to the Mesozoic Era. Phil. Trans. R. Soc., Lond. B. Biol. Sci. 341:243-252.

Wing, S. L., J. Alroy, and L. J. Hickey. 1995. Plant and mammal diversity in the Paleocene to early Eocene of the Bighorn Basin. Palaeogeogr. Palaeoclimatol. Palaeoecol. 115:117-156.

Wing, S.L., T. M. Bown, and J. D. Obradovich. 1991. Early Eocene biotic and climatic change in interior western North America. Geology 19:1189-1192.

Wolfe, J. A. 1978. A paleobotanical interpretation of Tertiary climates in the northern hemisphere. Amer. Sci. 66:694-703.

Wolfe, J. A. 1979. Temperature parameters of humid to mesic forests of eastern Asia and relation to forests of other regions of the northern hemisphere and Australasia. Prof. Papers U.S. Geol. Surv. 1106:1-37.

Wolfe, J. A. 1993. A method of obtaining climatic parameters from leaf assemblages. Bull. U.S. Geol. Surv. 2040:1-73.

Wolfe, J. A., and R. Z. Poore. 1982. Tertiary marine and nonmarine climatic trends. Pp. 154-158 in Climate in Earth History, National Academy Press, Washington, D.C.

Wood, H. E., R. W. Chaney, J. Clark, E. H. Colbert, G. L. Jepsen, J. B. Reeside, Jr., and C. Stock. 1941. Nomenclature and correlation of the North American continental Tertiary. Bull. Geol. Soc. Amer. 52:1-48.

Zachos, J. C., W. A. Berggren, M.-P. Aubry, and A. Mackenson. 1992. Isotope and trace element geochemistry of Eocene and Oligocene foraminifers from Site 748, Kerguelen Plateau. Proc. Ocean Drilling Program, Scientific Results 120:839-854.

Zachos, J. C., T. J. Bralower, and E. Thomas. 1993. An early Cenozoic stable isotope record from western Pacific Site 865: Implications for the stability of tropical SST. Abst. Geol. Soc. Amer. 1993:354.

Plant Response to Multiple Environmental Stresses: Implications for Climatic Change and Biodiversity

IRWIN N. FORSETH

Associate Professor, Department of Botany, University of Maryland, College Park

Currently it is estimated that about 25,000 of the more than 250,000 species of plants on Earth are classified as extinct, endangered, or vulnerable (Prance, 1990). Causes of past extinctions and present endangerments encompass a variety of activities associated with the growing human population, including browsing and overgrazing, land clearing for agriculture, deforestation, collection of rare or valuable species, loss of interactive organisms, and the introduction of aggressive exotic species of plants. Documentation of the effects of deforestation or invasion of alien species has occurred. However, the assessment of human-caused climatic changes on present and future plant biodiversity levels is a more difficult undertaking.

Anthropogenic sources of tropospheric gases such as carbon dioxide (CO_2), methane (CH_4), chlorofluorocarbons (CFCs), sulfur dioxide (SO_2), and nitrogen oxides (NO_x) have been rising dramatically since the start of the Industrial Revolution (Bolin, 1991). These gases have the ability to absorb infrared radiation and reradiate it back to Earth. The "greenhouse effect" that this process causes is well established in the atmospheric sciences. In fact, it is calculated that without the preindustrial concentration of these trace gases, average global temperatures would be approximately 33°C lower than they are presently (Bolin, 1991; Schneider, 1993). Thus, the increase in concentration of these gases has resulted in predictions of an increase in mean global temperatures. Predictions of this effect range from 1.5 to 4.5°C if CO_2 levels double over the next century (Schneider, 1993).

Many of the projections about future global climates come from general circulation models (GCMs). These models have variations and uncertainties in their projections and in their ability to model the Earth's climate. However, they do provide plausible regional scenarios of climatic change that ecologists can use to examine community and ecosystem responses (Schneider, 1993). All models show substantial changes in climate when CO_2 is doubled. Most models project greater temperature increases in midlatitude, temperate regions and in midcontinental regions, relative to overall global means (Schneider, 1993). In addition, many of these models predict that changes in regional precipitation patterns will occur, with decreases in midlatitude areas. These areas are currently major crop-producing regions of the world. A final prediction of many GCMs is that doubling CO_2 concentrations in the atmosphere may lead to increased occurrences of extreme events, such as major storms, long droughts, severe cold spells, or prolonged heat waves (Schneider, 1993). Because of the vulnerability of small populations to extreme climatic events, these latter occurrences are of special concern to ecologists who study biodiversity.

LESSONS FROM THE PAST

Paleoecologists study the way biological communities have responded to environmental changes in the past. Davis (1989) lists four valuable insights gained from paleoecological research that are important to research on global change. The first insight is that species respond individually to climate. For example, over the last 18,000 years, species composition of forests in North America has changed considerably (Davis, 1981; Webb, 1987). Some forest communities have reached their present species composition only within the last 2,000 years (Davis, 1981). Thus, it is unrealistic to assume that whole communities or biomes will change in concert in response to climatic change.

A second insight is that responses of species to climatic change often occur with time lags (Davis, 1989). These lags vary among species, and those with long time constants may be particularly vulnerable to rapid climatic changes. With the rapid changes predicted for the next 200 years, the capacity for most species to disperse and establish new populations in suitable regions will be overwhelmed (Davis and Zabinski, 1992). For example, Davis (1990) examined the fossil record of eastern hemlock (*Tsuga canadensis*) pollen over the last 10,000 years in North America. The frontier of the species moved northward and westward at an average rate of 20-25 km per century (Davis, 1981, 1990). Using current predictions of temperature change, temperature isotherms will move northward at a rate of 300 km over the next 100 years (Davis, 1990). Even the most rapid rate of 50 km per century documented for hemlock in eastern upper Michigan (Davis et al., 1986) is too slow to keep up with this rate of temperature change. In the past, *T. canadensis* has expanded from outlying populations that are 50-100 km past the main boundary of the population (Davis, 1990). However, such populations may not play as large a role in future expansion of the species due to a reduction in their number from logging and other forms of human disturbance. Other impediments to the response of *T. canadensis* to climatic change are: a reduction in the strength of seed sources due to human-caused decreases in hemlock abundance throughout its range; a loss of old growth forests, which provide dead logs that are sites for germination and establishment of hemlock seedlings; and increased herbivory by dense populations of deer that reduce hemlock reproduction (Davis, 1990).

A third relevant insight is that disturbance regimes will change as climate changes. For example, in the White Mountains of New Hampshire, wind always has been an important source of disturbance near the treeline (Spear, 1989). However, at low elevations, fire was the most important source of disturbance up to 7,000 years ago, when more mesic conditions caused it to be supplanted in importance by windstorms (Davis, 1985). Changes in disturbance regime may produce a larger change in vegetation than would have resulted from the effects of climatic change alone (Davis, 1989).

The final insight is that multiple environmental changes will be important and produce ecosystems that have no modern analogs (Davis, 1989). Evidence for this comes from the increasing divergence seen in plant assemblages from present-day counterparts as we go further into the past. Thus, analogy with present-day and past assemblages of species will have limited utility for prediction of future ecosystems. Rather, a functional understanding of the responses of individual species to multiple impacts will be needed to predict vegetation responses (Davis, 1989).

RESPONSE TO CO_2

Plants have strong responses to virtually every aspect of global change. For example, many plants increase their photosynthetic carbon gain and growth in response to increased CO_2 concentration (Woodward et al., 1991). However, there are key differences among species of plants in their response to increased CO_2. Plants with the photosynthetic pathway called C_4 metabolism (because the first products of CO_2 fixation are four carbon acids) are already saturated by CO_2 at current atmospheric levels. Thus, they show little, if any, increase in photosynthesis as CO_2 concentrations increase (Woodward et al., 1991). In contrast, plants with the C_3 photosynthetic pathway increase photosynthetic rates up to concentrations of 1,000 parts per million (ppm) CO_2 and beyond (Pearcy and Ehleringer, 1983). These plants typically show enhanced growth rates along with their increased photosynthetic performance.

There are also differences in response to increased CO_2 levels associated with plant life history. Hunt et al. (1991) examined the vegetative growth responses of 25 native herbaceous species of widely varying ecology to increased CO_2 concentrations. The response to a doubling of CO_2 concentration varied from no increased growth to increases up to 3.66 times that found in ambient CO_2 concentrations. Species that were classified as having a "competitive" (C) life history (sensu Grime, 1979) showed the greatest enhancements, while those naturally found in stressful habitats (i.e., low nutrient or saline habitats) (S life history), and those found in highly disturbed habitats (R or ruderal life history) showed no enhancement in growth. Hunt et al. (1991) concluded that vegetative communities of intermediate productivity, those that contain a combination of suppressed competitors and more stress-tolerant species, may experience dramatically altered species composition under the eutrophying effects of increased CO_2 concentrations.

Plants may increase the efficiency with which they use other resources, such as water and nitrogen, when grown under increased levels of CO_2 (Morison and Gifford, 1984). Both CO_2 fixed in photosynthesis and water lost through transpiration must pass through pores in the leaf epidermis called stomata. Generally, stomata close in response to increased CO_2 concentrations. This stomatal closure would reduce transpiration without a concomitant reduction in photosynthesis. Hence, water-use efficiency (the ratio of photosynthesis to transpiration) may increase (Woodward et al., 1991). Countering this effect is the response of leaf temperature to stomatal closure. With reduced stomatal apertures, leaf temperatures generally rise, resulting in increased transpiration from the leaf. Thus, it is not always possible to predict the exact water-use response of plants to increased CO_2. However, some plants have been found to perform better under limited water conditions when CO_2 concentrations are higher than ambient (Marks and Strain, 1989).

RESPONSE TO TEMPERATURE

Plant response to temperature is quite variable. In terms of photosynthesis, many plants have very broad temperature optima (Berry and Björkman, 1980), maintaining high photosynthetic rates across a large temperature range. Additionally, photosynthesis in many plants has the ability to acclimate to changes in temperature that occur over the span of days to weeks (Berry and Björkman, 1980). Thus, it is more likely that extreme conditions, such as unexpected frosts or extremely hot temperatures, are more important limits to plant distribution than is photosynthetic response to an increase in the mean ambient temperature of a few degrees.

RESPONSE TO WATER

Water availability, either through low soil water levels or through high evaporative demands, is the single most important environmental parameter limiting plant distribution and productivity on a world-wide basis (Schulze, 1986). There is great variability in the ability of different species of plants to survive periods of drought. Some species are effective drought avoiders, dropping leaves

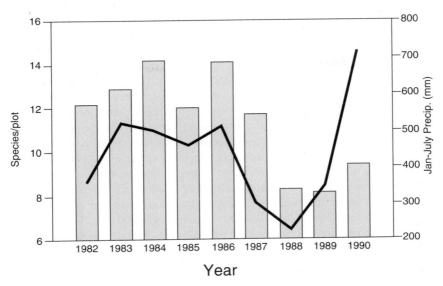

FIGURE 12-1 Species per plot (bars) in four old fields at Cedar Creek Natural History Area from 1982-1990. Also plotted is precipitation for January through July (line). In 1988, precipitation was at a 50-year low. Data are from Tilman and El-Haddi (1992).

or closing stomata in order to reduce the incidence of water stress. Others are capable of osmotic adjustment, which lowers the water potential at which stomata close and photosynthesis ceases. One particularly illuminating study on the potential effects of drought on biodiversity was that performed by Tilman and El-Haddi (1992) at the University of Minnesota. Using long-term records of the presence of species in permanent plots, they were able to examine the effect of rare severe drought on species richness. Local species richness in four different grassland fields fell an average of 37% during the drought year of 1988 (Figure 12-1). Even though plant biomass and precipitation recovered in 1989 and 1990, species richness did not recover. The authors attributed this to either a lack of available propagules or a failure of young seedlings to reestablish local populations within plots containing other established species (i.e., they were recruitment limited). There were differences between the chance that a species was lost and its abundance prior to 1988. In perennial grasses, forbs, legumes, and woody species, the predrought abundance was negatively correlated with loss from plots. These results show that some assemblages of rare species are especially vulnerable to extinction during environmental stress, and have grave implications for the effects of global climatic change on biodiversity.

RESPONSE TO CO$_2$ AND WATER

Plants rarely encounter single stresses in their environment. Rather, they are faced with multiple environmental stressors, cooccurring in various combinations and at various intensities at different times during growth. The effects of these multiple environmental stresses can be far from additive. Some may even counter the effect of others, e.g., increased CO$_2$ concentrations may reduce the effect of drought by reducing plant water loss and increasing water-use efficiency. Thus, the ability of plants to respond to future global changes will be complicated by the interaction of environmental variables. There have been many studies of the response of plants to single factors, such as CO$_2$. Fewer studies have been conducted under multiple environmental stresses, or in combination with biotic interactions such as competition. Examples of studies that did include multiple factors include Bazzaz and Carlson (1984), who observed that competitive interactions changed with CO$_2$ concentration and water availability. Under well-watered, high CO$_2$ conditions, C$_3$ plants were competitively superior. However, under water-limited conditions, C$_4$ plants were competitively superior, regardless of CO$_2$ concentration. In contrast, Marks and Strain (1989) found that under elevated levels of CO$_2$ *Aster pilosus*, an early successional C$_3$ weed, was competitively superior to *Andropogon virginicus*, a C$_4$ grass that usually displaces *A. pilosus* in the successional sequence, regardless of water supply. Another study that involved different levels of CO$_2$ along a moisture gradient was conducted by Miao et al. (1992). These authors found that the magnitude of gray birch and red maple seedling response to elevated CO$_2$ was contingent on soil moisture levels (Figure 12-2). Elevated CO$_2$ modified the overall pattern of response of these species to the soil moisture gradient. Red maple was enhanced preferentially at wetter ends of the gradient than was gray birch. Hence, the niche overlaps between the species were reduced.

These studies illustrate that experiments combining various CO$_2$ levels with gradients of other resources can result in species-specific, nonintuitive results. However, even these experiments lack the scale of time and area needed to effectively predict ecosystem responses to climatic change.

RESPONSE TO INCREASES IN UV-B RADIATION

In addition to their effects as greenhouse gases, CFCs have the capability to catalytically destroy stratospheric ozone. Stratospheric ozone is the primary atmospheric attenuator of ultraviolet-B (UV-B) radiation (280-320 nm) penetrating the Earth's atmosphere. For every 1% decrease in stratospheric ozone, we may see approximately a 2% increase in UV-B radiation at the Earth's surface. Plant response to UV-B radiation runs the gamut from little effect to large reductions in photosynthesis and growth (Teramura, 1983). An elegant, long-term study conducted by Barnes et al. (1988) illustrates the type of subtle effects that in-

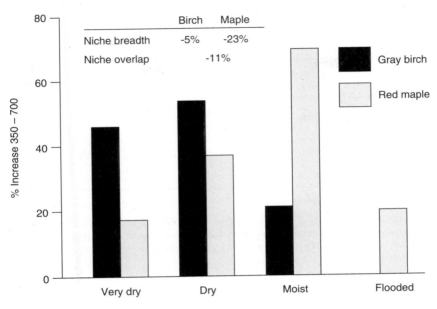

FIGURE 12-2 Percentage increase in biomass of gray birch (solid bars) and red maple (shaded bars) when grown in 700 ppm CO_2 versus 350 ppm CO_2. Plants were grown in different moisture conditions, as indicated on the graph. Also shown is the degree of niche breadth for moisture when plants were grown in 700 ppm CO_2 compared to 350 ppm CO_2. Niche overlap represents the reduction in overlap in moisture regimes for the two species when grown in 700 ppm CO_2 compared to 350 ppm CO_2. From Miao et al. (1992).

creased UV-B may have on species mixtures. The authors found that supplemental UV-B radiation had little effect on wheat or wild oat plants grown in monoculture. However, enhanced UV-B did alter the competitive balance between the two species studied, especially during wet years (Table 12-1). The mechanism of this effect was a change in the amount of leaf area with height that the two species showed when grown in mixtures. The authors concluded that changes in competitive balance in terrestrial plants may be a more sensitive indicator of solar UV augmentation than changes in total biomass of plants.

There have been several studies of the interaction of UV-B radiation with other environmental variables, such as light level, nutrient availability, and water stress. In general, the effects of UV-B radiation are reduced under water stress and low nutrient availabilities, and enhanced under low ambient light levels (Sullivan and Teramura, 1989; Tevini and Teramura, 1989). Recently, several studies have examined the interaction between UV-B radiation and increased CO_2 concentrations. In several species of crops, the effects of increased CO_2 con-

TABLE 12-1 Results of a Four-Year Study of Competition Between Wheat (*Triticum aestivum*) and Wild Oat (*Avena fatua*) Exposed to Enhanced UV-B Radiation

Year	Precipitation (mm)	Crowding Control	Coefficient UV-B
1981	17	0.94	1.28
1982	37	1.16	1.39[a]
1983	17	1.36	1.11
1984	91	1.24	1.82[a]
1985	43	1.66	1.65
1986	3	1.28	1.44

[a]Significantly different at $p < 0.05$.

NOTE: Enhancements of UV-B ranged from that expected from a 16% reduction in stratospheric ozone to that expected after a 40% reduction in stratospheric ozone. Any value of crowding coefficient above 1.0 indicates superior competitive ability for wheat.

SOURCE: Barnes et al. (1988).

centrations were eliminated or reduced by supplemental levels of UV-B radiation (Teramura et al., 1990; Ziska and Teramura, 1992). In *Pinus taeda* (loblolly pine), supplemental UV-B radiation not only reduced the growth enhancements caused by high levels of CO_2 but also altered allocation patterns to above-ground versus below-ground organs (Sullivan and Teramura, 1994). The authors concluded that this may have ramifications for seedling establishment, growth, and competitive interactions of this species in the field.

RESPONSES OF NATIVE VERSUS EXOTIC SPECIES

An important aspect of the response of native plants to global change is the potential for altered interactions with introduced species. For example, in the southeastern United States, two weedy vine exotics are especially successful, Kudzu (*Pueraria lobata*) and Japanese honeysuckle (*Lonicera japonica*). These two species have several properties that may allow them to spread even more in future environments. Since they are exotics to North America, they have few natural enemies. They are weedy, with very good colonization abilities. They are vines, with large allocation of carbon to above-ground stems and leaves. They respond very strongly to increased levels of CO_2 with increased elongation and above-ground growth (Sasek and Strain, 1988, 1989, 1991). Presently, their northern distribution limits are set by low winter temperatures and the date of the first hard fall frost. As temperatures and CO_2 levels rise, they may become even more aggressive competitors in their spread northward. Competition between these weedy aggressive exotics and rare native species—already endangered and placed under stress by high temperatures and increased periods of drought—may further reduce the biodiversity of native floras.

CONCLUSIONS

Human activities such as land clearing and burning of fossil fuel have caused a continuing increase in the concentration of greenhouse gases, especially CO_2 in the atmosphere. These gases are long-lived enough to remain in high concentrations in the atmosphere for the next century or more, even with immediate reductions in their emission. Projections of their effects on global climate include increased ambient temperatures, increased ambient UV-B radiation, changes in precipitation patterns and amounts, and increased occurrences of extreme weather events. Plants have strong responses to all of these predicted changes. Of primary importance is the fact that species of plants all respond individually and in different degrees to environmental conditions. Hence, the composition and distribution of species within present-day communities are likely to change. Altered competitive interactions and climatic conditions are bound to lead to loss of more species. This response may be exacerbated by the continuing spread of aggressive weedy exotics. So far, research has been concentrated at the level of responses of species to single environmental variables. New studies incorporating multiple environmental factors and multiple species have shown complex, species-specific results. Larger, longer-term studies of this type are needed before we can predict accurately the extent and nature of global change on plant biodiversity.

REFERENCES

Bazzaz, F. A., and R. W. Carlson. 1984. The response of plants to elevated CO_2. I. Competition among an assemblage of annuals at two levels of soil moisture. Oecologia 62:196-198.

Barnes, P. W., P. W. Jordan, W. G. Gold, S. D. Flint, and M. M. Caldwell. 1988. Competition, morphology and canopy structure in wheat (*Triticum aestivum* L.) and wild oat (*Avena fatua* L.) exposed to enhanced ultraviolet-B radiation. Func. Ecol. 2:319-330.

Berry, J., and O. Björkman. 1980. Photosynthetic response and adaptation to temperature in higher plants. Ann. Rev. Plant Physiol. 31:491-543.

Bolin, B. 1991. Man-induced global change of climate: The IPCC findings and continuing uncertainty regarding preventive action. Envir. Conserv. 18:297-303.

Davis, M. B. 1981. Quaternary history and the stability of forest communities. Pp. 132-153 in D. C. West, H. H. Shugart, D. B. Botkin, eds., Forest Succession: Concepts and Application. Springer-Verlag, N.Y.

Davis, M. B. 1985. History of the vegetation on the Mirror Lake watershed. Pp. 53-65 in G. E. Likens, ed., An Ecosystem Approach to Aquatic Ecology: Mirror Lake and its Environment. Springer-Verlag, N.Y.

Davis, M. B. 1989. Insights from paleoecology on global change. Bull. Ecol. Soc. Amer. 70:222-228.

Davis, M. B. 1990. Climatic change and the survival of forest species. Pp. 99-110 in G. M. Woodwell, ed., The Earth in Transition. Patterns and Processes of Biotic Impoverishment. Cambridge University Press, Cambridge, England.

Davis, M. B., K. D. Woods, S. L. Webb, and R. P. Futuyma. 1986. Dispersal versus climate: Expansion of *Fagus* and *Tsuga* in the upper Great Lakes region. Vegatatio 67:93-103.

Davis, M. B., and C. Zabinski. 1992. Changes in geographical range resulting from greenhouse warming: Effects on biodiversity in forests. Pp. 297-308 in R. Peters and T. Lovejoy, eds., Global Warming and Biological Diversity. Yale University Press, New Haven, Conn.

Grime, J. P. 1979. Plant Strategies and Vegetation Processes. John Wiley and Sons, N.Y.

Hunt, R., D. W. Hand, M. A. Hannah, and A. M. Neal. 1991. Response to CO_2 enrichment in 27 herbaceous species. Func. Ecol. 5:410-421.

Marks, S., and B. R. Strain. 1989. Effects of drought and CO_2 enrichment on competition between two old-field perennials. New Phytol. 111:181-186.

Miao, S. L., P. M. Wayne, and F. A. Bazzaz. 1992. Elevated CO_2 differentially alters the responses of co-occurring birch and maple seedlings to a moisture gradient. Oecologia 90:300-304.

Morison, J. I. L., and R. M. Gifford. 1984. Plant growth and water use with limited water supply in high CO_2 concentrations. II. Plant dry weight, partitioning and water use efficiency. Aust. J. Plant Physiol. 11:375-384.

Pearcy, R. W., and J. R. Ehleringer. 1983. Comparative ecophysiology of C_3 and C_4 plants. Plant Cell Envir. 7:1-13.

Prance, G. T. 1990. Flora. Pp. 387-391 in B. L. Turner II, W. C. Clark, R. W. Katz, J. F. Richards, J. T. Mathews, and W. B. Meyer, eds., The Earth as Transformed by Human Action: Global and Regional Changes in the Biosphere Over the Past 300 Years. Cambridge University Press, N.Y.

Sasek, T. W., and B. R. Strain. 1988. Effects of carbon dioxide enrichment on the growth and morphology of kudzu (*Pueraria lobata*). Weed Sci. 36:28-36.

Sasek, T. W., and B. R. Strain. 1989. Effects of carbon dioxide enrichment on the expansion and size of kudzu (*Pueraria lobata*) leaves. Weed Sci. 37:23-28.

Sasek, T., and B. Strain. 1991. Effects of CO_2 enrichment on the growth and morphology of a native and an introduced honeysuckle vine. Amer. J. Bot. 78:69-75.

Schneider, S. H. 1993. Scenarios of global warming. Pp. 9-23 in P. M. Kareiva, J. G. Kingsolver, R. B. Huey, eds., Biotic Interactions and Global Change. Sinauer Associates, Sunderland, Mass.

Schulze, E.-D. 1986. Carbon dioxide and water vapor exchange in response to drought in the atmosphere and in the soil. Ann. Rev. Plant Physiol. 37:247-274.

Spear, R. W. 1989. Late-Quaternary history of high-elevation vegetation in the White Mountains of New Hampshire. Ecol. Monogr. 59:125-151.

Sullivan, J. H., and A. H. Teramura. 1989. Field study of the interaction between solar ultraviolet-B radiation and drought on photosynthesis and growth in soybean. Plant Physiol. 92:141-146.

Sullivan, J. H., and A. H. Teramura. 1994. The effects of UV-B radiation on loblolly pine. 3. Interaction with CO_2 enhancement. Plant Cell Envir. 17:311-317.

Teramura, A. H. 1983. Effects of ultraviolet-B radiation on the growth and yield of crop plants. Physiol. Plantarum 58:415-427.

Teramura, A. H., J. H. Sullivan, and L. H. Ziska. 1990. Interaction of elevated ultraviolet-B radiation and CO_2 on productivity and photosynthetic characteristics in wheat, rice, and soybean. Plant Physiol. 94:470-475.

Tevini, M. M., and A. H. Teramura. 1989. UV-B effects on terrestrial plants. Photochem. Photobiol. 50:479-487.

Tilman, D., and A. El-Haddi. 1992. Drought and biodiversity in grasslands. Oecologia 89:257-264.

Webb, T., III. 1987. The appearance and disappearance of major vegetational assemblages: Long-term vegetational dynamics in eastern North America. Vegatatio 69:177-187.

Woodward, F. I., G. B. Thompson, and I. F. McKee. 1991. The effects of elevated concentrations of carbon dioxide on individual plants, populations, communities and ecosystems. Ann. Bot. 67:23-38.

Ziska, L. H., and A. H. Teramura. 1992. CO_2 enhancement of growth and photosynthesis in rice (*Oryza sativa*): Modification by increased ultraviolet-B radiation. Plant Physiol. 99:473-481.

PART IV

UNDERSTANDING AND
USING BIODIVERSITY

From bacteria to butterflies,
understanding is the key to wise use.

CHAPTER
13

Names: The Keys to Biodiversity

F. CHRISTIAN THOMPSON

Research Entomologist, Systematic Entomology Laboratory,
Plant Sciences Institute, Beltsville Agricultural Research Center,
United States Department of Agriculture, Washington, D.C.

Besides biodiversity, the one thing that all the chapters included in this volume have in common is scientific names. These names form the essential language, the means we use to communicate about biodiversity. To avoid a Tower of Babel, a common system of nomenclature is required: a system that is effective and efficient (and at minimal cost). Presented below are the essential aspects of this language for biodiversity and a discussion of where we are in respect to their implementation.

The long-term conservation of biodiversity can be achieved only through the approach used by the Instituto Nacional de Biodiversidad (INBio)—"save it," "characterize it," and "sustainably use it" (Janzen, Chapter 27 of this volume). Characterization requires that we have a language with which to communicate about biodiversity: a way of describing it, so that we all know what we are talking about and that we are talking about the same things. How do we characterize biodiversity? The first step is to name its components. Biodiversity is divisible into three levels: ecological, taxonomic, and genetic. Of these levels, taxonomic diversity is critical because taxa are the units that contain genetic diversity and are the units that make up ecological diversity. Since taxa are the core of biodiversity, names for taxa are the most critical component of any language of biodiversity.

A UNIVERSAL LANGUAGE OF BIODIVERSITY

What are names? Names are tags. Tags are words, short sequences of symbols that are used in place of something complex which would require many more words to describe. Hence, tags save time and space. Instead of a long description, we use a short tag. A scientific name differs from a common name in that the scientific name is a unique tag. In other languages, there may be multiple tags for the same thing. Imagine the various words in English that are used to describe *Homo sapiens*. In computer (database) jargon, data elements that are used to index information are termed keys, and keys that are unique are called primary keys. Scientific names are primary keys. The word "key" has another meaning in English, which is "something that unlocks something." Scientific names are those critical keys that unlock biosystematic information, all that we know about living organisms. Scientific names are tags that replace descriptions of objects or, more precisely, concepts based on objects (specimens). Scientific names are unique, there being only one scientific name for a particular concept, and each concept has only one scientific name.

Scientific names are more than just primary keys to information. They represent hypotheses. To systematists, this is a trivial characteristic that sometimes is forgotten and thereby becomes a source of confusion later. To most users, this is an unknown characteristic that prevents them from obtaining the full value from scientific names. If a scientific name were only a unique key used for storing and retrieving information, it would be just like a social security number. *Homo sapiens* is a unique key used to store and retrieve information about man, but that key also places the information about man into a hierarchical classification. Hierarchical classifications allow for the storage, at each node of the hierarchy, of information common to the subordinate nodes. Hence, redundant data, which would be spread throughout a nonhierarchical system, are eliminated.

Biological classifications, however, do more than just hierarchically store information. Given that one accepts a single common (unique) history for life and that our biological classifications reflect this common history in their hierarchical arrangement, then biological classifications allow for prediction: they allow us to predict that some information stored at a lower hierarchical node may belong to a higher node, that is, is common to all members of the more inclusive group. These predictions take the form of the following: if some members of a group share a characteristic that is unknown for other members of the same group, then that characteristic is likely to be common to all members of the group.

So scientific names are tags, unique keys, hierarchical nodes, and phylogenetic hypotheses. Thus, systematists pack a lot of information into their names and users can get a lot from them.

Scientific names are hypotheses, not proven facts. Systematists may and frequently do disagree about hypotheses. Hypotheses, which in systematics range from what is a character to what is the classification that best reflects the history of life, are always prone to falsification, and, hence, change. Disagreements about classification can arise from differences in paradigm or information.

Systematists use different approaches to construct classifications, such as cladistic versus phylogenetic versus phenetic methods. Given the same set of data that underlies a given hierarchy, cladists can derive classifications that are different from those derived by the phenetists (Figure 13-1). Even among cladists, there can be differences as to the rank (genus, family, order, etc.) and thereby the hierarchical groups used. These are disagreements based on paradigm. There can be disagreement about the hypotheses that underlie the information used to construct the classifications, such as what are the characters. Disagreement can arise among systematists because they use different information. While disagreements will affect the ability to predict, they need not affect the ability to retrieve information.

The desirable attribute that must be preserved to ensure complete access to information across multiple classifications is uniqueness. Our scientific nomenclature must guarantee that any scientific name that is used in any classification is unique among all classifications. This can be assured by having two primary keys. Unfortunately, having two keys increases the overhead of our information systems. Most systematists and *all* users want to avoid this problem by

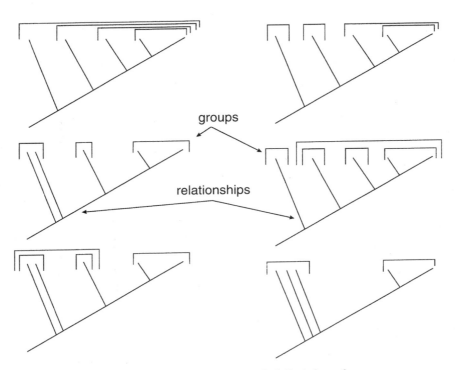

FIGURE 13-1 Multiple classifications for identical cladistic hypotheses.

mandating that there be only *one* classification. Although in theory there is only one correct classification, as there was only one history of life, in reality there have been multiple classifications in the past, there may be multiple classifications in use today, and there will be multiple classifications in the future. That is the price of scientific progress, of the increase in our knowledge of the world. If information is to be retrieved across time, if we want to extract information stored under obsolete classifications, and if we want to avoid dictating "the correct" classification, then we need a nomenclatural system that supports two unique keys.

The two keys for our language of biodiversity are the *valid* name and the *original* name. The valid name is the correct name for a concept within a classification; the original name is the valid name in the classification in which it was proposed. Valid names may be different among classifications, but the original name is invariant across classifications (Table 13-1). Valid names are the best names to use because they provide the full value of scientific names. These are the names that provide a basis for prediction. The original name is useful only for retrieval of information across multiple classifications. Although valid and original names may be and frequently are the same, users must know the differences between them. Specifically, they need to know that a valid name is a powerful tool for inference, that a valid name provides for prediction of unknown attributes of the organism that bears the name. But they must understand that there may be multiple valid names in the literature or in use, and that valid names represent hypotheses that may change as our knowledge is tested and improved. Most importantly, if there are multiple valid names in use, then the users must be aware that there are conflicting scientific hypotheses being advocated and that they must select the name that best serves their purpose. If users do not want to decide, do not want to use classifications to organize and synthesize their information, then they may use the original name to index their information, being assured that it will always be a unique key.

There are other problems today with our classifications: synonymy (having two names for the same concept) and homonymy (having the same name for

TABLE 13-1 Multiple Classifications and Primary Keys to Information

Year	Valid Name	Original Name	Authority
1776	*Musca balteata*	*Musca balteata*	De Geer
1822	*Syrphus balteatus*	*Musca balteata*	Meigen
1843	*Scaeva balteatus*	*Musca balteata*	Zetterstedt
1917	*Episyrphus balteatus*	*Musca balteata*	Matsumura
1930	*Epistrophe balteata*	*Musca balteata*	Sack
1950	*Stenosyrphus balteatus*	*Musca balteata*	Fluke
Today	*Episyrphus balteatus*	*Musca balteata*	Vockeroth

different concepts). These problems are, however, largely due to ignorance. If we knew all names and their types and could agree on what are species, then by applying the rules of nomenclature we immediately could eliminate all problems of synonymy and homonymy. Homonymy is eliminated by the rule of uniqueness. Synonymy is addressed by the rules of typification, which tie a physical instance of a concept to a name. Synonymy is resolved by logic of circumscription and the convention of priority (or usage). The name of a concept is the name affixed to one and only one of the types that falls within its circumscription. The name used is determined by which name is the oldest (priority) or most widely used (usage). The specific rules for resolving homonymy and synonymy, as well as for the proper formation and documentation of names, are our Codes of Nomenclature (International Botanical Congress, 1994; International Commission on Zoological Nomenclature, 1985; Sneath, 1992). These rules, however, do not address the problem of multiple classifications or ignorance of the universe of applicable names and their typification.

There is one final problem, the species problem. This is the problem of the basic unit of information or data. The basic unit of information for nomenclature is the species (or more precisely the species-group, which includes the category of subspecies). The problem is that the species is not a single element of datum, but consists of information—data derived from specimens that have been identified as belonging to that species. Mistakes can be made during this identification, which is another hypothesis. Information ultimately is not derived from species, but from specimens. Information management in biodiversity really begins with data management of specimens. The problems of specimen-based data management are not intractable, but are readily addressed by the use of bar-codes, another form of unique keys.

The species problem is also one of circumscription, the definition of the limits of a taxon. A group with the same name and type may be more or less inclusive, depending on the characters used to define its limits. Zoologists differ from botanists in not considering circumscription to be a problem because, minimally, all identically named taxa have at least some characteristics in common. The problem of how much is held in common, therefore, is best resolved by enumeration of the included taxa or specimens. The history of circumscription can be tracked by use of an additional key that uniquely identifies the person who defined the limits and the date of that action. It is sufficient for our purposes to know that data based on specimens always will be summarized into information units based on species and that all such information should be based on specimens.

THE REAL WORLD: DIVERSITY AND DISPERSION

The All Taxa Biodiversity Inventories (ATBIs), taxasphere, national biological surveys, etc., are means of addressing the problem of characterization of biodiversity, the most important step in conserving biodiversity. To characterize, a language is needed. To characterize biodiversity, all our available resources will be needed because the task is immense (World Conservation Monitoring Centre, 1992). The resources for characterizing biodiversity are diverse and dispersed. The only way the job is going to get done is by forming partnerships, by working together. That requires communications that can accommodate the diversity and dispersion.

If there were only one classification, if that classification were controlled by one person, and if all information were stored in one system at one place under that classification, then there would be one source for answers to questions about biodiversity. Unique and comprehensive information systems are powerful tools. In reality, however, things are different. Different classifications exist and resources are dispersed. To share resources, therefore, different classifications must be understood. To utilize distributed resources, a universal communication system that allows multiple classifications is needed. Scientific nomenclature provides the basis for this system: a set of unique keys to traverse the world of distributed databases, to find information anywhere on our future information superhighways. However, to make the system work, a universal data dictionary is required. Such a dictionary would allow users with any name to find the keys to unlock information about biodiversity. A universal data dictionary requires that the problems of synonymy and homonymy are solved and ensures that all classifications and names are accommodated.

PROGRESS AND PROMISE

The present language of biodiversity is binominal nomenclature that was introduced by Carolus Linnaeus, a Swedish professor of natural history. This system was the direct result of an earlier governmental biodiversity project. The Swedish Crown had some far-flung possessions and wanted to know what use could be made of them. They sent Linnaeus to investigate, to survey what today is called biodiversity, and to write a report characterizing what was found, with recommendations on how to use it. At the time, there was only a binary system of nomenclature: one word for the genus, with the species being described by a series of adjectives. Given the diversity Linnaeus found, he did not want to waste time repeating the long strings of adjectives that were required to characterize the biodiversity. Because the characterizations were in his flora of Sweden, he used a combination of the genus name and a single word (an epithet) for each species to form a unique key to those descriptions (see Stearn, 1957, for more details). The system was an immediate success. Linnaeus codified the system, built and maintained a universal information database for all names (his *Systema Naturae*, 1758), and trained a cadre of students to carry on his work. The students dispersed and converted others. But because no one could be *the* master except Linnaeus, they divided nature up. There were to be no more

Systema Naturae. At first, there were a series of *Systemae* for parts of nature, maintained by the students as the authorities for this kingdom or that area, but authorities in many countries quickly became involved. Two hundred years later, systematists cannot tell society how many organisms have been described or what are all their names.

What can systematists do? They can cooperate and, as a whole, recreate what Linnaeus started: a system of nature. They can agree on and follow a set of standards for nomenclature; ensure that those standards are adequate for our informational needs; eliminate the chaos of the past, first by gathering together the names that are today dispersed across a vast sea of literature, second by putting a limit on searching the past, and finally by accepting what is found after a reasonable search. They can ensure that the chaos will not return by requiring wide dissemination or registration of new names.

Where are systematists today in accomplishing these tasks? On standards, we are almost there. For bacteria, there is already a modern system of nomenclature (Ride, 1991; Sneath, 1986). For zoology and botany, we will have one shortly, and a start has been made on a universal code for all life (Hawksworth, 1994). Information about proposed changes in the botanical code are published regularly in the journal *Taxon*, and those for the zoological code in *Bulletin of Zoological Nomenclature*. A new draft version of the *International Code of Zoological Nomenclature* is now available for comment and can be adopted as soon as a year hence. With the acceptance of this new *Code*, the problems of name changes due to nomenclature will be eliminated. Some systematists are afraid that the new *Code* will be used to enforce the acceptance of one classification, rather than allow for a diversity of them. Such fears are unfounded because the new draft will preserve the "freedom of taxonomic thought and action." The draft includes new requirements stating that zoologists must document properly their classifications (implicit typification will be required for all names) and publish them where the whole community can evaluate them. This will be achieved by a registration system for new names. Bacteriology already requires this and botany has adopted it for the future. The new draft will be simpler to use, since the strict requirement of Latin grammar will be eliminated. Stability and universality will be enhanced by allowing zoologists to balance usage and priority more effectively in the determination of valid names. Finally, the new draft will provide the means to certify names and the associated nomenclatural data en masse, so as to free zoologists of the burdens of historical searches.

The last 250 years have left scientific names scattered across the most diverse array of media possible. No other science requires its practitioners to be responsible for such a mess. Scientists are expected to know the common and current hypotheses. They should not be required to know what was printed 200 years ago, distributed, and subsequently forgotten and largely lost. Systematists must deal with such ancient history, regardless of whether the concept had been published previously, was rediscovered subsequently, or was invalid.

That forgotten name need not even be in the domain of the systematist's expertise to cause problems.

The solution for systematics is simple: change the rules of nomenclature. This is being done. Then make a reasonable effort to gather together all the existing names and associated data and accept those names and data as correct. That will free systematists of the burden of history for nomenclature's sake.

The Systematic Entomology Laboratory of the U.S. Department of Agriculture has embarked on a voyage to do just this for insect names. We have proposed to the entomological community that together we develop a comprehensive database of names of terrestrial arthropods. We christened our ship BIOTA (Biosystematic Information on Terrestrial Arthropods). The most immediate and highest priority is to document all the names of terrestrial arthropods known to occur in North America. This represents the official adoption of the goal of the Entomological Collections Network, originally proposed by Miller (1992). We expect to reach this point within 2 years. We already have accumulated nearly 100,000 names and have commitments from cooperating specialists for another 40,000. This nomenclatorial database will include the essential keys, both the valid name and the original name for each species, that our specialists have recognized. The specialists will be identified in the data record, and minimal classificatory data, such as subfamily, family, order, and class, will be included. All synonyms, homonyms (all available names, sensu Zoological Code [International Commission on Zoological Nomenclature, 1985]) and common invalid combinations (names valid under other classifications) and misidentifications will be included. This information will provide the community with the necessary keys to the biodiversity of arthropods.

BIOTA also includes more comprehensive cataloging efforts, such as the Biosystematic Database of World Diptera. These efforts will include data on typification of names. The resultant catalogs will be submitted to the appropriate user-community for review and eventually to the International Commission of Zoological Nomenclature for certification. The Diptera database project has been endorsed by the International Congresses of Dipterology and is overseen by a committee of the Council for those Congresses (the council is a scientific member of the International Union of Biological Sciences). The family-group names, those names that may apply to the higher levels of classification, are nearing completion as the result of a 50-year effort by Curtis W. Sabrosky, one of our U.S.D.A. specialists. Some 4,296 names have been documented. Genus-group names have been entered and shortly will be distributed to specialists. Some 17,271 names were found, representing perhaps 8,000 valid genera. Names of species-groups are being entered now, with some 45,994 already in the database, perhaps 40% of the world total. All names for the flies of the Nearctic region have been entered and will be published shortly (28,890 names for 19,562 species and 2,356 genera).

Similar efforts are underway or already completed for other major groups of organisms (Bisby, 1994). Given continued support, we are likely to see the problems of names solved by the year 2000. Nomenclature no longer will be an impediment to efforts to characterize our biodiversity. We then will have a set of keys to what we know of biodiversity and a language capable of effectively and efficiently incorporating new information and accommodating diverse keys to that information.

SERVICE TO SOCIETY

Names are the keys to biodiversity, but what does one do when one has no name, only a specimen? How does one discover the proper name for a specimen? If it is a bird, one can use a field guide, like Peterson's (1980), to identify it. If not, one asks an expert. If it is an insect, one will find those experts at the Systematic Entomology Laboratory (SEL), which identifies more insects than any other organization. But identification has costs: when experts identify specimens, they are not building classifications nor describing new biodiversity. Realizing the existing shortage of experts to classify specimens, SEL sought new approaches to relieve their experts of the burdens and distractions of identification.

The obvious answer was found within the question: if one does not need experts to identify birds, why does one need them to identify flies? Peterson (1980) proved that, if users were presented with the critical characters (his field marks) in a graphic way, then users readily could identify birds. Many of the field marks used in Peterson's field guides were known to Linnaeus 200 years ago, but Linnaean descriptions are difficult for users to understand. Compare, for example, the plate of common ducks in Peterson's field guide (Peterson, 1980:51) and the corresponding page from Linnaeus' *Systema Naturae* (Linnaeus, 1758: 126) (Figure 13-2). As good as Peterson field guides are, they, like the traditional identification key, are rather inflexible. To identify, for example, a pintail, one must first know that it is a duck, a freshwater dabbler, or one must thumb through the pages of the field guide until the appropriate picture is found. With a computerized identification system, the user can select the most obvious character, such as the long tail. Then the computer would list the dozen or so species that have long tails. The user could ask the computer what are the best characters to discriminate among these species, and the computer would respond with a ranked list of the most useful characters, some of which may be head color, body shape, habitat, and geography.

A computerized identification system that builds on and extends the visual approach of Peterson has been produced. The Fruit Fly Expert System allows users to identify some 200 fruit flies, including all the important species of pests (Thompson et al., 1993). The system uses data encoded in the standard DELTA (DEscriptive Language for TAxonomy; Pankhurst, 1991) format, so other data sets can be prepared easily. The system is extremely flexible. Many taxa can be

126   AVES ANSERES. Anas.

Glau-   23. A. iridibus flavis, capite griſeo, collari albo. *Faun.*
cion.   *ſvec.* 104.
         Glaucion. *Bell. av.* 33: *b. Aldr. orn. l.* 13. *c.* 38.
         Will. orn. 231. *Raj. av.* 144.
         Habitat in Europæ *maritimis.*

Penelo- 24. A. cauda acutiuſcula ſubtus nigra, capite brunneo,
pe.   fronte alba. *Fn. ſvec.* 105.
         Penelope. *Geſn. av.* 108. *Aldr. orn. l.* 19. *c.* 38. *Jonſt.*
         av. l. 49. *Will. orn.* 288. *t.* 72. *Raj. av.* 146.
         Anas fiſtularis. *Alb. av.* 2. *p.* 88. *t.* 99.
         Habitat in Europæ *maritimis &* paludibus.

acuta.   25. A. cauda acuminata elongata ſubtus nigra, occipite
         utrinque linea alba.
         Anas cauda cuneiformi acuta. *Fn. ſvec.* 96.
         Anas cauda acuta. *Geſn. av.* 121. *Will. orn.* 289. *t* 73.
         Raj. av. 147. *Alb. av.* 2. *p.* 84. *t.* 94.
         Habitat in Europæ *maritimis.*

A

MARSH DUCKS
(Dabblers)

♀

♂

COMMON PINTAIL

♀

♂

AMERICAN WIGEON

♀

♂

B   EURASIAN WIGEON

FIGURE 13-2   Examples of identification aids. A. *Systema Naturae* (1758); B. Peterson's
Field Guide to the Birds (1980).

209

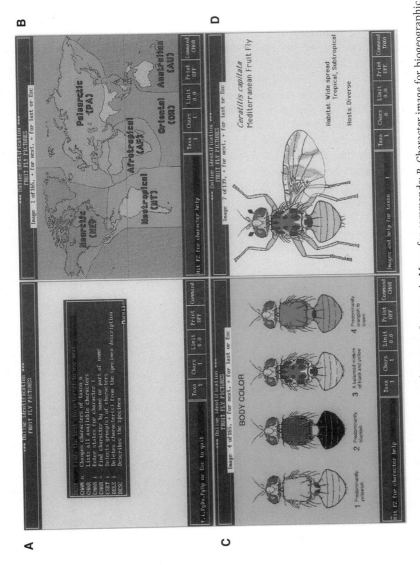

FIGURE 13-3 Computer screens from the Fruit Fly Expert System. A. Menu for commands; B. Character image for biogeographic regions; C. Character image for body color; D. Taxon image for Mediterranean fruit fly, *Ceratitis capitata*.

eliminated immediately by restricting the data set according to geographic location (Figure 13-3B) or other biological data. Any character can be chosen in any order, or the computer can list the best characters based on their ability to separate the remaining taxa. Characters are illustrated and multiple states are allowed (Figure 13-3C). This speeds the identification process in two ways, by enabling direct comparison of images with the specimen and by reducing the total number of decisions that must be made, because more than the traditional two alternatives can be evaluated efficiently at one time. Characters and computer commands are explained in help files that can be accessed at any time. Computer commands are either selected from menus or entered directly (Figure 13-3A). How closely a specimen must match characters (the error limit) can be set, so more matches must be made before a taxon is rejected. Errors, once detected, can be corrected easily without stepping through all the characters again. The identification can be verified easily. The computer can generate a complete description and image of any taxon (Figure 13-3D), list only the differences between the specimen and another taxon or between any two taxa, or generate a list of all the diagnostic characters for a particular taxon. The Expert System is not a panacea: unusual specimens, those outside the domain of the data set or with distorted features, still will have to be sent to systematists. Systematists are still needed, but by relieving them of most routine identifications, they can be more productive, and users will get their identifications faster.

CONCLUSION

The Systematic Entomology Laboratory has been and is committed to the characterization of biodiversity to help society develop an understanding of biodiversity and the ability to use it wisely. We are building classifications, a set of keys to enable us to better communicate about biodiversity. We also are committed to developing tools, such as expert systems and biosystematic information databases, to allow users to obtain these keys (names) and to know what are the best keys (valid names).

REFERENCES

Bisby, F. 1994. Global master species databases and biodiversity. Biol. Int. 29:33-38.
Hawksworth, D. L. 1994. Developing the bionomenclatural base crucial to biodiversity programmes. Biol. Int. 29:24-32.
International Botanical Congress. 1994. International Code of Botanical Nomenclature. Regnum Vegetabile 131. Koeltz Scientific Books, Königstein, Germany. 328 pp.
International Commission on Zoological Nomenclature. 1985. International Code of Zoological Nomenclature, third ed. International Trust for Zoological Nomenclature, London. 338 pp.
Linnaeus, C. 1758. Systema Naturae, tenth ed. L. Salvii, Stockholm. 824 pp.
Miller, S. E. 1992. Specimen databases and the lack of standard nomenclature: A proposal for North American insects. Insect Coll. News 7:7-8.

Pankhurst, R. J. 1991. Practical Taxonomic Computing. Cambridge University Press, Cambridge, England. 202 pp.

Peterson, R. T. 1980. A Field Guide to the Birds. A Completely New Guide to All the Birds of Eastern and Central North America. Houghton Mifflin, Boston. 384 pp.

Ride, W. D. L. 1991. Justice for the living. A review of bacteriological and zoological initiatives in nomenclature. Pp. 105-122 in D. L. Hawksworth, ed., Improving the Stability of Names: Needs and Options. Regnum Vegetabile 123. Koeltz Scientific Books, Königstein, Germany. 358 pp.

Sneath, P. H. A. 1986. Nomenclature of Bacteria. Pp. 36-48 in W. D. L. Ride and T. Younès, eds., Biological Nomenclature Today. IRL Press, Eynsham, England. 70 pp.

Sneath, P. H. A., ed. 1992. International Code of Nomenclature for Bacteria. 1990 Revision. American Society for Microbiology, Washington, D.C. 232 pp.

Stearn, W. T. 1957. An introduction to the *Species Plantarum* and cognate botanical works of Carl Linnaeus. In C. Linnaeus, Species Plantarum (a facsimile of the first edition of 1753). The Ray Society, London. 176 pp.

Thompson, F. C., A. L. Norrbom, L. E. Carroll and I. M. White. 1993. The fruit fly biosystematic information database. Pp. 3-7 in M. Aluja and P. Liedo, eds., Fruit Flies: Biology and Management. Springer-Verlag, N.Y.

World Conservation Monitoring Centre. 1992. Global Biodiversity. Status of the Earth's Living Resources. Chapman and Hall, London. 585 pp.

14

Systematics: A Keystone to Understanding Biodiversity

RUTH PATRICK

Francis Boyer Chair of Limnology, Academy of Natural
Sciences of Philadelphia, Pennsylvania

Systematics is the science by which species are recognized. Understanding and analyzing species and their relationships is essential in order to evaluate biodiversity.

Many of the aspects of biosystematics that are now being explored are based on understanding the similarities, differences, and the chemicals associated with such similarities and differences. Unless we know the relationship of taxa, it is very difficult to determine the significance of a single chemical difference. Systematics also is very important to determine the potential distribution of a chemical of interest. In other words, similar species are liable to contain similar chemical compounds and behave in similar ways.

Most children are curious about the natural world around them. They like to explore waterbugs and how they move or the pleasant odors of plants and the brightness of their colors. Children are innately classifiers, putting together those things that they think are similar. These basic desires to classify and recognize differences between organisms often has led to a deeper understanding of what is the species.

Unfortunately, when a child enters the elementary grades this interest sometimes is lost, often because their teachers do not have any interest in the natural world and do not excite the child's curiosity about how flowers grow or where animals live and why. Thus, in childhood, this great interest in the natural world often is lost temporarily as the student studies other topics such as mathematics, literature, and history.

Academic institutions that have collections of organisms should take a special interest in encouraging the interests of children. Some of our most effective systematists started their careers in sorting specimens in museums. This may seem like a dull task, but it is exciting if you are doing it for the first time and are just beginning to learn how to classify plants or animals.

This interest in the organization of the natural world should be encouraged, and the faculties throughout high school and college should be available to whet the interest of the young student. Often this is done by having a young scholar spend time sorting collections. It is very important that the collection manager take time to instruct the young person as to why they are doing what they do and how one tells differences.

Museums and colleges with a systematics program and collections available for student use offer an excellent opportunity to give the basic training necessary to understand systematics. Museums particularly are able to take young high school students and encourage them in their great interest in some particular group of organisms such as insects or flowers or molluscs. Often these young students then go on to college and make a great success in a career having to do with biodiversity or ecology.

It is the opportunity to handle specimens and to observe them critically with the help of a mentor that encourages a budding interest in a career of worthwhile research. For example, with a relatively small grant extending over 3 years that was provided by the Pew Charitable Trusts, a committee consisting of Dr. Peter Raven, Missouri Botanical Garden; Dr. Edward O. Wilson, Harvard University; Dr. Thomas Lovejoy, the Smithsonian Institution; Dr. Phil Humphrey, University of Kansas; and myself organized a program that has started the careers of some 25 students in the field of systematics.

In this program, we invited 18 institutions to submit programs to train young, beginning college students or students halfway through their college careers in systematics. The requirement was that each student must have a mentor who is willing to devote at least 1 day a week or its equivalent to helping them in their work. Each student was to have a special project that would result in a seminar or a publishable paper at the end of their training. Their training usually extended over 1.5 years.

The main requirement was that the students learn the modern techniques for carrying out the studies of systematics. As a result of this effort carried out by Cornell University, the Philadelphia Academy of Natural Sciences, the Smithsonian Institution, and the University of Kansas, 25 students have achieved a basic knowledge in systematics and are well on their way to writing a thesis in their chosen subject in graduate school.

These systematic programs have been in fish, insects, reptiles, paleobotany, and plants (particularly angiosperms). Several publishable papers have resulted from these studies, usually written with the mentor, and, most importantly, these students have acquired a basic interest in systematics and have chosen it for their academic careers. The college students were taught how to compare and contrast species according to certain traits. They have learned how to use

Students collecting specimens in the field.

computers and to do simple statistical analyses to objectively evaluate differences in the analysis of the specimens. These students have learned cytology and how to prepare slides or smears to identify chromosome numbers and understand the importance of genetic differences which enabled them to differentiate species of plants and animals. The student should be taught how to compare and contrast species according to certain traits. Learning to use the computer and to do simple statistical analysis is a very important part of this training. The students should learn techniques for studying the chromosome numbers and gene differences in plants and animals.

Field trips are very important in the training of students. It is most important for the student to realize that a species is not a single specimen but a collection of specimens, in other words a population. This population, in the natural world, will undoubtedly show variance, and being able to distinguish between those types of variances that are ephemeral or that are rather superficial and those that are essential to the functioning of the organism is important.

It is most important that each student work closely with a mentor. At frequent intervals the student should be asked to define clearly the similarities and the differences between two species in question. Many techniques should be used to determine whether this difference is significant.

Today we have computer techniques that enable us to compare many different factors in a relatively short period of time. A few years ago this was a very labor-intensive process. We also have developed various mathematical formulae which help us to see whether differences are significant. However, I caution that using a computer to tell differences alone is not sufficient. One needs to examine the species and determine if the differences found by different types of study correspond with those determined by the computer.

Computers also enable us to examine genetic differences and thus determine whether some of the differences we find are truly significant from a reproductive standpoint or whether they are frivolous variations.

The question "What is a species?" has long been an intriguing question and continues to be. It is very important that we use as many techniques as possible to determine what are the differences between species and how significant they are. Which ones are stable? Which ones are ephemeral?

We have changed our analyses of what is a species considerably over the last 30 years. It used to be largely based on visual differences. Today we also are able to analyze for functional differences and cytological differences that are not so easily seen when one encounters a species in the field.

As a result of this increased complexity in the science of systematics, one needs to have a well-equipped laboratory, large collections to work on, and a mentor who is wise and willing to be easily available to the student. Access to cytologists, geneticists, and their methodology also is important. It takes a great deal more money and more time to become an effective systematist today than it did when Linnaeus or Bailey or Gray or Parker were alive.

Systematics as a science is a necessity for understanding biodiversity, whether one is seeking new chemicals for medicine or food, whether one is trying to determine the effectiveness of various fibers for clothing and shelter, or whether one is looking for energy-producing substances such as coal and oil. The science of systematics and knowing what are the organisms that produced the substance that you desire is important.

Systematics is not only important to the business person and to the theoretical biological scientist, but also to the regional planner and environmentalist for the maintenance of our natural environment. We now know that biodiversity is a keynote to naturalness, and that this biodiversity must be maintained if the important functions of the production of oxygen, the transfer of nutrients through the ecosystem, the disappearance of wastes by digestion by organisms, and the availability of many species for man's use are to continue.

Unfortunately, interest in systematics is dwindling. In many groups of organisms, there are only one or two specialists in the world today. Some groups have no living specialists concerned with their identification. For these reasons, the training of systematics should be a number one priority of our country in order to supply the most nutritious food at the least cost and with the least impact on the environment, to be at the forefront in the development of new medicines to improve human health, and to maintain a natural environment suitable for recreation. Providing many more scholarships and assistantships to students pursuing a career in systematics would go a long way toward achieving these goals.

CHAPTER

15

Biodiversity and Systematics: Their Application to Agriculture[1]

DOUGLASS R. MILLER

Research Entomologist, Systematic Entomology Laboratory

AMY Y. ROSSMAN

Research Leader, Systematic Botany and Mycology Laboratory

Agricultural Research Service, U.S. Department of Agriculture, Beltsville, Maryland

Systematics and biodiversity are the threads that form and hold the fabric of agriculture together. Agricultural development is increasingly dependent on the interactions and knowledge base of systematics and biological diversity. Biodiversity itself is the grist for the agricultural mill—the germplasm. Systematics is the explorer, describer, organizer, and predictor of biological diversity. Agriculture is a primary user of the products of the interactions between biodiversity and systematics. This chapter examines the history of this interaction and shows how current and future practices in agriculture increasingly require more detailed knowledge of the world's biota. This chapter also focuses on a series of examples demonstrating the importance of systematics and biodiversity in agriculture and suggests a solution to a serious dilemma that faces agriculture today and in the future.

HISTORY AND THE FUTURE

Before the advent of pesticides, herbicides, and inorganic fertilizers, the need to understand the biology and biological requirements of agricultural organisms was of considerable importance. Control strategies for insect pests, for example, depended on knowledge of life histories so that agricultural practices could be used to circumvent high pest population levels. With the development of chemical pesticides and fertilizers came the complacent attitude that these products would solve the world's agricultural problems, and there was no longer a need to know the identity and interactions of the organisms that comprised

[1]Portions of this chapter were previously published in *Bioscience* 45:680-686.

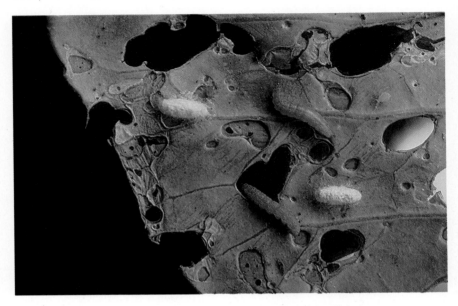

The diamondback moth larva, Plutella xylostella *(L.) (Plutellidae), is a minor pest on cabbage and other plants in the mustard family (Cruciferae).*

agricultural systems. For example, the solution to a problem with the Japanese beetle (*Popillia japonica* Newman) defoliating a vineyard was to spray the grapes with DDT. Solutions to problems with parasitic helminths relied heavily on antihelminthic compounds, with little regard for understanding the identity or bionomics of the organisms involved. In agricultural fields where production was limited by low nitrogen, the solution was to apply high levels of inorganic nitrogen and not to worry about the presence or effects on microorganisms such as mycorrhizal fungi. Rachael Carson's book *Silent Spring* (1962) drew attention to the devastating effects that pesticides could have on the environment. Entomologists, parasitologists, mycologists, and others rapidly became aware of the adaptability of pestiferous species to chemical control agents with their remarkable capacity to develop resistance. Suddenly, it became obvious that, once again, it was important to know the identities of the complex of organisms that occurred, or should occur, in agricultural fields, livestock, orchards, and backyards, and to discover how they interact.

The most important organisms in agricultural systems are the most poorly understood in the context of biodiversity and systematics, including fungi, free-living and plant-parasitic nematodes, other microorganisms, endoparasitic helminths, insects, mites, and other arthropods. Even for many important species of pests, superficial understanding of the group is a major impediment to solving important problems in agriculture. In collaborative research between the

Systematic Entomology Laboratory of the U.S. Department of Agriculture and the Department of Entomology of the University of Maryland, it has been demonstrated that the tobacco budworm complex of pestiferous moths (*Heliothis virescens* complex) encompasses 13 species, not 5 as previously thought (Poole et al., 1993). It is unbelievable that the group containing the second most serious agricultural pest in the Western Hemisphere (Poole et al., 1993) is so poorly known. Although this research summarizes all available information on the species that are represented by specimens in collections, it probably does not encompass even a majority of the species in the complex. Only through comprehensive inventories of the biodiversity of Mexico, Central America, and South America will currently undescribed species become known to science and society. Only then will agricultural scientists know the extent of the threat posed by this group of moth pests and be prepared to develop strategies against potential new introductions.

AGRICULTURAL DILEMMA

The agricultural community now finds itself in a difficult situation. It increasingly espouses environmentally positive approaches such as biological control, sustainable agriculture, and pest management, but lacks a comprehensive understanding of many of even the most common organisms that form the foundation for these approaches. Further, destruction of native ecosystems for new agricultural lands (Myers, 1988) eliminates the organisms upon which agriculture must depend for its future. Ehrlich (1988) points out that it is the "less cuddly" organisms that are the most important for the future of human existence.

In the following sections, we discuss several of the areas of agriculture that are strongly dependent on products from biodiversity and systematics.

Biological Control

Biological control programs mandate detailed knowledge of both the pests to be controlled and the natural enemies that prey on them (Rosen, 1977). Is it any wonder that programs in biological control are difficult to implement when so little is known about the systematics of the organisms involved? In general, the first step is to seek out systematic information known about the target pest and its natural enemies, and, based on this information, to survey the natural enemies in the field.

An example of insufficient systematic knowledge that caused the delay of a biological control program concerns a mealybug that was decimating the coffee plantations of Kenya (Le Pelley, 1968). The mealybug was misidentified for many years as *Planococcus citri* (Risso) and later *P. lilacinus* (Cockerell). Discovery, evaluation, and introduction of the natural enemies of these species had no effect on the pest and was a waste of time and resources. Ultimately, a systematist was asked to study the mealybug, who described it as a new species that formerly was restricted to Uganda. Natural enemies were located in Uganda, introduced to Kenya, and within a few years the pest became rare.

A more recent example concerns whiteflies that occur in the United States and are part of the *Bemisia tabaci* complex. Considerable confusion centers around the status of the sweet potato whitefly or type A, which has occurred in the United States for nearly 100 years, and type B, or the silverleaf whitefly, which apparently has been introduced recently (Perring et al., 1993). There are two camps of thought on the status of the silverleaf taxon; some believe that it is a separate species from the sweet potato whitefly (Perring et al., 1993), and have described it as *B. argentifolii* (Bellows and Perring, 1994), and others contend that it is the same (Campbell et al., 1993). This distinction is important, since it is unclear where the silverleaf whitefly is indigenous and therefore where to search for effective natural enemies. If the *Bemisia* fauna of the world already were known and monographed and the associated natural enemies were studied as part of biological diversity research, then the biological control community could implement its programs with less delay and farmers would not suffer millions of dollars in damage while systematists and biocontrol explorers provide needed information.

A similar dilemma prevails when knowledge of biodiversity in the United States is lacking. Because of insufficient information on North American biota, it is not uncommon for a species of natural enemy to be introduced even though it already occurs here. In insects this happens because only about half of the species that occur in the United States are known (Kosztarab and Schaefer, 1990), and there is no single source or database that provides information on species that compose the insect fauna. In other circumstances, repeated introductions occur because of insufficient systematic knowledge. For example, in the 1970s a parasitic wasp, *Elachertus argissa* (Walker) was introduced from Austria and Italy to control a moth, the larch casebearer (Ryan et al., 1975, 1977). In a revision of *Elachertus*, Schauff (1985) demonstrated that the species already occurred in the United States under the name of a synonym *E. proteoteratis* Howard. He also found that the species parasitizes a wide range of moth hosts and suggested that it was unlikely to be an effective biological control agent.

In the future, biological control will have an increasing importance in pest control. With the reality of bioengineered organisms, it now becomes possible to alter a pest organism so that it is beneficial rather than harmful (Freeman and Rodriguez, 1993). Development of pesticide-resistant natural enemies also adds an innovative new dimension to biological control (Hoy, 1982). Such strategies will be much more effective if they can draw on the genes of the millions of natural enemies that currently exist in the pool of biological diversity, but are unknown to science.

Integrated Pest Management

Integrated pest-management strategies depend heavily on knowing the identity of the organisms that occur in the crop system, understanding the complex interactions among these organisms, introducing biological control agents that are compatible with management schemes, understanding the ecological dynamics of the system, and developing minimally disruptive methods for managing the pests without diminishing the value of the agricultural product. Two examples demonstrate the importance of information on biodiversity in urban pest-management systems in Maryland. *Euonymus japonica* was a common ornamental shrub in most areas of the state, but in recent years it has become less common because it is killed by the euonymus scale when not treated with insecticides. A systematist in Korea and a researcher at the Beltsville Agricultural Research Center discovered a lady beetle that controls the scale (Drea and Carlson, 1987). University of Maryland Extension personnel have conducted further research on the beetle and now use it as part of their pest-management strategies. Results of collaborative research among a systematist, biocontrol specialist, and extension scientist have reduced pesticide usage for euonymus scale control and have provided the means for *E. japonica* to again become common (Davidson, personal communication, 1993).

Another example involves the control of certain moth borers in ornamental trees. Until recently, the only control for these pests has been to drench the bark with a pesticide at the time that the larva bores through the bark. Once the larva successfully penetrates living plant tissue below the bark, chemical control is no longer possible; this leaves only a small window of time when even insecticides can kill these pests. In recent years, considerable research has been conducted on exploiting the biodiversity of entomophagous nematodes as biological control agents in agroecosystems. One of the first major successes was with *Steinernema carpocapsae* in controlling tree borers (Davidson et al., 1992). As part of urban pest-management strategies, these nematodes are sprayed on infested trees whenever damage is detected, and they control tree borers in about 80% of the cases. The discovery and use of this nematode has reduced pesticide usage and enhanced the likelihood of control at many times of the year. One can only wonder what other benefits might be available if the biodiversity and classification of nematodes were more completely understood.

Sustainable Agriculture

Sustainable agriculture depends on detailed systematic knowledge of the organisms that occur in agroecosystems, and attempts to maximize the use of the native biota and natural resources and to minimize the use of nonindigenous organisms and external influences such as inorganic chemicals. Even in areas such as the Beltsville Agricultural Research Center, where agricultural research has been under way for more than 50 years, the native biota is only partially known. An example involves a study on the genus *Trichogramma*, a group of small wasps that primarily parasitize moth eggs, and have been useful in biological control. Over a period of 3 years, unparasitized moth eggs were placed around a field of corn or soybeans to determine the diversity of species of *Trichogramma*. After exposure to the indigenous wasp fauna for 3-5 days, the eggs were brought into the laboratory and parasites were allowed to emerge. Fifteen different species of *Trichogramma* were reared from the moth eggs, including nine that were new to science (Thorpe, 1984). The identity and role of these wasps and other organisms in the agroecosystem are crucial bits of information for successful implementation of sustainable agriculture programs. Even in areas of the United States where the biota is perceived as being well known, the biodiversity of the small and diverse organisms such as small wasps requires considerable study before strategies such as integrated pest management and sustainable agriculture can be truly successful.

Pest Introductions and Quarantine

Estimates of cumulative loss for adventive species in the United States up to 1991 were $96,944,000,000 (Office of Technology Assessment, 1993). This figure apparently does not include the cost of control of pest species, which would more than double the amount. A total of over 4,500 nonindigenous species has been introduced into the United States, and this figure includes only species that have been detected and classified (Office of Technology Assessment, 1993). Clearly, introduction of unwanted species into the United States is a major problem! It is the difficult task of the Animal and Plant Health Inspection Service (APHIS), of the U.S. Department of Agriculture to exclude these organisms. Unfortunately, the knowledge base and tools to assess biodiversity that this organization needs to make better-informed decisions are not always available. APHIS generally does not allow plant-associated organisms identified as pest risks into the country if they currently do not occur here. With few exceptions, the general corollary is: if an organism already occurs in the United States, no action will be taken against it, even if it is intercepted at a port of entry. The supposition then is that APHIS knows the identity of all of the organisms that occur in the United States. Furthermore, APHIS, or its cooperators, must be able to identify all organisms that are intercepted at ports of entry. APHIS also has the responsibility of determining if organisms unwanted by other countries occur in areas in the United States that export agricultural products to that country. Clearly, the information required by APHIS is derived from systematic research and mandates a world-wide knowledge of the Earth's agriculturally important biodiversity.

Some examples of the impact of systematic knowledge on APHIS decisions are as follows. A smut associated with wheat imported into Canada from the

European corn borer larvae, Ostrinia nubilialis, *Hübner (Crambidae), a species introduced from Europe, are a major pest of corn in the United States.*

United States was initially determined to be *Tilletia indica*, a high-risk exotic pathogen. A ban on the import of wheat from the United States to Canada was suggested. A systematist was asked to study the material and correctly identified the smut as *T. barclayana*. This species occurs on rice and apparently became a contaminant on the wheat in question when it was stored in a warehouse that previously contained rice. This identification was confirmed with isozyme analysis and a potentially costly international incident was averted (Palm, personal communication, 1993). A second example involves a nematode, *Nematodirus battus*, which was first discovered in North America and the Western Hemisphere in 1985 (Hoberg et al., 1985). Since its original description from sheep in Great Britain, this nematode has been one of the most pathogenic parasites of lambs in Great Britian and Europe in spite of concerted control efforts (Rickard et al., 1989). Consequently, in recognition of the potential economic impact related to this parasite, *N. battus* was the only nematode of sheep rated by APHIS as a serious foreign agent of animal disease. The parasite is cryptic, and because of an unusual life cycle, outbreaks of disease generally occur only after populations of high density have developed over several years (Rickard et al., 1989). Accurate identification is confounded because disease is not associated with adult parasites, and both parasitic larvae and adults are morphologically similar to other less pathogenic species. However, due to the efforts of systematists in

Oregon and Maryland, populations of this nematode were detected soon after first introduction into North America (Hoberg et al., 1986). Surveillance programs, measures of control, and extensive epidemiological studies were initiated to limit the potential impact of this newly introduced nematode (Hoberg et al., 1986; Rickard et al., 1989). Aside from this highly visible parasitic helminth, there has been a history of introductions of nematode parasites with exotic bovid hosts from Africa, camelids from South America, and cervids from Europe (e.g., Rickard et al., 1993). Many of these parasites represent potential threats to wild and domestic ruminants in North America, further highlighting the importance of systematics within the context of cosmopolitan groups, and the necessity to develop accurate inventories of the world fauna.

Plant Germplasm

In the area of plant germplasm, botanists have come a long way in expanding science's understanding of species of crops and the world's vascular plant flora, but, even in this relatively well-known group of organisms, much remains to be done. The now famous examples of discoveries by plant systematists of important but nondescript species of tomatoes and corn document the mind-boggling potential that plants hold for the future of agricultural crops. Millions of dollars have been added to the tomato industry through increased levels of soluble solids derived from an inconspicuous species of Andean tomato, and disease resistance has been added to the genome of cultivated corn from a nearly extinct species discovered in Mexico (Iltis, 1988). With recent developments in genetic engineering, nearly any species of plant has the potential to contribute genes of importance in enhancing agricultural crops. Ethnobotanical data suggest that of the approximately 250,000 species of plants (Wilson, 1988), more than 7,000 species have been used for human benefit as food (Ehrlich and Wilson, 1991). Understanding the diversity and systematics of these plants will add significantly to the goal of developing a much broader spectrum of agricultural products and will enhance sustainable approaches without destroying forests and more natural environments.

THE BRUSH-FIRE APPROACH TO AGRICULTURAL SYSTEMATICS

Under current conditions of small budgets and reduced personnel, resources devoted to systematics and biodiversity are woefully small and stretched to the limit (House of Lords, 1991). The impact is that each of the small number of agricultural systematists must serve many roles, including those of researchers, identifiers, curators, database managers, illustrators, and technicians. The best strategy to support the biodiversity and systematic needs of agriculture is to direct resources toward the groups of organisms that are most likely to cause future agricultural problems or toward the beneficial groups that show promise of enhancing agricultural systems, and undertake monographic research on these groups. Unfortunately, there rarely is time or resources to take a coordinated approach, and this leaves agriculture with a systematic support system that is patched together with chewing gum and rusted wire. In most instances, agricultural problems are solved using the brush-fire approach of reacting when an emergency arises, undertaking a narrow research program on the pest causing the problem, hoping to find a solution to the immediate problem, and going on to the next agricultural brush fire. This approach does not solve problems of the future and does not add significantly to the development of a predictive classification system. In fact, in many situations, systematic analysis of a single species removed from the context of phylogeny and biogeography and not carefully integrated into the classification system detracts rather than adds useful information.

An example of a circumstance where comprehensive research followed a brush-fire solution involved a pest in Africa. An unknown mealybug was attacking cassava in West Africa and was costing farmers $1.4 billion each year (Anonymous, 1986). A systematist described the species (*Phenacoccus manihoti*) as new and suggested that natural enemy exploration be undertaken in Central or South America (Matile-Ferrero, 1977). A mealybug identified as *P. manihoti* was discovered in northern South America and several of its parasites were imported to Africa. Unfortunately, none of the biological control agents were effective, and a mealybug systematist was asked to study the South American material. The systematist determined that the mealybug from northern South America actually was a second species different from *P. manihoti* (Cox and Williams, 1981). True *P. manihoti* was located further south and effective parasites were discovered and successfully introduced into West Africa (Herren and Neuenschwander, 1991). After this brush fire was put out, the International Fund for Agricultural Development offered substantial financial support for a study on the mealybugs of South America. A recently published book (Williams and Grana de Willink, 1992) serves as a first step towards understanding the diverse mealybug fauna of the area and partially prepares the world for the emergence of the next South American mealybug pest.

SYSTEMATICS: THE PREDICTOR OF BIODIVERSITY

Classification systems serve as the knowledge base from which predictions about organisms are made. Too frequently this predictability is taken for granted as an innate capacity of the human mind, but in fact it is part of the classification process. A simple example involves a person who sees a hornet sitting on someone's arm. The prediction is that the hornet will sting the arm if disturbed or crushed. Such a prediction could be made in any part of the world where hornets occur. As long as the hornet is a female, the prediction will have a high level of predictive accuracy. Similar, but more sophisticated predictions can be made from other classification systems. For example, recent research on the *Trichoderma*-like fungi used in biological control has demonstrated that a species previously placed in *Gliocladium* (*G. virens*) shares many characters with the genus *Trichoderma* and should be placed in the latter genus. This is important because many species of *Trichoderma* have mycotoxins that are useful for the biological control of plant-pathogenic ascomycetes, rather than basidiomycetes, as is the case for true species of *Gliocladium* (Rehner and Samuels, 1994). The sexual state of *T. virens* is *Hypocrea gelatinosa* or a closely related species, and it is predicted that other closely related sexual states of *Hypocrea* will prove to be valuable agents of biological control similar to *T. virens*. Scientists already have found high levels of mycotoxins in isolates of *Hypocrea*.

A second example comes from research carried out by Farrell and Mitter (1990) at the University of Maryland. They hypothesized that the leaf beetles in the genus *Phyllobrotica* and their *Scutellaria* plant hosts coevolved, and they supported this hypothesis by showing close congruence between the phylogenies of the beetle and host. One species of beetle had no information on host association, so Farrell and Mitter predicted the host species based on the phylogenies of the beetles and the plant; this prediction proved to be correct.

Another example involves an anticancer compound that was screened from extracts of the plant *Maytenus buchananii* collected from a small population in Kenya. Because the species would be eliminated if further collections were made, a systematist was consulted to provide information on related species. Based on the classification system of the genus, the systematist predicted that *M. rothiana* in India would most likely have the desired compound. This prediction proved to be correct (Shands and Kirkbride, 1989).

Specialists in biological control often ask systematists where to look for natural enemies and which species are most closely related to the pest that is the subject of the project. For example, the sugar beet leafhopper, *Circulifer tenellus*, was originally placed in the genus *Eutettix*, which occurs in South America. Exploration for natural enemies in South America was unsuccessful. A systematist examined the species and placed it in *Circulifer*, which is of Old World origin. Natural enemies subsequently were located in the Mediterranean area (Rosen, 1977).

THE SOLUTION

The question is where do we go from here? How can we describe and classify the components of Earth's biodiversity and find ways to preserve, enjoy, and use it sustainably? An initiative from the systematics community called *Systematics Agenda 2000: Charting the Biosphere* (Systematics Agenda 2000, 1994:1) proposes "to discover, describe, and classify the world's species." If we can walk on the moon or search for life in outer space, we can fully explore life on Earth. We know that life occurs here, but we have only an inkling of its diversity, grandeur, and wonderment. The benefits to humanity will be enormous. Imagine if all natural enemies of the silverleaf whitefly were known, a biological control program could be implemented more quickly and millions of dollars might be saved. APHIS would have information on all of the organisms that occur in the United States and could make better-informed quarantine decisions at ports of entry. Species of *Trichogramma* would be classified, and sustainable farming would benefit from systematic knowledge of all organisms that interact to maintain and stabilize the system. Crop plants would be enhanced by genes from plants that currently are unknown or are poorly known. High-risk adventive species would be well known and their distribution would be monitored so that some of the billions of dollars that might be spent on these organisms in the future could be saved.

Other chapters of this volume discuss the significant progress that systematics has made in the recent past. Some larger-sized organisms are well known, and even groups of smaller-sized organisms have a knowledge base many times larger than that which existed at the turn of the century. But progress is too slow if the currently suggested rates of extinction are correct. If we are to understand and benefit from major parts of the extant biota, we must make an all-out effort now. Through computer technology, imaging systems, morphometric analysis, molecular techniques, and recent theoretical innovations, the systematics community is poised and capable of carrying out the *Systematics Agenda 2000* initiative. Armed with comprehensive knowledge of the components of Earth's biological resources, we will be better able to maintain high levels of agricultural productivity and reduce environmental impact so that future generations can continue to enjoy the mysteries and aesthetics of complex biological systems.

REFERENCES

Anonymous. 1986. The 1 billion a year mealybug brings in more projects. CAB Int. News (October):6-7.

Bellows, T. S., Jr., and T. M. Perring. 1994. Description of a new species of *Bemisia*. Ann. Entomol. Soc. Amer. 87:195-206.

Campbell, B. C., J. E. Duffus, and P. Baumann. 1993. Detemining whitefly species. Science 261:1333.

Carson, R. 1962. Silent Spring. Houghton Mufflin, Boston. 368 pp.

Cox, J. M., and D. J. Williams. 1981. An account of cassava mealybugs with a description of a new species. Bull. Entomol. Res. 71:247-258.

Davidson, J. A., S. Gill, and M. J. Raupp. 1992. Controlling clearwing moths with entomopathogenic nematodes: The dogwood borer case study. J. Arboriculture 18:81-84.

Drea, J. J., and R. W. Carlson. 1987. The establishment of *Chilocorus kuwanae* in eastern United States. Proc. Entomol. Soc. Wash. 84:821-824.

Ehrlich, P. R. 1988. The loss of diversity: Causes and consequences. Pp. 21-27 in E. O. Wilson and F. M. Peter, eds., BioDiversity. National Academy Press, Washington, D.C.

Ehrlich, P. R., and E. O. Wilson. 1991. Biodiversity studies: Science and policy. Science 253:758-762.

Farrell, B. D., and C. Mitter. 1990. Phylogenies of insect/plant interactions: Have *Phyllobrotica* leaf-beetles and the Lamiales diversified in parallel? Evolution 44:1389-1403.

Freeman, S., and R. J. Rodriguez. 1993. Genetic conversion of a fungal plant pathogen to a non-pathogenic, endophytic mutualist. Science 260:75-78.

Herren, H. R., and P. Neuenschwander. 1991. Biological control of cassava pests in Africa. Ann. Rev. Entomol. 36:257-283.

Hoberg, E. P., G. L. Zimmerman, and L. G. Rickard. 1985. *Nematodirus battus*: A review of recent studies relative to the development of surveillance and control programs. Proc. Eighty-ninth Ann. Meeting U.S. Animal Health Assoc. 1985:420-431.

Hoberg, E. P., G. L. Zimmerman, and J. R. Lichtenfels. 1986. First report of *Nematodirus battus* in North America: Redescription and comparison to other species. Proc. Helminthol. Soc. Wash. 53:80-88.

House of Lords. 1991. Systematic biology research. Select Committee on Science and Technology. HL Paper 22-I. HMSO, London. 234 pp.

Hoy, M. A. 1982. Genetics and genetic improvement of the Phytoseiidae. Pp. 72-89 in M. A. Hoy, ed., Recent Advances in Knowledge of the Phytoseiidae. Division of Agricultural Science, Publ. 3284, University of California, Berkeley. 92 pp.

Iltis, H. H. 1988. Serendipity in the exploration of biodiversity: What good are weedy tomatoes? Pp. 98-105 in E. O. Wilson and F. M. Peter, eds., BioDiversity. National Academy Press, Washington, D.C.

Kosztarab, M., and C. W. Schaefer. 1990. Conclusions. Pp. 241-247 in M. Kosztarab and C. W. Schaefer, eds., Systematics of the North American Insects and Arachnids: Status and Needs. Information Series 90-1, Virginia Agricultural Experiment Station, Blacksburg. 247 pp.

Le Pelley, R. H. 1968. Pests of Coffee. Longmans, Green and Company, Ltd., London. 590 pp.

Matile-Ferrero, D. 1977. Une cochenille nouvelle nuisible au manioc en Afrique équatoriale, *Phenacoccus manihoti* n. sp. Ann. Soc. Entomol. Fr. (new series) 13:145-152.

Myers, N. 1988. Tropical forests and their species: Going, going...? Pp. 28-35 in E. O. Wilson and F. M. Peter, eds., BioDiversity. National Academy Press, Washington, D.C.

Office of Technology Assessment. 1993. Harmful Non-indigenous Species in the United States. OTA-F-565. U.S. Government Printing Office, Washington, D.C. 391 pp.

Perring, T. M., A. D. Cooper, R. J. Rodriguez, C. A. Farrar, and T. S. Bellows, Jr. 1993. Identification of a whitefly species by genomic and behavioral studies. Science 259:74-77.

Poole, R. W., C. Mitter, and M. Huettel. 1993. A revision and cladistic analysis of the *Heliothis virescens* species group with a preliminary morphometric analysis of *Heliothis virescens*. Department of Information Services, No. 4, Mississippi Entomological Museum. 51 pp.

Rehner, S. A., and G. J. Samuels. 1994. Taxonomy and phylogeny of *Gliocladium* analysed from nuclear large subunit DNA sequences. Mycol. Res. 98:625-634.

Rickard, L. G., E. P. Hoberg, J. K. Bishop, and G. L. Zimmerman. 1989. Epizootiology of *Nematodirus battus*, *N. filicollis*, and *N. spathiger* in Western Oregon. Proc. Helminthol. Soc. Wash. 56:104-115.

Rickard, L. G., E. P. Hoberg, N. M. Allen, G. L. Zimmerman, and T. M. Craig. 1993. *Spiculopteragia spiculoptera* and *S. asymmetrica* from red deer (*Cervus elaphus*) in Texas. J. Wildlife Dis. 29:512-515.

Rosen, D. 1977. The importance of cryptic species and specific identifications as related to biological control. Pp. 23-35 in J. A. Romberger, Biosystematics in Agriculture. Beltsville Symposium on Agricultural Research. Allanheld, Osmun and Co., Montclair, N.J. 340 pp.

Ryan, R. B., W. E. Bousfield, R. E. Denton, R. L. Johnsey, L. F. Pettinger, and R. F. Schmidtz. 1975. Additional releases of larch casebearer parasites for biological control in the western United States. U.S.D.A. Forest Service Research Note PNW-242, Pacific Northwest Forest Range Experiment Station, Portland, Oreg. 7 pp.

Ryan, R. B., W. E. Bousfield, R. E. Denton, R. L. Johnsey, L. F. Pettinger, and R. F. Schmidtz. 1977. Releases of recently imported larch casebearer parasites for biological control in the western

United States, including *Agathis pumila*. U.S.D.A. Forest Service Research Note PNW-290, Pacific Northwest Forest Range Experiment Station, Portland, Oreg. 4 pp.

Schauff, M. E. 1985. Taxonomic study of the Nearctic species of *Elachertus* Spinola. Proc. Entomol. Soc. Wash. 87:843-858.

Shands, H. L., and J. H. Kirkbride. 1989. Systematic botany in support of Agriculture. Symbolae Botanicae Upsalienses 28:48-54.

Systematics Agenda 2000. 1994. Systematics Agenda 2000: Charting the Biosphere. Technical Report. Systematics Agenda 2000, a Consortium of the American Society of Plant Taxonomists, the Society of Systematic Biologists, and the Willi Hennig Society, in cooperation with the Association of Systematics Collections, N.Y. 34 pp.

Thorpe, K. W. 1984. Seasonal distribution of *Trichogramma* species associated with a Maryland soybean field. Envir. Entomol. 13:127-132.

Williams, D. J., and M. C. Grana de Willink. 1992. Mealybugs of Central and South America. CAB International and Cambridge University Press, Cambridge, England. 635 pp.

Wilson, E. O. 1988. The current state of biodiversity. Pp. 3-18 in E. O. Wilson and F. M. Peter, eds., BioDiversity. National Academy Press, Washington, D.C.

CHAPTER
16

Snout Moths: Unraveling the Taxonomic Diversity of a Speciose Group in the Neotropics

M. ALMA SOLIS

Research Scientist, Systematic Entomology Laboratory,
U.S. Department of Agriculture, Washington, D.C.

Biodiversity studies often ignore the insects, one of the most important and diverse groups of organisms on this planet. With estimates of 5-30 million species of insects, the simple request to provide a scientific name can be nearly impossible for insect taxonomists, in strong contrast with the capability of vertebrate taxonomists who work with smaller and better-known taxa. Within the insects, butterflies have been regarded as good subjects for biodiversity studies because there are relatively fewer species and they are taxonomically well known. Butterflies constitute about 15% of the 140,000+ species of Lepidoptera (approximated from Heppner, 1991); the rest are moths. However, it would be a mistake to utilize only less diverse groups in biodiversity studies, because the more speciose groups such as moths may have a greater impact on the Earth's sustainable resources. Moths have many positive attributes for biodiversity studies: they are found in almost all habitats and niches, possess many specialized behaviors, are good indicators of areas of endemism, show rapid responses to environmental disturbance, can be sampled easily with quantitative methods, and have many taxa that are readily identifiable (Miller and Holloway, 1991).

The Pyraloidea, or snout moths, are endowed with all the attributes for investigations in biodiversity. Although they contain over 16,000 described species (Table 16-1) and at least 16,000 remain to be described (Munroe, personal communication, 1994), a project to study the taxonomic diversity of a smaller subgroup within the pyraloids is possible and will advance our knowledge and communication about this economically important group of moths. The Pyraloidea, like every group of organisms, has had many unique biological and historical influences that affect its present taxonomic status. This is a spe-

TABLE 16-1 Subfamilies of the Pyraloidea and Numbers of Species for the Western Hemisphere

Subfamilies	World Total[a]	Neotropical[b]	Costa Rica
Pyralidae			
Pyralinae	900[c]	23	2
Chrysauginae	437	378	48
Epipaschiinae	572[d]	282	65
Galleriinae	261	43	3
Phycitinae	2,629	341	SPS
Total	4,927	1,105	118
Crambidae			
Midilinae	47[e]	47[e]	8
Linostinae	3	3	1
Musotiminae	166	79	16
Scopariinae	479	30	26
Nymphulinae	716	218	40
Odontiinae	367	84[e]	17
Glaphyriinae	200	161[e]	30
Evergestinae	137	24	3
Pyraustinae	7,381	1,450	363
Schoenobiinae	169	72	8
Cybalomiinae	55[e]	9	0
Crambinae	1,877	433	SPS
Wurthiinae	7	0	0
Noordinae	6	0	0
Cathariinae	1	0	0
Total	11,611	2,534	512
Total for Pyralidae and Crambidae	16,538	3,639	630

[a]Heppner (1991), unless otherwise indicated.
[b]Munroe et al. (1995).
[c]Solis and Shaffer (manuscript).
[d]Solis (1992, 1993).
[e]Munroe and Solis (in press).

NOTE: SPS=still at preliminary stage.

cific report on the initiation, management strategies, and progress of a study on the taxonomic diversity of the Pyraloidea of Costa Rica.

SNOUT MOTHS AND BIODIVERSITY

Snout moths are primarily tropical in distribution, but occur world-wide, including most oceanic islands. They are known from low and middle elevations in the tropics to the high arctic. While many pyraloids may have broad continental and even world-wide distributions, there are many subgroups that are endemic to islands and continental areas and that contain examples of extreme disjunction (Clarke, 1986; Munroe and Mutuura, 1971; Zimmerman, 1958). Neotropical pyraloids comprise about 20% of the world fauna

Colorful snout moth larva feeding on host plant.

(Table 16-1). The largest subfamily is the Pyraustinae, and to date 363 species have been delineated in Costa Rica. There are a few small subfamilies found only within the Western Hemisphere, such as the Midilinae and Linostinae. Others, such as the Chrysauginae and Glaphyriinae, are found primarily in the Neotropics and have only a few species in the Neartic, Oriental, or Palearctic. All but three small subfamilies of the Pyraloidea occur in Costa Rica.

Pyraloid caterpillars have diverse habits. They consume dried or decaying plant or animal matter, wax in bee and wasp nests, and living plants. Some are known to be inquilines in ant nests (some Galleriinae), predators of scale insects (some Phycitinae), and aquatic scavengers in flowing water (some Nymphulinae). The Nymphulinae, with over 700 species, is the only taxon in Lepidoptera with truly aquatic caterpillars. Snout moth caterpillars are associated with non-vascular and vascular plants, feeding on both gymnosperms and angiosperms. They demonstrate plant specificity at different taxonomic levels. For example, the Midilinae are known to bore in the plant family Araceae, and the genus *Diaphania* Stephens is known to feed only on Cucurbitaceae. Snout moth caterpillars are intimately associated with their plant hosts either as external or concealed feeders. They have developed a broad array of concealed feeding strategies: folding, rolling, webbing or tying of leaves; making tunnels or tubes of silk or frass; mining leaves; and boring into stems, roots, buds, and fruits.

Because of the close association of the immature stages with plants, moths in general have shown a rapid response to environmental disturbance (Holloway and Barlow, 1992). Changes in plant communities can be monitored easily and studied efficiently because adult moths can be sampled in a qualitative and quantitative manner with lights, such as mercury-vapor lamps placed in front of a white sheet or a Robinson-pattern light trap (Barlow and Woiwood, 1990; Holloway et al., 1990; Robinson and Tuck, 1993). Although the Pyraloidea have been included in many qualitative diversity studies (i.e., a list of snout moths from a particular area), they have been included only recently in quantitative studies. Barlow and Woiwod (1990) utilized pyraloids to compare the seasonal and overall diversity of two lowland sites in Sulawesi. Robinson and Tuck (1993) compared the diversity of primitive moths, specifically delineating subfamilies in the Pyraloidea, in three vegetational areas of Borneo.

Pyraloids can be used in diversity studies because many are large enough to be sorted visually to species and are readily identifiable using collections such as the Natural History Museum in London or the National Museum of Natural History in Washington, D.C. The Pyraloidea consist of two families. The Pyralidae are the smaller group, with only about 5,000 species and 5 subfamilies, and the Crambidae are the larger group, with over 11,000 species and 15 subfamilies (Table 16-1). Some groups of pyraloids are brightly colored, large in individual size, small in numbers of species, and have been well collected. The Midilinae (Munroe, 1970) and *Cliniodes* Guenee (Odontiinae) (Munroe, 1964) are over 2.5 cm in wing length; have distinctive wing patterns and colors, such as white, orange, and yellow; and have had a history of taxonomic work and illustration. In contrast, many pyraloid groups are not visually pleasing and are small in individual size and large in numbers of species. For example, the Glaphyriinae and Scopariinae are small in size and have a homogeneous color pattern; the Pyralinae in Africa and Nymphulinae in southeast Asia are speciose in poorly collected areas. Nevertheless, some pyraloid groups have been treated recently in revisionary or phylogenetic studies, such as the Phycitinae (Shaffer, 1976) and Epipaschiinae (Solis, 1993), or in geographic studies, such as the Phycitinae (Heinrich, 1956; Neunzig, 1986; 1990), Pyraustinae and related subfamilies (e.g., Munroe, 1972, 1973a, 1976).

SNOUT MOTHS AND AGRICULTURAL DIVERSITY

Many pyraloids are pests to a wide array of crops and stored products. Some pyraloids rank among the most destructive pests to graminaceous crops such as corn, sugarcane, and rice. Table 16-2 illustrates the diversity of commodities affected by pyraloids and lists species that have been targets of concentrated world-wide efforts of control with parasites and predators. Although not listed, each genus may include closely related species that are also pests. For example, *Chilo* Zincken includes 41 species in the world, and although *C. suppressalis* (Walker) is one of the worst pests on rice, there are other species of *Chilo* that are pests: two on rice, five on sugarcane, and five on corn (Bleszynski, 1970). Hundreds more pyraloids are minor pests to many major fruits and vegetables (e.g., tomatoes, beans, squash, cabbage, celery) and of forest trees, both conifers and hardwoods, where they feed on foliage and fruits. Members of *Etiella* Zeller, *Cadra* Walker, and *Plodia* Guenée are major pests of stored products (e.g., grains, dried fruit, nuts) and have been dispersed world-wide.

Although the impact of snout moths on the human condition is based on their adverse effect, it is these same characteristics that have prompted testing of species of pyraloids for the biocontrol of noxious weeds. The most striking

TABLE 16-2   Commodities and Pyraloid Targets of Major Biocontrol Efforts

Commodity	Pest
Rice	*Chilo suppressalis* (Walker)
	Scirpophaga incertulas (Walker)
Sugarcane	*Diatraea saccharalis* (Fabricius)
	Hedylepta accepta (Butler)
Corn	*Ostrinia nubilalis* (Hübner)
Coconut	*Tirathaba complexa* (Butler)
Lima beans	*Etiella zinckenella* (Treitschke)
Oranges	*Amyelois transitella* (Walker)
Teak	*Hapalia machaeralis* (Walker)
Karoo	*Loxostege frustralis* Zeller
Banana	*Nacoleia octasema* (Meyrick)

SOURCE: Clausen (1978).

success story is *Cactoblastis cactorum* (Berg)(Phycitinae). It was introduced into many countries for the control of *Opuntia* Mill, the prickly pear cactus. In 1787, *Opuntia* was introduced into Australia and, by 1925, it had infested more than 24,000,000 hectares. But by 1934, after introduction of *C. cactorum*, 90% of the *Opuntia* stands were eliminated and the land restored for agricultural use (Clausen, 1978). Some pyraloids, including species in the Nymphulinae with aquatic caterpillars, have been investigated for the control of aquatic weeds such as water hyacinth, hydrilla, and alligatorweed (Habeck and Allen, 1974).

The study of the taxonomic diversity of pest species and their relatives is important for biocontrol studies, primarily for the importation of effective parasites and predators, and in the case of noxious weeds, for the diversity of insects feeding on the plants for experimentation as control agents. As a result of their notoriety as pests and control agents, some pyraloids have been well studied taxonomically, for example in the Schoenobiinae (Common, 1960; Heinrich, 1938; Lewvanich, 1981), Crambinae (Bleszynski, 1970; Dyar and Heinrich, 1927), and Pyraustinae (Munroe, 1973b; Mutuura and Munroe, 1970). In addition there have been studies of snout moth guilds on economically important host plants such as legumes (Neunzig, 1979), cacti (Heinrich, 1939), and aquatic plants (Center et al., 1982).

BIODIVERSITY IN COSTA RICA: AN OPPORTUNITY

In 1989, I initiated a project to study the taxonomic diversity of the Pyraloidea of Costa Rica. A combination of the following historical and ongoing events allowed a taxonomic diversity project in Central America, and specifically in Costa Rica: creation of the Instituto Nacional de Biodiversidad (INBio) in Costa Rica, organization of the largest neotropical pyraloid collection in the world at the National Museum of Natural History, a group of pyraloid taxonomists interested in biodiversity research, and publication of a checklist of snout moths for the entire neotropical region by Munroe et al. (1995). Two prerequisites for a biodiversity study of an area are thorough collections from the study area and detailed taxonomic research. Large collections with long series of specimens from the entire range of the species are absolutely necessary to study variation and permit accurate identification of the species. Taxonomic research demands knowing the names and identities of all the species that have been described of a particular group from the area. This list, in turn, allows the taxonomist to solve basic problems such as synonymy and recognition of new species.

The decision to undertake a study of the diversity of neotropical snout moths with Costa Rica as a taxonomic center was fueled in part by the creation of INBio (Gamez et al., 1993), because it fulfills one of two important criteria, that is, thorough collections in the area of study. The Institute was created to document the country's flora and fauna in a creative and organized way (Janzen et al., 1993). INBio, through the efforts of parataxonomists, has extensive collections of Pyraloidea from national parks and conservation areas in Costa Rica. Parataxonomists are Costa Ricans trained by INBio to collect and prepare biological organisms near their homes, and they are providing a thorough survey of Costa Rica in a relatively short period of time. The parataxonomists deposit the material at INBio where moth specimens are first labeled: a traditional insect label and a bar-code label. The bar-code label represents information about locality, collector, date of collection, and a unique number for each specimen. The label is read by a bar-code laser light and placed into a computer database which is then made available to taxonomists working on the material. The inventory manager of pyraloids at INBio sorts, organizes, sends material to taxonomists, and enters information (e.g., the scientific name determined by a taxonomist) into a database (Figure 16-1). Another important consideration with respect to collections is the Pyraloidea collection at the National Museum of Natural History. Although it is the largest neotropical pyraloid collection in the world, in 1989 over two-thirds was unsorted material and the remaining one-third was based on nineteenth century taxonomy. Over the past 4 years, a massive sorting and reorganization of the snout moth collection has made the material more accessible for taxonomic research and general retrieval of data.

Munroe (personal communication, 1990) estimated the total snout moth fauna in Costa Rica to be 2,000 species. To date, over 600 species from Costa Rica (Table 16-1) have been identified. Because of the size of the group, it was necessary to make decisions concerning the type of research required to answer specific taxonomic problems in each of the 17 subfamilies occurring in Costa

LOCALE	PERSONNEL	PRODUCT

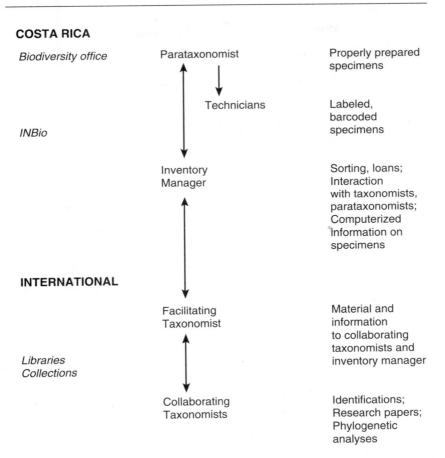

COSTA RICA

Biodiversity office   Parataxonomist   Properly prepared specimens

Technicians   Labeled, barcoded specimens

INBio

Inventory Manager   Sorting, loans; Interaction with taxonomists, parataxonomists; Computerized information on specimens

INTERNATIONAL

Facilitating Taxonomist   Material and information to collaborating taxonomists and inventory manager

Libraries Collections

Collaborating Taxonomists   Identifications; Research papers; Phylogenetic analyses

FIGURE 16-1 Simplified flow chart of material and information between Costa Rica and the international taxonomic community.

Rica (Table 16-1). Some subfamilies (e.g., Glaphyriinae) or certain genera (e.g., *Neurophyseta* Hampson in Musotiminae) with moths that are individually very small, and where very few species have been described from Central America, were targeted for preliminary taxonomic research papers primarily to describe the fauna. This approach has been found to be fruitful. For example, the Glaphyriinae of Costa Rica comprise about 19% of the total neotropical glaphyriine fauna. Approximately 27% of the glaphyriine species in Costa Rica are new; specifically, 22 species had been described previously and 8 are new spe-

Snout moths are a diverse, economically important group.

cies. While this proportion of new species may seem high, it was expected because the moths are small and had not been well collected. Although the work continues in all subfamilies, the Phycitinae and Crambinae are the only two speciose groups that are still in the preliminary stage of identification. The Pyraustinae, one of the largest subfamilies with about 1,500 species in the Neotropics and over 350 in Costa Rica, are still a major area of research because they have a number of genera, such as *Omiodes* Guenée, which are riddled with problems of synonymy and species complexes.

The goal of this project is to produce a handbook for the identification of pyraloid moths in Costa Rica. But to solve many of the taxonomic problems, such as synonymies, complexes, and discovery and description of new species, research papers prior to the handbook would be more efficient, appropriate, and useful. The first step in a study of taxonomic diversity is to know the names of all the species that have been described from the area. In the field of taxonomy, this is known as a checklist or a catalog derived from the literature. To be taxonomically informative and useful for future systematic studies, a list of the species should be based on a study with world-wide scope. This is especially true for snout moths because many genera are pantropical. Fortunately, such a study has been conducted by Munroe et al. (1995) for the neotropical region, based on world-wide generic concepts; this study provides the basis for the study of a geographical subset within the neotropical region. A checklist is just that, a list of names. What a taxonomist really needs is the identity, or morphological criteria, for assigning a name to a particular species of snout moth. Technically, this and many other taxonomic problems are solved by studying and dissecting type specimens, a single specimen on which a species name and its description is based. Dissection and comparison of Costa Rican species to type specimens help to maintain nomenclatural stability and enhance world-wide communication. Type specimens of species usually are deposited by the taxonomist who described them in collections of museums, such as the National Museum of Natural History in Washington, D.C., and the the Natural History Museum in London. Although both are large repositories for type specimens of the Pyraloidea, type specimens also can be found in many other museums throughout the world.

A checklist also provides the basis for efficiently solving the common problems of classification and indirectly aids in the discovery and description of new species. In any group that has not been revised recently, there are usually a large number of synonyms, i.e., many names for the same species described from different countries over the years by different people. For example, the same species could have been described by Möschler from Puerto Rico, by Dyar from Panama, by Schaus from Costa Rica, by Druce from Mexico, by Amsel from Venezuela, and by Dognin from Ecuador. It was common practice in the last century to describe new species from different countries without communicating with museums, other workers, or the literature to determine if the species previously had been described. Another problem is that of a species complex or one name for what is really many species. This occurs usually because the moths externally look the same and no one has studied them in detail over their entire distribution. Discovery of new species is an automatic byproduct of studying the entire fauna taxonomically, not regionally. If the identity of all previously described species is known, a species that is undescribed will be recognized as such. The actual description of all new species is not necessary or feasible at this point, and the decision of which new species will be described is based on a biological need to unravel complexes of species and to provide information about groups or taxa that have been historically underrepresented in the taxonomic literature.

BENEFITS OF RESEARCH ON PYRALOIDEAN BIODIVERSITY

The ultimate benefit of the Costa Rican project is transfer of information within and from the field of taxonomy to all levels of society via several different vehicles such as literature and education. The project will summarize information held in libraries, collections, and by taxonomists and place this knowledge at the fingertips of ecologists, conservationists, quarantine officers, and farmers in Central America. It will also be the first step to "taxonomic self-sufficiency" of pyraloid moths for Central American countries (Janzen et al., 1993). This self-sufficiency will be accomplished by producing accurate scientific names and detailed characters that allow individuals to identify and communicate specifically and comparatively about the diversity of organisms on a global basis. The most tangible product that will reach a diverse audience is a handbook for the identification of pyraloid moths in Costa Rica with photographs, the correct scientific name, distribution, and taxonomic and biological notes on each species. Analysis of the distribution of Costa Rican species throughout the Western Hemisphere also will be included and will form the foundation for other kinds of studies in other areas of Central America.

Transfer of information via education has occurred at all levels of this project. In 1990, a group of parataxonomists participated in a month-long training course on how to collect, identify, and prepare pyraloids. The inventory manager at INBio is encouraged and able to learn via research projects with the taxonomists and serves as a link between the parataxonomist and taxonomist, as well as with a wide variety of other users of such information in Costa Rica. Two others working on this project have Ph.D.s in systematics, but have little or no experience working with Pyraloidea, and are learning about pyraloids as they conduct collaborative research. Education and hands-on experience, from the parataxonomists to Ph.D.s in systematics, have been important components in the transfer of information and have made it possible for the project to progress more rapidly than if a single taxonomist proceeded alone to conduct a diversity study of a speciose group of moths.

CONCLUSION

The keys to the transfer of information about a speciose group of insects in a short period of time (e.g., 10 versus 30 years) are collections and taxonomists. The Costa Rican project became a reality through the study, development, and organization of pyraloid collections throughout the world. In turn, the Pyraloidea collections at the National Museum of Natural History in Washington, D.C., and INBio will become more accessible and useful for phylogenetic research and other biological studies. In addition, all other collections throughout the world involved in this project will have been improved. Taxonomic refinement and computerization of pyraloid collections will provide an impetus for studies of diversity and faunistic composition of snout moths in other tropical areas of Central America. Finally, the most difficult aspect, but the most important, has been the transfer of information from taxonomists (unaffiliated or affiliated in some other capacity with other organizations) who have volunteered their time, and from the past work of taxonomists that we now build on and use to answer some of our present questions.

REFERENCES

Barlow, H. S., and I. P. Woiwood. 1990. Seasonality and diversity of Macrolepidoptera in two lowland sites in the Dumoga-Bone National Park, Sulawesi Utara. Pp. 167-172 in W. J. Knight and J. D. Holloway, eds., Insects and the Rain Forests of South East Asia (Wallacea). The Royal Entomological Society of London, London.

Bleszynski, S. 1970. A revision of the world species of *Chilo* Zincken (Lepidoptera: Pyralidae). Bull. Brit. Mus. (Nat. Hist.) Entomol. 25(4):101-195.

Center, T. D., J. K. Balciunas, and D. H. Habeck. 1982. Descriptions of *Sameodes albiguttalis* (Lepidoptera: Pyralidae) life stages with key to lepidoptera larvae on water hyacinth. Ann. Entomol. Soc. Amer. 75(4):471-279.

Clarke, J. F. G. 1986. Pyralidae and Microplepidoptera of the Marquesas Archipelago. Smithsonian Contr. Zool. 416:1-485.

Clausen, C. P., ed. 1978. Introduced parasites and predators of arthropod pests and weeds: A world review. Agriculture Handbook U.S. 480:1-545.

Common, I. F. B. 1960. A revision of the Australian stem borers hitherto referred to *Schoenobius* and *Scirpophaga* (Lepidoptera: Pyralidae, Schoenobiinae). Aust. J. Zool. 8(2):307-347.

Dyar, H. G., and C. Heinrich. 1927. The American moths of the genus *Diatraea* and allies. Proc. U.S. Natl. Mus. 71:1-48.

Gamez, R., A. Piva, A. Sittenfeld, E. Leon, J. Jimenez, and G. Mirabelli. 1993. Costa Rica's Conservation Program and National Biodiversity Institute (INBio). Pp. 53-68 in R. Gamez et al., eds., Biodiversity Prospecting. World Resources Institute, Washington, D.C.

Habeck, D. H., and G. E. Allen. 1974. Lepidopterous insects as biological control agents of aquatic weeds. Pp. 107-113 in Proceedings of the Fourth International Symposium on Aquatic Weeds, Vienna, Austria.

Heinrich, C. 1938. Moths of the genus *Rupela*. Pyralididae. Schoenobiinae. Proc. U.S. Natl. Mus. 84:355-388.

Heinrich, C. 1939. The cactus-feeding phycitinae: A contribution toward a revision of the American Pyralidoid moths of the family Phycitidae. Proc. U.S. Natl. Mus. 86:331-413.

Heinrich, C. 1956. American moths of the subfamily Phycitinae. Bull. U.S. Natl. Mus. 207:1-581.

Heppner, J. B. 1991. Faunal regions and the diversity of Lepidoptera. Trop. Lepid. 2:1-85.

Holloway, J. D., and H. S. Barlow. 1992. Potential for loss of biodiversity in Malaysia, illustrated by the moth fauna. Pp. 293-311 in A. A. Kadir and H. S. Barlow, eds., Pest Management and the Environment in 2000. CAB International and Agricultural Institute of Malaysia, Kuala Lumpur.

Holloway, J. D., G. S. Robinson, and K. R. Tuck. 1990. Zonation in the Lepidoptera of northern Sulawesi. Pp. 153-166 in W. J. Knight and J. D. Holloway, eds., Insects and the Rain Forests of South East Asia (Wallacea). The Royal Entomological Society of London, London.

Janzen, D., W. Hallwachs, J. Jimenez, and R. Gamez. 1993. The role of parataxonomists, inventory managers, and taxonomists in Costa Rica's National Biodiversity Inventory. Pp 223-254 in Gamez et al., eds., Biodiversity Prospecting. World Resources Institute, Washington, D.C.

Lewvanich, A. 1981. A revision of the Old World species of *Scirpophaga* (Lepidoptera: Pyralidae). Bull. Brit. Mus. (Nat. Hist.) Entomol. 42(4):185-298.

Miller, S. E., and J. Holloway. 1991. Priorities for conservation research in Papua New Guinea: Nonmarine invertebrates. Pp. 44-58 in M. Pearl, B. Bechler, A. Allison, and M. Taylor, eds., Proceedings of Conservation and Environment in Papua New Guinea: Establishing Research Priorities Symposium. Papua New Guinea and Wildlife Conservation International, Papua New Guinea.

Munroe, E. 1964. Four new species of *Cliniodes* Guenée, with notes on the genus and some relatives and segregates (Lepidoptera: Pyralidae). Can. Entomol. 96:529-538.

Munroe, E. 1970. Revision of the subfamily Midilinae (Lepidoptera: Pyralidae). Mem. Entomol. Soc. Can. 74:1-94.

Munroe, E. 1972. Pyraloidea. Pyralidae. Pp. 1-250 in R. B. Dominick, C. Edwards, D. C. Ferguson, J. G. Franclemont, R. W. Hodges, and E. G. Munroe, eds., The Moths of America north of Mexico, Fasc. 13.1A-1B. E.W. Classey, Ltd., and The Wedge Entomological Research Foundation, London.

Munroe, E. 1973a. Pyraloidea. Pyralidae. Pp. 253-304 in R. B. Dominick, C. Edwards, D. C. Ferguson, J. G. Franclemont, R. W. Hodges, and E. G. Munroe, eds., The Moths of America north of Mexico, Fasc. 13.1C. E.W. Classey, Ltd., and The Wedge Entomological Research Foundation, London.

Munroe, E. 1973b. A supposedly cosmopolitan insect: The celery webworm and allies, genus *Nomophila* Hübner (Lepidoptera: Pyralidae: Pyraustinae). Can. Entomol. 105:177-216.

Munroe, E. 1976. Pyraloidea. Pyralidae. Pp. 1-150 in R. B. Dominick, T. Dominick, D. C. Ferguson, J. G. Franclemont, R. W. Hodges, and E. G. Munroe, eds., The Moths of America north of Mexico, Fasc. 13.2A-B. E.W. Classey, Ltd., and The Wedge Entomological Research Foundation, London.

Munroe, E., and A. Mutuura. 1971. Geographical Distribution of Pyraustinae (Lepidoptera: Pyralidae) of temperate East Asia. Trans. Lepid. Soc. Jap. 22:1-6.

Munroe, E., V. O. Becker, J. C. Shaffer, M. Shaffer, and M. A. Solis. 1995. Pyraloidea. Pp. 34-105 in J. B. Heppner, ed., Atlas of Neotropical Lepidoptera Checklist: Part 2. Association for Neotropical Lepidoptera, Gainesville, Fla.

Munroe, E., and M. A. Solis. In press. Pyraloidea. In N. Kristensen, ed., Handbook of Zoology. Walter de Gruyter and Company, Berlin.

Mutuura, A., and E. Munroe. 1970. Taxonomy and distribution of the European Corn Borer and allied species: Genus *Ostrinia* (Lepidoptera: Pyralidae). Mem. Entomol. Soc. Can. 71:1-112.

Neunzig, H. H. 1979. Systematics of immature phycitines (Lepidoptera: Pyralidae) associated with leguminous plants in the southern United States. Bull. U.S. Dept. Agriculture 1589:1-119.

Neunzig, H. H. 1986. Pyraloidea, Pyralidae. Pp. 1-112 in D. R. Davis, T. Dominick, D. C. Ferguson, J. G. Franclemont, R. W. Hodges, E. G. Munroe, and J. A. Powell, eds., The Moths of America North of Mexico, Fasc. 15.2. The Wedge Entomological Research Foundation, Washington, D.C.

Neunzig, H. H. 1990. Pyraloidea, Pyralidae. Pp. 1-165 in D. R. Davis, T. Dominick, D. C. Ferguson, J. G. Franclemont, E. G. Munroe, and J. A. Powell, eds., The Moths of America North of Mexico, Fasc. 15.3. The Wedge Entomological Research Foundation, Washington, D.C.

Robinson, G. S., and K. R. Tuck. 1993. Diversity and faunistics of small moths (Microlepidoptera) in a Bornean rainforest. Ecol. Entomol. 18:385-393.

Shaffer, J. 1976. A revision of the neotropical Peoriinae (Lepidoptera: Pyralidae). Syst. Entomol. 1:281-331.

Solis, M. A. 1992. Check list of the Old World Epipaschiinae and the related New World genera *Macalla* and *Epipaschia*. J. Lepid. Soc. 46(4):280-297.

Solis, M. A. 1993. A phylogenetic analysis and reclassification of the genera of the *Pococera* complex (Lepidoptera: Pyralidae: Epipaschiinae). J. N.Y. Entomol. Soc. 101(1):11-83.

Zimmerman, E. C. 1958. Pyraloidea. Pp. 1-456 in Insects of Hawaii, Vol. 8., University of Hawaii Press, Honolulu.

CHAPTER 17

Phylogeny and Historical Reconstruction: Host-Parasite Systems as Keystones in Biogeography and Ecology

ERIC P. HOBERG

Research Zoologist and Associate Curator, U.S. National Parasite Collection, Agricultural Research Service, Biosystematics and National Parasite Collection Unit, U.S. Department of Agriculture, Beltsville, Maryland

"Parasites furnish information about present day habitats and ecology of their individual hosts. These same parasites also hold promise of telling us something about host and geographical connections of long ago. They are simultaneously the product of an immediate environment and a long ancestry reflecting associations of millions of years. . . . Eventually there may be enough pieces to form a meaningful language which could be called parascript—the language of parasites which tells of themselves and their hosts both of today and yesteryear." (Harold Manter, 1966:70).

Biodiversity represents the complex interaction of phylogeny, ecology, geography, and history as determinants of organismal evolution and distribution. Accordingly, our perception of biotic diversity is a function of scale with respect to populational, genealogical, and ecological identity (Eldredge, 1992; Ricklefs, 1987), spatial relationships (Barrowclough, 1992), and temporal duration (Brooks and McLennan, 1991). Conceptually, biodiversity includes a range of micro- to macroassociations extending from the enumeration of taxa and elucidation of interactions within contemporary ecosystems to the recognition of ancestral areas, regions of endemism, and significant centers of organismal evolution. At one end of this continuum is historical biogeography, which encompasses the study of pattern and process in the distribution of organisms, and historical ecology, which is involved with macroevolutionary process in community development. Thus, with phylogenetic analyses, a record of geological change and knowledge of ecological interactions, hypotheses may be posed about production of biological diversity and evolution of community structure within a rigorous macroevolutionary context (Brooks and McLennan, 1991). In this framework, parasites constitute elegant indicators of the historical and eco-logical development, temporal longevity, current health, and prospects for con-tinuity of biotas.

Current research in biodiversity has focused on regions already profoundly influenced by anthropogenic perturbations and predicted to be strongly affected by climatic change (Ehrlich and Wilson, 1991; Soulé, 1991; Wilson, 1988). How-ever, many of these approaches to faunal assessment have been limited in scope taxonomically, geographically, and temporally. Recent attempts in assessment of biodiversity, although covering a range of local, biomic, and global scales (Barrowclough, 1992), often have concentrated on modern communities and on taxonomically restricted groups, thus obscuring a broader historical and eco-logical context. There has been a notable, but often necessary, focus that has led research to be largely centered on birds, mammals, selected groups of arthro-pods, and vascular plants in an often nondimensional framework that has em-phasized contemporary biotic associations.

Most components of the world's vertebrate and invertebrate fauna are para-sitized, and the helminth parasites of these hosts (Platyhelminthes, Nematoda, Acanthocephala; Table 17-1) can provide a new dimension to understanding ecological interactions, patterns of distribution, and the complex history of many geographic regions and biotas. An interesting perspective on parasites is sug-gested by recent evaluations of global biodiversity (Barrowclough, 1992; Ehrlich and Wilson, 1991; Erwin, 1988; Wilson, 1988) which noted that there may be in excess of 30-100 million species, primarily represented by arthropods. How-ever, if the ubiquitous nature of zooparasitic nematodes among vertebrates and invertebrate hosts is considered, along with the realization that many will be host-specific in their distributions, it is clear that parasites represent a substan-tial facet of biodiversity that yet has to be evaluated in detail. Even with the small fraction of parasitic helminths so far described, a wealth of data on life

TABLE 17-1   Species Diversity of the Major Phyla of Helminthic Parasites of Vertebrates: The Numbers of Described Species

Platyhelminthes	
Monogenea	>5,000
Digenea	>5,000
Eucestoda	~5,000
Nematoda	>12,000
Acanthocephala	~700

Perspectives on Diversity
45,000 species of potential vertebrate hosts
~4,000 species of tapeworms are described
~ 20% of ~850 species of elasmobranchs have been examined
hundreds of species of Tetraphyllidea (Eucestoda) are described

history and distribution exists for application in ecological and historical assessments of biotas. Such information, although traditionally of interest only to parasitologists, is now considered to directly complement and augment knowledge derived solely from the study of free-living organisms on which parasites are dependant. Because of the unique insights gained from parasitological studies about evolution of ecological interactions and community structure, parasitology is becoming recognized as an integral constituent of biodiversity research programs (Brooks, 1985; Brooks and McLennan, 1991, 1993). Biodiversity studies thus are promoting a revitalization of systematics and a renewed appreciation of parasitology as one of the most integrative of the biological disciplines.

HISTORICAL RECONSTRUCTION

Historical reconstruction deals with estimation of organismal diversification, distributional history, ecological interactions, and continuity of community structure over evolutionarily significant time frames. It is the study of pattern and process in the origin and development of biotas emphasizing the historical components of diversity. The basic facets of such research involve an integration of phylogeny (Hennig, 1966; Wiley, 1981; Wiley et al., 1991), historical biogeography (Cracraft, 1982; Nelson and Platnick, 1981; Wiley, 1988a), and historical ecology (Brooks, 1985) in the construction of a macroevolutionary framework to understand biological evolution (Brooks and McLennan, 1991; Brooks et al., 1992). The organization of past environments and the faunal and floral assemblages that they constituted is of critical interest to this research program. Primary issues emphasize how communities (biotas) have been structured by climatological, geological, and biotic factors and the utilization of historical analysis as a key to understanding attributes of contemporary diversity patterns (Brooks et al., 1992; Cracraft, 1980).

Utilization of phylogenetic methodologies allows researchers to develop hypotheses about the history of biotic associations (Brooks, 1985). By focusing on patterns of species distribution and speciation (largely an allopatric process), these hypotheses examine the relationship of diversification within and among clades as a function of past and present environments (Brooks et al., 1992). Analyses of the historical sequence of addition of species and longevity within communities allows identification of elements that are residents (as a function of vicariance) or colonizers (via dispersal) of specific biotas. As a logical extension, it becomes possible to consider the nature and duration of ecological associations, stability of ecological interactions, and the evolution and maintenance of specific life history traits.

The significance of historical reconstruction, and the disciplines subsumed within this program, to current approaches in biodiversity assessment resides in the observation that the past is the key to the present. Historical reconstruction allows identification of historically important centers of diversification (ancestral areas), provides a predictive framework to assess the importance of specific habitats, geographic regions, and biotas and areas of critical genealogical and ecological diversity (Brooks and McLennan, 1991; Erwin, 1991; Vane-Wright et al., 1991). Identification of the most vulnerable members of a community subsequently is based on phylogenetic distinctiveness (Vane-Wright et al., 1991), degree of endemism (Brooks et al., 1992; Erwin, 1991) and unique historical ecological associations (Brooks et al., 1992). Recognition of these facets in turn allows us to make more reliable predictions about the impacts of natural and anthropogenic perturbations on ecosystem structure and stability (Brooks et al., 1992). In this regard, it is becoming increasingly evident that parasites constitute probes for the elucidation of complex linkages among organisms across the genealogical, ecological, spatial, and temporal scales that characterize biodiversity.

PARASITES AND HISTORICAL RECONSTRUCTION

Associations of hosts and parasites generally are not random and tend to reflect some degree of stability indicative of long-term (evolutionary) relationships. This is not a recently recognized phenomenon among parasitologists, many of whom have investigated the predictable nature of parasite-host assemblages (for review, see Klassen, 1992a). For example, Krabbe (1869) and Fuhrmann (1909) were among the earliest biologists to note the narrow distributions of specific cestode taxa among phylogenetically related taxa of birds. It was von Ihering (1891, 1902), however, who first utilized helminthic parasites in a zoogeographic context. His recognition of an old (Gondwanan) connection between South America and New Zealand formed the conceptual foundation for parasitology as an integral facet of studies in historical biogeography and coevolution (Brooks and McLennan, 1993; Klassen, 1992a).

The rich and longstanding tradition of an interplay between parasitology, coevolution, and biogeography is reflected in a series of rules that outline the determinants of the parasite-host relationship (Brooks and McLennan, 1993; Klassen, 1992a). Among the most prominent is Fahrenholz's Rule, which states that parasite phylogeny mirrors host phylogeny; and Manter's Rules that deal with host-specificity and its relationship to coevolution and aspects of historical biogeography. Manter (1966) believed that parasites could serve as keystones for understanding the history of biotas because he recognized the importance of helminths as phylogenetic, ecological, and biogeographic indicators of their host groups (Brooks and McLennan, 1993). This message was implicitly understood among parasitologists, based on the empirical evidence of host-specificity, complex life cycles that were dependant on specific and predictable trophic relationships (Figure 17-1), and definable geographic distributions for many assemblages (Brooks and McLennan, 1993; Klassen, 1992a). Although these generalities were recognized and codified, testing of their universality was limited until the

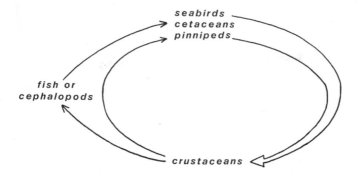

FIGURE 17-1   Representation of a complex parasite life cycle dependant on trophic link-ages, showing the putative cycles for a number of genera of tetrabothriid cestodes that infect marine homeotherms.   Adult tapeworms occur in seabirds (*Tetrabothrius*), ceta-ceans (*Tetrabothrius, Trigonocotyle,* and *Priapocephalus*), and pinnipeds (*Anophryo-cephalus*); larvae or metacestodes develop in crustaceans as intermediate hosts. Larval parasites also become available to seabirds and marine mammals via fish or cephalopods that serve as paratenic or ecologically-transported hosts. In this life cycle, transmission is dependant on specific predator-prey relationships.

advent of phylogenetic methodologies (Hennig, 1966) and their application to historical analysis of host-parasite systems (Brooks, 1979a, 1981; Brooks and McLennan, 1991, 1993).

Conceptual Foundations

Host and geographic ranges of parasites historically are constrained by ge-nealogical and ecological associations (Brooks, 1979a, 1981, 1985) that can be examined within a phylogenetic context (Brooks and McLennan, 1991, 1993). This framework allows us to examine the origin, temporal continuity, and dis-tribution of a parasite-host assemblage with respect to alternative, but not mu-tually exclusive, hypotheses for coevolution or colonization (Mitter and Brooks, 1983).

Coevolution refers to "association by descent" and may be reflected in vicariant geographical patterns for parasites or in ancestor-descendant relation-ships for hosts and parasites (Brooks and McLennan, 1993). Coevolutionary hy-potheses are corroborated by congruence between area relationships (i.e., con-gruence in the distributional relationships of multiple parasite clades across a definable geographic region) and parasite phylogenies, or between host-parasite phylogenies. Predictions that follow from coevolutionary hypotheses include a protracted association for hosts and parasites, a high degree of cospeciation and coadaptation, and possible recognition of numerical and phylogenetic relicts

(Brooks and Bandoni, 1988; Brooks and McLennan, 1991, 1993; Hoberg, 1992). Although host-specificity may be observed in coevolutionary associations, it is a phenomenon that should not be considered to reflect the temporal extent of a host-parasite relationship (Brooks, 1985; Hoberg, 1986). Congruence of biogeographic patterns among parasite clades permits identification of general area relationships that are indicative of faunal assemblages that have been influenced by the same physical and biotic processes as determinants of distribution. Such general patterns are employed to recognize historically important regions of ongoing organismal evolution and areas of endemism that are indicative of relictual communities.

Alternatively, uncovering incongruencies between parasite and host phylogenies or area relationships highlights the components of a biota that have been structured by colonization (host-switching). Predictions based on a colonization hypothesis include a similarity in host-trophic ecology, geographically delimited faunas, and associations of variable temporal extent and degree of cospeciation or coadaptation as determined by the time frame for colonization of an area or host clade (Brooks and McLennan, 1991; Hoberg, 1986, 1992, 1995). Within the context of coevolution or colonization then, the temporal duration of an assemblage may be elucidated with respect to host-parasite distribution, historical biogeography of hosts, and aspects of regional history and physical geology (Hoberg, 1986; Hoberg and Adams, 1992).

Historical evaluations of host-parasite associations have been termed "parascript studies" in recognition of the messages conveyed by parasites to investigations of evolution, ecology, and biogeography (Brooks and McLennan, 1993; Manter, 1966). Such studies are minimally based on an integration of genealogical hypotheses for parasites and hosts, distributional patterns, and geological history (Brooks and McLennan, 1993; Wiley, 1988b). Typically, data from parasites are optimized onto a host tree or area cladogram to provide an historical context for hypotheses about the development of specific parasite-host assemblages. However, even in the absence of robust hypotheses for the host group, substantial historical ecological and biogeographic conclusions may be derived from reconstruction of area and host relationships based on assessments of multiple groups of parasites (Brooks, 1981, 1988), a process termed Brooks Parsimony Analysis (Wiley, 1988b; Brooks and McLennan, 1991). Indeed, phylogenetic studies for a variety of helminth groups, formulated solely on the characters intrinsic to parasites, have provided parasitologists with a means to independently evaluate the history of host taxa or geographic regions for which there is currently a paucity of information (Brooks and McLennan, 1993).

Current Research Programs

Parasites constitute "biodiversity probes" that can be applied directly to questions of contemporary diversity and the historical development of community structure (Brooks and McLennan, 1993; Brooks et al., 1992; suggested by D. R. Brooks in Gardner and Campbell, 1992a). The dual phenomena of host-parasite evolution (coevolution, colonization, or host-switching) and faunal distribution (endemism and elucidation of ancestral areas for biotas) are keystones to examining patterns of diversity. In addition, complex life cycles of helminths are strongly correlated with intricate foodwebs and are dependant on specific ecological interactions and climatological conditions (Figure 17-1). Dependance on a series of intermediate (usually invertebrate), paratenic (invertebrate and vertebrate), and definitive hosts (vertebrates) indicates that each species of parasite exquisitely represents an array of organisms within a community and tracks broadly and predictably across many trophic levels (Brooks, 1985; Brooks and McLennan, 1993; Hoberg, 1986, 1992, 1995). Thus, knowledge of the evolution of parasite-host assemblages provides direct estimates of the history of ecological associations and community development and is indicative of the temporal continuity of trophic assemblages (e.g., Hoberg, 1987). Not surprisingly, studies of helminthic parasites in terrestrial, aquatic, and marine systems have shown these organisms to be elegant markers of contemporary and historical ecological relationships, biogeography, and host phylogeny (Table 17-2; and reviewed in Brooks and McLennan, 1993).

Parasites are thus critical to the recognition of historically important centers of organismal diversification and the spatial and temporal continuity of biotas. At a contemporary level, a predictive framework, with parasites as indicators, exists for elucidating the impacts of natural or anthropogenic perturbations to faunas and ecosystems. It then becomes possible to examine the influence of faunal introductions, extirpations, extinctions, and habitat alterations within a rigorous framework employing parasites as ecological indicators.

The range of explicitly phylogenetic and historical studies of parasites spans the time frame from the Mesozoic era through Recent time, and covers a full spectrum of habitats and taxa of vertebrates (Table 17-2). These studies have indicated that coevolution is not a universal phenomenon and, interestingly, that entire faunas have originated by host-switching and subsequent coevolution—e.g., colonization of pelagic marine birds and mammals by tetraphyllidean cestodes of elasmobranchs in the Tertiary (Hoberg, 1987). This reinforces the importance of the dual components of parasite evolution, guild associations, and evolutionary time in the development of biotas and the utility of parasites in historical reconstruction. These concepts can be examined in greater detail using specific examples from freshwater, terrestrial, and marine communities.

Freshwater Rays, Parasites and the Amazon: A Pacific Origin

Among the elasmobranchs, only stingrays of the family Potamotrygonidae are restricted in their distributions to riverine habitats and occur only in the major drainages of eastern South America. They were traditionally considered

TABLE 17-2 Key Papers Using Helminthic Parasites in Historical Reconstruction in Terrestrial, Aquatic and Marine systems[a]

Parasites/Hosts	Region	Age	Historical Concepts[b]
Terrestrial Systems			
Platt (1984)			
Nematoda/Cervidae	Holarctic	Pliocene-Pleistocene	C,B
Glen and Brooks (1985, 1986),			
Brooks and McLennan (1993)			
Nematoda/Primates	Africa	Late Tertiary	C,B
Gardner (1991)			
Nematoda/Rodentia	Neotropical	Tertiary	C,B,E
Gardner and Campbell (1992a,b)			
Cestoda/Marsupials	Gondwanan	Cretaceous	C,B,E
Hoberg and Lichtenfels (1994)			
Nematoda/Artiodactyla	Cosmopolitan	Tertiary	C,B
Aquatic Systems			
Von Ihering (1891, 1902)			
Turbellaria/Crustacea	Gondwanan	Cretaceous	B
Brooks (1977), Brooks and McLennan (1993)			
Digenea/Anura	Gondwanan	Cretaceous	C,B
Brooks and Overstreet (1978)			
Digenea/Crocodilia	Pangean	Mesozoic	C,B
Brooks (1978a, 1978b)			
Eucestoda/Fish, Amphibians	Pangean	Mesozoic	C,B
Brooks (1979b, 1980),			
Brooks and O'Grady (1989)			
Helminths/Crocodilia	Pangean	Mesozoic	C,B
Brooks et al. (1981),			
Brooks and McLennan (1991, 1993),			
Brooks (1992)			
Helminths/Stingrays	Neotropical (Amazon)	Cretaceous	C,B,E
Bandoni and Brooks (1987a)			
Amphilinidea/Teleosts	Pangean	Mesozoic	C,B
Klassen and Beverley-Burton (1987)			
Monogenea/Siluriforms	Nearctic	Cretaceous	C,B,E
Klassen and Beverley-Burton (1988)			
Monogenea/Centrarchids	Nearctic	Pliocene-Pleistocene	C,B
Boeger and Kritsky (1988),			
Van Every and Kritsky (1992)			
Monogenea/Cypriniforms	Neotropical (Amazon)	Pliocene-Pleistocene	C,B
MacDonald and Brooks (1989)			
Digenea/Turtles, Snakes	Nearctic	—	C
Beverley-Burton and Klassen (1990)			
Monogenea/Fishes	Laurasian	Cretaceous	C,B
Platt (1992)			
Digenea/Freshwater Turtles	Pangean	Triassic	C,B

TABLE 17-2 *Continued*

Parasites/Hosts	Region	Age	Historical Concepts[b]
Marine Systems			
Manter (1966)			
Digenea/Fishes	—	—	B,E
Hoberg (1986, 1992)			
Eucestoda/Seabirds	Holarctic	Pliocene-Pleistocene	C,B,E
Bandoni and Brooks (1987b)			
Gyrocotylidea/Holocephala	Pangean	Mesozoic	C,B
Brooks and Deardorff (1988)			
Helminths/Elasmobranchs	Gondwanan	Cretaceous	C,B
Hoberg and Adams (1992),			
Hoberg (1992, 1995)			
Eucestoda/Pinnipedia	Holarctic	Pliocene-Pleistocene	C,B,E
Klassen (1992b)			
Mongenea/Tetraodontiforms	Indo-Pacific, Caribbean	Tertiary	C,B

[a]Additional studies cited in Brooks (1988) and Brooks and McLennan (1993).
[b]Historical concepts: C=coevolution; B=historical biogeography; E=historical ecology.

to be of marine origin, having ascended each river system from the Atlantic basin in the late Tertiary (see Brooks, 1992; Brooks et al., 1981). However an alternative hypothesis suggesting relatively archaic origins from faunas present in the eastern Pacific was based on the study of multiple taxa of helminthic parasites typical of these rays (Brooks and Deardorff, 1988; Brooks et al., 1981).

This alternative hypothesis was based on the discovery that the sister-groups for some helminths inhabiting potamotrygonids occur in marine rays in the Pacific, not in the Atlantic (Brooks and Deardorff, 1988; Brooks and McLennan, 1991). When the phylogenetic hypotheses for helminths of potamo-trygonids were expressed as an area cladogram, an historical geological or vicariant backbone sequentially linking faunas in the upper Amazon, Parana, Orinoco, and Magdalena drainages of eastern South America was revealed (Figure 17-2). This sequence corresponds to the geological history of the region beginning with the southern Andean orogeny in the late Cretaceous. Notably, the ancient Amazon flowed into the Pacific until the Miocene, when it was blocked by the northward trend of the Andean orogeny, and later developed into a major inland sea prior to flowing into the developing Atlantic basin (Brooks, 1992). Thus, entrapment and subsequent diversification of the ancient Amazonian fauna coincided with the uplifting of the Andes (Brooks, 1992; Brooks et al., 1981, 1992).

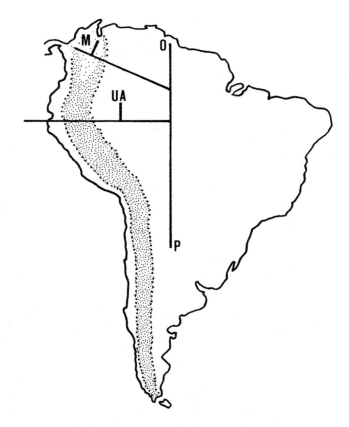

FIGURE 17-2 Area relationships and history of the helminthic parasite fauna of the potamotrygonid freshwater stingrays endemic to South America (after Brooks and McLennan, 1993; Brooks et al. 1981). This area cladogram shows the vicariant backbone for the origins and diversification of the helminth fauna, with an ancestral region in the eastern Pacific, initial entrapment in the upper Amazon (UA), and sequential vicariance and diversification in the Parana (P), Orinoco (O), and Magdalena (M) drainages. The stippled region depicts the present position of the Andes Mountains, which developed by orogeny from the south to the north starting in the Cretaceous. The secondary history of dispersal for the fauna among adjacent river basins is not depicted, but is shown in Brooks and McLennan (1993).

A Pacific origin for the freshwater stingrays and many of their parasites is compatible with similar hypotheses for the derivation of freshwater anchovies and needlefish in the Amazon basin (Brooks, 1992). However the occurrence of a riverine/lacustrine parasite fauna with marine affinities is dependant on the availability of suitable intermediate and definitive hosts to support specific life cycles. In this instance, suitable molluscan and arthropodan species must have been isolated with the stingrays over an extended period of time in estuarine systems that coincided with the Andean orogeny (Brooks, 1992). This leads to the striking conclusion that a sizable component of current diversity in freshwater systems of the Neotropics might have been derived from marine ancestors in the eastern Pacific in the late Cretaceous to Tertiary (Brooks, 1992). Thus, historical biogeographic and ecological analysis of parasites of potamotrygonids highlights the unique nature of what is now a nonrenewable constituent of biodiversity in tropical freshwater communities (Brooks et al., 1992).

Tapeworms and Transantarctic Marsupials

Similar archaic origins also have been suggested for the linstowiid cestode faunas of marsupials and some monotremes in the Neotropics and Australia (Gardner and Campbell, 1992a,b). For example, some cestodes of didelphid marsupials in South America have a limited range in the Yungas-Chaco ecotone of Bolivia, a region considered to contain a relictual biota. Phylogenetic analysis of the known species of *Linstowia* corroborates this contention, suggesting ancient transantarctic connections for hosts and parasites extending from the late Cretaceous (Figure 17-3; also see Gardner and Campbell, 1992a).

This fauna has implications for understanding long-term ecological stability of habitats in South America and Australia. Life cycles of species of *Linstowia* are complex, requiring both an arthropodan intermediate and a mammalian final host. This implies that specific life history associations and patterns of transmission that are currently evident existed prior to the geological separation of South America, Antarctica, and Australia in the late Mesozoic to

The mouse opossum, a South American marsupial.

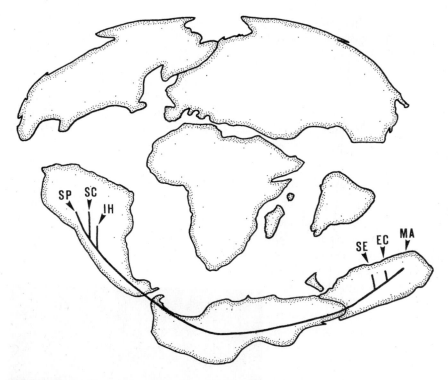

FIGURE 17-3 The transantarctic distributional history for *Linstowia* from didelphid marsupials and monotremes in South America and Australia is summarized (based on Gardner and Campbell, 1992a). This area cladogram depicts the current distribution and phylogenetic relationships of the fauna that developed following vicariance of an ancestral host-parasite assemblage in the Cretaceous. The following species are depicted: in South America, IH=*Linstowia iheringi*, SC=*L. schmidti*, and SP=a currently unnamed species; in Australia, SE=*L. semoni*, EC=*L. echidnae*, and MA=*L. macrouri*. The paleoreconstruction of the southern Gondwanan continental land masses is based on Dietz and Holden (1976).

early Tertiary (Gardner and Campbell, 1992a). Cestode faunas continue to exist in endemic regions that have remained intact over the past 60 million years. It is apparent that such historically important biotas can be identified using a combination of information on current distribution and life cycles in conjunction with phylogenetic analyses of a fauna. In this case, the occurrence of a particular parasite-host assemblage has provided extensive information about ecological interactions across a temporal scale from the contemporary to the archaic.

Seabirds, Pinnipeds, and Pleistocene Marine Refugia

Although much of the historical research using host-parasite phylogeny has involved faunas with relatively ancient origins, often structured by vicariant (usually tectonic) processes (Table 17-2), it is also possible to examine associations that have relatively recent derivations. For example, consider the congruent and synchronic patterns in distribution and speciation that have been postulated for phylogenetically disparate groups of tapeworms that parasitize pinnipeds and seabirds in the Holarctic (Hoberg, 1986, 1992, 1995; Hoberg and Adams, 1992). Combined results of phylogenetic analyses of tetrabothriid cestodes of the genus *Anophryocephalus*, a largely host-specific group inhabiting phocids and otariids, and dilepidid cestodes of the genus *Alcataenia* among the Alcidae (principally in puffins, murres, guillemots, and some auklets) yielded a general area relationship or pattern for host and parasite diversification in the North Pacific basin, North Atlantic, and adjoining areas of the Arctic. Although the host groups are relatively old (Miocene), the parasite groups are not, with host-parasite associations having developed via colonization of pinnipeds and alcids over the last 3 million years (since the late Pliocene and Pleistocene) (Hoberg, 1992). In the absence of a phylogenetic reconstruction, this is not an obvious conclusion.

Two primary areas of diversity are recognized for these groups: (1) a putative ancestral area for both faunas in the North Atlantic sector of the Arctic basin that coincides with the current distribution of basal members of both assemblages; and (2) a region of secondary diversification for hosts and parasites in the North Pacific (Figure 17-4). This general pattern resulted from early vicariance of a Holarctic fauna across the Beringian region, followed by radiation in the North Pacific with subsequent range expansion into the Arctic basin and Atlantic for alcids, phocids, and their respective parasites. Diversification during the late Pliocene and Pleistocene was tied to climatic factors. During glacial maxima, refugial habitats were distributed in marginal zones of the North Pacific, Sea of Okhotsk, Aleutian Islands, and Arctic basin as a function of eustatic reduction in sea level. These changes resulted in the isolation of small populations of hosts and parasites, establishing the initial conditions for either peripheral isolates or microvicariance speciation (e.g., Lynch, 1989; Wiley, 1981). During interstadials (warm periods between glacial advances), ranges of hosts and parasites expanded, increasing the potential for host-switching. The parasite fauna in this region thus is characterized by a pattern of sequential colonization, vicariance, and speciation of peripheral isolates that is driven by the fluctuating geographic ranges of the definitive host groups in response to refugial expansion and contraction (Hoberg, 1995). Although the phylogenetic histories for parasites and hosts were highly incongruent (as would be expected for instances of host-switching), there was a general synchrony in speciation events during radiation of these phylogenetically unrelated tapeworms in marine birds

FIGURE 17-4 Historical summary of the congruent and synchronic relationships of marine parasite faunas among pinnipeds and alcids in the Holarctic during the late Pliocene and Pleistocene showing distributional history (solid lines), current geographical limits for the faunas (dashed lines delimiting the Pacific and Atlantic basins), and emergent continental shelf during glacial maxima (stippled regions)(after Hoberg, 1992; also see Hoberg, 1986, 1995; Hoberg and Adams, 1992). (1) Putative ancestral area for *Alcataenia* and *Anophryocephalus* in seabirds and pinnipeds, respectively (ca. 3.0 to 3.5 million years ago). (2) Early Holarctic distributions for host-parasite assemblages attained by ca. 2.5 to 3.0 million years ago. (3) Initial submergence of Beringia (ca. 3.0 million years ago) was followed by range expansion for hosts and parasites into the North Pacific and subsequent colonization and radiation in refugial habitats during the Pleistocene. (4) Secondary Holarctic ranges for some species of *Alcataenia* and *Anophryocephalus* were attained during the Quaternary. (5) Diversification of *Anophryocephalus* continued with the colonization of sea lions during the Quaternary.

and mammals (Hoberg, 1992; Hoberg and Adams, 1992). Given the complexity of these parasite distributions, the high degree of biogeographic congruence indicates that these systems have been strongly influenced by the same geological and climatological factors.

These host-parasite assemblages are indicative of specific ecological associations via food webs that have been maintained since at least the late Pliocene (Hoberg, 1992, 1995). Trophic interactions, food habits, and foraging behavior also appear to have influenced the potential for colonization among feeding guilds where hosts exploited a common prey source (Hoberg, 1984a,b, 1986, 1987). Although ecologically cohesive and characterized by a high level of host-specificity, it is apparent that these faunas are not relicts of a previous biota, but were structured by host-switching, geographical colonization, and cyclic climatological fluctuations that characterized the late Pliocene and Pleistocene (Hoberg, 1992, 1995). In other words, the dramatic climatic changes associated with the Pliocene-Pleistocene glaciations promoted speciation, not extinction, in these Holarctic marine faunas. These marine parasite-host systems thus provide a counterintuitive view of diversification at high latitudes over the past 3 million years.

CONCLUSIONS: PARASITES AS HISTORICAL PROBES FOR BIODIVERSITY

Parasites are elegant indicators of historical ecological associations and biogeography. The complexities of parasite life cycles, dependant on a series of intermediate, paratenic, and definitive hosts, indicate that each species of helminth represents a broader array of organisms within a community. Thus, knowledge of the evolution of a parasite-host assemblage can provide direct estimates of the history of ecological associations and is indicative of the continuity of trophic assemblages through time. The genealogical and biogeographic histories of parasitic helminths provide considerably more information than that available only from the evaluation of phylogenetic hypotheses for free-living organisms alone (e.g., Erwin, 1991; Stiassny, 1992). This means that parasites are admirably suited for a role as historical and contemporary biodiversity probes (Brooks and McLennan, 1994; Brooks et al., 1992; Gardner and Campbell, 1992a) and for augmenting the development of conservation strategies through recognition of regions of critical diversity and evolutionary significance.

Considered from a conservation standpoint, the phylogenetic history of parasite-host assemblages allows direct predictions about the age and duration of specific faunal associations, identification of regions of endemism and evolutionary "hot spots," and the historical structure of ecosystems and communities. These data can be applied to complex decisions in conservation biology through recognition of distinctive clades (e.g., Stiassny, 1992; Vane-Wright et al., 1991), geographic regions of past (and future) evolutionary importance (e.g.,

Brooks et al., 1992; Erwin, 1991), and unique historical ecological associations (Brooks and McLennan, 1991, 1994).

The focus of Brooks et al. (1981) and Gardner and Campbell (1992a) on helminthic faunas of South American elasmobranchs and mammals highlights the importance of phylogenetic studies of parasites in identifying significant areas of endemism in the Neotropics. Succinctly stated by Brooks and McLennan (1994:23): "Areas of endemism are important because they have been the focus of biodiversity production in the past and thus may be `hot spots' of evolutionary potential for the future." In this regard, the distinctive community associated with the Yungas-Chaco ecotone of Bolivia indicates that this region could merit reserve status (Gardner and Campbell, 1992a). In addition, the drainages of the upper Parana River, Amazon River, and Magdalena River support vital faunal assemblages, represented by unique parasite and elasmobranch faunas, which are currently isolated and ". . . represent good compromises between the need to protect and the need to develop . . ." (Brooks et al., 1992:58).

Historical studies of parasite-host associations also have applications in outlining potential climatological determinants of biodiversity. It has been predicted that communities in the Arctic may respond dramatically to environmental alteration driven by changes in global climate (e.g., Danks, 1992; Douglas et al., 1994). Historically, patterns of diversification and biogeography among seabirds, pinnipeds, and parasites across the Holarctic have been strongly influenced by environmental fluctuations during the Quaternary (Hoberg, 1986, 1995). Applied as an analog, altered patterns of distribution among contemporary parasite assemblages in the Arctic may serve as indicators of natural and anthropogenically driven perturbations in climate.

The past is the key to the present, where historical reconstruction involving parasites contributes to a predictive framework for discovering the interaction of biotic communities, environment, and climate. As a consequence, parasites constitute powerful tools and represent keystones to be applied to questions of the origin, maintenance, and distribution of organismal diversity.

ACKNOWLEDGMENTS

The conceptual foundations for this paper were developed through discussions over the past several years with D. R. Brooks, G. J. Klassen, D. Siegel-Causey, and D. A. McLennan. Critical and insightful evaluations of the manuscript by D. A. McLennan, A. Y. Rossman, D. R. Miller, and J. R. Lichtenfels were greatly appreciated.

REFERENCES

Bandoni, S. M., and D. R. Brooks. 1987a. Revision and phylogenetic analysis of the Amphilinidea Poche, 1922 (Platyhelminthes: Cercomeria: Cercomeromorpha). Can. J. Zool. 65:1110-1128.

Bandoni, S. M., and D. R. Brooks. 1987b. Revision and phylogenetic analysis of the Gyrocotylidea Poche, 1926 (Platyhelminthes: Cercomeria: Cercomeromorpha). Can. J. Zool. 65:2369-2389.

Barrowclough, G. F. 1992. Systematics, biodiversity and conservation biology. Pp. 121-143 in N. Eldredge, ed., Systematics, Ecology and the Biodiversity Crisis, Columbia University Press, N.Y.

Beverley-Burton, M., and G. J. Klassen. 1990. New approaches to the systematics of the ancyrocephalid Monogenea from Nearctic freshwater fishes. J. Parasitol. 76:1-21.

Boeger, W. A., and D. C. Kritsky. 1988. Neotropical Monogenea. 12. Dactylogyridae from *Serrasalmus nattereri* (Cypriniformes, Serrasalmidae) and aspects of their morphological variation and distribution in the Brazilian Amazon. Proc. Helminthol. Soc. Wash. 55:188-213.

Brooks, D. R. 1977. Evolutionary history of some plagiorchoid trematodes of anurans. Syst. Zool. 26:277-289.

Brooks, D. R. 1978a. Evolutionary history of the cestode order Proteocephalidea. Syst. Zool. 27:312-323.

Brooks, D. R. 1978b. Systematic status of proteocephalid cestodes from reptiles and amphibians in North America with descriptions of three new species. Proc. Helminthol. Soc. Wash. 45:1-28.

Brooks, D. R. 1979a. Testing the context and extent of host-parasite coevolution. Syst. Zool. 28:299-307.

Brooks, D. R. 1979b. Testing hypotheses of evolutionary relationships among parasites: The Digeneans of crocodilians. Amer. Zool. 19:1225-1238.

Brooks, D. R. 1980. Revision of the Acanthostominae Poche, 1926 (Digenea: Cryptogonimidae). Zool. J. Linn. Soc. 70:313-382.

Brooks, D. R. 1981. Hennig's parasitological method: A proposed solution. Syst. Zool. 30:229-249.

Brooks, D. R. 1985. Historical ecology: A new approach to studying the evolution of ecological associations. Ann. Missouri Bot. Garden 72:660-680.

Brooks, D. R. 1988. Macroevolutionary comparisons of host and parasite phylogenies. Ann. Rev. Ecol. Syst. 19:235-259.

Brooks, D. R. 1992. Origins, diversification, and historical structure of the helminth fauna inhabiting neotropical freshwater stingrays (Potamotrygonidae). J. Parasitol. 78:588-595.

Brooks, D. R., and S. M. Bandoni. 1988. Coevolution and relicts. Syst. Zool. 37:19-33.

Brooks, D. R., and T. L. Deardorff. 1988. *Rhinebothrium devaneyi* n. sp. (Eucestoda: Tetraphyllidea) and *Echinocephalus overstreeti* Deardorff and Ko (1983) (Nematoda: Gnathostomatidae) in a thorny back ray, *Urogymnus asperrimus*, from Enewetak Atoll, with phylogenetic analysis of both species groups. J. Parasitol. 74:459-465.

Brooks, D. R., and D. A. McLennan. 1991. Phylogeny, Ecology and Behavior. A Research Program in Comparative Biology. University of Chicago Press, Chicago. 434 pp.

Brooks, D. R., and D. A. McLennan. 1993. Parascript: Parasites and the Language of Evolution. Smithsonian Institution Press, Washington, D.C. 429 pp.

Brooks, D. R., and D. A. McLennan. 1994. Historical ecology as a research programme: Scope, limitations, and the future. Pp. 1-27 in P. Eggleton and R. Vane-Wright, eds., Phylogenetics and Ecology. Linnean Society Symposium Series No. 17, Academic Press, London.

Brooks, D. R., and R. T. O'Grady. 1989. Crocodilians and their helminth parasites: Macroevolutionary considerations. Amer. Zool. 29:873-883.

Brooks, D. R., and R. M. Overstreet. 1978. The family Liolopidae (Digenea), including a new genus and two new species from crocodilians. Int. J. Parasitol. 8:267-273.

Brooks, D. R., T. B. Thorson, and M. A. Mayes. 1981. Fresh-water stingrays (Potamotrygonidae) and their helminth parasites: Testing hypotheses of evolution and coevolution. Pp. 147-175 in V. A. Funk and D. R. Brooks, eds., Advances in Cladistics. New York Botanical Garden, N.Y.

Brooks, D. R., R. L. Mayden, and D. A. McLennan. 1992. Phylogeny and biodiversity: Conserving our evolutionary legacy. Trends Ecol. Evol. 7:55-59.

Cracraft, J. 1980. Biogeographic patterns of terrestrial vertebrates in the southwest Pacific. Palaeogeogr. Palaeoclimatol. Palaeoecol. 31:353-369.

Cracraft, J. 1982. Geographic differentiation, cladistics and vicariance biogeography: Reconstructing the tempo and mode of evolution. Amer. Zool. 22:411-424.

Danks, H. V. 1992. Arctic insects as indicators of environmental change. Arctic 45: 159-166.

Dietz, R. S., and J. C. Holden. 1976. The breakup of Pangea. Pp. 126-137 in J. T. Wilson, ed., Continents Adrift and Continents Aground. Freeman and Company, San Francisco.

Douglas, M. S. V., J. P. Smol, and W. Blake, Jr. 1994. Marked post-18th century enviromental change in high-Arctic ecosystems. Science 266:416-419.

Ehrlich, P. R., and E. O. Wilson. 1991. Biodiversity studies: Science and policy. Science 253:758-762.

Eldredge, N. 1992. Where the twain meet: Causal intersections between the genealogical and ecological realms. Pp. 1-14 in N. Eldredge, ed., Systematics, Ecology, and the Biodiversity Crisis. Columbia University Press, N.Y.

Erwin, T. L. 1988. The tropical forest canopy: The heart of biotic diversity. Pp. 123-129 in E. O. Wilson and F. M. Peter, eds., BioDiversity. National Academy Press, Washington, D.C.

Erwin, T. L. 1991. An evolutionary basis of conservation strategies. Science 253:750-752.

Fuhrmann, O. 1909. Die Cestoden der Vögel. Zoologisch Jahrb. 10: 1-232.

Gardner, S. L. 1991. Phyletic coevolution between subterranean rodents of the genus Ctenomys (Rodentia: Hystricognathi) and nematodes of the genus Paraspidodera (Heterakoidea: Aspidoderidae) in the neotropics: Temporal and evolutionary implications. Zool. J. Linn Soc. 102:169-201.

Gardner, S. L., and M. L. Campbell. 1992a. Parasites as probes for biodiversity. J. Parasitol. 78:596-600.

Gardner, S. L. and M. L. Campbell. 1992b. A new species of Linstowia (Cestoda: Anoplocephalidae) from marsupials in Bolivia. J. Parasitol. 78:759-799.

Glen, D. R., and D. R. Brooks. 1985. Phylogenetic relationships of some strongylate nematodes of primates. Proc. Helminthol. Soc. Wash. 52:227-236.

Glen, D. R., and D. R. Brooks. 1986. Parasitological evidence pertaining to the phylogeny of the hominoid primates. Biol. J. Linn. Soc. 27:331-354.

Hennig, W. 1966. Phylogenetic Systematics. University of Illinois Press, Urbana. 263 pp.

Hoberg, E. P. 1984a. Alcataenia fraterculae sp. n. from the horned puffin, Fratercula corniculata (Naumann), Alcataenia cerorhincae sp. n. from the rhinoceros auklet, Cerorhinca monocerata (Pallas), and Alcataenia larina pacifica ssp. n. (Cestoda: Dilepididae) in the North Pacific basin. Ann. Parasitol. Hum. Comp. 59:335-351.

Hoberg, E. P. 1984b. Alcataenia longicervica sp. n. from murres, Uria lomvia (Linnaeus) and Uria aalge (Pontoppidan), in the North Pacific basin, with redescriptions of Alcataenia armillaris (Rudolphi, 1810) and Alcataenia meinertzhageni (Baer, 1956) (Cestoda: Dilepididae). Can. J. Zool. 62:2044-2052.

Hoberg, E. P. 1986. Evolution and historical biogeography of a parasite-host assemblage: Alcataenia spp. (Cyclophyllidea: Dilepididae) in Alcidae (Charadriiformes). Can. J. Zool. 64: 2576-2589.

Hoberg, E. P. 1987. Recognition of larvae of the Tetrabothriidae (Eucestoda): Implications for the origin of tapeworms in marine homeotherms. Can. J. Zool. 65:997-1000.

Hoberg, E. P. 1992. Congruent and synchronic patterns in biogeography and speciation among seabirds, pinnipeds and cestodes. J. Parasitol. 78:601-615.

Hoberg, E. P. 1995. Historical biogeography and modes of speciation across high latitude seas of the Holarctic: Concepts for host-parasite coevolution among the Phocini (Phocidae) and Tetrabothriidae (Eucestoda). Can. J. Zool. 73:45-57.

Hoberg, E. P., and A. M. Adams. 1992. Phylogeny, historical biogeography, and ecology of Anophryocephalus spp. (Eucestoda: Tetrabothriidae) among pinnipeds of the Holarctic during the late Tertiary and Pleistocene. Can. J. Zool. 70:703-719.

Hoberg, E. P., and J. R. Lichtenfels. 1994. Phylogenetic systematic analysis of the Trichostrongylidae (Nematoda) with an initial assessment of coevolution and biogeography. J. Parasitol. 80:976-996.

Klassen, G. J. 1992a. Coevolution: A history of the macroevolutionary approach to studying host-parasite associations. J. Parasitol. 78:573-587.

Klassen, G. J. 1992b. Phylogeny and biogeography of ostraciin boxfishes (Tetraodontiformes: Ostraciidae) and their gill parasitic *Haliotrema* sp. (Monogenea: Ancyrocephalidae): A study in host-parasite coevolution. Ph.D. dissertation, University of Toronto, Toronto. 366 pp.

Klassen, G. J., and M. Beverley-Burton. 1987. Phylogenetic relationships of *Ligictaluridus* spp. (Monogenea: Ancyrocephalidae) and their ictalurid (Siluriformes) hosts: an hypothesis. Proc. Helminthol. Soc. Wash. 54:84-90.

Klassen, G. J., and M. Beverley-Burton. 1988. North American fresh water ancyrocephalids (Monogenea) with articulating haptoral bars: Host-parasite coevolution. Syst. Zool. 37: 179-189.

Krabbe, H. 1869. Bidrag til Kundskab om Fugelenes Baendelorme. Dansk Videns. Selsk. Skr., 5. Raekle Naturvidesnskabelig og Mathematisk 8:251-368.

Lynch, J. D. 1989. The gauge of speciation: On the frequencies of modes of speciation. Pp. 527-553 in D. Otte and J. A. Endler, eds., Speciation and its Consequences. Sinauer Associates, Sunderland, Mass.

MacDonald, C. A., and D. R. Brooks. 1989. Revision and phylogenetic analysis of the North American species of *Telorchis* Luehe, 1899 (Cercomeria: Trematoda: Digenea: Telorchidae). Can. J. Zool. 67:2301-2320.

Manter, H. W. 1966. Parasites of fishes as biological indicators of recent and ancient conditions. Pp. 59-71 in J. E. McCauley, ed., Host Parasite Relationships. Oregon State University Press, Corvallis.

Mitter, C., and D. R. Brooks. 1983. Phylogenetic aspects of coevolution. Pp. 65-98 in D. J. Futuyma and M. Slatkin, eds., Coevolution. Sinauer Associates, Sunderland, Mass.

Nelson, G., and N. I. Platnick. 1981. Systematics and Biogeography: Cladistics and Vicariance. Columbia University Press, N.Y.

Platt, T. R. 1984. Evolution of the Elaphostongylinae (Nematoda: Metastrongyloidea: Protostrongylidae) parasites of cervids (Mammalia). Proc. Helminthol. Soc. Wash. 51:196-204.

Platt, T. R. 1992. A phylogenetic and biogeographic analysis of the genera of Spirorchinae (Digenea: Spriorchidae) parasitic in freshwater turtles. J. Parasitol. 78:616-629.

Ricklefs, R. E. 1987. Community diversity: Relative roles of local and regional processes. Science 235:167-171.

Soulé, M. E. 1991. Conservation: Tactics for a constant crisis. Science 253:744-749.

Stiassny, M. L. 1992. Phylogenetic analysis and the role of systematics in the biodiversity crisis. Pp. 109-120 in N. Eldredge, ed., Systematics, Ecology, and the Biodiversity Crisis. Columbia University Press, N.Y.

Van Every, L. R., and D. C. Kritsky. 1992. Neotropical Monogenoidea. 18. *Anacanthorus* Mizelle and Price, 1965 (Dactylogyridae, Anacanthorinae) of piranha (Characoidea, Serrasalmidae) from the central Amazon, their phylogeny, and aspects of host-parasite coevolution. J. Helminthol. Wash. 59:52-75.

Vane-Wright, R. E., C. J. Humphries, and R. J. Williams. 1991. What to protect: Systematics and the agony of choice. Biol. Conserv. 55:235-254.

von Ihering, H. 1891. On the ancient relations between New Zealand and South America. Proc. New Zealand Inst. 24:431-445.

von Ihering, H. 1902. Die Helminthen als Hilsmittel der zoogeographischen Forschung. Zool. Anziger. 26:42-51.

Wiley, E. O. 1981. Phylogenetics: The Theory and Practice of Phylogenetic Analysis. John Wiley and Sons, N.Y. 439 pp.

Wiley, E. O. 1988a. Vicariance biogeography. Ann. Rev. Ecol. Syst. 19:513-542.

Wiley, E. O. 1988b. Parsimony analysis and vicariance biogeography. Syst. Zool. 37:271-290.

Wiley, E. O., D. Siegel-Causey, D. R. Brooks, and V. A. Funk. 1991. The Compleat Cladist: A Primer of Phylogenetic Procedures. University of Kansas Museum of Natural History Press, Lawrence. 158 pp.

Wilson, E. O. 1988. The current state of biological diversity. Pp. 3-18 in E. O. Wilson and F. M. Peter, eds., BioDiversity. National Academy Press, Washington D.C.

CHAPTER
18

Comparative Behavioral and Biochemical Studies of Bowerbirds and the Evolution of Bower-Building

GERALD BORGIA

Professor, Department of Zoology, University of Maryland, College Park

Unprecedented human-induced changes in the environment are causing a rapid loss of species and their habitats. Common responses to the loss of bio-diversity include habitat conservation and increased efforts to understand the origin and maintenance of biodiversity. While there is no doubt that these are critical activities, it is unclear the extent to which these and related conservation approaches can stem the tide of extinctions. Global threats such as climatic warming and ozone depletion place all species on the planet at risk and compli-cate attempts to determine the level of threat for particular species or communi-ties.

There must be an intensive effort to collect information about species before they go extinct or are relegated to reserves in degraded habitats. Otherwise, we never will have detailed information about the life history, specialized adapta-tions, social behavior, or relationships with other species for the vast majority of species that go extinct. For those left close to extinction in degraded habitats, in zoos, or in preserves, our ability to understand their ecological and evolutionary relationships to other species and the surrounding habitat will be greatly com-promised.

Wilson (1992) describes "unmined riches" locked in the diverse and poorly known biotas. He offers recent discoveries of natural seed stocks and the use of secondary compounds from plants as pharmaceutical agents as examples of such riches. But biologists have been less emphatic about the unmined intellectual resources that are lost with extinction. As species go extinct and habitats are degraded, the opportunity to use natural communities as sources of information about basic physiological, ecological, and evolutionary processes is lost forever.

As with basic research in other areas, the natural historical sciences also offer prospects for unforeseen intellectual and economically important discoveries.

The grim prospect of the loss of much biodiversity is tempered somewhat by the availability of many well-trained experts in a variety of natural historical fields, by the maturation of these fields in which intellectually significant issues have been identified, and the explosion of new tools to decode the information locked up in natural systems. Molecular methods for phylogenetic reconstruction, methodologies for comparative studies, and automated equipment that allows collection of data on diverse sorts of organisms all are becoming available and are being put to use in field studies. It is a sad coincidence that, as we are gaining sophisticated tools for exploring the natural history of organisms, there is a dramatic loss of species that can be studied. We are at a critical time when we have the opportunity to collect some of the most detailed and useful natural history information before many species go extinct. A broadly based biodiversity initiative that emphasizes both conservation and natural historical studies directed at species in still vibrant populations must be considered the only appropriate response to the loss of biodiversity.

EVOLUTION OF BOWERBIRDS

In this Chapter, I present information from a comparative study of bowerbirds (Ptilonorhynchidae) on the likely causes of the evolution of bowers. This work has been motivated by an interest in the causes of mate choice in species with extreme displays. It serves to illustrate the importance of studies that involve comparisons between species.

Bowerbirds occur across the Australio-Papuan region and are unique in that males build structures on the ground called bowers that appear to function in mate attraction and related activities. The first 7 years of my work on bowerbirds focused on the satin bowerbird (*Ptilonorhynchus violaceus*). Previously, there had been no detailed quantitative study of any aspect of bowerbird behavior. After those 7 years, I felt that I had a good understanding of typical bowerbird behavior. This belief was shaken after a preliminary study of spotted bowerbirds (*Chlamydera maculata*) that showed fundamentally different patterns of courtship and male interactions from those observed in satin bowerbirds. Since then, my students and I have studied nine additional species of bowerbirds.

As an example demonstrating the value of comparative studies involving a large number of species, this work shows that even among closely related species there can be very large differences in behavior. One cannot characterize a group like bowerbirds based on studies of single species, and there is no typical species that fully represents this group. Information from numerous species often is needed to understand the evolution of complex traits like bower-building.

Male satin bowerbird displaying at bower.

Mate selection in species with elaborate male display traits was a topic central to Charles Darwin's (1859, 1871) seminal writings on sexual selection. Prominent in his discussion of sexual selection were the bowerbirds and their unique behavior of building bowers. Bowers typically are made of sticks. In some species, bowers can reach 1.5 m high and are built near display courts decorated with more than 2,000 decorations. These elaborate decorated structures frequently have been described as one of the wonders of the animal world.

There are now numerous hypotheses explaining how extreme displays evolve, although clear answers remain elusive. "Good genes" models propose that extreme sexual displays function as indicators of male quality to females choosing mates (e.g., Andersson, 1982, 1986; Borgia, 1979; Hamilton and Zuk, 1982; Trivers, 1972; Zahavi, 1975). Alternatively, the runaway model (Fisher, 1930; Kirkpatrick, 1982; also see Lande, 1981) posits that female preferences can produce greatly elaborated male display traits without providing enhanced vigor to offspring. Other models include: passive choice (Andersson, 1982; Parker, 1983), intrasexual signaling (Halliday, 1978; LeCroy et al., 1980), proximate benefits (e.g., protection to females provided by well-constructed bowers; Borgia et al., 1985) and innate preferences (Burley, 1985; Kirkpatrick, 1987; Ryan et al., 1990).

Recent empirical studies have shown that elaborate displays in polygynous species are typically not single traits, but a complex sets of traits (Andersson, 1989; Borgia, 1985; Gibson et al., 1991; McDonald, 1989; Møller and Pomiankowski, 1993; Prum, 1990; Zuk et al., 1990). In bowerbirds, bowers and decorations are part of a generalized display that includes plumage, acoustical, and dancing elements directed at females during courtship. Recent studies indicate that females prefer males with well-built and well-decorated bowers (Borgia, 1985; Borgia and Mueller, 1992), indicating that an important current function of bowers and decorations is to attract mates. The large number of models for explaining elaborated display traits, the lack of specific predictions from some of these models, and the complex array of traits involved in the display of polygynous species have made it difficult to develop clear-cut conclusions about the evolution of extreme male displays. Elsewhere it has been suggested (Borgia et al., 1985) that bowers may function in female assessment of male quality as sires (good genes), as a protective device for females being courted (proximate benefit), or as a structure with no direct functional significance outside the context of sexual selection (runaway or latent preference).

Comparative studies of the evolution of traits among related species can provide critical information about the sequential evolution of the components of display and their initial and derivative functions (Basolo, 1990; Brooks and McLennan, 1991; Kusmierski et al., 1993; Prum, 1990).

Several recent developments have made it possible to carry out a detailed comparative analysis of the function of bowers. First, the use of remote-controlled cameras aimed at bowers where males display and mate has allowed my students, field assistants, and me to simultaneously monitor all activity at as many as 30 bowers through an entire mating season. Because bowers are widely separated and the mating periods may last several months, it would have been impossible otherwise to obtain detailed information from numerous bowers. Second, the advent of molecular techniques has made it possible to build a phylogeny of bowerbirds based on mitochondrial DNA sequence information. The independent derivation of this phylogeny makes it possible to infer the historical pattern of the evolution of display traits by using maximum parsimony methods to map the distribution of particular male display traits and their occurrence in ancestors onto the phylogeny. By combining these techniques, my students, collaborators, and I have been able to gather detailed information on the display traits of bowerbirds and map this information onto an independently derived phylogeny of the bowerbirds.

Building bowers is restricted to the family Ptilonorhynchidae, which is made up of six genera. One genus, the catbirds (*Ailuroedus*, three species), is monogamous and does not clear a display court. Members of the remaining five genera are polygynous, clear a court, and typically build a bower. There are two major designs of bowers. Maypole bower-builders decorate a sapling with sticks. Males of species in the genus *Amblyornis* (four species) typically decorate a single spire surrounded by a mossy circular court, and two species build a hut-like dome that covers part of the court. *Prionodura newtoniana* builds a two-spired structure with a cross perch connecting the spires. Avenue-builders (three genera, eight species) build a two-walled structure with a display court near the end of the bower. Two other species, toothbilleds (*Scenopoeetes dentirostris*)

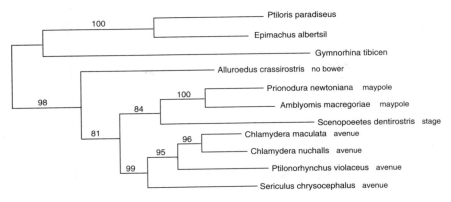

FIGURE 18-1 Phylogeny of the bowerbirds based on information from mt-DNA sequences taken from Kusmierski et al. (1993). The topmost grouping represents two Birds of Paradise (paradise riflebird and the brown sickelbill) and the Australian magpie. The bowerbirds are represented in three clades represented by the monogamous catbirds (*Ailuroedus*) that build no bower, the maypole-builders (including the toothbill bowerbird, *Scenopoeetes dentirostris*), and the avenue-builders. Recent results (Kusmierski, personal communication, 1995) indicate that Archbold's bowerbird (*Archboldia papauensis*) is a member of the clade that builds maypoles. Like *Scenopoeetes* in the same clade, the most probable pattern of evolution is that Archbold's bowerbird has lost its tendencies to build bowers and these losses are independent. Numbers at nodes represent the percentage of times a group occurred in 1,000 bootstrap replicates.

and Archbold's (*Archboldia papuensis*), clear display courts but do not build a characteristic bower.

A phylogeny based on sequences of mitochondrial DNA from the cytochrome-b gene, developed in cooperation with Kusmierski and Crozier (Kusmierski et al., 1993), indicates that there was an initial separation of lineages that led to the monogamous catbirds and the remaining polygynous species. Later, a separation in the clade that led to bower-builders produced a lineage that led to the avenue-builders, and another that produced the maypole-builders (*Archboldia* and *Prionodura*) and the two polygynous species that do not build bowers (*S. dentirostris* and *A. papuensis*) (Figure 18-1). Because building bowers is the dominant pattern in both lineages and it occurs in no other avian family, it most likely evolved once, preceding the split of the ancestors of the avenue- and maypole-builders.

HYPOTHESES FOR THE EVOLUTION OF BOWERS

Many of the models of the evolution of extreme displays can be used to develop specific models for the origin and evolution of bowers. For example, the placement of sticks in an incipient bower could have been the result of an arbitrary female preference (as part of a runaway or latent preference model). Such a hypothesis, however, posits no particular functional role for bowers apart from mate attraction and fails, by itself, to account for the variation in form of bowers. Evidence of a function for bowers, especially if it is consistent among different types of bowers, would weigh against this hypothesis.

Another possibility is that bowers could have functioned to provide protection from various threats (Borgia et al., 1985). Sources of threat include predators that might attack males and females, marauding males that force copulations on females while they view the court owner's courtship, and court owner himself forcing copulations on females not ready to mate.

The hypotheses of predation and marauding-males are not well supported. There is no evidence of predation on females or males while males are displaying on courts in more than 100,000 hours that cameras monitored the display courts of males. This is especially relevant because, in most species, males are not protected from predators by the bower during courtship. In addition, females are not protected in some types of bowers, including the open bowers of MacGregor's bowerbird (*Amblyornis macgregoriae*), which may be similar to ancestral bowers, and those of the streaked bowerbird (*A. subalaris*), where the male occupies the covered part of the bower during courtship, opposite to what is predicted by the female protection hypothesis (Sejkora and Borgia, in preparation). Last, females change their behavior from being very reluctant to stand outside the bower before copulation to being quite willing to stand there afterward. The observations that many types of bowers do not confer protection, that the sexes stand in the wrong place, and that the female's behavior changes after copulation also fail to support the hypothesis that bowers provide protection from marauding males.

The remaining hypothesis, protection from the courting male, is not limited by these difficulties. Species from the two clades that build bowers show generally similar patterns of how they use bowers during courtships that lead to copulations. In both lineages, males display facing the female. The bower enhances the female's ability to escape unwanted matings by blocking the male's direct path to her and forcing him to run around the wall of the bower or maypole to mate. The increased distance that males must travel to reach a female enhances her ability to escape unwanted matings.

Several issues emerge in attempts to evaluate this hypothesis. First, why should males build a structure that decreases their prospects of forced copulation? If males are programmed to maximize their reproduction and forced copulations can contribute to their reproductive success, it at first seems counterintuitive for males to build a structure that limits this type of reproductive benefit. A plausible solution is that building bowers may offer a compensating benefit that outweighs losses from forced copulation. If the reduced threat of forced copulation causes more females to be attracted to courts with bowers than those without, then gains from increased visitation by females could more than compensate for the lost ability to force copulations with females. Female attraction might arise for several reasons, as discussed above.

Second, how do we test the hypothesis that females prefer males with bowers that function to protect them from forced copulation? It is rather tricky to infer the function of a trait at its evolutionary origin, given no fossils or other record of the ancestral form of the bower. Information from observations of the different species of bowerbirds and knowledge of their evolutionary relationships can be used to establish likely hypotheses. The criteria that would strongly support a hypothesis for an incipient function of bowers are: (1) the proposed function should be consistent with the design of the likely ancestral bower, (2) this function should be significant across all types of bowers, and (3) species that do not build bowers should show alternative solutions to the problem solved by the bower. There is no guarantee that even the correct answer will meet all three criteria. It is possible, for example, that bowers could have taken on a variety of secondary functions that have replaced the incipient function of the bower.

In the following section, I present descriptions of courtship for four species, two with bowers and two without. Examples from two different types of bowers, avenue and maypole, illustrate how these bowers are used in courtship. The species described here characterize a modal type in their clades. For each, there are related species that show widely divergent behaviors but which are consistent with the use of the bower for protection.

Display in the Satin Bowerbird: An Avenue-Builder

Satin bowerbirds (*Ptilonorhynchus violaceus*) are representative of species that build avenues. They occur in rain forests that fringe eastern Australia. The bowers of satin bowerbirds have two walls of sticks separated by a central avenue where females stand and then crouch when they are courted. The bower is aligned in a north-south direction with a decorated display court on the north end. The male displays while facing the female from the display court with a decoration held in his beak.

The females visiting the bower and the court typically alight in the vegetation on the south side of the bower and then move directly into the avenue of the bower. Initial vocalizations consist of numerous guttural chortles and squeaks that progress into a typical call sequence that consists of an initial mechanical buzzing followed by mimicry of a kookaburra (*Declio gigas*), a Lewin's honeyeater (*Meliphaga lewinii*), and less frequently, a crow (*Corvus coronoides*) (Loffredo and Borgia, 1986). The buzzes in the mechanical calls occur in conjunction with rapid movements by the male across the north entrance of the bower accompanied by flicks of one or both of his wings.

Females signal their willingness to copulate by lowering from an upright stance to a crouch. To copulate, the male circles to the opposite side of the bower to mount the female. After a brief (3-second) copulation, the female shakes intermittently in or near the bower for up to several minutes and then leaves. Females usually visit the bowers of several males, but mate with only one. The average courtship lasts slightly more than 4 minutes.

In satin bowerbirds and other avenue-builders, most courtships end with the female leaving from the north entrance of the bower as the male moves from the court toward the south entrance in his attempt to copulate. Only 9% of satin bowerbird's courtships are successful, although the most attractive males mate in 25% of their courtships. There is a significant relationship between the number of decorations and the mating success of males, indicating that it is important for females to enter the bower and see the display in order to assess the males.

Display in Macgregor's Bowerbird: A Maypole-Builder

Macgregor's bowerbird (*A. macgregoriae*) occurs at high elevations in the mountain ranges across central and eastern New Guinea. Its bower is a simple maypole, a sapling decorated with sticks and moss. Commonly, the sapling is rather thin and sticks are placed nearly horizontally, increasing the diameter around the sapling to approximately 25 cm. The low part of the maypole and the floor of the round court surrounding it are covered with a fine compressed moss mat that rises to form a circular rim approximately 40 cm from the may-

The bower of MacGregor's bowerbird.

pole. Many small decorations are used on the court, including insect parts and seed pods. Regurgitated fruit pulp is hung near the ends of the sticks of the maypole bower. Numerous large woody black fungi are arrayed on the rim of the court and on nearby logs.

Females arriving for courtship often land on the maypole, move down it, and then hop onto the court. The male usually is already present on the court and may have been calling prior to the female's arrival. The male positions himself on the opposite side of the maypole from the female with his chest pressed up against it and with his head plume concealed. He calls and, as the female moves around the maypole, he makes a counter move so as to keep the bower between him and her. Calling and counter moves continue for approximately 4 minutes before the male increases the intensity of display by expanding his bright orange head plume and violently shaking his head from side to side. The side-to-side shaking is associated with a rapid foot movement that appears to counterbalance the rapid movement of his head. Seen from the female's side of the bower, this display creates rapid orange flashes on each side of the maypole. After several bouts of head-shaking, the male moves around the bower toward the female in an attempt to copulate with her.

Although the shape of avenue and maypole bowers is fundamentally different, there is a striking similarity between maypole bowers and those of avenue-builders in the way in which the bower is used to separate the male and female. In each case, the male develops a prolonged courtship display. He watches the female and, when she signals her readiness to mate by crouching, he moves around the bower to mount her. If the female is not prepared to copulate, the bower serves as a dodge that allows her to leave from the opposite side from where the male is approaching.

Display in Toothbilled Bowerbirds: No Bower with Leks

Toothbilled bowerbird courtship is very different from that of other species of bowerbirds. Males clear courts but do not build bowers. Male courts are close together and aggregated into a lek (a group of displaying males not associated with resources needed by females). Males interact with frequent loud calls, with dominant males interrupting the calls of males on adjacent courts. Unlike other bowerbirds, they use large objects (e.g., leaves) as decorations on their courts. Courts surround several trees of small diameter, the bases of which have been cleaned meticulously of debris.

Females arrive on the court and stand very still, as if waiting for the male. Toothbill courtship is very brief. The male aggressively mounts the female with little or no display after she arrives on the court. The average time of courtship is 3.8 seconds. However, copulations are prolonged and violent relative to the brief and more cooperative behavior seen in other species. During these copulatory bouts, the male continues to display with characteristic low buzzing calls and wing beats. Females leave the court immediately after mating without the prolonged flapping of wings that is characteristic of other species.

By far, toothbill males spend less time on their court than any other species, and it appears that the adjustments in their display and mating behavior reflect an especially high susceptibility to predation while they are on the ground. The aggressive nature of the courtship suggests that females might not be fully prepared to copulate when they arrive, but the need to reduce time on the ground has caused males to attempt to speed this process.

The loud vocal interactions of males on courts and the use of large decorations suggest that females can evaluate male qualities such as dominance before they arrive on the court. At lek centers, males that preliminary studies show are dominant in vocal interactions have the highest mating success. This correlation suggests that females are choosing mates. Observations by Frith and Frith (1993) and our group show that males hiding behind a tree on their court sometimes call to females on nearby perches. However, we have not seen this calling lead to copulation.

In toothbills, if male calling interactions and large leaves allow females to choose mates before they arrive on the courts, then bowers may not be necessary. The female already may have selected the male before arriving at his court, so the prospect of forced copulation is not threatening. The capture of females by males indicates that forced copulations are possible in bowerbirds. The hiding of males behind trees during calling displays suggests a situation analogous to the initial condition of bower evolution in which females seek some protection from the courting male.

Archbold's Bowerbird: No Bower and No Lek

Archbold's bowerbird (*Archboldia papuensis*) is the other polygynous species that does not build a bower. Male Archbold's clear a large display court overlain with a mat of ferns where males place a variety of decorations, including snail shells, dark fruit, beetle wings, and King of Saxon Bird-of-Paradise (*Pteridophora alberti*) head plumes. Typically, small decorations are in piles near the fringes of the display court and are arrayed on the limbs overhanging the court. Male Archbold's bowerbirds neatly drape limbs that cross up to 1.5 m above the display court with a uniform curtain of flowerless orchid vines that nearly touch the display court and subdivide it. The cumulative array of the curtains provides a rather dramatic visual effect.

Male Archbold's bowerbirds are large and uniformly black, with a bright yellow head crest that extends from between the eyes, over the top of the head, to the neck. The crest covers more area on the bird's head than the crests in species of *Amblyornis*, but the individual plumes are much shorter.

Courtship in Archbold's bowerbird also is unique. Males chase females around the court. Occasionally the female pauses, and the male stops near her and attempts to approach her with his body pressed close to the court. If the male is successful in approaching her, he faces toward the female with his head near the ground. He emits a chattering call and rapidly moves his head with slight side-to-side movements. If the female does not move after the initial frontal display, he slowly moves behind her while maintaining a position near the ground and then rises up to copulate. Copulation is brief and lasts only 3 seconds, as is typical for all bowerbirds except toothbills.

In Archbold's bowerbird, courtship is not constrained to a particular site, as it is in species with bowers. The preliminary phase of courtship involves chases about the court, and the large size of the court may accommodate these chases. The walls may function in constraining the direction females can move.

Male Archbold's bowerbirds have evolved an alternative solution to the problem of copulating with females in the absence of a bower. Unlike toothbills, they do not interact over long distances or have leks, and they do not attempt to grab and copulate females by force. Like most other bowerbirds, they have prolonged courtship and frequent female rejections, suggesting that choice occurs on the display court. The low position of males during courtship does not compromise the female's ability to escape the courting male, even when he is nearby.

One explanation for the loss of bower-building in Archbold's bowerbird may be related to the widely ranging displays that females use to test males. In most species of bowerbirds, males can be directly compared because they compete by stealing their competitor's decorations and destroying their bowers. Visiting females may assess male competitive ability by the quality of his display (Borgia, 1985; Borgia et al., 1985). In *Archboldia*, the bowers are spread very far apart, and the possibility of male interaction is low. The frequent chases may be a means by which females test male athleticism. Elsewhere I have suggested that intense displays which span large areas in male spotted bowerbirds function similarly for female assessment (Borgia and Mueller, 1992; Borgia, 1995). In *Archboldia*, where the bowers of males are spread apart in small forest islands, there may be a similar need for males to demonstrate fitness in athletic rather than interactive components of display. Comparisons among all of the species of bowerbirds that I have studied indicate that the two species with interbower distances of >700 m (*Archboldia* and *Chlamydera maculata*) have especially large display courts and male displays that range over these courts, whereas the displays of the remaining nine species with bowers that are closer together have relatively small courts and male display is restricted to these courts ($\chi^2=3.14$, df=10, p=0.02).

A MODEL FOR THE EVOLUTION OF BOWERS

The hypothesis that protection from the courting male is important in the evolution of bowers is supported by the patterns of evolution in bowers and its consistency with the diversity of types of bowers. The ancestor to the lineage that led to modern species that build bowers probably displayed on a decorated ground court. Females favored courts with a natural barrier, such as a sapling, that separated them from males during courtship because it allowed them to approach the male and closely observe his display and decorations while still retaining the ability to leave if not stimulated by the display. Males who enhanced this barrier, e.g., placed sticks around the sapling and enlarged its diameter, offered females a safer vantage point for observing display. Males could gain from this elaboration by exploiting the female preference for mating in a protected environment. Increased female visitation and lessened threat during courtship contributed to an overall increase in matings over what might be achieved by forced copulations. Gains for females from the avoidance of forced copulations might include eschewal of genetically inferior males and reduction of direct physical costs (e.g., parasite transmission and time lost in remating). Remating of females which have been forced to copulate with other males would lower the value of forced copulations to court owners and may have caused males to shift efforts toward attraction of females.

The simple bower described above is similar to the bower built by MacGregor's bowerbird. Once the tendency to use sticks to build a bower evolved, however, it was possible to build a bower that functioned in female protection but did not require a central sapling. The loss of dependence on the use of saplings could have allowed males more freedom in selecting sites for bowers, in orienting their decorations, and in displaying their decorations. The ancestors of avenue-builders probably added a second barrier because it oriented females toward illuminated parts of the court where males could concentrate their decorations on a well-lit stage. The orientation of bowers in a north-south direction, the consistent placement of most decorations on the north side, and the clearing of leaves over display courts support this hypothesis.

This hypothesis is consistent with all types of bower-construction and our observations of how bowers are used in courtship. In both avenue and maypole bowers, males are forced to run around a barrier in order to reach the rear of the female where they can copulate. The delay caused by this extra traveling time gives females an opportunity to escape males that are unattractive to them. The behavior of the two species that do not build bowers also is consistent with the protection hypothesis. In toothbilleds, there is no bower, but because females appear to move to the ground for copulation only after they make their mating decision, they leave the court without mating proportionately less often than females in other species (Borgia, in preparation). In Archbold's bowerbird, by lying close to the ground as they approach females, males are not threatening as forced copulators.

CONCLUSIONS

We rarely can be sure about the evolutionary origins of a trait, but we can use information from comparisons among extant species to formulate reasonable hypotheses. In the case of the hypothesis that bowers provide protection from courting males, the great diversity of behaviors among species of bowerbirds with very consistent elements within species gives surprisingly strong support for this hypothesis of the origins of bower-building. This work has the added benefit of suggesting an important role for models of proximate benefit in explaining elaborated male traits. This work could not have been accomplished had the number of species of bowerbirds available to study been limited by extinctions. This places an immediate imperative on carrying out detailed comparative studies of behavior before there are large reductions in numbers of species. It also suggests that attempts to preserve representative species may not be productive because there are no typical species.

ACKNOWLEDGMENTS

This research was supported by funds from the National Science Foundation (BNS 85-10483 and BSR 89-11411), from the Graduate Dean and Dean of Life Sciences, University of Maryland, and from the University of Maryland Computer Science Center. The New South Wales and Queensland National Parks, the Australian Bird and Bat banding scheme, and the Papua New Guinea Wildlife and Conservation Department provided permits. R. Crozier, J. Dimuda, G. Harrington, I. Hayes, J. Hayes, N. Hayes, J. Lauridsen, M. J. Littlejohn, J. Kikkawa, N. Raga, J. Hook, and M. and J. Turnbull provided various important forms of support. C. Depkin, D. Bond, A. Day, and J. Morales served as team leaders. More than 100 volunteer assistants provided excellent help in the field. This paper is dedicated to the memory of Iris, Jack, and Ned Hayes, who were great friends and critics.

REFERENCES

Andersson, M. 1982. Female choice and runaway selection. Biol. J. Linn. Soc. 17:375-93.
Andersson, M. 1986. Evolution of condition-dependent ornaments and mating preferences: Sexual selection based on viability differences. Evolution 39:967-1004.
Andersson, S. 1989. Sexual selection and cues for female choice in leks of Jackson's widowbird *Euplectes jacksoni*. Behav. Ecol. Sociobiol. 25:403-410.
Basolo, A. 1990. Female preference predates the evolution of the sword in swordtail fish. Science 228:340-344.
Borgia, G. 1979. Sexual selection and the evolution of mating systems. In M. Blum and A. Blum, eds., Sexual Selection and Reproductive Competition. Academic Press, N.Y.
Borgia, G. 1985. Bowers as markers of male quality. Test of a hypothesis. Anim. Behav. 35:266-271.
Borgia, G., and U. Mueller. 1992. Bower destruction, decoration stealing, and female choice in the spotted bowerbird (*Chlamydera maculata*). Emu 92:11-18.
Borgia, G., S. Pruett-Jones, and M. Pruett-Jones. 1985. Bowers as markers of male quality. Zeit. für Tierpsychol. 67:225-236.
Brooks, D. R., and D. A. McLennan. 1991. Phylogeny, Ecology, and Behavior: A Research Program in Comparative Biology. University of Chicago Press, Chicago.

Burley, N. 1985. The organization of behavior and the evolution of sexually selected traits. Pp 22-44 in P. A. Gowaty and D. W. Mock, eds., Avian Monogamy. American Ornithologists' Union, Washington, D.C.

Darwin, C. 1859. On the Origin of Species. John Murray, London.

Darwin, C. 1871. The Descent of Man and Selection in Relation to Sex. John Murray, London.

Fisher, R. A. 1930. The Genetical Theory of Natural Selection. Clarendon Press, Oxford, England.

Frith, C., and D. Frith. 1993. Courtship display in the Tooth-billed Bowerbird Scenopoeetes dentirostris (Ptilonochynchidae). Australian Bird Watcher 11:103-113.

Gibson, R., J. Bradbury, and S. Vehrencamp 1991. Mate choice in lekking sage grouse revisited: The roles of vocal display, female site fidelity, and copying. Behav. Ecol. 2: 165-180.

Halliday, T. 1978. Sexual selection and mate choice. Pp. 180-213 in J. R. Krebs and N. D. Davis, eds., Behavioural Ecology: An Evolutionary Approach. Sinauer Associates, Sunderland, Mass.

Hamilton, W., and M. Zuk. 1982. Heritable true fitness and bright birds: A role for parasites? Science 218:384-387.

Kirkpatrick, M. 1982. Sexual selection and the evolution of female choice. Evolution 36:1-12.

Kirkpatrick, M. 1987. Sexual selection by female choice in polygynous animals. Ann. Rev. Ecol. Syst. 18:43-70.

Kusmierski, R., G. Borgia, R. Crozier, and B. Chan. 1993. Molecular information on bowerbird phylogeny and the evolution of exaggerated male characters. J. Evol. Biol. 6:737-752.

Lande, R. 1981. Models of speciation by sexual selection on polygenic traits. Proc. Natl. Acad. Sci. 78:3721-3725.

LeCroy, M., A. Kulipi, and W. Peckover. 1980. Goldie's bird of paradise: Display, natural history, and traditional relationships of people to the bird. Wilson Bull. 92:298-391.

Loffredo, C., and G. Borgia. 1986. Sexual selection, mating systems, and the evolution of avian acoustical displays. Amer. Nat. 128:733-794.

Maynard Smith, J. 1991. Theories of sexual selection. Trends Ecol. Evol. 6:146-151.

McDonald, D. 1989. Correlates of male mating success in a lekking bird with male-male cooperation. Anim. Behav. 37:1007-1022.

Møller, A. P., and A. Pomiankowski. 1993. Why have birds got multiple sexual ornaments? Behav. Ecol. Sociobiol. 32:167-176.

Parker, G. A. 1983. Mate quality and mating decisions. Pp. 141-166 in P. Bateson, ed., Mate Choice. Cambridge University Press, Cambridge, England.

Prum, R. O. 1990. Phylogenetic analysis of the evolution of display behavior in the Neotropical manakins (Aves: Pipridae). Ethology 84:202-231.

Ryan, M. J., J. H. Fox, W. Wilczynski, and A. S. Rand. 1990. Sexual selection for sensory exploitation in the frog Physalaemus pustulosus. Nature 343:66-67.

Trivers, R. L. 1972. Parental investment and sexual selection. Pp. 136-179 in B. G. Campbell, ed., Sexual Selection and the Descent of Man. Aldine Press, Chicago.

Wilson, E. O. 1992. The Diversity of Life. Belknap Press, Cambridge, Mass. 424 pp.

Zahavi, A. 1975. Mate selection—a selection for a handicap. J. Theor. Biol. 53:205-214.

Zuk, M., R., Thornhill, and J. D. Ligon. 1990. Parasites and mate choice in red jungle fowl. Amer. Zool. 30:235-244.

PART V

BUILDING TOWARD A SOLUTION:
NEW DIRECTIONS AND APPLICATIONS

Coral reefs: marine biodiversity writ large.

CHAPTER
19

Microbial Biodiversity
and Biotechnology

RITA R. COLWELL
President, University of Maryland Biotechnology Institute,
College Park, Maryland

Biodiversity and biotechnology are strongly interrelated and interdependent. With respect to microbial diversity, microorganisms have been used in a variety of ways in the food industry. A good example is the dairy industry, in which microorganisms have been used to produce butter, yogurt, and cheese. Genetic modification of key microorganisms used in the manufacture of these foods has been accomplished to improve flavor, texture, and taste of the final product. Thus, the enormous diversity of the Earth provides the foundation for biotechnology, not only in food production, but also in agriculture to improve soil fertility, in the pharmaceutical industry as a source of antimicrobial and antitumor agents, and in production of compounds of value in the chemical industry, e.g., dextran, a bacterial polymer.

Despite the popular press' view of biotechnology, it is not a field that was discovered only recently. In fact, it enjoys a long history of contribution to the well-being of humans, since leavening of bread and the fermentation of grapes to make wine were activities practiced in China and Egypt centuries ago. However, following on the steps of the Industrial Revolution, development of large-scale fermentation processes, coupled with major discoveries in genetics providing an understanding of genes and chromosomes and the molecular basis of heredity, the explosive developments in modern biotechnology were made possible during the past 20 years.

The new "era" of biotechnology includes a myriad of innovations based on recombinant DNA. In addition to recombinant DNA, the tools of molecular biology include protoplast fusion and production of hybridomas to provide monoclonal antibodies. The value of these tools is that they make possible the

279

production of biological hybrids otherwise not possible by classical mating and selection. Gene cloning also provides the capacity to introduce genes into organisms to provide a new function for that organism. An example of this is the luminescent tobacco plant, which results when the lux gene (a gene or genes responsible for production of enzymes involved in bioluminescence) has been cloned from a bioluminescent organism, usually a microorganism, and inserted into the genetic structure of the host plant. The lux gene is inserted next to a gene whose function is to be studied so that both genes are activated at once and the plant glows whenever the adjacent gene produces its product. This has allowed major advances in our knowledge of the structure and function of genes. Furthermore, biotechnological products such as the "pomato," a result of protoplast fusion of the tomato and potato, have advanced our basic knowledge of manipulating the properties of organisms for human benefit, including plants of agricultural importance. Beside these novel results, which have contributed to our basic knowledge of how organisms function, the more apparent benefits of biotechnology, such as the production of human insulin through genetic manipulation of easily cultured microorganisms, sometimes have been overshadowed.

Faced with burgeoning human populations, many government leaders expect the biotechnology "revolution" to provide a solution for the ever-increasing global demand for food and to resolve human medical and epidemiological issues. Unfortunately, there is a significant distance between this hope and actual fulfillment. Nevertheless, without the contributions of biotechnology to agriculture and medicine, it is unlikely that these problems will be overcome.

This paper provides a brief description of the biotechnology industry, similarly reviews the biodiversity of microorganisms, examines the significance of this biodiversity for a sustainable biosphere, and evaluates the significance of microbial biodiversity for the further development and application of biotechnology for human needs.

GROWTH OF THE BIOTECHNOLOGY INDUSTRY

Major sectors of the biotechnology industry currently include agriculture, food processing, industrial chemicals, and pollution control. When environmental biotechnology, which includes pollution control and bioremediation (the restoration of a damaged or disturbed ecosystem to a functional ecological state through the activities of living organisms), begins to fulfill its promise, the biotechnology industry will increase both its markets and its economic diversity significantly.

Biotechnology companies in the United States now number well over 1,300. Projections for the growth of the biotechnology industry made in the early 1980s were that, by 1990, the industry would be valued at approximately $5-7 billion. In fact, the Annual Report on the Biotechnology Industry (Burrill and Lee, 1993) noted that, in the United States alone, biotechnology sales were $10-12 billion in

1993. Sales of the U.S. biotechnology industry are projected to be about $100 billion by the year 2035.

Funding for marine biotechnology has increased in the United States from a negligible amount of less than $1 million in 1983 to approximately $100 million in 1994 (Zilinskas et al., 1994). In Japan, investment in marine biotechnology is estimated to be in the range of $1 billion during the past decade.

Strong interactions between universities and the industry have developed in recent years. Components of technology transfer between universities and the biotechnology industry include intellectual property sharing, technology marketing, formation of start-up companies, and establishment of incubator facilities. Basic research undertaken in universities provides the knowledge necessary for understanding the diversity of organisms (especially microorganisms) and is the basis for the biotechnology industry, providing critical understanding of genetic and cellular mechanisms and thereby allowing application of biotechnological principles to the improvement of the human condition. The well educated and highly trained work force supplied by universities provides the technical staff, enables technology transfer, and ensures success of the industry.

BIODIVERSITY OF MICROORGANISMS

The driving forces of microbial (bacterial, viral, viroid, filamentous fungal, yeast, microalgal, and protozoan) diversity include the genetic constitution of these organisms, the environment in which they are found, and ecological interactions with other components of the biosphere. The result is an extraordinary richness of microbial diversity, most of which remains to be explored.

Microorganisms inhabit virtually every ecological "nook and cranny" in the biosphere. Because of their very small size, direct observations of species diversity of microorganisms in natural environments rarely can be made unless an unusual abundance of a single species occurs, producing a characteristic texture of growth (e.g., that found in geothermal springs). New biotechnological methods, however, are beginning to reveal a vast, previously unrecognized realm of microbial life that has tremendous ecological and medical importance. Of the above microorganisms, this paper emphasizes the spectacular functional diversity of bacteria and viruses as they relate to biotechnological developments.

Bacteria exhibit significant diversity in size and morphology. Most planktonic bacteria in lakes and oceans that have been cultured successfully are coccoid or spherical cells, approximately 0.2 to 0.6 μ in diameter, or short, rod-shaped cells about 1 to 3 μ in length and 0.2 to 0.6 μ in diameter. Many species of bacteria demonstrate unusual shapes and forms (e.g., helical, coiled, triangulate, etc.). Other bacteria show unusual biochemical and biomechanical properties, such as the magnetotactic bacteria, which possess intracellular magnetic particles allowing the cells to orient to the Earth's magnetic poles. Although many bacterial species can be grown easily in culture, the vast majority either are not easily cultured or are unculturable by present methods. Consequently, we have little idea of the true morphological and functional diversity of all microorganisms extant on the Earth.

Viruses are intracellular parasites containing only one type of nucleic acid, either RNA or DNA, and usually a protein coat. Following invasion of the host cell, most viruses incorporate themselves directly into the DNA or RNA of the host cell's genetic machinery. They then rely on their host cell to provide part or all of the materials for replication and expression of the viral genetic information. Unlike host cell genetic elements, however, some viruses also can exist functionally as extracellular particles (e.g., as seen in the viruses that attack bacteria). Although these viruses are capable of surviving for long periods of time in the external environment, their ecology and reproduction is linked intimately with their host. They may serve as a vector in the transfer of genetic information from one host to another. Viruses are especially valuable in human medicine, i.e., gene therapy, in serving as the "carrier" of a fully functional gene that can correct an inborn error of metabolism by gene substitution or replacement.

Viruses are far more common and ecologically important than previously thought. For example, in water samples collected from the Chesapeake Bay, viruses (observed with the electron microscope) are more abundant than bacteria in late summer and early autumn (Wommack et al., 1992). Recent studies also indicate that there are more viruses than bacteria in the open ocean. Many of these viruses attack bacteria, and may represent a significant control agent for the ubiquitous bacterial flora in the environment, and they may control algal abundance as well. We are only beginning to comprehend the ecological significance and diversity of these submicroscopic organisms in the environment.

Another aspect of microbial diversity, not readily recognized, is the relatively limited number of species that have been described and the even greater number of estimated species not yet known. Only 3,000-4,000 species of bacteria have been described (Hawksworth and Colwell, 1992). It has been estimated that there may be as many as 300,000 species of bacteria, but more likely the number is closer to 3,000,000. The number of species of viruses has been estimated to be approximately 5,000, but only about 500 have been described.

About 17% of known species of fungi have been cultured, but less than 1% of known fungal species are available in the world's culture collections; described species represent only 1-3% of the total estimated species of fungi (Hawksworth and Colwell, 1992). Thus, the bulk of undescribed species are microorganisms (bacteria, viruses, viroids, filamentous fungi, yeasts, microalgae, and protozoans).

THE ECOLOGICAL IMPORTANCE OF MICROBIAL BIODIVERSITY

Microorganismal diversity provides essential ecological services for the biosphere by regulating the composition of the atmosphere, controlling the structure and fertility of the soil, and regulating agricultural pests.

Microbial activities comprise vital links in the chain of geochemical events that occur when nutrient elements are cycled. For example, in the carbon cycle, methanogenic microorganisms can influence the climate by producing methane gas, which is a major greenhouse gas. Although not uniquely produced by microorganisms, carbon dioxide, also a greenhouse gas, is produced by respiration and heterotrophic decomposition by many bacteria and fungi. Plankton and some species of bluegreen bacteria produce dimethyl sulphide, a compound that promotes cloud formation and rainfall. The evidence suggests that microorganisms may affect climate through cloud formation. In marine ecosystems, some algae and cyanobacteria produce dimethyl sulphide (DMS) in large quantities. DMS is volatile and is readily oxidized in the atmosphere to form first dimethyl sulfoxide and then sulfate which acts as a nucleating agent in water droplet formation. Thus, the more DMS that is produced, the more water droplets formed, the more clouds produced, and the greater the rainfall.

In addition to the effects of carbon dioxide and methane in the atmosphere, nitrifying and denitrifying microorganisms produce nitrous oxide which photochemically reacts with ozone, contributing to events that admit increasing concentrations of ultraviolet radiation to the Earth's atmosphere.

Photosynthetic organisms fix carbon dioxide and release oxygen to the atmosphere. Microbial respiration replenishes atmospheric carbon dioxide and reduces significant amounts of atmospheric oxygen to water and other compounds. Some bacterial species oxidize elemental nitrogen, which ultimately is returned to the atmosphere by bacteria that reduce the nitrogen in nitrate compounds to the elemental state. The bacteria in soil oxidize methane to carbon dioxide and produce volatile compounds of sulfur, phosphate, and nitrate that enter the atmosphere.

Microbial diversity is intimately related to soil structure and function. Bacteria generally occur at concentrations of $\geq 10^9$ cells/gram of soil. Soil governs the productivity of plants and, therefore, the sustainability of agriculture, forestry, and natural ecosystems. Some of the best soil is formed in grassland pastures, where bacteria are associated with root material and are attached to clay particles (Lynch and Poole, 1979). In most cases, these bacteria are responsible for transforming and cycling carbon, nitrogen, phosphorus, iron, and sulfur in the soil and for the manner in which aggregates and clumps of soil are formed. In well-drained soil sustained by a healthy bacterial flora, much of the space between soil aggregates is filled with air. Since oxygen is necessary for metabolism in plant roots, this aerated structure of the soil is necessary for soil productivity. Unfortunately, if certain microbial species become dominant in the soil, the system can become anaerobic (lacking in oxygen). Also, some bacteria produce gums and cements (Margulis et al., 1989) that can block pores in the soil. Both of these effects can decrease soil fertility. It is essential to understand how the biodiversity of microbes in soil maintains agricultural productivity.

Another example of the diverse ecological activities of bacteria is their ability to control insects. Many insects carry a microbial flora on their surface and in their gut. Populations of microorganisms pathogenic for the insect may develop if the insect is injured. Bacteria also produce chemical compounds that adversely affect insect growth. Thus, manipulation of microbial populations provides a mechanism by which agricultural pests can be controlled.

In summary, microbial biodiversity represents the foundation of a sustainable biosphere and is fundamental to sustainable agriculture. The activities of microorganisms, in the aggregate, and the diversity of species, most of which still remains undescribed, provide a rich source of genetic variation for application to biotechnology.

THE SIGNIFICANCE OF MICROBIAL
BIODIVERSITY FOR BIOTECHNOLOGY

Recently, the bacterium *Agrobacterium tumefaciens* has been used to genetically engineer plants. *Agrobacterium tumefaciens* carries a plasmid that normally causes plant disease. However, by inserting specific genes into the bacterial plasmid, desirable characteristics can be transferred to the plant genome where the plasmid DNA is integrated. Thus, herbicide resistance, pesticide activity, and designer fruits and vegetables can be produced.

In the marine sphere, many of the unique compounds of potential medical value that are retrieved from the tissues of a range of marine invertebrate and vertebrate organisms are produced by marine bacteria. An example of a compound produced by marine microorganisms is tetrodotoxin, a toxin found in the fatty tissues and reproductive organs of the puffer fish, and originally thought to be produced by the fish itself. Tetrodotoxin is a very powerful analgesic (about 300,000 times more powerful than cocaine) that resembles procaine in its ability to inhibit transmission of nerve cells. Tetrodotoxin has been used in Japan to treat pain in neurogenic leprosy and cancer. Thus, tetrodotoxin is a compound produced by bacteria found on the puffer fish and, more recently, bacteria associated with benthic invertebrates, as well. Tetrodotoxin is one example of a compound that may have wide pharmaceutical value in the future.

Diversity "hot spots," such as the Great Barrier Reef off the coast of Australia, provide a wide array of known and unknown microorganisms. The diversity of corals and other marine invertebrates on these reefs have long been appreciated, but some of their bacterial symbionts have been described only recently.

Tunicates (Didemnidaceae), i.e., sea squirts, produce compounds (didemnins) that have antileukemic properties. Collections of sponges (Sclerospongia) from Caribbean coral reefs also have yielded bacterial species that produce unusual antimicrobial compounds. Cross sections of the sclerosponge revealed that 50% of the sponge mass is comprised of procaryotic structures concluded to be bacteria. Most of these bacterial species have not yet been cultured (Santavy and Colwell, 1990). Of those cultured, several have been found to produce compounds of potential pharmacological value. In general, marine sponges are widely known for production of chemicals that deter predators or even cause ill effects in divers exposed to them; many of these compounds may be produced by the diverse bacteria living within the tissues of the sponge.

In another application of microbial diversity to biotechnology, an enzyme active at high temperatures was prepared relatively recently from *Thermus aquaticus*, a microorganism which had been isolated from the hot springs of Yellowstone National Park many years ago. Using this enzyme, the polymerase chain reaction (PCR) was developed, an enormously powerful molecular genetic tool that cuts and splices DNA and allows detailed examination of gene sequences. Application of PCR in criminology is already well known.

An important aspect of microbial diversity in biotechnology is utilization of microorganisms as "janitors" of the planet, i.e., in bioremediation. Microorganisms are useful in restoring habitats to a functional ecological state because they are capable of degrading pollutant compounds, purifying water and soil in the process. Genetically engineering microorganisms to degrade toxic compounds more rapidly and completely offers great promise for bioremediation and biorestoration in the future.

RELATION OF LOSS OF MACROORGANISMAL BIODIVERSITY TO MICROBIAL ECOLOGY AND DIVERSITY

While loss of diversity obviously results in lost products and lost markets, an insidious and often unrecognized result of disturbed environments and biodiversity loss is the effect on human health. Deforestation, for example, not only creates problems of environmental degradation, but enhances the potential for epidemics of new and unpredictable diseases that may be devastating for human populations. An example is an epidemic that occurs when native or introduced macroorganisms reach high densities in the course of extreme population fluctuations associated with the disturbed environmental conditions, such as destruction of native vegetation for logging or farming, weather changes related to the El Niño Southern Climatic Oscillation, or anthropogenic effects on atmospheric warming and precipitation. For example, deforested areas may allow massive increases in mosquito populations, leading to malaria outbreaks in humans. Another example is that of rodents, which harbor microorganisms that are pathogenic for humans but occur in sufficiently low frequency, when rodent populations are low, that the risk of human disease also is relatively low. When rodent population densities reach unusually high levels, however, serious outbreaks of disease caused by viruses that are carried by rodents may occur in humans. In other instances, a microorganism normally found in rodents may shift hosts, once rodent populations reach critical densities, and cause a pathogenic epidemic among humans in the area. The recent Hanta virus outbreak in the southwestern United States is but one example of such phenomena (Epstein et al., 1993; Epstein, 1995). The Hanta virus is carried by a deer mouse, but can be transmitted to humans in certain circumstances. A proliferation of deer mice associated with a change in weather patterns resulted in outbreaks of the virus in humans who inhaled dust contaminated with dried mouse urine; these outbreaks resulted in a number of deaths.

Thus, the biodiversity of microbial populations, the biology of which is not understood, may impact human populations adversely as global populations continue to increase and environmental disturbances become more common. Research is needed to document and understand not only the biodiversity and population dynamics of macroorganisms in disturbed conditions, but also the biology and diversity of associated microorganisms, to avert such epidemics.

Cholera offers a useful example of disease outbreaks whose biology and relationship to natural environments needs to be understood in order to minimize adverse effects on human populations. Beginning in 1991, cholera outbreaks began devastating Latin America. Initially, approximately 285,000 cases occurred in Peru, with about 108,000 hospitalizations and 3,000 deaths. The disease subsequently moved to Ecuador and then to Colombia. Coastal areas were the first impacted. The migration pattern of the disease can be related to disturbances in the weather and in the pattern of upwelling in the East Pacific Ocean caused by the El Niño Southern Oscillation Event (Epstein et al., 1993). The epidemic strain of cholera, *Vibrio cholerae 01*, has been demonstrated to be associated with zooplankton that occur in coastal waters in Asia (Colwell and Huq, 1994), providing an explanation for the greater intensity of the disease in coastal areas of South America. A useful biotechnological product of these studies was a highly specific detection agent, a monoclonal antibody prepared against *V. cholerae 01*, allowing detection of *V. cholerae 01* in environmental samples. A diagnostic kit, now manufactured commercially, permits detection in the field so that the organism can be tracked in the environment (Colwell et al., 1992).

CONCLUSIONS

There is much yet to be learned about microorganisms in natural ecosystems. An improved understanding of microbial community structure, ecology, and population genetics is needed, as well as more information on microbial diversity and interactions of microorganisms at the species level. A workable species concept for microorganisms is lacking, but nucleic acid sequencing, a highly promising development in microbial systematics, may offer a pragmatic approach to the definition of microbial species. However, information for cataloging species and developing databases is just beginning to be collected at the level necessary for reproducible definition and description of species, i.e., for *polyphasic* definition of species.

Microbial diversity already has contributed significantly to biotechnology, but until a full understanding of microbial diversity and microbial interactions is gained, the benefits for biotechnology will not be realized completely. Microbial diversity, as a significant component of overall biological diversity, plays a major role in maintaining human health and sustaining the well-being of the environment. In any prospective search for compounds of medical or agricultural value, whether in a tropical forest of Latin America, in Maryland soil, or in off-shore Atlantic or Pacific ocean water, the microbiological resources that remain to be discovered undoubtedly will enrich the lives of the human race, and will reveal the intricate interlocking mechanisms of biodiversity that underlie the well-being of humans and the balance of our biosphere.

REFERENCES

Burrill, G. S., and K. B. Lee, Jr. 1993. Ernst and Young's Eighth Annual Report on the Biotechnology Industry: Biotech 94 Long-Term Value, Short-Term Hurdles. Ernst and Young, San Francisco. 98 pp.

Colwell, R. R., and A. Huq. 1994. Vibrios in the environment: Viable but nonculturable *Vibrio cholerae*. Pp. 117-134 in I. K. Wachsmuth, P. A. Blake, and O. Olsvik, eds., *Vibrio cholerae* and Cholera. American Society of Microbiology, Washington, D.C.

Colwell, R. R., J. A. K. Hasan, A. Huq, L. Loomis, R. Siebeling, M. Torres, S. Galvez, S. Islam, and D. Bernstein. 1992. Development and evaluation of a rapid, simple, sensitive, monoclonal antibody-based coagglutination test for direct detection of *Vibrio cholerae* 01. FEMS Microbiol. Lett. 97:215-220.

Epstein, P. R. 1995. Emerging diseases and ecosystem instability: New threats to public health. Amer. J. Pub. Health 85:113-117.

Epstein, P. R., T. E. Ford, and R. R. Colwell. 1993. Marine ecosystems. The Lancet 342:132-135.

Hawksworth, D. L., and R. R. Colwell, eds. 1992. Biodiversity amongst microorganisms and its relevance. Biodiv. Conserv. 1:221-345.

Lynch, J. M., and N. J. Poole. 1979. Microbial Ecology: A Conceptual Approach. John Wiley and Sons, Inc., N.Y.

Margulis, L., J. O. Corliss, M. Melkonian, and D. J. Chapman, eds. 1989. Handbook of Protoctista. Jones and Bartlett Publishers, Boston. 914 pp.

Santavy, D. L., and R. R. Colwell. 1990. Comparison of the symbiont bacterial community associated with the Caribbean sclerosponge *Ceratoporella nicholsoni* and ambient seawater. Mar. Ecol. Prog. Ser. 67:73-82.

Wommack, K. E., R. T. Hill, M. Kessel, E. Russek-Cohen, and R. R. Colwell. 1992. Distribution of viruses in the Chesapeake Bay. Appl. Envir. Microbiol. 58:2965-2970.

Zilinskas, R. A., R. R. Colwell, D. W. Lipton, and R. Hill. 1994. The Global Challenge of Marine Biotechnology. National Sea Grant College Program Publication. Maryland Sea Grant College, College Park. 330 pp.

CHAPTER
20

The Impact of Rapid Gene Discovery Technology on Studies of Evolution and Biodiversity

CAROL J. BULT, JUDITH A. BLAKE, MARK D. ADAMS, OWEN WHITE,
GRANGER SUTTON, REBECCA CLAYTON, AND ANTHONY R. KERLAVAGE
*Science Research Faculty and Staff, The Institute for
Genomic Research (TIGR), Gaithersburg, Maryland*

CHRIS FIELDS
*Scientific Director, National Center for Genome
Resources (NCGR), Santa Fe, New Mexico*

J. CRAIG VENTER
Director, The Institute for Genomic Research (TIGR), Gaithersburg, Maryland

Topics of critical concern for biodiversity include the development of standards to assess levels and distributions of diversity and the creation of a global network of interoperable databases to manage, analyze, and distribute large amounts of information to researchers and policymakers. The technical challenges confronted by researchers working to map and sequence the human genome (National Center for Human Genome Research, 1990) have led to new strategies for rapid gene discovery and informatics (Fields, 1992; Kerlavage et al., 1993; Venter et al., 1992) that we, the authors, are applying to large-scale molecular assessment of biodiversity. In this chapter, we describe our research efforts in areas relevant to biodiversity assessment; specifically, we outline the development of (1) multiple molecular markers for identification and classification of species, (2) new algorithms for multiple sequence alignment, and (3) databases that link taxonomic, phylogenetic, geographic, and molecular data.

WHAT IS RAPID GENE DISCOVERY?

Rapid gene discovery is the process of "tagging" transcribed genes by obtaining partial sequences of cloned DNA copies of many randomly selected messenger RNA (mRNA) transcripts. In the laboratory, the isolated mRNAs are reverse transcribed into cDNA (complementary DNA) that is then cloned and sequenced. The partial cDNA sequences, which are typically 200–400 basepairs (bp) in length, are called Expressed Sequence Tags (ESTs; Adams et al., 1991). Single-pass, automated sequencing of ESTs from randomly selected clones from cDNA libraries permits the rapid identification of genes expressed in cells, tissues, or whole organisms (Adams et al., 1991, 1992, 1993a,b; Khan et al., 1992; McCombie et al., 1992; Okubo et al., 1992; Waterston et al., 1992). The genes that ESTs "tag" are putatively identified by evaluation of the degree of similarity between the nucleotide and amino acid translations of an EST sequence and previously described DNA and protein sequences from public sequence databases (e.g., GenBank, EMBL, SwissProt). Significant similarity is evaluated using computer-assisted algorithms such as BLAST (Altschul et al., 1990) and BLAZE (Brutlag et al., 1993).

Automated single-pass DNA sequencing is more than 98.5% accurate, on average, for up to 400 bp per sequencing reaction (Adams et al., 1993a). For abundant ESTs, we observe an average of seven- to eight-fold redundancy, which provides an additional measure of quality assurance for sequence accuracy. The EST approach to gene discovery differs from previous methods in that it does not rely on screening for and sequencing of full-length cDNA clones. We estimate that the EST approach reduces the costs (in time and materials) associated with gene discovery by 2-3 orders of magnitude.

In 1993, the sequencing core at The Institute for Genomic Research (TIGR) generated up to 300,000 bp from 1,000 DNA templates per day (Adams et al., 1994). In comparison, the submission of sequence data to GenBank from all other DNA sequencing labs combined averaged just over 309,000 bp per day (based on submissions to GenBank from January to July 1993). Our current maximum sequencing throughput is 500,000 bp per day; with modifications of existing hardware and software, throughput is expected to reach 1 million bp of sequence each day within the next 3 to 5 years. Since the genome size of many microorganisms is in the range of 2-4 million bp, it is feasible that the complete genomes of some organisms will be sequenced in a single week in the near future.

ESTs were first used for rapid identification of genes expressed in the human brain (Adams et al., 1991, 1992, 1993a,b). Beginning in late 1992, researchers at TIGR extended the EST approach to include other human tissues and organs to determine patterns of gene expression during human development. As of April 1994, over 200,000 ESTs have been sequenced from 300 human cDNA libraries that were constructed from 37 distinct tissues and organs (Adams et al., 1995). Over 40,000 unique genes are represented by these EST sequences, including more than half of the estimated 60,000 genes in the human genome. The results of the human EST sequencing project provide the most comprehensive picture of gene expression patterns during human development to date.

When human ESTs are compared to existing public sequence databases, approximately 33% find exact matches among sequences already published,

2% identify (by similarlity of nucleotide sequences) potential new members of existing gene families (genes that arise via duplication events), 7% identify (by similarity of amino acid sequences) potential human homologs of genes from other species, and 57% do not match any published DNA or protein sequence. The gene identification rates are comparable to those obtained for ESTs or genomic sequences from *Escherichia coli* (a common enteric bacterium), *Saccharomyces cerevisiae* (baker's yeast), and *Caenorhabditis elegans* (a nematode), and from human chromosomal DNA sequencing (Adams et al., 1993a). We use the term *isolog* to describe ESTs that either have a significant but nonexact match to a nucleotide sequence within the same species or whose only significant match is to a protein sequence from a different species. This nomenclature distinguishes identity based solely on similarity of sequences from explicit hypotheses of common ancestry associated with the term *homolog*.

Although still a new approach to genome characterization, EST studies are proving to be a rich source of data, not only for gene discovery, but also for comparative evolutionary analyses. EST projects have proved useful for identifying potential homologs, identifying new members of gene families, genome nucleotide composition analysis (White et al., 1993), gene mapping (Polymeropoulos et al., 1993), and the analysis of synteny (genes that are located on the same chromosome) (Helentjaris, 1993).

MOLECULAR DATA IN SYSTEMATICS

Although the use of "informational macromolecules" (Zuckerlandl and Pauling, 1965) for systematics has become commonplace, molecular approaches to systematics have been criticized recently as not providing the degree of phylogenetic resolution that had been predicted. This assessment is premature given the limited sampling of possible molecular markers that have been tested empirically (Doyle, 1993). A brief survey of the relevant literature reveals that only about 40 molecular markers (nuclear and organellar combined) have been used for phylogenetic analyses to date. The small subunit of ribosomal DNA (rDNA), the largest subunit of ribulose bisphosphate decarboxylase (rbcL), and cytochrome-b are by far the most commonly employed markers in molecular systematic studies. Molecular phylogenetic studies to date have relied on single exemplars of large speciose groups and on the information content of one or two genes, research strategies that can lead to a distorted representation of evolutionary history (Lecointre et al., 1993). Overall, the potential of molecular data for systematic studies has barely been tapped (Doyle, 1993).

The development of new molecular markers has proceeded at a relatively slow rate because the process has depended in large part on retrospective analyses of gene sequences present in public sequence databases (Friedlander et al., 1992). The sequences currently available in these databases are biased in kinds and numbers of genes and species represented, making it difficult to design markers that are applicable across a wide range of species or to distinguish between orthologs (genes related by speciation events) and paralogs (genes related by gene duplication events) (Fitch, 1970). Distinguishing between orthologs and paralogs is a critical concern for molecular systematists because the evolutionary histories of species and genes do not always coincide (Fitch, 1970; Page, 1993; Sanderson and Doyle, 1992).

ESTs as a Tool for Marker Development

Two qualities of ESTs make them a valuable tool for designing molecular markers. First, the rapid generation of large numbers of candidate markers means that a diverse pool of sequences can be screened in a short period of time. Second, ESTs provide sensitive probes for identifying potential gene homologs across species. The population of cDNAs represented by EST sequences are biased only in that the transcripts present in a cDNA library will represent the gene expression patterns of a particular cell, tissue, organ, or lifecycle. Many of the transcripts we identify by ESTs are "housekeeping" genes essential for basic cellular function (e.g., elongation factor-1a, actin, ribosomal proteins, etc.). Housekeeping genes often contain regions that are highly conserved and, therefore, are good candidates as markers for tracking deep evolutionary divergences.

EST sequences can be used to identify potential homologs in much the same way that DNA hybridization probes identify homologs on Southern blots. ESTs have the advantage of being more selective as probes, because direct comparison of sequences reveals differences not detectable even with quantitative hybridization. ESTs are also more sensitive as probes because the protein translation of an EST can be used to detect similarity at the level of amino acids even when the match for nucleotide sequences is low. For example, we identified a novel human very-low-density-lipoprotein (VLDL) receptor via a protein match (87% similarity over 39 amino acid residues) of a translated EST to a rabbit VLDL receptor (Adams et al., 1993a). Detection of the novel human VLDL receptor by standard hybridization procedures was not possible because of the many nucleotide differences between the probe and the gene sequences of the novel human receptor.

Recent findings suggest that it will not always be necessary to isolate full-length cDNAs to confirm an identification based on an EST match. Because most of the cDNA clones in a library will be represented by many different ESTs, the complete cDNA sequence can be determined by assembling multiple overlapping ESTs. For example, the complete cDNA of a new human isolog of the yeast *sui1* translation initiation factor gene was assembled using overlapping, independent EST sequences (Fields and Adams, 1994).

In addition to the dozen or so EST projects world-wide that focus on humans, EST sequencing currently is under way in a number of laboratories for such diverse organisms as *Arabidopsis thaliana* (mouse-ear cress; Newman et al., 1994), *Brassica napus* (oilseed rape; Kwak et al., 1994), *Pinus taeda* L. (loblolly pine; Kinlaw et al., 1994), *Ceanorhabiditis elegans* (nematode; McCombie et al., 1992; Waterston et al., 1992), *Mus musculus* (domestic house mouse; Hoog, 1991), *Saccharomyces cerevisiae* (baker's yeast; Weinstock, personal communication, 1994), *Oryza sativa* L. (rice; Rice Genome Newsletter, 1992), *Zea mays* (maize; Shen et al., 1994), *Pyrococcus furiosus* (Robb, personal communication, 1993), *Plasmodium faciparum* (malaria parasite; Reddy et al., 1993), and *Macrops eugenii* (tammar wallaby; Collet and Joseph, 1994) . The National Center for Biotechnology Information (NCBI) maintains and distributes EST data via its EST database (dbEST; Boguski et al., 1993). As of November 1994, there were over 60,000 EST sequences from 24 different species in dbEST. (Information on how to use the dbEST server can be obtained by sending an electronic mail message with the word "HELP" as the text to est_report@ ncbi.nlm.nih.gov.)

The phylogenetic breadth represented by the organisms listed above is still quite limited and will not provide data adequate for testing many of the fundamental questions in molecular evolution (Sogin, 1991). However, we have begun several collaborative EST projects to encompass a broader sampling of phylogenetic diversity. These data will be invaluable as a starting point for designing new molecular markers and for addressing basic questions in molecular evolution, such as the extent and significance of ancient conserved regions (Green et al., 1993), the nature of the last universal ancestor (Forterre et al., 1993), the patterns and processes of gene and protein evolution, and the identification of orthologs and paralogs.

An Annealing Algorithm for Multiple Sequence Alignment

The critical starting point in any systematic analysis is the hypothesis of common ancestry for each character in the analysis (i.e., character homology). For morphological characters, hypotheses of homology are based on some understanding of common developmental processes. For molecular data, a multiple sequence alignment serves as the basis for hypotheses of homology at each nucleotide or amino acid position. Alignments are critical for the analysis of EST data, as the putative identification of ESTs are made on the basis of sequence comparisons to genes of known function. Most of the alignment algorithms available commercially and in the public domain rely on clustering by sequence similarity. With these algorithms, sequences are compared to each other and a matrix of similarity scores is generated. The two sequences having the closest similarity scores are aligned and a consensus sequence is calculated. The consensus sequence is then aligned with the next closest sequence, and the process is repeated until all of the sequences are aligned. A major limitation of this "greedy" alignment method is that, once a pairwise alignment is made, regions of low similarity are locked in place for all subsequent iterations. As a result, suboptimal alignment is perpetuated in some regions when sequences having varying degrees of similarity are compared.

In collaboration with colleagues at Maspar Computer, Inc., we have developed an "annealing" alignment algorithm that does not use the distance metric approach described in the preceding paragraph and that can take advantage of the processing power of a parallel computer (Sutton and Busse, 1993). The annealing process starts by searching for pairwise regions of high similarity. Only regions of high similarity are aligned initially for each pairwise comparison, leaving regions of lower similarity free to find a better alignment in subsequent pairwise comparisons. Each iteration of the algorithm is made at progressively lower match stringency until a stable alignment is achieved. As with distance metric algorithms, scoring matrices for amino acid replacements can be used to optimize matches when highly divergent protein sequences are compared. The annealing alignment method is superior to distance metric approaches in aligning sequences that are significantly different in length or that contain long gaps.

Databases and Biodiversity

A current limiting factor in research on molecular systematics is not the generation of data, but the effective management, analysis, and distribution of data (Fields, 1992; Kerlavage et al., 1993). Researchers need access to a federated system of databases linking diverse data resources that can answer basic and applied questions about biodiversity (National Research Council, 1993). Three databases developed at TIGR to support our research programs provide a model for the development of seamless links between diverse information resources: the Expressed Sequence Tag Database (ESTDB; Kerlavage et al., 1993), the Expressed Gene Anatomy Database (EGAD; Fields et al., 1993), and the Sequences, Sources, Taxa database (SST; Bult et al., 1994). The steady-state acquisition and analysis of EST sequence data at TIGR is supported by a relational database, ESTDB, together with a suite of custom-built and public domain tools for analysis and user-interfaces. We have developed EGAD to link EST sequences with other relevant data, including allelic differences, map location, gene expression, cellular role, and biochemical function. SST is designed to support large-scale systematics and gene discovery projects. SST links data on DNA and protein sequences with information on specimens, collections, and taxonomy, and will serve to (1) document the use of specimens from curated collections in comparative molecular analyses, and (2) facilitate the proper documentation of all taxa used in molecular studies.

The relational structure of SST and EGAD allows complex queries that are not possible with any existing public database such as: "Return a list of all taxa collected in Panama between 1988 and 1993 for which sequence data are available, together with names of the genes sequenced from each organism and voucher specimen locations," or "Which cell-surface receptors expressed in human embryos have sequence isologs in *Drosophila* or *Caenorhabiditis*?" Thus, SST and EGAD will answer a wide range of biological questions that currently can be addressed only by compiling a large amount of data by hand. We expect that the ability to easily correlate the types of data handled by EGAD and SST will lead to more complex questions being addressed both experimentally and retrospectively. Together, ESTDB, EGAD, and SST support research efforts in gene discovery, molecular systematics, and population-level genetic diversity. EGAD and SST are being developed as publicly accessible resources as part of a federated system of interoperable databases. Demonstrations of these databases are available via the World Wide Web at http://www.tigr.org.

APPLYING RAPID GENE DISCOVERY TECHNOLOGY TO BIODIVERSITY

Many concepts of biodiversity exist, and what is meant by "assessing biodiversity" evades precise definition. At the very least, the term *biodiversity* incorporates genetic and ecological variation, through space and time, of individuals within populations as well as monophyletic (sensu Hennig, 1966) groups of taxa. But biodiversity also is used to describe the numbers of species in defined geographic areas (species richness). What then is an appropriate measure of biodiversity that can be used by researchers and policy makers? Is it allelic richness within individuals, populations, species, or larger groups? The number of populations per species or of species per monophyletic group? The numbers of species per hectare? How do measures based on organismal taxonomy relate to measures based on sequence similarity considered independently of species origin, and how do these different measures relate to patterns of ecological variation and to the biology of an organism? The key to addressing these related but quite distinct questions is sampling of both markers and species. The importance of sampling is illustrated by work under way at TIGR on microbial diversity. We have found high levels of sequence variation in the small subunit of rDNA both within and between conspecific strains in a broad sample of eubacterial genera (Clayton et al., 1995). This is an unexpected result given that regions of the small subunit of rDNA are conserved from eubacteria to vertebrates. Are these differences due to sequencing error? Or contamination of a bacterial culture? Or to different mechanisms of rDNA evolution in bacteria? Or to misidentification of bacterial species? Sequencing error and contamination are relatively straightforward to detect, but to address the other possibilities will require a suspension of preconceived ideas of molecular and biological diversity and a focus on collection of data and clear, unbiased analysis.

We are working to scale-up and automate the DNA extraction, amplification, and sequencing steps associated with molecular systematics so that multiple exemplars and multiple molecular markers can be analyzed simultaneously with the same accuracy and efficiency that we now achieve in sequencing ESTs. Our current research focuses on the use of molecular markers to identify monophyletic groups of organisms (i.e., organisms that descend from a common ancestor) and their geographic distributions. Future research will include measuring allelic variation at the population level. We use the term "Phylogenetic Species Tags" (PSTs) to describe the sequence-based phylogenetic trees that are generated from large-scale, molecular-based analyses. Just as ESTs are the starting point for further characterization of a particular gene and its biological function and cellular role, PSTs represent a first pass at identification and classification of taxa. As a tool for assessing biodiversity, PSTs are a form of molecular triage to assist in identifying taxonomic priorities for conservation efforts (Stiassny, 1992) and to identify unique species (Novacek, 1992). For example, molecular surveys of specific taxonomic groups using suites of multiple genetic markers could be used to compare geographic areas for evidence of loss of species diversity or as a tool to monitor the reestablishment of species and population diversity following remediation efforts.

The power of the molecular approach is not in the technology per se, but is in the comparative methodologies that serve as the basis for phylogenetic systematics (Farris, 1983; Funk and Brooks, 1990). Its success depends not only on the volume of data that can be generated, but on the development of molecular markers appropriate for a given research question; the establishment of baseline data on molecular variation; networks of databases that link molecular, geographic, ecological, and morphological data; and close collaborations with scientists at collections-based research institutions world-wide.

At the National Forum on BioDiversity held in Washington, D.C., in 1986, E.O. Wilson declared that "the magnitude and control of biological diversity is not just a central problem of evolutionary biology; it is one of the key problems of science as a whole" (Wilson, 1988:14). The authorship of this chapter constitutes evidence that we share this belief, since it includes individuals with expertise in systematics, evolutionary biology, molecular biology, computational biology, computer science, and protein biochemistry. The technological and informational advances associated with the Human Genome project have created the infrastructure necessary for exploring molecular variation on a large scale. The union of the technology with the power of the comparative method will undoubtedly lead to intriguing new insights into evolution and biodiversity assessment far into the twenty-first century.

REFERENCES

Adams, M. D., and 83 other authors. 1995. Initial assessment of human gene diversity and expression patterns based upon 83 million nucleotides of cDNA sequence. Nature 377(Suppl.):3-174.
Adams, M. D., M. Dubnick, A. R. Kerlavage, R. Moreno, J. M. Kelley, T. R. Utterback, J. W. Nagle, C. Fields, and J. C. Venter. 1992. Sequence identification of 2,375 human brain genes. Nature 355:632-634.
Adams, M. D., J. M. Kelley, J. D Gocayne, M. Dubnick, M. H. Polymeropoulos, H. Xiao, C. R. Merril, A. Wu, B. Olde, R. F. Moreno, A. R. Kerlavage, W. R. McCombie, and J. C. Venter. 1991.

Complementary DNA sequencing: Expressed sequence tags and human genome project. Science 252:1651-1656.

Adams, M. D., A. R. Kerlavage, C. Fields, and J. C. Venter. 1993a. 3,400 new expressed sequence tags identify diversity of transcripts in human brain. Nature Genetics 4:256-267.

Adams, M. D., A. R. Kerlavage, J. M. Kelley, J. D. Gocayne, C. Fields, C. M. Fraser, and J. C. Venter. 1994. A model for high-throughput automated DNA sequencing and analysis core facilities. Nature 368:474-475.

Adams, M. D., M. B. Soares, A. R. Kerlavage, C. Fields, and J. C. Venter. 1993b. Rapid cDNA sequencing (expressed sequence tags) from a directionally cloned human infant brain cDNA library. Nature Genetics 4:373-380.

Altschul, S. W. Gish, W. Miller, E. Meyers, and D. Lipman. 1990. A basic local alignment search tool. J. Mol. Biol. 215:403-410.

Boguski, M. S., T. M. J. Lowe, and C. M. Tolstoshev. 1993. dbEST- database for "expressed sequence tags." Nature Genetics 4:332-333.

Brutlag, D. L., J. P. Dautricourt, R. Diaz, J. Fier, B. Moxon, and R. Stamm. 1993. BLAZE: An implementation of the Smith-Waterman comparison algorithm on a massively parallel computer. Comput. Chem. 17:203-207.

Bult, C. J., J. A. Blake, A. Kerlavage, A. Glodek, W. FitzHugh, O. White, G. Sutton, L. FitzGerald, M. W. Chiu, M. Adams, R. Clayton, J. C. Venter, and C. Fields. 1994. The Expressed Gene Anatomy (EGAD) and Sequences, Sources, Taxa (SST) databases: Integrated biological databases to support research in gene discovery and evolution (abstract). P. 19 in The Second International Conference on the Plant Genome, San Diego, California. Scherago International, Inc., N.Y.

Clayton, R. A., G. Sutton, P. S. Hinkle, C. Bult, and C. Fields. 1995. Intraspecific variation in small subunit ribosomal RNA in prokaryotes. Int. J. Syst. Bacteriol. 45:595-599.

Collet, C. and R. Joseph. 1994. The identification of nuclear and mitochondrial genes by sequencing randomly chosen clones from a marsupial mammary gland cDNA library. Biochem. Gen. 32:181-190.

Doyle, J. J. 1993. DNA, phylogeny, and the flowering of plant systematics. BioScience 43:380-389.

Farris, J. S. 1983. The logical basis of phylogenetic analysis. Pp. 7-36 in N. I. Platnick and V. A. Funk, eds., Advances in Cladistics, Vol 2. Columbia University Press, N.Y.

Fields, C. 1992. Data exchange and inter-database communications in genome projects. Trends Biotechnol. 10:58-61.

Fields, C., and M. D. Adams. 1994. Expressed sequence tags identify a human isolog of the *sui1* translation initiation factor. Biochem. Biophys. Res. Comm. 198:288-291.

Fields, C., L. FitzGerald, O. White, A. Glodek, P. Hinkle, C. Bult, M. Adams, J. C. Venter, and A. Kerlavage. 1993. The Expressed Gene Anatomy Database (abstract). P. 32 in Genome Sequencing and Analysis Conference V, Hilton Head, South Carolina. Mary Ann Liebert, Inc., N.Y.

Fitch, W. 1970. Distinguishing homologous from analogous proteins. Syst. Zool. 19:99-113.

Forterre, P., N. Benachenhou-Lahfa, F. Confalonieri, M. Duguet, C. Elie, and B. Labedan. 1993. The nature of the last universal ancestor and the root of the tree of life, still open questions. BioSystems 28:15-32.

Friedlander, T., J. Regier, and C. Mitter. 1992. Nuclear gene sequences for higher level phylogenetic analysis: 14 promising candidates. Syst. Biol. 41:483-490.

Funk, V. A., and D. R. Brooks. 1990. Phylogenetic systematics as the basis of comparative biology. Smithsonian Contr. Bot. No. 73. Smithsonian Institution Press, Washington, D.C.

Green, P., D. Lipman, L. Hillier, R. Waterston, D. States, and J-M. Claverie. 1993. Ancient conserved regions in new gene sequences and the protein databases. Science 259:1711-1715.

Helentjaris, T. 1993. Implications for conserved genomic structure among plant species. Proc. Natl. Acad. Sci. 90:8308-8309.

Hennig, W. 1966. Phylogenetic Systematics. University of Illinois Press, Urbana.

Hoog, C. 1991. Isolation of a large number of novel mammalian genes by a differential cDNA library searching strategy. Nucleic Acids Res. 19:6123-6127.

Kerlavage, A. R., M. D . Adams, J. C. Kelley, M. Dubnick, J. Powell, P. Shanmugam, J. C. Venter, and C. Fields. 1993. Analysis and management of data from high-throughput expressed sequence tag projects. Pp. 585-594 in Proceedings of the Twenty-Sixth Annual Hawaii International Conference on Systems Sciences. The Institute of Electrical and Electronics Engineers, Inc., Los Alamitos, Calif.

Khan, A. S., A. S. Wilcox, M. H. Polymeropoulos, J. A. Hopkins, T. J. Stevens, M. Robinson, A. K. Orpana, and J. M. Sikela. 1992. Single pass sequencing and physical and genetic mapping of human brain cDNAs. Nature Genetics 2:180-185.

Kinlaw, C. S., C. Baysdorfer, D. E. Harry, S. M. Gerttula, A. T. Groover, J. M. Lee, M. E. Devey, and D. B. Neale. 1994. Partial sequence analysis of loblolly pine cDNAs and identification of mapped genes (abstract). P. 42 in The Second International Conference on the Plant Genome, San Diego, California. Scherago International, Inc., N.Y.

Kwak, J. M., Y. S. Park, M. S. Soh, and H. G. Nam. 1994. Generation of expressed sequence tags of Brassica plants (abstract). P. 52 in The Second International Conference on the Plant Genome, San Diego, California. Scherago International, Inc., N.Y.

Lecointre, G., H. Philippe, H. L. Van Le, and H. Le Guyader. 1993. Species sampling has a major impact on phylogenetic inference. Mol. Phylogen. Evol. 2:205-224.

McCombie, W. R., M. D. Adams, J. M. Kelley, M. G. FitzGerald, T. R. Utterback, M. Khan, M. Dubnick, A. R. Kerlavage, J. C. Venter, and C. Fields. 1992. Caenorhabditis elegans expressed sequence tags identify gene families and potential gene homologues. Nature Genetics 1:124-131.

National Center for Human Genome Research. 1990. Understanding Our Genetic inheritance: The U.S. Human Genome Project, The First Five Years (FY 1991-1995). NIH Pub. 90-1580. National Institutes of Health, Bethesda, Md.

National Research Council. 1993. A Biological Survey for the Nation. National Academy Press, Washington, D.C. 205 pp.

Newman, T. , F. J. de Bruijn, P. Green, K. Keegstra, H. Kende, L. McIntosh, J. Ohlrogge, N. Raikhel, S. Somerville, M. Thomashow, E. Retzel, and C. Somerville. 1994. Genes galore: A summary of methods for accessing results from large-scale partial sequencing of anonymous Arabidopsis cDNA clones. Plant Physiol. 106:1241-1255.

Novacek, M. 1992. The meaning of systematics and the biodiversity crisis. Pp. 101-108 in N. Eldredge, ed., Systematics, Ecology, and the Biodiversity Crisis. Columbia University Press, N.Y.

Okubo, K., N. Hori, R. Matoba, T. Niiyama, A. Fukushima, Y. Kojima, and K. Matsubara. 1992. Large scale cDNA sequencing for analysis of quantitative and qualitative aspects of gene expression. Nature Genetics 2:173-179.

Page, R. D. M. 1993. Genes, organisms, and areas: The problem of multiple lineages. Syst. Biol. 42:77-84.

Polymeropoulos, M. H., H. Xiao, J. M. Sikela, M. Adams, J. C. Venter, and C. R. Merril. 1993. Chromosomal distribution of 320 genes from a brain cDNA library. Nature Genetics 4:381-386.

Reddy, G. R., D. Chakrabarti, S. M. Schuster, R. J. Ferl, E. C. Almira, and J. B. Dame. 1993. Gene sequence tags from Plasmodium falciparum genomic DNA fragments prepared by the genease activity from mung bean nuclease. Proc. Natl. Acad. Sci. 90:9867-9871.

Rice Genome Newsletter. 1992. Volume 1, Number 1. Yuzo Minobe, ed. Rice Genome Research Program. Genome Research Program of the Ministry of Agriculture, Forestry and Fisheries of Japan, Tokyo.

Sanderson, M. J., and J. J. Doyle. 1992. Reconstruction of organismal phylogenies from multigene families: Paralogy, concerted evolution, and homoplasy. Syst. Biol. 41:4-17.

Shen, B., N. Carneiro, I. Torres, B. Stevenson, C. Baysdorfer, R. Ferl, and T. Helentjaris. 1994. Partial sequencing and mapping of two cDNA libraries in maize (abstract). P. 62 in The Second International Conference on the Plant Genome, San Diego, California. Scherago International, Inc., N.Y.

Sogin, M. L. 1991. Early evolution and the origin of eukaryotes. Curr. Opin. Gen. Dev. 1:457-463.

Stiassny, M. 1992. Phylogenetic analysis and the role of systematics in the biodiversity crisis. Pp. 109-120 in N. Eldredge, ed., Systematics, Ecology, and the Biodiversity Crisis. Columbia University Press, N.Y.

Sutton, G., and T. Busse. 1993. An annealing algorithm for multiple sequence alignment on the MASPAR parallel computer (abstract). P. 44 in Genome Sequencing and Analysis Conference V, Hilton Head, South Carolina. Mary Ann Liebert, Inc. Publishers, N.Y.

Venter, J. C., M. D. Adams, A. Martin-Gallardo, W. R. McCombie, and C. Fields. 1992. Genome sequence analysis: Scientific objectives and practical strategies. Trends Biotechnol. 10:8-11.

Waterston, R., C. Martin, M. Craxton, C. Huynh, A. Coulson, L. Hillier, R. Durbin, P. Green, R. Shownkeen, N. Halloran, M. Metzstein, T. Hawkins, R. Wilson, M. Berks, Z. Du, K. Thomas, J. Thierry-Mieg, and J. Sulston. 1992. A survey of expressed genes in *Caenorhabditis elegans*. Nature Genetics 1:114-123.

White, O., T. Dunning, G. Sutton, M. Adams, J. C. Venter, and C. Fields. 1993. A quality control algorithm for DNA sequencing projects. Nucleic Acids Res. 21:3829-3838.

Wilson, E. O. 1988. The current state of biological diversity. Pp. 3-18 in E.O. Wilson and F. M. Peter, eds., BioDiversity. National Academy Press, Washington, D.C.

Zuckerlandl, E., and L. Pauling. 1965. Molecules as documents of evolutionary history. J. Theor. Biol. 8:357-366.

Initial Assessment of Character Sets from Five Nuclear Gene Sequences in Animals[1]

TIMOTHY P. FRIEDLANDER
Research Associate, Center for Agricultural Biotechnology,
University of Maryland Biotechnology Institute

JEROME C. REGIER
Director, Center for Agricultural Biotechnology,
University of Maryland Biotechnology Institute

CHARLES MITTER
Associate Professor, Department of Entomology,
University of Maryland, College Park, Maryland

There is growing agreement that, because any single sequence alone may be misleading, molecular systematic inferences can rest securely only on concordant results from multiple independent sequences (Miyamoto and Cracraft, 1991). For nucleotide sequences, most inferences in animals have been based on either the mitochondrial genome or the nuclear ribosomal gene family. The analyses described herein are directed at documenting the informativeness of additional nuclear gene sequences.

The selection of phylogenetically informative sequences from the nuclear genome is not trivial. Current methods of inference deal most effectively with point substitutions and simple insertion/deletion events within independently evolving orthologous sequences. Problems such as distinguishing orthologs from paralogs or detecting nonindependent evolution are often intractable. Even data on point substitutions are hard to interpret when divergence is great and multiple hits are common. For these reasons, most of the nuclear genome probably is not useful for any given systematic question. For example, nontranscribed regions, which constitute the vast majority of the nuclear genome, are unlikely to be informative about higher-level taxonomic relationships. The

[1]This publication is reprinted with slight modifications from Friedlander, T. P., J. C. Regier, and C. Mitter. 1994. Phylogenetic information content of five nuclear gene sequences in animals: Initial assessment of character sets from concordance and divergence studies. Syst. Biol. 43:511-525.

difficulty of a priori selection of likely informative sequences currently limits the exploitation of nuclear genes for phylogenetic studies.

In a previous study, we delimited 14 protein-encoding nuclear genes whose sequences are likely to contain interpretable phylogenetic information, largely in the form of point substitutions (Friedlander et al., 1992). The criteria for their selection included appropriate levels of sequence conservation for deep taxonomic splits, as inferred from comparisons of published sequences, and desirable features of gene structure. In particular, these genes are present in just one or a few copies, simplifying identification of orthologous comparisons. They each contain over 1,000 basepairs of fairly uniformly evolving coding regions, hence many potential characters. They are free of internal repetitive elements or obvious nucleotide bias that would complicate sequence alignments (homology statements) and analysis.

In the current study, we have sought to gauge more directly the phylogenetic information content of five promising genes for animal phylogenetics. They are dopa decarboxylase (DDC, EC 4.1.1.26), phosphoenolpyruvate carboxykinase (PEPCK, EC 4.1.1.32), the nonrepeating portion of the largest subunit of RNA polymerase II (POL II, EC 2.7.7.6), elongation factor-2 (EF-2), and elongation factor-1α (EF-1α, orthologs of the F1 gene in *Drosophila melanogaster* [Walldorf et al., 1985]). All available, alignable animal sequences for these genes (as of 1993), representing a range of divergence times from about 10 million (Catzeflis et al., 1992; Jacobs et al., 1989) to more than 550 million years ago (Bowring et al., 1993; Conway Morris, 1993), were examined for phylogenetic informativeness in three ways.

First, the sequences were mapped onto accepted phylogenies strongly supported by previous evidence (Figure 21-1). This permitted the assessment of character support for each clade and the temporal partitioning of characters by their times of divergence. Mapping also permitted identification of homoplasious characters.

Second, the sequences were analyzed by parsimony and the resulting minimum-length trees (Figures 21-2 to 21-6) compared with the accepted phylogeny (Figure 21-1). Character sets consisting of amino acids, of total nucleotides, and of nucleotides from each of the three codon positions were analyzed separately. Characters were partitioned in this manner because nucleotides within protein-encoding sequences evolve at different rates (Fitch, 1980; Li and Graur, 1991:67-98), and thus are likely to be maximally informative at different taxonomic levels. We collectively refer to the analyses listed in this and the previous paragraph, comprising evaluation of new sequences in light of a previously well-established phylogeny, as a "concordance test." Such tests provide the strongest available validation of the phylogenetic utility of a candidate gene.

A third, less direct but readily obtainable, predictor of phylogenetic informativeness is pairwise sequence divergence. Over the lowest part of its range, pairwise divergence should be related to the number of informative characters.

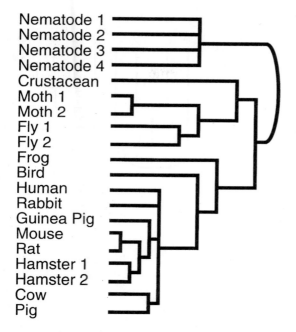

Nematode 1
Nematode 2
Nematode 3
Nematode 4
Crustacean
Moth 1
Moth 2
Fly 1
Fly 2
Frog
Bird
Human
Rabbit
Guinea Pig
Mouse
Rat
Hamster 1
Hamster 2
Cow
Pig

FIGURE 21-1 Accepted phylogenetic relationships for all taxa used in this study. Species names are listed in Table 21-1. Curved branches leading to the nematode in this and subsequent figures denote uncertain placement with respect to arthropods and vertebrates.

At higher levels, however, these differences can underestimate the actual number of substitutions to an ever-increasing degree due to "multiple hits" (Kimura, 1982; Saitou, 1989; Shoemaker and Fitch, 1989), with concomitant loss of phylogenetic information. Under the Jukes-Cantor model (Jukes and Cantor, 1969), for example, when observed divergence is approximately 0.55, the actual average number of substitutions per site is 1.0, while divergence of 0.75 corresponds to complete saturation, equivalent to comparisons between random sequences. This model, in conjunction with empirical saturation plots, provides heuristic guidelines for the onset of saturation. These guidelines, in combination with the phylogenetic concordance studies, permit a first estimate of the taxonomic level over which each character set in the five genes will be maximally informative (Cracraft and Helm-Bychowski, 1991).

MATERIALS AND METHODS

Data

Nucleotide and amino acid sequences of animal genes were accessed through the GenBank/EMBL and SWISS-PROT data banks. Sequences were aligned first using the GAP program in the University of Wisconsin Genetics Computer Group's software package (Devereux et al., 1984) and then by eye in multiple sequence comparisons. Sequence regions of uncertain alignment (and therefore

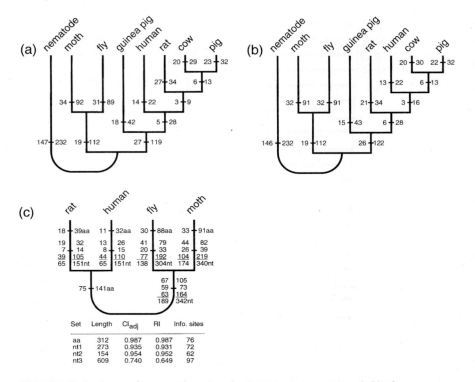

FIGURE 21-2 Concordance study using the DDC gene sequences. (a,b) The two most parsimonious trees for DDC amino acid sequences. Minimum and maximum branch lengths under any character optimization rule are mapped onto appropriate branches. Tree lengths (L)=652, consistency indices with autapomorphies removed (CI=CI$_{adj.}$)= 0.827, retention indices (RI)=0.689, numbers of informative characters for the aligned data set=145-158 (ambiguous because of missing data). (c) Most parsimonious tree for DDC nucleotide sequences (identical to test phylogeny). Minima and maxima for both amino acid and nucleotide substitutions are mapped, the latter by codon position (top to bottom: aa, nt1, nt2, nt3, all nucleotides). Tree statistics are shown.

homology), mostly insertion/deletion neighborhoods, were removed before analysis. Excluded portions amount to no more than 15% of the total coding regions in any of the five genes. For POL II only, the carboxy-terminal 20% was excluded from the outset because of multiple, internally repeating sequences with problematical alignment across species (Suzuki, 1990). For DDC, the nematode nucleotide sequence was excluded because it could not be aligned with any confidence with the other four available sequences.

The five genes studied, the species in which they were examined, sequence accession numbers, portions of sequences aligned in the analyses, and total numbers of aligned sites are listed in Table 21-1. EF-1α sequences from one fly (F2 gene, Walldorf et al., 1985) and one bee (probable F2 ortholog; Walldorf and Hovemann, 1990) were excluded from this study owing to nonorthology with other EF-1α sequences. The accepted (and, when different, most parsimonious) cladograms for all character sets and test taxa are shown in Figures 21-2 to 21-6.

Test Phylogeny

The relationships depicted in Figure 21-1 among the taxa used in this study, representing some of the most securely established relationships of those in any organismal groups, are supported by morphology and in most cases by multiple other lines of evidence. Thus, they provide a benchmark against which the utility of new character sets can be assessed. These groups include tetrapods and their subgroups, the Amniota and Mammalia (Benton, 1990); within mammals, Artiodactyla (cow and pig) and rodents; and within rodents, the subgroups hystricognaths (guinea pig) and myomorphs (rat and mouse, hamsters) (Luckett and Hartenberger, 1985). There is evidence that guinea pigs may not be rodents (Graur et al., 1991). The remaining mammalian relationships are controversial and left unresolved. Monophyly of the holometabolous insects and their subgroups, Diptera (flies) and Lepidoptera (moths), is not in doubt (Kristensen, 1991). Monophyly of the Arthropoda has been questioned, but a closer relationship of Crustacea to insects than to either Nematoda or Vertebrata

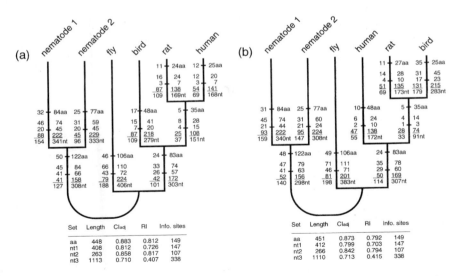

FIGURE 21-3 Concordance study using the PEPCK gene sequences. Character mapping as in Figure 21-2. Tree statistics are shown. (a) Most parsimonious tree for PEPCK amino acid and nucleotide position 1 and 2 datasets. (b) Most parsimonious tree for PEPCK nucleotide position 3.

FIGURE 21-4 Concordance study using the POL II gene sequences. Most parsimonious tree for POL II sequence data sets (identical to test phylogeny). Character mapping as for Figure 21-2. Tree statistics are shown.

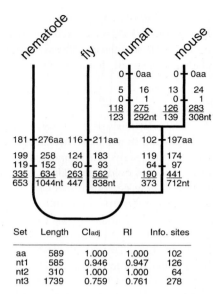

Set	Length	Cl_adj	RI	Info. sites
aa	589	1.000	1.000	102
nt1	585	0.946	0.947	126
nt2	310	1.000	1.000	64
nt3	1739	0.759	0.761	278

FIGURE 21-5 Concordance study using the EF-2 gene sequences. Most parsimonious tree for EF-2 sequence data sets (identical to test phylogeny). Character mapping as for Figure 21-2. Tree statistics are shown.

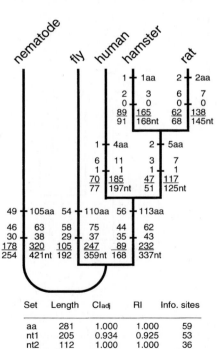

Set	Length	Cl_adj	RI	Info. sites
aa	281	1.000	1.000	59
nt1	205	0.934	0.925	53
nt2	112	1.000	1.000	36
nt3	1071	0.747	0.545	279

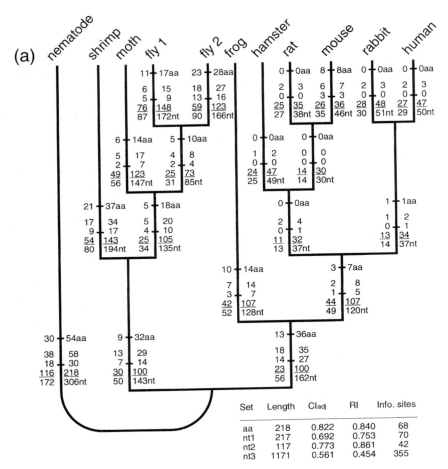

Set	Length	Cl$_{adj}$	RI	Info. sites
aa	218	0.822	0.840	68
nt1	217	0.692	0.753	70
nt2	117	0.773	0.861	42
nt3	1171	0.561	0.454	355

FIGURE 21-6 Concordance study using the EF-1α gene sequences. Character mapping as for Figure 21-2. Tree statistics are shown. (a) Test phylogeny for EF-1α sequences. (b) Strict consensus tree of the two most parsimonious trees for amino acid sequences, with uncertain placement of the mouse. There are either seven or eight characters mapping to the mouse branch, and either two or three characters mapping to the mammal branch. (c) Strict consensus tree of the two most parsimonious trees for nucleotides in codon position 1, with uncertain placement of the rabbit. (d) Strict consensus tree of six most parsimonious trees for nucleotides in codon position 2, with uncertain placement of flies within insects and of rodents and the rabbit within mammals. Numbers assigned to various branches vary as shown. (e) Most parsimonious tree for nucleotides in codon position 3.

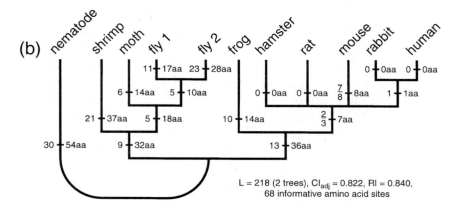

L = 218 (2 trees), Cl$_{adj}$ = 0.822, RI = 0.840,
68 informative amino acid sites

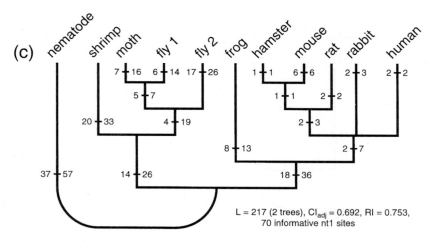

L = 217 (2 trees), Cl$_{adj}$ = 0.692, RI = 0.753,
70 informative nt1 sites

FIGURE 21-6   Continued

has not (review in Eernisse et al., 1992). Monophyly of Nematoda has not been questioned, but the position of this phylum with respect to the other two groups (arthropods, vertebrates) is not resolved (Eernisse et al., 1992).

Phylogenetic Analyses

All analyses of parsimony were done with PAUP versions 3.0 (Swofford, 1991). Pairwise sequence divergences by gene and character set obtained with this program are listed in Tables 21-2 and 21-3. Values from 0.01 to 0.30 are italicized in Table 21-2 to discriminate more optimal values from more saturated ones (see Discussion below). Most parsimonious trees were obtained using the

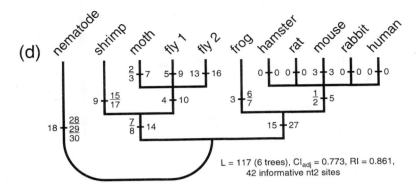

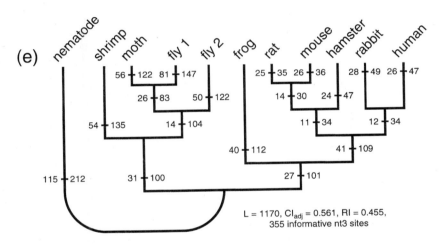

FIGURE 21-6 *Continued*

branch-and-bound or exhaustive analyses options. Tree lengths (L), consistency indices adjusted for uninformative sites (CI_{adj}), retention indices (RI), and informative characters are listed with each tree. In addition, the minimum and maximum number of synapomorphies under any character optimization are placed on branches of the accepted and most parsimonious trees.

RESULTS

Dopa Decarboxylase

Analysis of amino acid sequences from eight taxa (Table 21-1) yields two most parsimonious trees differing only in the position of the rat (Figures 21-

TABLE 21-1 Genes Examined, Animals Studied for Each, Sequence Accession Numbers, Numbers of Aligned Sites, and Sequence Portions Used in Concordance Studies

Genes	Accessions	No./position of aligned sites
Dopa decarboxylase		1,422 nt
Bos taurus (cow)	P27718	1-467 (aa only)
Caenorhabditis elegans (nematode)	Z11576	1-467 (aa only)
Cavia porcellus (guinea pig)	P22781	1-467 (aa only)
Drosophila melanogaster (fly)	X04661, M2411, X16802	1-1413
Homo sapiens (human)	M88700	1-1422
Manduca sexta (moth)	(Hiruma and Riddiford, 1990)	1-1416
Rattus norvegicus (rat)	M27716	1-1422
Sus scrofa (pig)	P80041	1-467 (aa only)
Phosphoenolpyruvate carboxykinase		1,641 nt
Ascaris suum (nematode 2)	L01787	124-372, 389-1176, 1213-1548, 1579-1693 1720-1761, 1784-1899
Drosophila melanogaster (fly)	Y00402	154-402, 418-1200, 1234-1569, 1597-1710, 1735-1776, 1801-1917
Gallus gallus (bird)	M14229	79-327, 340-1122, 1156-1491, 1525-1638, 1663-1704, 1726-1842
Haemonchus contortus (nematode 1)	M76494	61-309, 325-1107, 1144-1479, 1510-1623, 1651-1692, 1714-1830
Homo sapiens (human)	L05144	79-327, 340-1122, 1156-1491, 1525-1638, 1663-1704, 1726-1842
Rattus norvegicus (rat)	K02299, K03243-K03248	79-327, 340-1122, 1156-1491, 1525-1638, 1663-1704, 1726-1842
RNA polymerase II (largest subunit)		4,360 nt
Caenorhabditis elegans (nematode)	M29235	19-468, 508-588, 595-1644, 1648-2730, 2761-3774, 3781-4444
Drosophila melanogaster (fly)	M27431	16-102, 108-471, 505-585, 589-1227, 1240-1650, 1654-3471, 3480-3758, 3771-4434
Homo sapiens (human)	X63564	28-114, 120-483, 529-609, 613-1251, 1264-1674, 1678-3495, 3505-3783, 3796-4459
Mus musculus (mouse)	M12130, M14101	28-114, 120-483, 529-609 613-1251, 1264-1674, 1687-3504, 3514-3792, 3805-4468

TABLE 21-1 *Continued*

Genes	Accessions	No./position of aligned sites
Elongation factor-2		2,409 nt
Caenorhabditis elegans (nematode)	M86959	1-267, 313-618, 625-743, 841-2559
Drosophila melanogaster (fly)	X15805	1-267, 289-594, 601-717, 817-2535
Homo sapiens (human)	M30456, X51466	1-267, 277-582, 592-708, 859-2577
Mesocricetus auratus (hamster)	M13708	1-267, 277-582, 592-708, 859-2577
Rattus norvegicus (rat)	Y07504	1-267, 277-582, 592-708, 859-2577
Elongation factor-1a (F1 ortholog)		1,374 nt
Artemia species (brine shrimp)	J01165, X00546, X03349	1-1374
Bombyx mori (moth)	D13338	1-1374
Cricetulus longicaudatus (hamster)	D00522	1-1374
Drosophila melanogaster (fly 1)	X06869	1-1374
Homo sapiens (human)	X03558	1-1374
Mus musculus (mouse)	X13661	1-1374
Onchocerca volvulus (nematode)	M64333	1-1374
Oryctolagus cuniculus (rabbit)	X62245	1-1374
Rattus norvegicus (rat)	X61043	1-1374
Rhynchosciara americana (fly 2)	X66131	154-1374
Xenopus laevis (frog)	M25504	1-1374

2a,b). Both trees are concordant with the expected topology from Figure 21-1. Approximately one-third of the 471 aligned sites are potentially informative. Least divergent pairs of taxa are found within mammals (d_{aa}~0.11, Table 21-2) and represent evolutionary splits of 65 million years ago or less (Novacek, 1982). The moth/fly divergence, estimated at approximately 275 million years ago (Kukalova-Peck, 1991), has a d_{aa} of ~0.26. Mammal/insect divergences (>550 million years ago, d_{aa}~0.39) are less than those for either mammal/nematode or insect/nematode comparisons (d_{aa}~0.60).

All nucleotide character sets recover the test phylogeny for the four taxa whose sequences could be aligned; 16% of the 1,422 aligned nucleotide sites are informative (Figure 21-2c). Nucleotides in the first two codon positions and amino acids analyzed for the same four taxa have RI and CI_{adj} above 0.900, while indices for nt3 are lower (e.g., RI=0.740). Pairwise nucleotide divergences vary widely with character set and taxonomic depth (Table 21-2). For example, nt2 is less divergent at all taxonomic depths than nt3, even within mammals. The within-mammal comparison may be the only comparison for the nt3 set that is not fully saturated (d_{nt3}=0.314).

TABLE 21-2 Pairwise Sequence Divergences by Gene and Character Set

Dopa decarboxylase

aa\nt1	nem	moth	fly	gpig	rat	human	cow
moth	0.607	—	*0.259*	0.350	0.344		
fly	0.604	*0.262*	—	0.352	0.340		
guinea pig	0.596	0.387	0.387	—			
rat	0.591	0.402	0.394	*0.134*	—	*0.095*	
human	0.587	0.391	0.385	*0.121*	*0.106*	—	
cow	0.598	0.402	0.404	*0.153*	*0.123*	*0.108*	—
pig	0.567	0.389	0.398	*0.146*	*0.134*	*0.102*	*0.110*

nt2\nt3	rat	human	fly	moth
rat	—	0.314	0.601	0.664
human	0.046	—	0.610	0.658
fly	0.211	0.215	—	0.626
moth	0.226	0.224	0.125	—

Phosphoenolpyruvate carboxykinase

aa\nt1	nem 2	nem 1	fly	bird	human	rat
nematode 2	—	*0.192*	0.324	0.314	*0.296*	0.303
nematode 1	*0.214*	—	0.353	0.316	0.307	0.305
fly	0.373	0.378	—	*0.285*	*0.287*	*0.293*
bird	0.376	0.375	*0.280*	—	*0.106*	*0.108*
human	0.361	0.361	*0.280*	0.129	—	*0.066*
rat	0.359	0.355	*0.290*	0.122	0.071	—

nt2\nt3	nem 2	nem 1	fly	bird	human	rat
nematode 2	—	0.580	0.614	0.669	0.644	0.614
nematode 1	*0.119*	—	0.618	0.623	0.640	0.623
fly	*0.243*	*0.247*	—	0.647	0.550	0.532
bird	*0.239*	*0.229*	*0.194*	—	0.495	0.486
human	*0.229*	*0.225*	*0.197*	*0.048*	—	0.351
rat	*0.221*	*0.219*	*0.196*	*0.049*	*0.018*	—

RNA polymerase II (largest subunit)

aa\nt1	mouse	human	fly	nem
mouse	—	*0.020*	*0.216*	*0.266*
human	0.000	—	*0.210*	*0.262*
fly	*0.219*	*0.219*	—	*0.263*
nematode	*0.265*	*0.265*	*0.275*	—

nt2\nt3	mouse	human	fly	nem
mouse	—	*0.276*	0.588	0.611
human	0.001	—	0.560	0.634
fly	*0.108*	*0.108*	—	0.617
nematode	*0.149*	*0.149*	*0.146*	—

Elongation factor-2

aa\nt1	nem	fly	human	hams	rat
nematode	—	*0.151*	*0.142*	*0.139*	*0.146*
fly	0.188	—	0.159	0.153	0.158

TABLE 21-2 *Continued*

Elongation factor-2—*continued*

human	*0.193*	*0.199*	—	*0.020*	*0.124*
hamster	*0.195*	*0.200*	*0.008*	—	*0.011*
rat	*0.195*	0.199	0.009	0.004	—

nt2\nt3	nem	fly	human	hams	rat
nematode	—	0.529	0.598	0618	0.605
fly	*0.083*	—	0.514	0.531	0.503
human	*0.092*	*0.091*	—	0.365	0.325
hamster	*0.092*	*0.091*	*0.002*	—	*0.283*
rat	*0.092*	*0.091*	*0.002*	*0.000*	—

Elongation factor-1α

aa\nt1	nem	shrimp	moth	fly 1	fly 2	frog	hams	rat	mouse	rabbit	human
nematode	—	*0.179*	*0.175*	*0.175*	0.199	*0.177*	*0.170*	*0.170*	*0.175*	*0.162*	*0.166*
shrimp	*0.180*	—	*0.096*	*0.098*	*0.128*	*0.153*	*0.151*	0.153	0.155	0.148	0.148
moth	*0.160*	0.101	—	*0.050*	*0.081*	0.131	0.124	0.133	0.131	0.129	0.129
fly 1	*0.185*	0.115	*0.059*	—	*0.081*	0.135	0.133	0.138	0.135	0.133	0.133
fly 2	*0.210*	*0.151*	*0.095*	*0.100*	—	0.167	0.162	0.170	0.170	0.170	0.165
frog	*0.167*	*0.164*	*0.144*	*0.155*	*0.176*	—	*0.039*	*0.041*	*0.048*	*0.039*	*0.037*
hamster	*0.155*	*0.158*	*0.128*	*0.144*	*0.174*	*0.038*	—	*0.009*	*0.015*	*0.015*	*0.017*
rat	*0.155*	*0.158*	*0.128*	*0.144*	*0.174*	*0.038*	*0.000*	—	*0.020*	*0.015*	*0.017*
mouse	*0.167*	*0.169*	*0.140*	*0.155*	*0.187*	*0.054*	*0.018*	*0.018*	—	*0.026*	*0.028*
rabbit	*0.155*	*0.158*	*0.128*	*0.142*	*0.171*	*0.041*	*0.002*	*0.002*	*0.020*	—	*0.011*
human	*0.155*	*0.158*	*0.128*	*0.142*	*0.171*	*0.041*	*0.002*	*0.002*	*0.020*	*0.000*	—

nt2\nt3	nem	shrimp	moth	fly 1	fly 2	frog	hams	rat	mouse	rabbit	human
nematode	—	0.602	0.608	0.685	0.598	0.569	0.595	0.615	0.621	0.617	0.591
shrimp	*0.096*	—	0.476	0.533	0.485	0.480	0.524	0.511	0.521	0.522	0.517
moth	*0.088*	*0.048*	—	0.443	0.463	0.500	0.507	0.485	0.488	0.517	0.502
fly 1	*0.103*	*0.055*	*0.026*	—	0.490	0.496	0.533	0.515	0.508	0.502	0.522
fly 2	*0.121*	*0.076*	*0.049*	*0.052*	—	0.530	0.547	0.564	0.558	0.539	0.527
frog	*0.101*	*0.096*	*0.085*	*0.094*	*0.111*	—	0.389	0.369	0.405	0.389	0.393
hamster	*0.103*	*0.092*	*0.079*	*0.090*	*0.115*	*0.017*	—	*0.183*	*0.173*	*0.207*	*0.192*
rat	*0.103*	*0.092*	*0.079*	*0.090*	*0.115*	*0.017*	*0.000*	—	*0.133*	*0.201*	*0.234*
mouse	*0.105*	*0.096*	*0.083*	*0.094*	*0.121*	*0.022*	*0.007*	*0.007*	—	*0.221*	*0.239*
rabbit	*0.103*	*0.090*	*0.076*	*0.087*	*0.113*	*0.017*	*0.002*	*0.002*	*0.009*	—	*0.164*
human	*0.103*	*0.090*	*0.076*	*0.087*	*0.113*	*0.017*	*0.002*	*0.002*	*0.009*	0.000	

NOTE: Within each matrix, the values for the first of the two character sets listed in the upper left corner are given below the diagonal, while those for the second are above the diagonal. Divergence values from 0.010 to 0.300 are italicized.

Phosphoenolpyruvate Carboxykinase

Of the four character sets tested for six taxa (Table 21-1), only the nt3 set failed to recover the test phylogeny (Figure 21-3a), grouping the rat with the bird instead of with the human (Figure 21-3b). However, no character set strongly supported one topology over the other.

TABLE 21-3   Selected Pairwise Sequence Comparisons
(Divergence) of Taxa by Gene and Character Set

Genes	Character set			
	nt1	nt2	nt3	aa
nematode/fly[a,b]				
DDC				0.604
PEPCK	0.337	0.245	0.616	0.376
POL II	0.263	0.146	0.617	0.275
EF-2	0.151	0.083	0.529	0.188
EF-1a	0.187	0.112	0.642	0.198
nematode/human[a,b]				
DDC				0.587
PEPCK	0.302	0.227	0.642	0.361
POL II	0.262	0.149	0.634	0.265
EF-2	0.142	0.092	0.598	0.193
EF-1a	0.166	0.103	0.591	0.155
fly/human[b]				
DDC	0.340	0.215	0.610	0.385
PEPCK	0.287	0.197	0.550	0.280
POL II	0.210	0.108	0.560	0.219
EF-2	0.159	0.091	0.514	0.199
EF-1a	0.149	0.100	0.525	0.157
human/rat[c]				
DDC	0.095	0.046	0.314	0.106
PEPCK	0.066	0.018	0.351	0.071
POL II	0.020	0.001	0.276	0.000
EF-2	0.024	0.002	0.325	0.009
EF-1a	0.023	0.006	0.237	0.011
moth/fly[b]				
DDC	0.259	0.125	0.626	0.262
EF-1a	0.066	0.038	0.453	0.077
hamster/rat[c]				
EF-2	0.011	0.000	0.283	0.004
EF-1a	0.012	0.004	0.178	0.009

[a]Nucleotide comparisons unavailable for DDC.

[b]Average values for nematodes for PEPCK and for flies for EF-1a.

[c]Mouse substituted for rat in POL II comparisons; average values for rat/mouse
for EF-1a.

One-fourth of 510 amino acid sites, and one-third of 1,641 nucleotide sequence sites, mostly in nt3, are potentially informative. Pairwise divergences
(Table 21-2) for the bird/mammal split (~300 million years ago; Benton, 1990)
are relatively small for amino acids and nt1+2 (0.13 and ~0.16 average) but large
($d_{nt3}=0.49$) at nt3. Divergence between the two mammals (~65 million years
ago) is about half as large (d_{aa}~0.07, d_{nt1+2}~0.04 average) as that for bird/mam-

mals in these characters. Divergence between the two nematodes is twice as large as that observed for bird/mammals (d_{aa}~0.21, d_{nt1+2}~0.16 average). Nt3 is saturated for all nematode/fly comparisons, almost saturated for bird/mammal comparisons, but less so for rat/human comparisons.

RNA Polymerase II (Largest Subunit)

All character sets recovered the mammal clade (Figure 21-4) among the four available sequences (Table 21-1). Approximately 10% of the alignable sites (1,495 aa/4,360 nt) were informative, and homoplasy was low for all character sets except nt3. Mouse/human (divergence ~65 million years ago) differences are entirely synonymous changes (Table 21-2, d_{aa}=0, d_{nt3}=0.276); all other nt3 divergence comparisons are nearly saturated (d>0.5). Nonsynonymous divergences between fly and mammal (>550 million years ago) are less than those between either taxon and nematode (d_{aa}~0.26 versus d_{aa}~0.33).

Elongation Factor-2

All character sets recover the expected phylogeny (Figure 21-5) for these five taxa (Table 21-1), although support for the rodent clade, as opposed to the mammal clade, is largely limited to nt3. Seven percent of 844 amino acid sites and 15% of 2,409 nucleotide sites were potentially informative. Homoplasy is low (RI>0.925) for all but nt3 (RI=0.545).

Amino acid differences (Table 21-2) within mammals (divergence ~65 million years ago) are few (d_{aa}<0.01). Synonymous site comparisons are saturated except within mammals, where d_{nt3} values range from 0.283 to 0.365. The nematode and fly are roughly equally divergent from mammals and from each other.

Elongation Factor-1α

Published EF-1α sequences are more numerous (11) and thus permit more taxonomic levels to be tested in concordance studies than the previous character sets (Figure 21-6a). Fifteen percent of 440 amino acid sites were potentially informative. Two most parsimonious trees were found, both concordant with the test phylogeny, but neither tree resolves how rodents are related to each other or to other mammals (Figure 21-6b).

One-third of 1,374 nucleotide sites were potentially informative. Nt1 yields two most parsimonious trees, neither of which resolves insect or rodent relationships correctly (Figure 21-6c). The rabbit shares a minimum of one change with either rodents or humans. Only two additional changes are required to recover the test phylogeny.

Six most parsimonious trees result from the analysis of nt2 (Figure 21-6d). Their strict consensus is concordant with the test phylogeny (Figure 21-6a), but not all minimum length trees group the two taxa of flies. Most phylogenetic trees group the human and rabbit and recover rodents. Some do both, but rodents are never resolved. The position of rabbits with regard to rodents and primates is controversial (Goodman et al., 1985; Li et al., 1990; Pesole et al., 1991).

One most parsimonious tree was found for nt3 (Figure 21-6e), one step shorter than the concordance tree. The former differs from the latter only by grouping one fly with the moth in preference to the other fly. All nucleotide sites analyzed together give this same result as equally parsimonious with the concordance tree (data not shown).

Least divergent pairs of taxa occur within mammals (~65 million years ago, $d_{aa} < 0.02$, Table 21-2). Maximum d_{nt3} within mammals is still comparatively small (<0.24). Frog/mammal divergences (~365 million years ago; Benton, 1990) are about twice as large as those for mammals. Between-insect comparisons (200-275 million years ago; Kukalova-Peck, 1991) are even greater ($d_{aa} \sim 0.10$, $d_{nt3} \sim 0.47$), followed by crustacean/insect, arthropod/vertebrate, and finally, all nematode comparisons ($d_{aa} \sim 0.17$, $d_{nt3} \sim 0.61$). Nt3 divergences appear saturated except within mammals (see below).

DISCUSSION

Our concordance tests on all available animal sequences provide evidence that these five genes contain substantial phylogenetic information. For each gene, parsimony analysis of amino acid and most nucleotide character sets recovered the test phylogeny (Figure 21-1), although with varying quantity and quality of support and resolution (Figures 21-2 to 21-6). While more evidence is needed, we predict that these five genes will prove widely applicable to phylogenetic studies.

The pairwise divergences in Table 21-3 illustrate that these five genes evolve at greatly different rates. DDC evolves most rapidly, followed by PEPCK. POL II, EF-2, and EF-1α are the slowest, and are well known for their extreme protein sequence conservation (Cammarano et al., 1992; Creti et al., 1991; Gropp et al., 1986; Iwabe et al., 1991; Rivera and Lake, 1992). Within each gene, nt3 is the fastest among sequence character sets, reflecting a preponderance of synonymous substitutions, while nt2 is slowest, presumably because all nt2 substitutions are nonsynonymous. Amino acids are intermediate, owing to the fact that amino acid changes may result from nucleotide changes in any codon position. Nt1 is also intermediate because changes can be both synonymous (leucine and arginine codons) and nonsynonymous. In our concordance studies, nonsynonymous substitutions and amino acid replacements largely agree with the test phylogenies, but are not fast enough to track relatively recent splits. Synonymous substitutions track recent splits, but are increasingly noisy with regard to deeper splits.

Given this variation in rates, it should be possible to preselect particular combinations of these genes and character sets which are most likely to be useful for resolving phylogenetic splits of a given temporal range. The current paucity of sequences prevents our concordance tests from specifying precise ranges of utility for these genes. However, a useful interim predictor of such ranges can be obtained from pairwise sequence differences. As a first approximation, we suggest divergences of 0.20 to 0.30 as a threshold, above which multiple hits are likely to lead tree construction astray. Plots of divergence against time and/or apparent transition/transversion ratios for a number of both mitochondrial and nuclear genes (e.g., Liu and Beckenbach, 1992; Mindell and Honeycutt, 1990; Shoemaker and Fitch, 1989) suggest that, above this level, an approach to saturation frequently becomes evident. This is also the point at which observed differences start to depart noticeably from actual ones in the Jukes-Cantor model (Jukes and Cantor, 1969), which predicts that our suggested interval corresponds to actual divergences of 0.23 to 0.38. Under models incorporating more constraints, including nucleotide bias and natural selection, underestimation of evolutionary divergence should be still more pronounced (Fitch, 1986). Above this level of divergence, suites of characters on long branches might converge.

On the other hand, the suggested threshold might be too conservative in some cases, because it ignores the ability of phylogenetic trees to reveal substitutions undetected by pairwise comparisons when there are taxa that branch off between the pair being compared (Saitou, 1989). Indeed, in several cases in our study, nt3 recovered the test phylogeny despite saturating divergence levels. At the other extreme, pairwise divergences less than 0.01, for example, will yield very few characters even from gene sequences hundreds of basepairs long. Thus, while the models for estimating actual divergence values from observed pairwise divergences provide heuristic guidelines, divergence levels corresponding to maximal phylogenetic utility will depend on the nature of the phylogenetic problem, including the pattern of taxon sampling, rates of evolution in different lineages, and the divergence times to be resolved.

Based on these divergence criteria and the estimated divergence times of our test taxa, we provide in Table 21-4 a first estimate of the time frames over which the different character sets for the five genes are likely to prove optimally informative with more extensive concordance testing. Perhaps the strongest practical implication of Table 21-4 is the importance of separate consideration of the different character sets within each gene (partitions of Bull et al., 1993). One advantage of such discrimination lies in avoiding faulty inferences. For example, for extremely conserved genes such as POL II, EF-2, and EF-1α, there is likely to be an intermediate time span in which no character set is appropriate, even though overall divergence levels might appear to lie within the useful range. This is because sites undergoing synonymous changes can become fully saturated before enough nonsynonymous changes have accumulated to provide

TABLE 21-4   Predicted Time Spans of Optimal Phylogenetic Utility for Nucleotide Position and Amino Acid Character Sets in Each Gene, Based on Concordance and Divergence Studies Among Test Taxa with Known Times of Separation

	Approximate Ages of Divergences			
	Cenozoic		Paleozoic	
Character sets	~35 MYA	50-65 MYA	275-365 MYA	~550 MYA
all nt	DDC PEPCK POL II EF-2 EF-1a	POL II EF-2 EF-1a		
nt1, nt2, aa		DDC PEPCK	PEPCK POL II EF-2 EF-1a	POL II EF-2 EF-1a
nt2 only			DDC	PEPCK

substantial phylogenetic information. We have evidence that this is so for EF-1α sequences applied to relationships among moths (unpublished observations).

Conversely, emphasizing character sets highlights a use for highly conserved protein-encoding genes that has gone largely unexploited. Because EF-1α is so conservative overall, it has been applied heretofore mostly to the very earliest divergences, e.g., those among major lineages of prokaryotes. Third position nucleotides in that same gene sequence, however, evolve rapidly enough to be useful at much lower taxonomic levels, with the advantage over introns or other rapidly evolving sequences that amino acid conservation should make both alignment and primer definition straightforward. Indeed, third position characters in all conservative, protein-encoding genes may be useful at lower taxonomic levels. We have preliminary evidence that, for EF-1α, nt3 sites will be highly informative about generic and subfamilial relationships in the moth family, Noctuidae (Cho et al., 1995).

ACKNOWLEDGMENTS

The authors gratefully acknowledge constructive comments on earlier versions by Drs. Grace Wyngaard and Michael Miyamoto and by two anonymous reviewers. We thank Soowon Cho for his contributions in the study of EF-1α sequences. This research was supported by funds from the National Science Foundation grant #DEB-9212669, the Center for Agricultural Biotechnology, the

U.S. Department of Agriculture-NRI CGP grant #90-37250-5482, and the Maryland Agricultural Experiment Station (MAES Scientific Article No. A6539, Contribution No. 8750).

REFERENCES

Benton, M. J. 1990. Phylogeny of the major tetrapod groups: Morphological data and divergence dates. J. Mol. Evol. 30:409-424.

Bowring, S. A., J. P. Grotzinger, C. E. Isachsen, A. H. Knoll, S. M.Pelechaty, and P. Kolosov. 1993. Calibrating rates of early Cambrian evolution. Science 261:1293-1298.

Bull, J. J., J. P. Huelsenbeck, C. W. Cunningham, D. L. Swofford, and P. J. Waddell. 1993. Partitioning and combining data in phylogenetic analysis. Syst. Biol. 42:384-397.

Cammarano, P., P. Palm, R. Creti, E. Ceccarelli, A. M. Sanangelantoni, and O. Tiboni. 1992. Early evolutionary relationships among known life forms inferred from elongation factor EF-2/EF-G sequences: Phylogenetic coherence and structure of the Archaeal domain. J. Mol. Evol. 34:396-405.

Catzeflis, F. M., J.-P. Aguilar, and J.-J. Jaeger. 1992. Muroid rodents: Phylogeny and evolution. Trends Ecol. Evol. 7:122-126.

Cho, S., A. Mitchell, J. C. Regier, C. Mitter, R. W. Poole, T. P. Friedlander, and S. Zhao. 1995. A highly conserved nuclear gene for low level phylogenetics: Elongation Factor-1α recovers morphology-based tree for heliothine moths. Syst. Biol. 12:650-656.

Conway Morris, S. 1993. The fossil record and the early evolution of the Metazoa. Nature 361:219-225.

Cracraft, J., and K. Helm-Bychowski. 1991. Parsimony and phylogenetic inference using DNA sequences: Some methodological strategies. Pp. 184-220 in M. M. Miyamoto and J. Cracraft, eds., Phylogenetic Analysis of DNA Sequences. Oxford University Press, N.Y.

Creti, R., F. Citarella, O. Tiboni, A. Sanangelantoni, P. Palm, and P. Cammarano. 1991. Nucleotide sequence of a DNA region comprising the gene for elongation factor 1α (EF-1α) from the ultra-thermophilic archaeote *Pyrococcus woesei*: Phylogenetic implications. J. Mol. Evol. 33:332-342.

Devereux, J., P. Haeberli, and O. Smithies. 1984. A comprehensive set of sequence analysis programs for the VAX. Nucleic Acids Res. 12:387-395.

Eernisse, D. J., J. S. Albert, and F. E. Anderson. 1992. Annelida and Arthropoda are not sister taxa: A phylogenetic analysis of spiralian metazoan morphology. Syst. Biol. 41:305-330.

Fitch, W. M. 1980. Estimating the total number of nucleotide substitutions since the common ancestor of a pair of homologous genes: Comparison of several methods and three beta hemoglobin messenger RNA's. J. Mol. Evol. 16:153-209.

Fitch, W. M. 1986. The estimate of total nucleotide substitutions from pairwise differences is biased. Phil. Trans. R. Soc. Lond. B 312:317-324.

Friedlander, T. P., J. C. Regier, and C. Mitter. 1992. Nuclear gene sequences for higher level phylogenetic analysis: 14 promising candidates. Syst. Biol. 41:483-490.

Goodman, M., J. Czelusniak, and J. E. Beeber. 1985. Phylogeny of primates and other eutherian orders: A cladistic analysis using amino acid and nucleotide sequence data. Cladistics 1:171-185.

Graur, D., W. A. Hide, and W.-H. Li. 1991. Is the guinea-pig a rodent? Nature 351:649-652.

Gropp, F., W. D. Reiter, A. Sentenac, W. Zillig, R. Schnabel, M. Thomm, and K. O. Stetter. 1986. Homologies of components of DNA-dependent RNA polymerases of Archaebacteria, eukaryotes and Eubacteria. Syst. Appl. Microbiol. 7:95-101.

Hiruma, K., and L. M. Riddiford. 1990. Regulation of dopa decarboxylase gene expression in the larval epidermis of the tobacco hornworm by 20-hydroxyecdysone and juvenile hormone. Dev. Biol. 138:214-224.

Iwabe, N., I. Kuma, H. Kishino, M. Hasegawa, and T. Miyata. 1991. Evolution of RNA polymerases and branching patterns of the three major groups of Archaebacteria. J. Mol. Evol. 32:70-78.

Jacobs, L. L., L. J. Flynn, and W. R. Downs. 1989. Neogene rodents of southern Asia. Pp. 157-177 in C. C. Black and M. R. Dawson, eds., Papers on Fossil Rodents, in honor of Albert Elmer Wood. Sci. Ser. No. 33, Los Angeles County Natural History Museum, Los Angeles.

Jukes, T. H., and C. R. Cantor. 1969. Evolution of protein molecules. Pp. 21-132 in H. N. Munro, ed., Mammalian Protein Metabolism, Vol. 3. Academic Press, N.Y.

Kimura, M. 1982. The neutral theory as a basis for understanding the mechanism of evolution and variation at the molecular level. Pp. 3-56 in M. Kimura, ed., Molecular Evolution, Protein Polymorphism and the Neutral Theory. Japan Scientific Societies Press, Tokyo, and Springer-Verlag, Berlin.

Kristensen, N. P. 1991. Phylogeny of extant hexapods. Pp. 125-140 in I. D. Naumann, P. B. Carne, J. F. Lawrence, E. S. Nielsen, J. P. Spradbery, R. W. Taylor, M. J. Whitten, and M. J. Littlejohn, eds., The Insects of Australia, second ed., Vol. 1. Division of Entomology, CSIRO. Cornell University Press, Ithaca, N.Y.

Kukalova-Peck, J. 1991. Fossil history and the evolution of hexapod structures. Pp. 141-179 in I. D. Naumann, P. B. Carne, J. F. Lawrence, E. S. Nielsen, J. P. Spradbery, R. W. Taylor, M. J. Whitten, and M. J. Littlejohn, eds. The Insects of Australia, second ed., Vol. 1. Division of Entomology, CSIRO. Cornell University Press, Ithaca, N.Y.

Li, W.-H., M. Gouy, P. M. Sharp, C. O'Huigin, and Y.-W. Yang. 1990. Molecular phylogeny of Rodentia, Lagomorpha, Primates, Artiodactyla, and Carnivora and molecular clocks. Proc. Natl. Acad. Sci. 87:6703-6707.

Li, W.-H., and D. Graur. 1991. Fundamentals of Molecular Evolution. Sinauer Associates, Sunderland, Mass.

Liu, H., and A. T. Beckenbach. 1992. Evolution of the mitochondrial cytochrome oxidase II gene among 10 orders of insects. Mol. Phylogen. Evol. 1:41-52.

Luckett, W. P., and J.-L. Hartenberger, eds., 1985. Evolutionary relationships among rodents: A multidisciplinary analysis. NATO ASI Series A, Life Sciences, Vol. 92. Plenum Press, N.Y.

Mindell, D. P., and R. L. Honeycutt. 1990. Ribosomal RNA in vertebrates: Evolution and phylogenetic applications. Ann. Rev. Ecol. Syst. 21:541-566.

Miyamoto, M. M., and J. Cracraft. 1991. Phylogenetic inference, DNA sequence analysis, and the future of molecular systematics. Pp. 3-17 in M. M. Miyamota and J. Cracraft, eds., Phylogenetic Analysis of DNA Sequences. Oxford University Press, N.Y.

Novacek, M. J. 1982. Information for molecular studies from anatomical and fossil evidence on higher eutherian phylogeny. Pp. 3-41 in M. Goodman, ed., Macromolecular Sequences in Systematic and Evolutionary Biology. Plenum Press, N.Y.

Pesole, G., E. Sbisa, F. Mignotte, and C. Saccone. 1991. The branching order of mammals: Phylogenetic trees inferred from nuclear and mitochondrial molecular data. J. Mol. Evol. 33:537-542.

Rivera, M. C., and J. A. Lake. 1992. Evidence that eukaryotes and eocyte prokaryotes are immediate relatives. Science 257:74-76.

Saitou, N. 1989. A theoretical study of the underestimation of branch lengths by the maximum parsimony principle. Syst. Zool. 38:1-6.

Shoemaker, J. S., and W. M. Fitch. 1989. Evidence from nuclear sequences that invariable sites should be considered when sequence divergence is calculated. Mol. Biol. Evol. 6:270-289.

Suzuki, M. 1990. The heptad repeat in the largest subunit of RNA polymerase II binds by intercalating into DNA. Nature 344:562-565.

Swofford, D. L. 1991. PAUP: Phylogenetic analysis using parsimony, Version 3.0. Illinois Natural History Survey, Champaign.

Walldorf, U., and B. T. Hovemann. 1990. *Apis mellifera* cytoplasmic elongation factor 1α (EF-1α) is closely related to *Drosophila melanogaster* EF-1α. FEBS Lett. 267:245-249.

Walldorf, U., B. Hovemann, and E. K. F. Bautz. 1985. *F1* and *F2*: Two similar genes regulated differently during development of *Drosophila melanogaster*. Proc. Natl. Acad. Sci. 82:5795-5799.

CHAPTER 22

Gap Analysis for Biodiversity Survey and Maintenance

J. MICHAEL SCOTT
Unit Leader

BLAIR CSUTI
Research Associate

National Biological Service, Idaho Cooperative Fish and Wildlife Research Unit, College of Forestry, Wildlife, and Range Science, University of Idaho, Moscow

Ehrlich (1988) postulated that reversing the loss of biodiversity will require a "quasi-religious transformation" of the way contemporary cultures view the value of human life and the intrinsic values of organic diversity. Even if that transformation were to occur today, we would be faced with a cruel reality: maintaining viable examples of every natural community, including the myriad of species they support, is the fundamental mechanism for conserving biodiversity (Noss and Cooperrider, 1994), but you cannot conserve biodiversity if you do not know where it is located. The elements of biodiversity, from genes and species to ecosystems, have distributions, but they have not been mapped at scales useful for developing a national biodiversity conservation and management strategy.

Centuries of scientific collectors have deposited tens of millions of specimens in the world's museums and herbaria. These form the foundation of our knowledge of species distributions, yet many areas and taxa remain poorly sampled. The Gap Analysis Program, a program of the National Biological Service, uses two relatively new technologies, satellite remote sensing and geographic information systems (GIS), to assist in the assessment of the status and distribution of several elements of biodiversity (Scott et al., 1993). While not a substitute for traditional biological surveys, we feel that gap analysis can provide a preliminary, landscape-scale assessment of the distribution of both species and ecosystem diversity in the United States that can be used to guide future field research and to provide a spatial framework for a preliminary national biodiversity conservation strategy. The gap analysis approach to biodiversity surveys holds promise for the rapid development of information on the distribution of several indicators of biodiversity in areas of the world that have been less well sampled.

Burley (1988) identified four steps in gap analysis: (1) identify and classify biodiversity, (2) locate areas managed primarily for biodiversity, (3) identify biodiversity that is un- or underrepresented in those managed areas, and (4) set priorities for conservation action. While these steps remain essential to gap analysis, the distribution of vegetation cover and species, gathered as a precursor to analysis, has considerable application to natural resource inventory and monitoring in and of itself.

HISTORY OF GAP ANALYSIS

Kepler and Scott (1985) used the distribution of endangered Hawaiian forest birds, gathered through field surveys (Scott et al., 1986), to perform a simple "gap analysis" on the island of Hawaii (Figure 22-1). They found little overlap between the distribution of endangered forest birds and the location of nature reserves. The Nature Conservancy of Hawaii and the U.S. Fish and Wildlife Service responded by establishing the Hakalau Forest National Wildlife Refuge in one of the areas where distribution of three endangered species overlapped.

Extensive field inventory, at the level of detail undertaken in Hawaii (Scott et al. 1986), is prohibitively expensive for large continental regions. To avoid these expenses, the Gap Analysis Program has developed methods to take advantage of currently available information to produce land cover and terrestrial vertebrate distribution maps (Scott et al., 1993). Other taxa, such as butterflies, can be added to data layers where sufficient information is available.

Pilot programs of gap analysis were initiated in Idaho in 1987 and in Oregon in 1988. By 1995, there were active programs in 36 states, and programs in 4 states—Utah, Idaho, Oregon, and Arizona—had been completed. Cooperation with state and federal agencies and private conservation groups has been an essential component of successful programs. Over 200 different public cooperators and private businesses now are involved in various state programs. Common needs, such as access to satellite imagery and databases on species, have encouraged cooperators to pool resources. While programs are state-based for administrative reasons, biodiversity analyses are best carried out for entire biological regions or at the national level. State-level information eventually will be merged to facilitate regional and national analyses.

CLASSIFYING AND MAPPING ECOSYSTEMS

Gap analysis requires three primary GIS data layers: vegetation cover, species range maps, and the location of land managed primarily for native species and natural ecosystem processes. Where available, information on other environmental factors, such as elevation, slope, aspect, soils, aquatic features, and

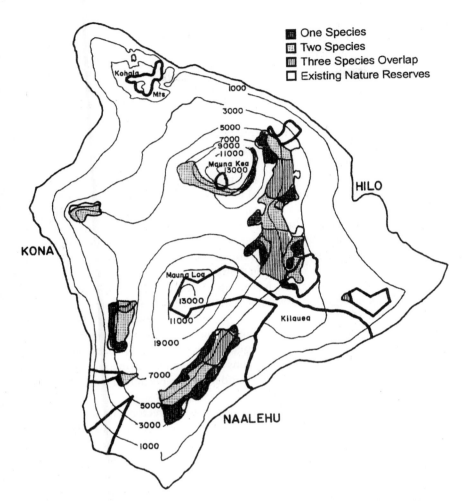

FIGURE 22-1 Ranges of four endangered forest birds on the island of Hawaii in comparison to the distribution of areas managed for biodiversity in 1982.

climate, can be used to improve the accuracy of maps of vegetation and species distributions. Additional information on socioeconomic attributes of landscapes (e.g., projected population trends, projected housing starts, ownership of land by state and federal agencies, zoning, etc.) can be examined to refine planning efforts for land use (Machlis et al., 1994).

Orians (1993:206) correctly pointed out that there is "no previously established, generally accepted taxonomy of habitats, communities, or ecosystems." Plant communities are the most visible component of ecosystems and have been widely considered to be acceptable surrogates for ecosystems (Austin, 1991; Austin and Margules, 1986). Gap analysis assumes that plant communities serve as integrators of many physical factors (type of soil, moisture regime, aspect, elevation, temperature) that interact at a site (Thomas, 1979). Floristic composition (described by the dominant or codominant species in the uppermost vegetation layer) provides a common denominator for description of plant communities. These communities then can be aggregated into any of a number of taxonomic schemes, such as that proposed by the United Nations Educational, Scientific, and Cultural Organization (1973).

In 1993, cooperating with the U.S. Geological Survey (USGS), the Environmental Protection Agency (EPA), and the National Oceanic and Atmospheric Administration (NOAA), the National Biological Service's Gap Analysis Program purchased complete LANDSAT Thematic Mapper satellite imagery for the conterminous United States. Imagery from one or two dates with favorable weather conditions in the years 1991-1993 will be made available to state programs and cooperators. The imagery will be preprocessed and archived by the U.S. Geological Survey's EROS Data Center in Sioux Falls, South Dakota. With cooperation from the EPA, USGS, and other groups, interpretation of this imagery will serve as the framework for a seamless national vegetation map.

LANDSAT Thematic Mapper imagery measures reflectance of 30 × 30 m pixels (picture elements) on the Earth's surface at seven wavelengths (see Scott et al., 1993, for details). Upland vegetation cover is mapped by aggregating pixels of similar reflectance into polygons using a minimum mapping unit of 100 hectares. Smaller minimum mapping standards have been used by most states to meet local needs.

Gap analysis programs use GIS software with vector data structure as a standard for vegetation mapping (ARC/INFO, Environmental Systems Research Institute, Inc., Redlands, California). Stands are delineated either by on-screen digitizing or by computer algorithms (Figure 22-2). Many plant communities are mapped as stands larger than 100 hectares. These usually contain considerable internal heterogeneity. An attribute file is created for each vegetation polygon containing information about primary, secondary, and tertiary plant communities within the polygon as well as other special features, such as the presence of small wetlands and other microhabitats. Wetlands larger than 40 hectares are mapped when identifiable from satellite imagery. Ultimately, 1:24,000 scale maps produced by the USGS National Wetlands Inventory will be available in digital form to represent this critical habitat element.

Although LANDSAT Thematic Mapper imagery provides a geographically consistent and repeatable spatial framework for vegetation mapping, identification of floristic dominant or codominant species often requires ancillary information, such as aerial photographs, low altitude airborne videophotography (Graham, 1993), existing large-scale vegetation maps, or field reconnaissance.

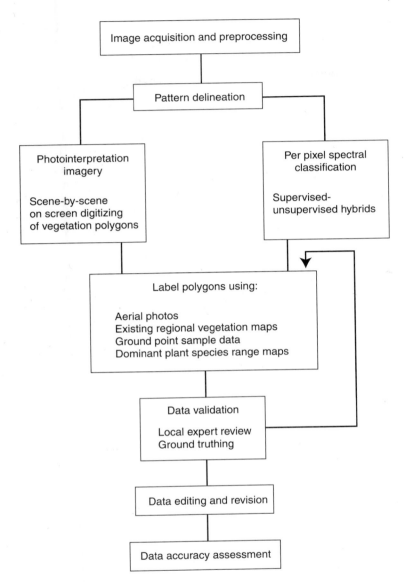

FIGURE 22-2 Flow chart showing steps in development of vegetation cover map from LANDSAT Thematic Mapper imagery and ancillary information. Two widespread approaches to pattern delineation are used by various states. Photointerpretation offers advantages for pattern recognition in complicated landscapes, while machine classification is more consistent and repeatable. National minimum mapping unit and labeling protocols must be followed regardless of the pattern delineation technique.

State gap analysis programs use a variety of available sources to label vegetation maps, keeping a record of the sources used for each polygon.

DISTRIBUTION MAPS OF SPECIES

Museum and herbarium specimens are the ultimate source of knowledge about the classification and distribution of species. Specimen locality records can be supplemented by reliable observations that are published in the literature or maintained in quality-controlled databases, such as the U.S. Fish and Wildlife Service's Breeding Bird Survey (see also Farr and Rossman, Chapter 31, and Umminger and Young, Chapter 32, in this volume). Continued scientific specimen collection is vital to more fully understand the distribution and variation among living organisms, but the rapid decline of many natural communities places urgency on development of a comprehensive biodiversity conservation strategy in the absence of a truly complete biological survey.

All distribution maps are statements of the probability of encountering a species in time and space. The distributional limits of many species fluctuate from year to year and may display long-term expansion or contraction. Within those distributional limits, populations appear and disappear from patches of suitable habitat (e.g., *Dipodomys deserti* in Joshua Tree National Monument, California; Miller and Stebbins, 1964) and may occur sporadically in unsuitable habitat. Specimen records are sparse even for many common species, and most collecting has taken place along transportation corridors, leaving vast areas unsurveyed. With some qualifications, these shortcomings can be overcome by interpolating and extrapolating the probable presence of a species in suitable environments between verified collection localities within the outer bounds of a species' range. Gap analysis uses this approach to develop distribution maps of species for which distributional and ecological data are readily available. To date, distribution mapping has focused on terrestrial breeding vertebrates, although distributional data for other well-studied groups, such as butterflies, are being gathered by some programs.

The distributional limits of each species are defined by specimen locality records or confirmed observations. They are represented as attributes of a geographic unit, in effect creating a rasterized distribution map. Raster data structure divides an area into units and assigns each a unique attribute (presence or absence of a species in this case). A 635 km² hexagonal grid array (White et al., 1992), developed to create a sampling framework for the EPA, is being used as the geographic unit of reference by several state programs, while the traditional county of occurrence is used by others.

The ancients recognized the relationship between species and natural communities (Morrison et al., 1992), and early concepts of the ecological niche had their foundation in natural history observations of the links between species and environmental features (Grinnell, 1914, 1917; Hutchinson, 1978). With certain qualifications, it is possible to relate the distribution of most species to that of plant communities in which they normally occur (Scott et al., 1993). Final maps of species' distributions are created by overlaying distributional limits with a vegetation map (Figure 22-3). A database indicating which plant communities provide suitable habitat for each species then identifies those vegetation

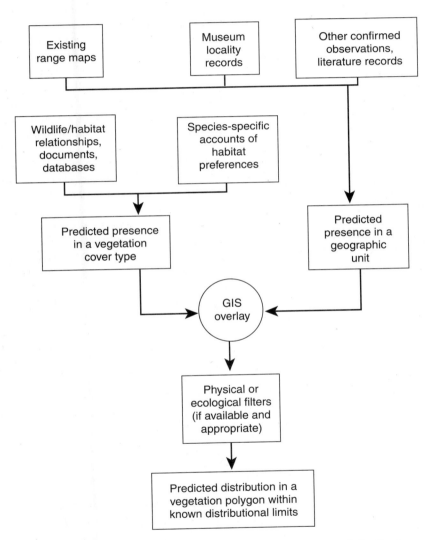

FIGURE 22-3   Flow chart showing steps in development of predicted distribution maps for breeding terrestrial vertebrates.

polygons within a species' range that probably are occupied (Figure 22-4). In this way, areas of unsuitable habitat are excluded from the predicted distribution. For example, species typical of coniferous forests are not predicted to occur in desert scrub or alpine fell fields.

The two most important limitations in this approach concern scale: (1) Many species will be present in a plant community only if certain microhabitat requirements also are present, and (2) many important habitat components are physical features (streams, cliff-faces, snags) that are too small to map. For this reason, species maps based on wildlife-habitat relationships can best be used "to predict the occurrence of species in general vegetation types and in environmental conditions across broad regions rather than at the scale of an individual stand" (Morrison et al., 1992:246). Put another way, it is virtually certain that California Thrashers (*Toxostoma redivivum*) occur in the Berkeley Hills, as they did in Grinnell's day (Grinnell, 1917), but there is less chance of encountering one in a particular 1-hectare stand of coyote bush (*Baccharis pilularis*) on any particular spring morning.

Maps of species distribution generated for gap analysis are intended to be used and validated at landscape scales ("kilometers in diameter," Forman and Godron, 1986:11), not at individual field sites (Edwards et al., 1996). The total number of species expected to occur in each vegetation stand can be displayed, creating a state map of species richness (Figure 22-5). Because of different biogeographic histories, areas of similar species richness in different ecoregions are likely to differ in species content. The richness maps used in any conservation evaluation will vary with the questions being asked (e.g., what is the distribution of endemic species, or what is the distribution of declining species of neotropical migrant birds?).

A further caveat concerns endangered, rare, or locally distributed species. Gap analysis does not predict that these species will occur other than at documented locations. Occurrences of virtually all the rarest and most endangered elements of biodiversity are tracked in databases of the Conservation Data Center which are established in all 50 states by The Nature Conservancy (Jenkins, 1988). The location of habitat for rare species can direct searches for additional populations, but only known occurrences, obtained from cooperating Conservation Data Centers, are used for developing a conservation strategy.

LOCATIONS OF AREAS MANAGED PRIMARILY FOR NATIVE SPECIES AND NATURAL ECOSYSTEM PROCESSES

As in the example from Hawaii (Figure 22-1), and others from Idaho (Figures 22-6 and 22-7), gap analysis compares the distribution of elements of biodiversity with that of areas in which the maintenance of biodiversity is a primary management goal. The boundaries of state and federal management units that meet this criterion are placed in a GIS format. State Conservation Data

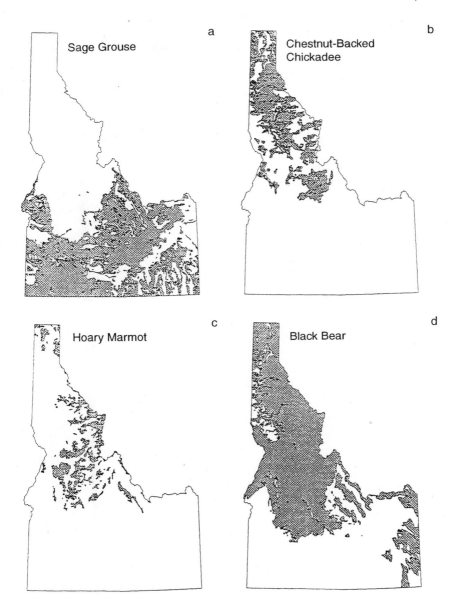

FIGURE 22-4 Predicted distribution of four species produced by intersection of occurrence in a geographic unit (counties, in this example) and polygons of suitable vegetation cover during the breeding season: (a) sage grouse (*Centrocerus urophasianus*), (b) chestnut-backed chickadee (*Parus rufescens*, (c) hoary marmot (*Marmota caligata*), (d) black bear (*Ursus americanus*).

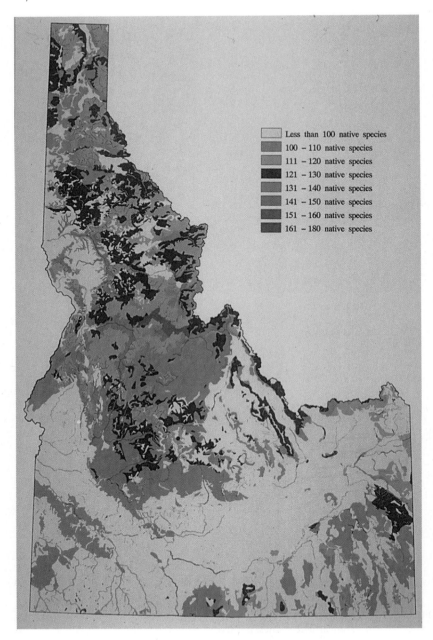

FIGURE 22-5 Species richness for native terrestrial vertebrates in Idaho.

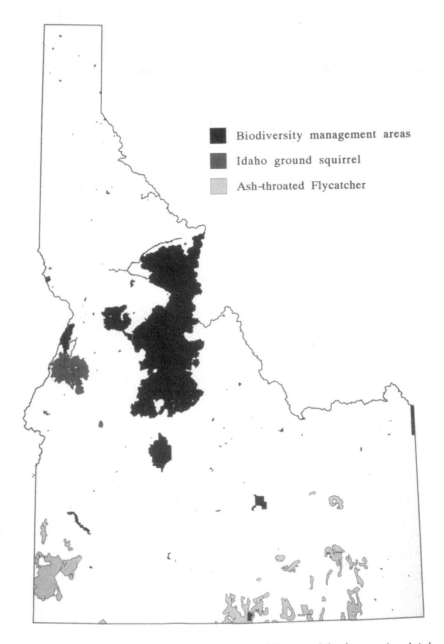

FIGURE 22-6 Distribution of Idaho ground squirrel (*Spermophilus brunneus*) and Ash-throated Flycatcher (*Myiarchus cinerascens*) in Idaho versus areas managed for long-term maintenance of natural communities.

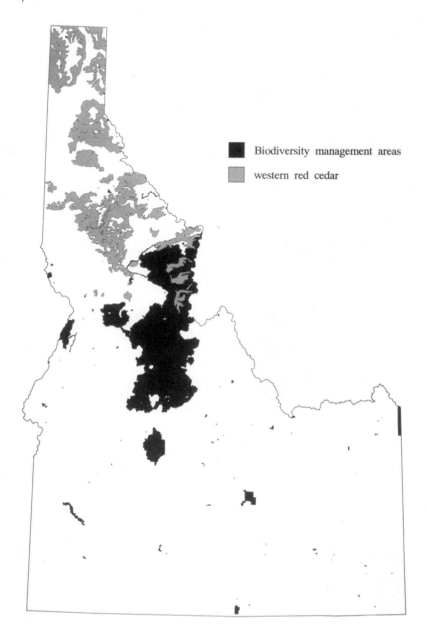

FIGURE 22-7 Distribution of western red cedar as a dominant type of cover in Idaho versus areas managed for long-term maintenance of biodiversity.

Centers and Natural Heritage Programs maintain files on each of these public management units, and their managed area files are the usual source for managed area boundaries for state programs. Examples of such areas include national parks, wilderness areas, research natural areas, and some national wildlife refuges (see Scott et al., 1993, for further details).

GAP ANALYSIS AS A CONSERVATION STRATEGY

It is impossible to manage for the long-term maintenance of biodiversity unless all the elements of biodiversity are represented in the areas to be managed in the first place (Margules et al., 1988). Gap analysis is a conservation evaluation technique (Margules, 1989; Usher, 1986) that identifies areas in which selected elements of biodiversity are represented. Once those areas are identified, other principles of conservation biology, such as population viability analysis, ecosystem patch dynamics, and habitat quality can be used to select specific sites and determine appropriate management area boundaries.

Noss (1987) and Noss and Cooperrider (1994) describe "coarse filter" and "fine filter" conservation strategies. Coarse filter strategies assume that most common species, including those of groups difficult to inventory, such as most invertebrates, will be represented in a reserve network that contains viable examples of all natural communities. Some species, especially those with restricted distributions, will be missed by the coarse filter. These species are captured by a fine filter that tracks the location of populations of individual species or rare natural communities.

The United States currently has no strategy to conserve biodiversity (Noss and Cooperrider, 1994). The most visible program to conserve biodiversity is the Endangered Species Act of 1973, a fine filter approach that protects one species at a time. The Endangered Species Act is a powerful tool with which to rescue species from the brink of extinction, but it is a reactive strategy that is in danger of being overwhelmed by growing numbers of species in peril and inadequate funding. Furthermore, many species that are not currently endangered will become so as their habitats are lost to human activities (Margules, 1989). The Endangered Species Act needs to be supplemented with a strategy to identify places that must be managed for their natural values if all communities and species are to persist (Scott et al., 1987; Tear et al., 1993). Like the Endangered Species Act, the ranking system of The Nature Conservancy's Conservation Data Centers provides a standard methodology for evaluating potential natural areas on the basis of threatened or endangered species or rare natural communities (Pearsall et al., 1986), but directs conservation action to elements of biodiversity already in peril.

Gap analysis provides a hierarchical approach to address conservation needs of both species and communities. Scott et al. (1987) called for an ecosystem approach to protecting biodiversity, supplemented by protecting species-rich areas for a variety of taxa. These areas should be selected to maximize their complementarity (see below). Finally, species not already captured in the network would be added on an individual basis.

Despite the contribution of biodiversity management areas to a conservation strategy, many mobile species and landscape processes require far more area than will ever be managed strictly for biodiversity (Brussard, 1991). The fate of these species will rest on the management of multiple-use lands surrounding nature reserves (Scott et al., 1990). Gap analysis data layers provide information about the context of areas being managed for different values, as well as opportunities to maintain connectivity between natural areas through landscape linkages (Csuti, 1991).

METHODS OF ANALYSIS: SELECTING COMPLETELY REPRESENTATIVE BIODIVERSITY MANAGEMENT NETWORKS

The simplest way to ensure that all ecosystems and species are represented in areas managed for biodiversity is to designate several areas for each species or ecosystem. Given sufficient time, local populations or ecosystems will experience catastrophic stochastic events. Species persist because populations elsewhere escape these events (Goodman, 1987). This strongly argues for multiple representation of species or ecosystems throughout their geographic range. Tear et al. (1993) suggest protecting at least three viable examples of each element of biodiversity (e.g., vegetation type, species) within each ecoregion.

Evidence from the study of island biogeography further suggests that only large areas will maintain anything like their original complement of species over evolutionary time. Barro Colorado Island, with an area of 16 km^2, has lost one-fourth of its species since its isolation in 1914 (Wilcox, 1980). Brown (1986) suggests that areas smaller than 500 km^2 are likely to lose more than half their species in a few thousand years.

There is a practical limit to the amount of land that any nation can manage primarily to maintain biodiversity (Pressey, 1990). The need for relatively large areas for management of biodiversity (e.g., >500 km^2) demands that those areas be maximally efficient in capturing all species and ecosystems. It simply is not possible to designate an appropriately large area (let alone several areas) for management of each and every species and ecosystem. By taking advantage of the fact that examples of many ecosystems occur in close proximity to one another and that species ranges overlap, it is possible to identify a subset of areas in which most elements of biodiversity are represented.

Pressey et al. (1993) articulate three principles for selecting reserve networks: (1) complementarity: the greatest efficiency in adding species or communities to a set of areas will be achieved if the areas are maximally different (or complementary); (2) flexibility: there are usually alternative areas that can add particular species or communities to a reserve network, therefore the selection process is somewhat flexible; and (3) irreplaceability: some elements of bio-diversity will occur only in one area, therefore these areas must be a part of any completely representative biodiversity management network—they are irre-placeable.

Considerable progress on quantitative approaches to efficient selection of reserve networks using iterative algorithms has been made in Australia (Bedward et al., 1992; Margules, 1989; Nicholls and Margules, 1993). A simple reserve selection algorithm identifies the area with the most species or types of vegeta-tion, then the area with the most species or types of vegetation not represented in the first choice, and so on. In some cases, it is more efficient to use an algo-rithm that selects the area with the rarest element first, then the area with the next rarest element that also contains the largest number of other elements, and so on (Pressey and Nicholls, 1989). Iterative algorithms can be extended to ensure that each element (species or ecosystem) is represented a number of times (once, twice, three times, etc.).

Similar results can be obtained if all possible combinations of two, three, four (and so on) areas are examined to identify those combinations that capture the most diversity at each step (Figure 22-8). This analysis (an example of an exact set coverage problem; Pennisi, 1993) identifies a family of areas, one of which is selected at each step. It also presents a more difficult computational problem and can be calculated for only a small number of steps for state or regional data sets. For example, for the state of Idaho, there are 25.3×10^{13} possible combinations of 389 hexagons taken 7 at a time.

While maximizing species richness (or diversity of types of vegetation) at each step in a selection process can lead to an efficient reserve network (Scott et al., 1987; Terborgh and Winter, 1983), some species may not occur in centers of richness. An area may contain relatively few species but still be a necessary part of a completely representative biodiversity management network due to the presence of species not found in other areas. Kareiva (1993), Prendergast et al. (1993), and Saetersdal et al. (1993) have pointed out that centers of species rich-ness for different groups, such as birds and butterflies, may not coincide (see also Robbins and Opler, Chapter 6 in this volume). The analysis of data layers is therefore a complex process, proceeding hierarchically from the community level of organization to complementary areas of high species richness and finally to species still not represented. Separate analyses for major taxonomic groups or ecoregions may identify priority areas with higher internal diversity than an analysis of all species throughout a political unit, where earlier choices are domi-nated by the taxon with the most species (i.e., birds) and later choices add only a few species but are otherwise redundant with earlier choices.

Because most species or plant communities occur in more than one area, the analysis itself becomes iterative. As an element of biodiversity occurs in more management areas, the priority of other areas which contain that element is reduced. Other factors can be used to assign priority among management areas,

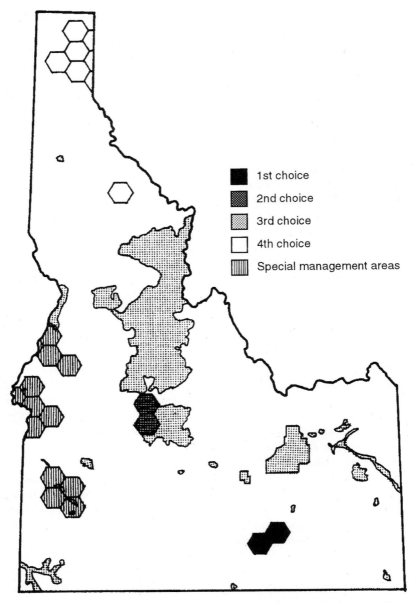

FIGURE 22-8   An exact set coverage algorithm identified four 635 km² areas in Idaho in which underrepresented species in existing reserves are predicted to occur. One choice from each of the four families will yield a set of four hexagons in which 97% of unprotected species are predicted to occur. Hexagon boundaries are arbitrary and are disregarded when designing boundaries for biodiversity management (Kiester et al, in press).

including threat, proximity to other areas (Nicholls and Margules, 1993), and taxonomic distinctness (Faith, 1994; Vane-Wright et al., 1991). Pressey and Bedward (1991) and Stoms (1994) have pointed out the influence of scale on biodiversity analysis. Determining the optimally sized subunits of a region for analysis remains an important research issue in conservation evaluation.

Predicting the biological value of priority areas for managing biodiversity is an efficient first step toward developing a conservation strategy. It must be followed by field reconnaissance to verify the value and condition of each area and to apply the principles of conservation biology to boundary delineation. Patterns of land ownership and economic activities affect the potential of managing any particular area solely for biodiversity and can be factored into the iterative process of building a reserve network.

AN EXAMPLE FROM IDAHO

The distribution of 119 types of vegetation and 357 species were mapped for the Idaho gap analysis. The representation of types of vegetation in managed areas is discussed elsewhere (Caicco et al., 1995). An analysis performed by A. R. Kiester and his colleagues (Kiester et al., 1996) determined that 49 terrestrial vertebrates were inadequately represented on existing managed areas in Idaho. Species were defined as inadequately represented on existing managed areas if they occurred less than three times on at least 10,000 contiguous hectares of land managed primarily for native species and natural ecosystems. The state was divided into 635 km^2 hexagonal sampling units following White et al. (1992). An exact set coverage algorithm identified 119 combinations of 4 hexagons in which 47 of 49 unprotected species were represented (Figure 22-8). Predicted distributions of unprotected species need to overlap hexagon boundaries only slightly to be assigned to a hexagon; therefore, actual areas that would support viable populations of these species will differ considerably from hexagonal sampling units. Analyses of the same data set using different definitions for unprotected species or different sample unit sizes are expected to identify somewhat different priority areas. The analysis presented here is one of many that can be integrated into a flexible conservation strategy for a completely representative network of areas that can be managed for the long-term maintenance of biodiversity.

ACKNOWLEDGMENTS

We thank Christopher R. Margules and Robert L. Pressey for their comments on Australian approaches to conservation evaluation. Bart R. Butterfield developed distribution maps for species in Idaho (Figures 22-5 and 22-6), and A. Ross Kiester carried out the exact set coverage analysis for Idaho that is illustrated (Figure 22-8).

REFERENCES

Austin, M. P. 1991. Vegetation: Data collection and analysis. Pp. 37-41 in C. R. Margules and M. P. Austin, eds., Nature Conservation: Cost Effective Biological Surveys and Data Analysis. CSIRO Publications, East Melbourne, Australia. 207 pp.

Austin, M. P., and C. R. Margules. 1986. Assessing representativeness. Pp. 45-67 in M. B. Usher, ed., Wildlife Conservation Evaluation. Chapman and Hall, London. 394 pp.

Bedward, M., R. L. Pressey, and D. A. Keith. 1992. A new approach for selecting fully representative reserve networks: Addressing efficiency, reserve design and land suitability with an iterative analysis. Biol. Conserv. 62:115-125.

Brown, J. H. 1986. Two decades of interaction between the MacArthur-Wilson model and the complexities of mammalian distributions. Biol. J. Linn. Soc. 28:231-251.

Brussard, P. F. 1991. The role of ecology in biological conservation. Ecol. Appl. 1:6-12.

Burley, F. W. 1988. Monitoring biological diversity for setting priorities in conservation. Pp. 227-230 in E. O. Wilson and F. M. Peter, eds., BioDiversity. National Academy Press, Washington, D.C.

Caicco, S. L., J. M. Scott, B. Butterfield, and B. Csuti. 1995. A gap analysis of the management status of the actual vegetation of Idaho (U.S.A.). Conserv. Biol. 9:498-511.

Csuti, B. 1991. Conservation corridors: Countering habitat fragmentation. Introduction. Pp. 81-90 in W. E. Hudson, ed., Landscape Linkages and Biodiversity. Island Press, Washington, D.C.

Edwards, T.C., Jr., E.T. Deshler, D. Fosler, and G.G. Molsen. 1996. Adequacy of wildlife habitat relation models for estimating spatial distributions of terrestrial vertebrates. Conserv. Biol. 10:263-270.

Ehrlich, P. R. 1988. The loss of diversity: Causes and consequences. Pp. 21-27 in E. O. Wilson and F. M. Peter, eds., BioDiversity. National Academy Press, Washington, D.C.

Faith, D. P. 1994. Genetic diversity and taxonomic priorities for conservation. Biol. Conserv. 68:69-74.

Forman, R. T. T., and M. Godron. 1986. Landscape Ecology. John Wiley and Sons, N.Y. 619 pp.

Goodman, D. 1987. How do any species persist? Lessons for conservation biology. Conserv. Biol. 1:59-62.

Graham, L. A. 1993. Airborne video for near-real-time vegetation mapping. J. Forestry 91:28-32.

Grinnell, J. 1914. An account of the mammals and birds of the Lower Colorado River Valley. Univ. Calif. Publ. Zool. 12:51-294.

Grinnell, J. 1917. The niche-relationships of the California thrasher. Auk 34:427-433.

Hutchinson, G. E. 1978. An Introduction to Population Ecology. Yale University Press, New Haven, Conn. 260 pp.

Jenkins, R. E., Jr. 1988. Information management for the conservation of biodiversity. Pp. 231-239 in E. O. Wilson and F. M. Peter, eds., BioDiversity. National Academy Press, Washington, D.C.

Kareiva, P. 1993. No shortcuts in new maps. Nature 365:292-293.

Kepler, C. B., and J. M. Scott. 1985. Conservation of island ecosystems. Int. Council Bird Preserv. Tech. Publ. No. 3:255-271.

Keister, A.R., J.M. Scott, B. Csuti, R. Noss, B. Butterfield, K. Sahr, and D. White. 1996. Conservation priortization using GAP data. Conserv. Biol. 10(4), in press.

Machlis, G. E., D. J. Forester, and J. E. McKendry. 1994. Biodiversity Gap Analysis: Critical Challenges and Solutions. Idaho Forestry, Wildlife, and Range Experiment Station Contr. No. 736. University of Idaho, Moscow. 61 pp.

Margules, C. R. 1989. Introduction to some Australian developments in conservation evaluation. Biol. Conserv. 50:1-11.

Margules, C. R., A. O. Nicholls, and R. L. Pressey. 1988. Selecting networks of reserves to maximize biological diversity. Biol. Conserv. 43:63-76.

Miller, A. H., and R. C. Stebbins. 1964. The Lives of Desert Animals in Joshua Tree National Monument. University of California Press, Berkeley. 452 pp.

Morrison, M. L., B. G. Marcot, and R. W. Mannan. 1992. Wildlife-Habitat Relationships: Concepts and Applications. University of Wisconsin Press, Madison. 343 pp.

Nicholls, A. O., and C. R. Margules. 1993. An upgraded reserve selection algorithm. Biol. Conserv. 64:165-169.

Noss, R. F. 1987. From plant communities to landscapes in conservation inventories: A look at The Nature Conservancy (USA). Biol. Conserv. 41:11-37.

Noss, R. F., and A. Y. Cooperrider. 1994. Saving Nature's Legacy: Protecting and Restoring Biodiversity. Island Press, Washington, D.C. 416 pp.

Orians, G. H. 1993. Endangered at what level? Ecol. Appl. 3:206-208.

Pearsall, S. H., D. Durham, and D. C. Eagar. 1986. Pp. 111-133 in M. B. Usher, ed., Wildlife Conservation Evaluation. Chapman and Hall, London. 394 pp.

Pennisi, E. 1993. Filling in the gaps. Sci. News 144:248-251.

Prendergast, J. R., R. M. Quinn, J. H. Lawton, B. C. Eversham, and D. W. Gibbons. 1993. Rare species, the coincidence of diversity hotspots and conservation strategies. Nature 365:335-337.

Pressey, R. L. 1990. Reserve selection in New South Wales: Where to from here? Aust. J. Zool. 26:70-75.

Pressey, R. L., and M. Bedward. 1991. Mapping the environment at different scales: Benefits and costs for nature conservation. Pp. 7-13 in C. R. Margules and M. P. Austin, eds., Nature Conservation: Cost Effective Biological Surveys and Data Analysis. CSIRO Publications, East Melbourne, Australia. 207 pp.

Pressey, R. L., C. J. Humphries, C. R. Margules, R. I. Vane-Wright, and P. H. Williams. 1993. Beyond opportunism: Key principles for systematic reserve selection. Trends Ecol. Evol. 8:124-128.

Pressey, R. L. and A. O. Nicholls. 1989. Application of a numerical algorithm to the selection of reserves in semi-arid New South Wales. Biol. Conserv. 50:263-278.

Saetersdal, M., J. M. Line, and H. J. B. Birks. 1993. How to maximize biological diversity in nature reserve selection: Vascular plants and breeding birds in deciduous woodlands, western Norway. Biol. Conserv. 66:131-138.

Scott, J. M., B. Csuti, J. D. Jacobi, and J. E. Estes. 1987. Species richness: A geographic approach to protecting future biological diversity. BioScience 37:782-788.

Scott, J. M., B. Csuti, and K. A. Smith. 1990. Playing Noah while paying the devil. Bull. Ecol. Soc. Amer. 71:156-159.

Scott, J. M., F. Davis, B. Csuti, R. Noss, B. Butterfield, C. Groves, H. Anderson, S. Caicco, F. D'Erchia, T. C. Edwards, Jr., J. Ulliman, and R. G. Wright. 1993. Gap analysis: A geographic approach to protection of biological diversity. Wildlife Monogr. 123:1-41.

Scott, J. M., S. Mountainspring, F. L. Ramsey, and C. B. Kepler. 1986. Forest bird communities of the Hawaiian islands: Their dynamics, ecology, and conservation. Studies in Avian Biology, No. 9. Cooper Ornithological Society, Lawrence, Kans. 431 pp.

Stoms, D. 1994. Scale dependence of species richness maps. Prof. Geogr. 46:346-358.

Tear, T. H., J. M. Scott, P. H. Hayward, and B. Griffith. 1993. Status and prospects for success of the Endangered Species Act: A look at recovery plans. Science 262:976-977.

Terborgh, J., and B. Winter. 1983. A method for siting parks and reserves with special reference to Colombia and Ecuador. Biol. Conserv. 27:45-58.

Thomas, J. W., ed. 1979. Wildlife Habitats in Managed Forests: The Blue Mountains of Oregon and Washington. U.S.D.A. Agriculture Handbook No. 553. U.S. Department of Agriculture, Washington, D.C. 512 pp.

United Nations Educational, Scientific, and Cultural Organization. 1973. International Classification and Mapping of Vegetation. UNESCO, Paris, France. 35 pp.

Usher, M. B., ed. 1986. Wildlife Conservation Evaluation. Chapman and Hall, London. 394 pp.

Vane-Wright, R. I., C. J. Humphries, and P. H. Williams. 1991. What to protect? Systematics and the agony of choice. Biol. Conserv. 55:235-254.

White, D., A. J. Kimerling, and W. S. Overton. 1992. Cartographic and geometric components of a global sampling design for environmental monitoring. Cart. Geogr. Info. Syst. 19:5-22.

Wilcox, B. A. 1980. Insular ecology and conservation. Pp. 95-117 in M. E. Soule and B. A. Wilcox, eds., Conservation Biology: An Evolutionary-Ecological Perspective. Sinauer Associates, Sunderland, Mass. 395 pp.

Conservation of Biodiversity
in Neotropical Primates

JAMES M. DIETZ

Associate Professor, Department of Zoology, University of Maryland, College Park

In 1986, the year of the National Research Council and Smithsonian Institution's National Forum on BioDiversity, compilations were published on the distribution and conservation status of most neotropical primates. These contributions, largely the effort of Russell Mittermeier (then of World Wildlife Fund, now of Conservation International), established regional and taxonomic priorities for conservation and research initiatives on neotropical primates (Mittermeier, 1986a,b; Mittermeier and Oates, 1985).

Since that time, considerable progress has been made in our understanding of the biology of neotropical primates and in the application of appropriate conservation methods to ensure that these species will survive to be considered in future biodiversity symposia. In this chapter, I select a few examples from the recent literature to illustrate progress in our understanding of neotropical primate biodiversity in the contexts of geographic distribution and taxonomy, habitat evaluation, studies on ecology, evolution and behavior, and conservation strategy. For each of these areas, I also identify important topics for future research and development.

SYSTEMATICS, PHYLOGENY, AND GEOGRAPHIC DISTRIBUTION

New World primates, called platyrrhines because of their flat noses and widely spaced nostrils, are thought to have evolved as forest dwellers, never descending to occupy the terrestrial niches widely used by Old World primates. The split between Old and New World primates took place some 40 million years ago (Rosenberger, 1992) and resulted in an extremely diverse array of platyrrhines at all taxonomic levels.

Most authorities recognize 16 extant platyrrhine genera, including *Ateles* (spider monkey), *Brachyteles* (muriqui), *Lagothrix* (wooly monkey), *Alouatta* (howler monkey), *Pithecia* (saki), *Chiropotes* (bearded sakis), *Cacajao* (uakari), *Aotus* (owl monkeys), *Callicebus* (titi monkey), *Cebus* (capuchin monkey), *Saimiri* (squirrel monkey), *Cebuella* (pygmy marmoset), *Callithrix* (marmosets), *Leontopithecus* (lion tamarins), *Saguinus* (tamarins), and *Callimico* (Goeldi's monkey) (review by Rosenberger, 1981).

In contrast with the general concordance of reviewers working at the genus level, considerable disagreement exists among authorities working at higher taxonomic levels (see reviews by Mittermeier and Coimbra-Filho, 1981, and de Vivo, 1991). Traditional classifications divided New World primates into "marmosets" or "callitrichids" and "nonmarmosets" or "cebids," in which the marmoset patterns of morphology were seen as primitive and those of cebids as advanced (e.g., Hershkovitz, 1977). In a 1986 revision, Ford reorganized recent platyrrhine genera using a cladistic approach. In 1992, Rosenberger of the National Zoological Park, Smithsonian Institution, launched a new synthesis based on phylogenetic relationships, adaptive change, and the fossil record. He recognized four monophyletic subfamilies in that classification. Callitrichids became callitrichines, closely related to squirrel monkeys and capuchins (Cebinae). Many features of callitrichines are now thought to have resulted from selection for small size and thus are derived and not ancestral or primitive traits.

Identification of the phylogenetic link between the Cebinae and Callitrichinae was made possible by the discovery of a new set of fossils at La Venta, Colombia, with morphological characteristics intermediate between those of the two subfamilies (Rosenberger, 1992). The rate of description of fossil platyrrhine genera has increased markedly in recent years. Of 16 known fossil genera, 7 were described since 1985 and 4 of these since 1990 (Fleagle and Rosenberger, 1990). Our understanding of phylogenetic relationships and extinctions of neotropical primates certainly will continue to improve if the trend toward increased description of fossil forms continues.

New advances in molecular genetic techniques now make it possible to objectively test hypotheses about the phylogenetic relationships among the platyrrhines. The Brazilian geneticists Maria Sampaio and Horácio Schneider examined sequences of DNA in nuclear globin genes and found close relationships between *Alouatta* and the atelines (*Ateles*, *Brachyteles*, and *Lagothrix*), between *Cebus* and *Saguinus*, and among *Pithecia*, *Chiropotes*, and *Cacajao* (Schneider et al., 1993), adding additional support to the phylogenetic model developed by Rosenberger.

Much of what we know about the diversity of neotropical primates at the species level is due to the painstaking descriptive detail provided in the published works of Hershkovitz (e.g., Hershkovitz, 1977). That author and a few others generally used pelage and cranial descriptions to classify the majority of extant species of neotropical primates. However, there has been considerable recent controversy about the criteria used in assigning species status to many of these forms (e.g., Forman et al., 1986, for *Leontopithecus* spp.; Rylands et al., 1993, for the Callitrichidae; Coimbra-Filho, 1991, for *Leontopithecus caissara*; and Meireles et al., 1992, for *Callithrix* spp.).

Geographic ranges of many forms have been altered by recent anthropogenic habitat modification. As a result, correlations between patterns of geographic distributions and morphological characteristics become increasingly difficult to interpret. The biological accuracy of taxonomic designations becomes particularly important because legislative protection and conservation funding for full species is significantly greater than those for subspecies (e.g., Ryder, 1986). In addition, conservation objectives of reintroduction and translocation of isolated populations are predicated on the conservation of genetic diversity within a species. If genetic management is to have its desired effect in situ, it is imperative that we be able to distinguish conspecifics from less closely related forms. Revision of neotropical primates at the species level should be given high priority, both in terms of contributions to primate conservation and basic research. Recently developed noninvasive genetic techniques such as sequencing of mitochondrial DNA extracted from hair follicles of museum specimens (e.g., Woodruff, 1993) may make it possible to resolve some of these controversies.

A group of golden lion tamarins in a Brazilian rainforest.

Living neotropical primates are found from central Mexico to northern Argentina and include some 68 species, comprising about one-third of all living species of primates (Mittermeier, 1986b; Mittermeier et al., 1992). In the mid-1980s, much emphasis was put on identification of major phytogeographic regions that contained relatively large numbers of primates, particularly rare, endangered, or endemic species (Mittermeier, 1986a,b; Mittermeier and Cheney, 1986). These analyses resulted in maps of priority areas for primate conservation and research. For example, the Brazilian Atlantic forest reportedly contained 6 genera and 15 species of primates, including 2 endemic genera and 9 endemic species, all of which were considered in danger of extinction (Mittermeier, 1986b).

In the past 3 years, four new species of primates were described in Brazil: *Cebus kaapori* (Queiroz, 1992), *Callithrix mauesi* (Mittermeier et al., 1992), and *Callithrix nigriceps* (Ferrari and Lopes, 1992), all from the Amazon region; and *Leontopithecus caissara* (Lorini and Persson, 1990), found not far from the large cities of Curitiba and São Paulo. The discovery of four unknown primates in such a short period of time, including one species in the backyard of one of the largest cities in the world, underscores how little we really know about the diversity of even well-studied taxa such as nonhuman primates and emphasizes the need for more basic research and surveys on primates in the Neotropics.

However, if conservation efforts on behalf of any species of primates are to succeed, it will be necessary to go beyond the identification of its geographic distribution and taxonomic relationships with other primates. Among other things, we will need a reliable estimate of how many individuals of that species remain in the wild and information about their population structure. Pioneers in Brazilian primatology, such as Coimbra-Filho, recognized decades ago that the geographic distribution of golden lion tamarins and other Atlantic forest primates had been reduced to small and degraded forest fragments (e.g., Coimbra-Filho, 1977).

Although biologists have been conducting censuses of neotropical primates for decades, lack of standardization of methods and the difficulty of observing primates in closed forest has made it difficult to assess the accuracy of the results of these surveys. Nonetheless, numbers of many neotropical primates appear to be decreasing. For example, in 1969, the Brazilian conservationist Alvaro Aguirre estimated that 2,000-3,000 muriquis (*Brachyteles arachnoides*) survived in Brazil's Atlantic forest (Brownlee, 1987). In the mid-1980s, the minimum number of muriquis known to exist in 10 locations was published as 240 (Mittermeier, 1986b). However, it was not until 3 years ago—when Kierulff (1993), a Brazilian graduate student, used play-backs of golden lion tamarin (*Leontopithecus rosalia*) vocalizations in virtually every forest that might contain these primates—that an accurate estimate of the number of individuals in the wild became known. This, the first exhaustive census for any neotropical primate, suggested that only 559 individuals in 103 groups remain in the wild and that the species is subdivided into 16 isolated subpopulations, 12 of these containing only 1 group. The accurate nature of this information removed any question about the degree of threat to this species and prompted significant changes in the focus of conservation and related research activities.

MONITORING CHANGES IN QUANTITY AND QUALITY OF HABITAT

In 1986, critical regions for conservation of neotropical primates were identified as huge phytogeographic areas such as the Amazon basin and "megadiversity countries" (Mittermeier, 1986b) that contained a large diversity of primate taxa. Since that date, our ability to pinpoint regions for conservation has improved markedly. In 1990, approximately 100 specialists on Amazonian fauna, flora, soils, and climate met in Manaus, Brazil, to produce a map of biologically important regions in the Amazon Basin based on criteria of endemism, rarity, and diversity. Important here is the diversity of specialists contributing information and the speed with which the results were disseminated.

Perhaps the most significant recent advance in our ability to quantitatively evaluate primate habitat is the combination of remote-sensing technology and geographic information system (GIS) analysis. A decade ago satellite images of regions of South America were expensive, difficult to acquire, and could be analyzed in only a very few locations. Currently, a variety of high-quality, inexpensive magnetic image data are available from several sources and cover virtually every region on Earth. Recent improvements and reduction in price of hand-held geographic position system (GPS) equipment now allow a researcher to precisely determine latitude and longitude for samples of habitat, thus simplifying the ground-truthing (verifying the interpretation of images obtained via remote observation devices in local habitats) of satellite data. Desk-top computer versions of GIS software make it possible to conduct sophisticated habitat analyses without acquiring specialized equipment. New techniques such as gap analysis (Scott et al., 1993, and Chapter 22, this volume), which evaluate the protection status of plant and animal communities by GIS overlay of distributional data on maps of existing protected areas, help conservationists to be proactive rather than reactive in their efforts to preserve biodiversity.

The accessibility of satellite data and new analytical techniques (Sader et al., 1990) resulted in three large steps forward in terms of establishing conservation and research priorities in the Neotropics. First, we now are able to quantify changes in vegetation patterns over time. For example, in the 1980s, widespread disagreement over rates of deforestation in the Brazilian Amazon hampered efforts at international collaboration in conservation (reviewed in Skole and Tucker, 1993). In the 1990s, independent reports by Brazilian and American scientists agreed that previous estimates were several-fold larger than actual rates of deforestation (Skole and Tucker, 1993). Since future treaties on biodiversity and climatic change are likely to require countries to limit their emissions of greenhouse gases, including those resulting from deforestation, estimates of deforestation based on satellite data acquire additional political and economic importance.

Second, we now can use satellite data to test hypotheses about the effects of anthropogenic factors on habitat quantity and quality. For example, we can use archival data to document habitat changes that followed the construction of a road into a forested area in the Amazon and subsequently use that information to predict how a proposed road might alter undisturbed forest in another region. Third, based on any criteria that can be mapped (e.g., forest type, socioeconomic variables, or distance from villages), it is now possible to quantitatively evaluate areas for creation as reserves, annexation to reserves, or as corridors between reserves (Figure 23-1). Remote sensing and GIS techniques

FIGURE 23-1 False color composite of data from Landsat's Thematic Mapper showing original boundary (black line) of Una Biological Reserve, Bahia state, Brazil, and forest tracts purchased and annexed to the Reserve through international collaborative efforts (green line). Dark brown represents areas covered by dense forest, yellow is secondary forest, blue is areas that have been recently cleared, and white and black are clouds and their shadows. Satellite data were acquired from Brazil's Instituto Nacional de Pesquisas Espaciais by WWF-Brazil. The image was produced by R. DeFries and D. Van Wie, Department of Geography, University of Maryland.

are becoming increasingly powerful research tools and should be included in the tool kit of any researcher or conservationist doing field work on neotropical primates.

FIELD STUDIES ON ECOLOGY, EVOLUTION, AND BEHAVIOR

Hundreds of scientific reports have been published in the past decade on the ecology and behavior of neotropical primates. I have selected a few examples to illustrate three categories of field research that have been particularly useful from a conservation perspective: field research using the comparative method, long-term demographic studies, and research addressing the effects of human disturbance factors on primate population size and structure.

Although the principles of evolutionary biology should be useful in explaining primate behavior, primates rarely have been at the forefront in this work (Richard, 1981). Cheney et al. (1986) identify several reasons why primates are rarely used to test evolutionary theory. Most species of primates are difficult to identify individually or even to observe under field conditions. Many species are so long-lived that accumulation of data on life histories may require decades of observation. In addition, the social behavior of many species is sufficiently complex that it becomes difficult to collect information on a sufficient number of individuals to obtain a complete picture of patterns of social organization (Dunbar, 1986). Seasonal and annual variation further complicates the design of field studies examining the effects of ecological variables or the adaptive significance of specific behaviors in tropical ecosystems. Finally, formulation of generalizations often is impeded by differences in researchers' methodology.

One approach that has proved useful in circumventing these obstacles is to ask how several primate taxa behave under the same circumstances and when the same observational methodology is used. Conclusions then may be inferred from differences between species. This comparative approach has the added advantage of focusing on an interesting question rather than on an interesting primate.

A good example of the comparative method applied to field research on primates is Terborgh's (1983) long-term study of the comparative ecology of five New World primates in Manu National Park, Peru. Comparisons across species allowed Terborgh to make inferences about, for example, the function of territorial defense, factors that influence size of the group, and the relationship between size of the group and the social system in primates. Another fruitful comparative study was conducted by Peres (1993) at a remote site in Amazonas, Brazil. Peres compared the feeding ecology and habitat use of 13 primate species for a period of 20 months. The comparative method allowed Peres to conclude that seasonal migrations by several species living in large groups were the result of "bottlenecks" in food availability in their habitats.

An interesting example of the comparative method is the comparison of wild- and captive-born reintroduced golden lion tamarins on private property adjacent to Poço das Antas Reserve, Rio de Janeiro, Brazil (Beck et al., 1991). Parallel studies of native and reintroduced tamarins allowed researchers to generalize about the effects of captivity on locomotion, communication, range use, foraging, and reproduction. Ninety-one tamarins were introduced between 1984 and 1991, and were closely monitored by local observers. Thirty-three survived until June, 1991, and 57 infants were produced. Deficits in food-finding, orientation, and locomotion, which ultimately caused most losses of reintroduced tamarins, were less severe in their wild-born offspring than in the originally introduced individuals (Beck et al., 1991).

A second approach that has proved useful both in terms of basic research and conservation application is long-term monitoring of demographic variables for a single primate population. During the past decade of study on golden lion tamarins in Poço das Antas Reserve, researchers individually marked 80% of the tamarins in the reserve and these became habituated to the presence of human observers. Continuous monitoring of all births, deaths, emigrations, and immigrations in 22 study groups allowed the researchers to compare average fecundity and survivorship for tamarins following specific behavior patterns, e.g., males that emigrate at an early age from the natal group versus those that remain as nonreproductive helpers (Baker, 1991; Baker et al., 1993; Dietz and Baker, 1993; Dietz et al., 1994a). The conservation implications of these data are discussed below.

Calculation of reproductive success under field conditions can be problematic, particularly for species of primates in which more than one male copulates with a reproductive female. Under these circumstances, new molecular genetic techniques, notably DNA fingerprinting, may allow paternity exclusion (identification of which males did not father the offspring in question) and thus direct determination of male reproductive success (reviewed in Martin et al., 1992).

It long has been thought that deforestation and hunting have negative effects on most primate populations. However, the effects of these and other anthropogenic factors rarely have been quantified for neotropical species. In recent reviews of the impact of habitat disturbance on a variety of species of primates, Johns (1991) and Johns and Skorupa (1987) reported that body size and degree of frugivory were negatively correlated with a species' probability of survival. Redford and Robinson (1991) surveyed the available literature and predicted that hunting by humans would cause primate population density to decrease as prey body mass increases. Freese et al. (1982) surveyed primate densities in Bolivia and Peru and suggested that predation by humans was the most important factor affecting primate densities. Peres (1990, 1991) used transect censuses of primates in hunted and unhunted areas to test hypotheses about the effect of hunting on primate densities. Peres found that body size alone largely determined the choice of primates hunted in Amazonian forests and that large species, such as woolly monkeys (*Lagothrix lagotricha*), often were hunted to extinction in areas near human settlements, even in the absence of habitat disturbance. Redford (1992) reached similar conclusions for a variety of large animals in neotropical forest. Two conclusions are suggested by these studies. First, conservation efforts need to focus on large-bodied primates with low rates of reproduction, even before habitat destruction begins. Second, past and present hunting pressure must be included in calculating population dynamics of large-bodied primates, even in areas of apparently intact habitat.

CONSERVATION STRATEGIES

The alarm that sounded in the mid-1980s about the immediate threat to most species of neotropical primates was answered in a number of ways. I have classified these strategies as information management and networking, application of scientific method, captive breeding, conservation education, and international collaboration.

Information Management and Networking

The advent of wide-spread use of electronic mail dramatically increased the speed and decreased the cost of international communication. Transfer of data files, even from remote field stations, has become routine in the past few years. Computerized bulletin boards are available on a number of relevant topics, including conservation biology, primates, ecology, the Neotropics in general, as well as several tropical countries.

Many international nongovernment conservation organizations (NGOs) support regional programs to develop and conduct conservation and research activities in neotropical countries. Some of these initiatives, e.g., the Golden Lion Tamarin Conservation Program in Brazil, which has been supported for 10 years by grants from the World Wildlife Fund and other NGOs, focus on conservation of primates and their habitat. The Nature Conservancy sponsors the establishment of Conservation Data Centers in 13 Latin American and Caribbean countries to aid in decision-making in conservation and sustainable development. Conservation International supports a Rapid Assessment Program in which teams of experts census diverse taxa of fauna and flora to set priorities for conservation action.

Several scientific journals and newsletters reporting on topics relevant to conservation of neotropical primates have appeared in the past decade, and some of these include articles or abstracts in Spanish or Portuguese. Examples include *Neotropical Primates* (newsletter of the IUCN/Species Survival Commission Primate Specialist Group), *CBSG News* (newsletter of the Captive Breeding Specialist Group), *Conservation Biology* (journal of the Society for Conservation Biology, Blackwell Press) and *Boletim da Sociedade Brasileira de Primatologia* (bulletin of the Brazilian Primatological Society).

Application of the Scientific Method

The past decade has seen a heartening increase in the generation of conservation theory and the application of rigorous scientific method and problem-solving techniques to conservation questions. Graduate degrees in conservation biology now are offered at many universities in the United States and Europe (reviewed in Jacobson, 1990, and Jacobson et al., 1995). The National Science Foundation now funds grant proposals in conservation and restoration biology.

Although the primary concern of conservationists is the survival of ecosystems, much of the theoretical and applied research in conservation biology is conducted at the population level. The link between population and ecosystem conservation is thought to be the appropriate selection of populations of "flagship species" that serve as the focus of conservation campaigns, "umbrella species" that have ranges large enough to include significant portions of natural ecosystems, and "keystone species" whose presence or absence determines the abundance and distribution of a number of other species. Appropriate selection of populations for conservation can conserve habitat and associated biodiversity effectively. The charismatic nature and ecological role of many neotropical primates makes them well-suited to be flagship species (reviewed in Dietz et al., 1994b).

A new approach to determining conservation action plans at the population level is the Population Viability Analysis or Population and Habitat Viability Analysis (PVA or PHVA; reviewed in Foose et al., 1995). This concept is predicated on the need to establish quantitative conservation goals for the population under study, e.g., a 95% probability of survival over a period of 100 years. Computer models such as VORTEX (Lacy, 1993) are used to estimate the probability of population extinction and time to extinction, based on the effects of stochastic events such as inbreeding, environmental variation, and catastrophes on population demographic variables. The probability of survival of multiple subpopulations in a metapopulation also may be estimated (Gilpin and Hanski, 1991). Examples of PHVA applied to four species of lion tamarins are presented in Seal et al. (1990).

The advantages of PHVA include the simultaneous treatment of several variables, the collaborative process by which the best available information is assembled for all interested parties, the quantitative nature of the results, and the versatility in evaluating management options. The shortcomings of PHVA are related to the assumptions implicit in the computer model and the effects of estimating unknown values. For example, the impact of environmental variation on fecundity and survivorship rarely is known for populations in nature. Small changes in the specified effects of environmental variation on reproduction or mortality can produce 10-fold differences in the predicted probability of extinction of the population.

Captive Breeding and Reintroduction

Not long ago, zoos were seen as consumers of primates and other animals, competing to assemble a collection of the rarest and most valuable examples for public exhibit. Recent changes in the objectives of modern zoological parks make these institutions important contributors to in situ conservation of neotropical primates. The principal objective of captive propagation now is seen as the reestablishment or reinforcement, not replacement, of wild populations (Ballou and Cooper, 1992). With over 13,000 specimens in zoological collections, primates are overrepresented relative to other mammalian orders. Nevertheless, only about 50% of primate taxa are represented in captivity (Magin et al., 1994). International management committees, organized to maximize retention of genetic diversity and minimize probability of extinction in captivity and the wild, are increasing for many vertebrate taxa, including primates.

A number of advisory groups work to coordinate the ex situ and in situ primate conservation activities of zoos. The Species Survival Commission (SSC), consisting of world experts on approximately 100 taxa or conservation topics, is the main source of advice to The World Conservation Union (IUCN) regarding the technical aspects of conservation of species. The SSC promotes action on the part of the international conservation community on behalf of species threatened with extinction and those important for human welfare. Two SSC Specialist Groups are particularly important for primate conservation: the Primate Spe-

A golden lion tamarin feeding.

cialist Group, comprised of about 200 experts working in the conservation of primates, and the Captive Breeding Specialist Group, whose objective is the conservation or establishment of viable populations of threatened species.

In 1992, a Taxon Advisory Group (TAG) to the American Association of Zoos and Aquaria (AAZPA) was formed to coordinate and facilitate captive breeding of New World Primates in support of in situ conservation efforts. Faunal Interest Groups (FIGs) coordinate conservation activities in zoos for fauna and flora from a specific geographic region, e.g., Brazil.

Studbooks, containing complete demographic records for a species in captivity, facilitate the scientific management of captive populations, e.g., by facilitating selection of appropriate individuals for pairings to reduce loss of genetic diversity (reviewed by Ballou and Lacey, 1995; Lande, 1995). Studbooks are maintained for 15 species of neotropical primates, and have been proposed for several others. For a few species, information on native populations is sufficient for inclusion in studbooks. For example, the studbook for golden lion tamarins soon will include all individuals in the species: captive, reintroduced, and native subpopulations.

The reintroduction of captive-born neotropical primates has been initiated for several species in the past few years. In addition to the reintroduction of golden lion tamarins mentioned above, reintroduction and monitoring of *Callithrix geoffroyi* (Geoffroyi's marmoset) in Espirito Santo state, Brazil, began in 1991 (Chiarello and Passamani, 1993). The costs and criteria for success in a primate reintroduction are reviewed by Kleiman et al. (1991). The IUCN-SSC Reintroduction Specialist Group and the AAZPA Reintroduction Advisory Group both were created in the past few years to facilitate in situ conservation through integration of wild and captive population management.

Conservation Education

The role of community education in directing the attitudes and behaviors of local peoples toward sustainable use of natural resources long has been recognized as an essential component of any viable conservation strategy. This is particularly true where charismatic species such as primates can be used as flagships for conservation efforts (Dietz et al., 1994b). However, only in the past few years have rigorous evaluation techniques been used to quantify the effects of community education in primate conservation projects.

Illustrating the value of quantitative evaluation of community education, educators conducted surveys before and after 2 years of activities related to the golden lion tamarin conservation project and documented a significant increase in the percentage of respondents who recognized golden lion tamarins in a photograph. Whereas 42.5% of these responses were attributed to project activities, only 6.9% reported seeing tamarins in the forest. The project activities that were mentioned most often included television (14.9% of responses) and radio

(6.9% of responses), a total effect roughly equivalent to all other project activities combined (Dietz and Nagagata, 1986; Dietz et al., 1994b). Graduate training in conservation education for South and Central American nationals is one of the best investments that can be made to ensure the future of neotropical primates.

International Collaboration

A final example illustrates progress in international collaboration. In this case, private lands were purchased and annexed to a Brazilian federal reserve too small to guarantee the survival of an endangered primate. A PHVA for golden headed lion tamarins (*L. chrysomelas*), conducted in the Una Biological Reserve, southern Bahia State, Brazil (the only protected area containing that primate), suggested that the probability of extinction of that primate was relatively high (Dietz et al., 1994c; Seal et al., 1990).

Based on the findings of the PHVA, a coalition of eight international conservation organizations conducted fund-raising efforts to buy out squatters and purchase land to annex to Una Reserve. Funds were transferred in several installments to Fundação Biodiversitas, a Brazilian NGO, which purchased lands at market price from local landowners and immediately conferred title to the Brazilian agency responsible for administration of the Reserve. To date, the area of the Reserve has been increased from 5,342 hectares to 7,059 hectares (Figure 23-1; Coimbra-Filho et al., 1993). Cooperation that spans disciplines, agencies, and nations is rare indeed, but if neotropical primates are to be conserved in nature it will be through collaborative efforts of this type in which the world's scientific and economic resources can be brought to bear on the problem.

REFERENCES

Baker, A. J. 1991. Evolution of the Social System of the Golden Lion Tamarin. Unpublished Ph.D. dissertation, University of Maryland, College Park.

Baker, A. J., J. M. Dietz, and D. G. Kleiman. 1993. Behavioural evidence for monopolization of paternity in multimale groups of golden lion tamarins. Anim. Behav. 46:1091-1103.

Ballou, J. D., and K. A. Cooper. 1992. Genetic management strategies for endangered captive populations: The role of genetic and reproductive technology. Symp. Zool. Soc. Lond. 64:183-206.

Ballou, J. D., and R. C. Lacy. 1995. Identifying genetically important individuals for management of genetic variation in pedigreed populations. In J. D. Ballou, M. Gilpin, and T. J. Foose, eds., Population Management for Survival and Recovery. Columbia University Press, N.Y.

Beck, B. B., D. G. Kleiman, J. M. Dietz, I. Castro, C. Carvalho, A. Martins, and B. Rettberg-Beck. 1991. Losses and reproduction in reintroduced golden lion tamarins *Leontopithecus rosalia*. Dodo 27:50-61.

Brownlee, S. 1987. These are real swinging primates. Discover (April): 67-77.

Cheney, D. L., R. M. Seyfarth, B. B. Smuts, and D. W. Wrangham. 1986. The study of primate societies. Pp. 1-8 in B. B. Smuts, D. L. Cheney, R. M. Seyfarth, R. W. Wrangham, and T. T. Struhsaker, eds., Primate Societies. University of Chicago Press, Chicago.

Chiarello, A. G., and M. Passamani. 1993. A reintroduction program for geoffroyi's marmoset, *Callithrix geoffroyi*. Neotrop. Primates 1(3):6-7.

Coimbra-Filho, A. F. 1977. Natural shelters of *Leontopithecus rosalia* and some ecological implications (Callitrichidae: Primates). Pp. 79-89 in D. G. Kleiman, ed., The Biology and Conservation of the Callitrichidae. Smithsonian Institution Press, Washington, D.C.

Coimbra-Filho, A. F. 1991. Sistemática, distribuição geográfica e situação atual dos símios brasileiros (Platyrrhini-Primates). Rev. Brasil. Biol. 50(4):1063-1079.

Coimbra-Filho, A. F., L. A. Dietz, J. C. Mallinson, and I. B. Santos. 1993. Land purchase for the Una Biological Reserve, refuge of the golden-headed lion tamarin. Neotrop. Primates 1(3):7-9.

de Vivo, M. 1991. Taxonomia de *Callithrix* Erxleben, 1777 (Callitrichidae, Primates). Fundação Biodiversitas, Belo Horizonte, Brazil.

Dietz, J. M., and A. J. Baker. 1993. Polygyny and female reproductive success in golden lion tamarins, *Leontopithecus rosalia*. Anim. Behav. 46:1067-1078.

Dietz, J. M., A. J. Baker, and D. Miglioretti. 1994a. Seasonal variation in reproduction, juvenile growth and adult body mass in golden lion tamarins *Leontopithecus rosalia*). Amer. J. Primatol. 34:115-132.

Dietz, J. M., L. A. Dietz, and E. Y. Nagagata. 1994b. The effective use of flagship species for conservation of biodiversity: The example of lion tamarins in Brazil. Pp. 32-49 in P. J. S. Olney, G. M. Mace, and A. T. C. Feistner, eds., Creative Conservation. Chapman and Hall, London.

Dietz, J. M., S. N. F. Sousa, and J. R. O. Silva. 1994c. Population structure and territory size in golden-headed lion tamarins, *Leontopithecus chrysomelas*. Neotrop. Primates 2(supl.):21-23.

Dietz, L. A., and E. Y. Nagagata. 1986. Programa de educação comunitária para a conservação do mico leão dourado *L. rosalia* (Linneaus, 1766): desenvolvimento e avaliação de educação como uma tecnologia para a conservação de uma espécie em extinção. Pp. 249-256 in M. T. de Mello, ed., A Primatologia no Brasil, Vol. 2. Sociedade Brasileira de Primatologia, Brasília, D.F., Brazil.

Dunbar, R. I. M. 1986. Demography and reproduction. Pp. 240-249 in B. B. Smuts, D. L. Cheney, R. M. Seyfarth, R. W. Wrangham, and T. T. Struhsaker, eds., Primate Societies. University of Chicago Press, Chicago.

Ferrari, S. F., and M. A. Lopes. 1992. A new species of marmoset, genus *Callithrix* Erxleben, 1777 (Callitrichidae, Primates), from western Brazilian Amazonia. Goeldiana Zool. 12:1-13.

Fleagle, J., and A. Rosenberger. 1990. The platyrrhine fossil record. J. Human Evol. 19:1-254.

Foose, T. J., B. de Boer, U. S. Seal, and R. Lande. 1995. Conservation management strategies based on viable populations. In J. D. Ballou, M. Gilpin, and T. J. Foose, eds., Population Management for Survival and Recovery. Columbia University Press, N.Y.

Ford, S. 1986. Systematics of the New World monkeys. Pp. 73-135 in D. Swindler, ed., Comparative Primate Biology, Vol. 1. Systematics, Evolution and Anatomy. Alan R. Liss, N.Y.

Forman, L., D. G. Kleiman, R. M. Bush, J. M. Dietz, J. D. Ballou, L. G. Phillips, A. F. Coimbra-Filho, and S. J. O'Brien. 1986. Genetic variation within and among lion tamarins. Amer. J. Physical Anthropol. 71:1-11.

Freese, C. H., P. G. Heltne, N. Castro, and G. Whitesides. 1982. Patterns and determinants of monkey densities in Peru and Bolivia, with notes on distributions. Int. J. Primatol. 3:53-90.

Gilpin, M., and I. Hanski. 1991. Metapopulation Dynamics: Empirical and Theoretical Investigations. Academic Press, London.

Hershkovitz, P. 1977. Living New World Monkeys, Vol. 1. University of Chicago Press, Chicago.

Jacobson, S. K. 1990. Graduate education in conservation biology. Conserv. Biol. 4(4):431-440.

Jacobson, S. K., E. Vaughan, and S. W. Miller. 1995. New directions in conservation biology: Graduate programs. Conserv. Biol. 9(1):5-17.

Johns, A. D. 1991. Forest disturbance and Amazonian primates. Pp. 115-135 in H. Box, ed., Primate Responses to Environmental Change. Chapman and Hall, London.

Johns, A. D., and J. P. Skorupa. 1987. Responses of rain-forest primates to habitat disturbance: A review. Int. J. Primatol. 8:157-191.

Kierulff, M. C. 1993. Avaliação das populações selvagens de mico-leão-dourado, *Leontopithecus rosalia*, e proposta de estratégia para sua conservação. Unpublished M.Sc. thesis. Federal University of Minas Gerais, Minas Gerais, Brazil.

Kleiman, D. G., B. B. Beck, J. M. Dietz, and L. A. Dietz. 1991. Costs of a re-introduction and criteria for success: Accounting and accountability in the Golden Lion Tamarin Conservation Program. Symp. Zool. Soc. Lond. 62:125-142.

Lacy, R. C. 1993. VORTEX: A computer simulation model for population viability analysis. Wildlife Res. 20:45-65.

Lande, R. 1995. Breeding plans for small populations, based on the dynamics of genetic variance in quantitative traits. In J. D. Ballou, M. Gilpin, and T. J. Foose, eds., Population Management for Survival and Recovery. Columbia University Press, N.Y.

Lorini, M., and V. G. Persson. 1990. Nova espécie de *Leontopithecus* Lesson, 1840, do sul do Brasil (Primates, Callitrichidae). Bol. Museu Nacional (Zoologia) 138:1-14.

Magin, C. D., T. H. Johnson, B. Groombridge, M. Jenkins, and H. Smith. 1994. Species extinctions, endangerment and captive breeding. Pp. 3-31 in P. J. S. Olney, G. M. Mace, and A. T. C. Feistner, eds., Creative Conservation. Chapman and Hall, London.

Martin, R. D., A. F. Dixson, and E. J. Wickings. 1992. Paternity in primates: Genetic tests and theories. Karger, Basel, Switzerland.

Meireles, C. M. M., M. I. Sampaio, H. Schneider, and M. P. C. Schneider. 1992. Protein taxonomy and differentiation in five species of marmosets (Genus *Callithrix* Erxleben, 1777). Primates 33(2):227-238.

Mittermeier, R. A. 1986a. Primate diversity and the tropical forest: Case studies from Brazil and Madagascar and the importance of megadiversity countries. Pp. 145-154 in E. O. Wilson and F. M. Peter, eds., BioDiversity. National Academy Press, Washington, D.C.

Mittermeier, R. A. 1986b. Primate conservation priorities in the neotropical region. Pp. 221-240 in K. Benirschke, ed., Primates: The Road to Self-Sustaining Populations. Springer-Verlag, N.Y.

Mittermeier, R. A., and D. L. Cheney. 1986. Conservation of primates and their habitats. Pp. 477-490 in B. B. Smuts, D. L. Cheney, R. M. Seyfarth, R. W. Wrangham, and T. T. Struhsaker, eds., Primate Societies. University of Chicago Press, Chicago.

Mittermeier, R. A., and A. F. Coimbra-Filho. 1981. Systematics: Species and subspecies. Pp. 29-109 in A. F. Coimbra-Filho and R. A. Mittermeier, eds., Ecology and Behavior of Neotropical Primates. Academia Brasileira de Ciências, Rio de Janeiro, Brazil.

Mittermeier, R. A., and J. F. Oates. 1985. Primate diversity: The world's top countries. Primate Conserv. 5:41-48.

Mittermeier, R. A., M. Schwarz, and J.M. Ayres. 1992. A new species of marmoset, genus *Callithrix* Erxleben, 1777 (Callitrichidae, Primates) from the Rio Maués, State of Amazonas, Central Brazilian Amazonia. Goeldiana Zool. 14:1-17.

Peres, C. A. 1990. Effects of hunting on western Amazonian primate communities. Biol. Conserv. 54:47-59.

Peres, C. A. 1991. Humbolt's woolly monkey decimated by hunting in Amazonia. Oryx 25(2):89-95.

Peres, C. A. 1993. Structure and spatial organization of an Amazonian terra firme forest primate community. J. Trop. Ecol. 9:259-276.

Queiroz, H. L. 1992. A new species of capuchin monkey, genus *Cebus* Erxleben 1777 (Cebidae, Primates), from eastern Brazilian Amazonia. Goeldiana Zool. 15:1-3.

Redford, K. H. 1992. The empty forest. BioScience 42(6):412-422.

Redford, K. H., and J. G. Robinson. 1991. Subsistence and commercial use of wildlife in Latin America. Pp. 6-23 in J. G. Robinson and K. H. Redford, eds., Neotropical Wildlife Use and Conservation. University of Chicago Press, Chicago.

Richard, A. 1981. Changing assumptions in primate ecology. Amer. Anthropol. 83:517-533.

Rosenberger, A. 1981. Systematics: The higher taxa. Pp. 9-27 in A. F. Coimbra-Filho and R. A. Mittermeier, eds., Ecology and Behavior of Neotropical Primates. Academia Brasileira de Ciências, Rio de Janeiro, Brazil.

Rosenberger, A. 1992. Evolution of New World monkeys. Pp. 209-216 in S. Jones, R. Martin, and D. Pilbeam, eds., The Cambridge Encyclopedia of Human Evolution. Cambridge University Press, Cambridge, England.

Ryder, O. A. 1986. Species conservation and systematics: The dilemma of subspecies. Trends Ecol. Evol. 1(1):9-10.

Rylands, A. B., A. F. Coimbra-Filho, and R. A. Mittermeier. 1993. Systematics, geographic distribution, and some notes on the conservation of the Callitrichidae. Pp. 11-77 in A. B. Rylands, ed., Marmosets and Tamarins: Systematics, Behaviour, and Ecology. Oxford University Press, Oxford, England.

Sader, S. A., T. A. Stone, and A. T. Joyce. 1990. Remote sensing of tropical forests: An overview of research and applications using non-photographic sensors. Photogram. Engineer. Remote Sens. 56(10):1343-1351.

Schneider, H., M. P. C. Schneider, I. Sampaio, M. L. Harada, M. Stanhope, J. Czelusniak, and M. Goodman. 1993. Molecular phylogeny of the New World Monkeys (Platyrrhini, Primates). J. Mol. Phylogeny Evol. 2(3):225-242.

Scott, J. M., F. Davis, B. Csuti, R. Noss, B. Butterfield, C. Groves, H. Anderson, S. Caicco, F. D'Erchia, T. C. Edwards, J. Ulliman, and R. G. Wright. 1993. Gap analysis: A geographic approach to protection of biological diversity. Wildlife Monogr. 123:1-141.

Seal, U. S., J. D. Ballou, and C. V. Padua. 1990. *Leontopithecus* Population Viability Analysis Workshop Report. Captive Breeding Specialist Group, Minneapolis, Minn.

Skole, D., and C. Tucker. 1993. Tropical deforestation and habitat fragmentation in the Amazon: Satellite data from 1978 to 1988. Science 260:1905-1910.

Terborgh, J. 1983. Five New World Primates. Princeton University Press, N.J.

Woodruff, D. S. 1993. Non-invasive genotyping of primates. Primates 34(3):333-344.

24

Using Marine Invertebrates to Establish Research and Conservation Priorities

JAMES D. THOMAS
Curator of Crustacea, Department of Invertebrate Zoology,
Smithsonian Institution, Washington, D.C.

Current methods and applications used to identify and select coral reefs for conservation efforts are seldom based on scientific methodology. Instead, protection efforts are focused on a series of coral reefs under direct or imminent threat of impact or alteration, resulting in a reactive policy approach. What is needed in light of current changes occurring in coral reef systems are programs with a strong preventative component designed to establish research and conservation priorities in coral reefs before they come under significant levels of threat. In this chapter, I outline inadequacies of current approaches to identifying and managing biodiversity in coral reefs, recommend new ways to establish selective criteria through taxonomic surveys and inventories, and provide an example from a coral reef system of exceptional biodiversity in the Madang Lagoon, Papua New Guinea. While not visually spectacular, this reef system houses remarkable levels of marine invertebrate biodiversity.

CORAL REEFS

Coral reefs have provided scientists with a rich source of facts and theory and have helped to forge fundamental views on evolution, biodiversity, and geology in the ocean realm. Scientists such as Darwin and Wallace recognized distinct patterns in biological distribution of coral reef organisms and attributed part of this pattern to geological events. With the acceptance of plate tectonic theory and accurate radiometric dating, these patterns have assumed biogeographic importance. Tectonic plate movements over fixed mantle hotspots produce accurate plate circuitry measurements (Yan and Kroenke, 1993), producing linear volcanic arcs and island archipelagos that encode a history of past geological events. According to Pandolfi (1992), a correlation exists between biogeographic pattern and geological history, and modern marine distribution patterns can best be interpreted by incorporating the geological history of the area under study.

Numerous hypotheses have been proposed to explain patterns of biodiversity on coral reefs. These include dispersal models (Kay, 1984), the Pacific plate vicariance theory (Springer, 1982), a "Pacifica" continental fragmentation theory (Nur and Ben-Avrahm, 1977), an expanding Earth theory (Carey, 1958, 1976), and a variety of ecological explanations (Vermeij, 1990). However, the systematic study of comparative levels of endemism in coral reef invertebrates to see which paradigm, or combination of paradigms, best explains biogeographical patterns remains to be engaged by the scientific community. Fundamental knowledge of biodiversity at the level of species is a prerequisite for such investigations.

Integrated studies of marine biodiversity are just beginning. To date, many taxonomic and most biogeographic studies of reef systems have been random and opportunistic, focusing on easily accessible islands. Taxonomic literature on tropical marine invertebrates is scattered, dealing mostly with large, spatially obvious components such as fish, scleractinian corals, and molluscs, thus creating a biological bias for larger organisms that may not be the best indicators

Fish swarm over a coral reef.

or measures of biodiversity (Thomas, 1992). Platnick (1992) refers to the condition of concentrating on the obvious spatial components as the "megafauna bias." The most informative invertebrate groups in terms of biogeography are those without a dispersive (pelagic) larval stage. Wide dispersal capabilities can mask small-scale distribution patterns. Thus, animals with restricted distributions more accurately reflect levels of endemism. Information on those groups of invertebrates with restricted distributions is needed to integrate the many competing hypotheses on distribution mechanisms.

While coral reefs contain the highest levels of biodiversity in any marine ecosystem, virtually nothing is known about possible extinctions or natural trends in biodiversity. World-wide reports of changes in coral reef systems are largely anecdotal and based primarily on ecological research. Human impact is suspected as a contributing factor in these changes, but the interaction of natural change versus human-induced impacts remains speculative (Roberts, 1988; Williams and Williams, 1990). The rate at which marine areas now are being designated for protected status as sanctuaries, refuges, preserves, parks, etc., creates a problem in setting priorities and selecting among the many candidates suggested for protection. The goal of maintaining native levels of biological diversity requires that knowledge regarding historical sources and levels of biodiversity be known so that competent decisions about research, conservation, and management action can be made.

DISTINGUISHING TYPES OF BIODIVERSITY

Above the generic level, the marine environment is more diverse than terrestrial systems (Ray and Grassle, 1991), yet most of what we know about biodiversity comes from studies in tropical moist forest systems. Biodiversity studies in marine environments, particularly the tropics, suffer from diffusion of scientific effort and an inability to identify critical sites based on levels of biodiversity. Present management and conservation efforts in marine systems are driven mainly by concepts other than biodiversity, e.g., species of special human interest, areas of spectacular natural beauty, and economics (Thomas, 1993). Such an approach places the highest priority on unusual areas with little regard to levels and trends of biodiversity. In some instances where biodiversity surveys have been conducted, emphasis is on "spatially obvious" organisms such as fish and corals that may not be the best indicators to identify the processes of environmental change (Angermeier and Karl, 1994).

Successful management of marine ecosystems must rely on relevant information about levels and trends of biodiversity, information that is almost entirely lacking in current management approaches. Biodiversity survey and inventory programs that are designed to identify and select those marine areas of highest scientific value are needed. Resources such as existing collections in natural history museums and coordination of ongoing taxonomic surveys and inventories need to be organized according to selective criteria. The limited number of active field systematists and taxonomists currently working in marine environments must be deployed effectively. Training programs to allow a wider dissemination of taxonomic expertise also are a critical element in a unified approach to marine biodiversity. Use of data from disciplines outside taxonomy and systematics must be incorporated into a marine biodiversity program for the twenty-first century. Information from fields such as plate tectonics, paleogeography, cladistics, ecology, and anthropology must be brought together in concert to fully understand what is happening in marine environments and what can be done to improve our understanding of system processes. At the administrative level, specific shifts in policy goals to include reliance on preventative rather than reactive management strategies must become a priority (Angermeier and Karl, 1994).

TAXONOMY AND BIODIVERSITY IN CORAL REEFS

Available taxonomic data tell us more about varying attention given to different groups of animals, the "taxonomy of taxonomists," than about the level of taxonomic knowledge (May, 1994). The information summarized by May (1994) illustrates the great disparity of attention received by different groups. Roughly one-third of taxonomists work on plants, while the remaining two-thirds split roughly equally between invertebrates and vertebrates. The estimated total number of species of vertebrates is 40,000; species of plants is 300,000; and species of invertebrates is about 1 million (with estimates up to 10 million; Grassle and Maciolek, 1992). Therefore, for every n taxonomists working on vertebrates, there are $0.1n$ taxonomists investigating plants and $0.01n$ taxonomists specializing in invertebrates. When we consider that a majority of invertebrate taxonomists study a single group, the insects, the great disparity within the current taxonomic work force that specialize in marine invertebrates becomes apparent. Estimates of millions of new species, with an estimated novelty rate of 40-80% for undescribed taxa, places an absolute accounting of all marine species outside the realm of possibility in any time frame that would make any significant difference in the current biodiversity crisis. I suggest that what is needed is sufficient effort directed toward taxonomic inventories of *selected* groups of bioindicators that are chosen by established taxonomic protocols. Such a selection process would target not only areas that are threatened or in crisis, but would incorporate objective scientific criteria to predict possible centers of evolutionary diversification that act as genetic sources for existing biodiversity (Thomas, 1992). Approaches that focus existing taxonomic expertise on areas that may be historical sources of biodiversity will help us understand how patterns of biodiversity at the level of species are maintained and replenished.

DOCUMENTING BIODIVERSITY IN TROPICAL MARINE ECOSYSTEMS

Despite extensive reports of large-scale change in coral reef systems, the scientific, conservation, and management community does not currently have the capability to investigate every region or problem that gains public attention.

Lack of comprehensive marine biodiversity protocols to identify areas of particular scientific, conservation, and management value have lead to a diffuse approach to the systematic investigation of coral reef biodiversity. Most research taking place now on reefs is ecological in nature, with little or no systematic effort directed toward taxonomic surveys and inventories of reef systems world-wide. This lack of focus is further compounded by the shortage of experienced field systematists and taxonomists. Training programs that would increase the number of personnel available for surveys and inventories on coral reefs, while frequently discussed, have resulted in few active programs. The need is especially acute for the marine invertebrates, particular groups of which are sensitive indicators of change in coral reefs (Thomas, 1993).

The scientific community must develop a process to assess biodiversity that uses selective criteria and sets priorities, in effect a form of "environmental triage." Every reef system cannot be investigated with current personnel and levels of funding. Difficult choices must be made that maximize existing personnel, equipment, and organizational structure. Numerous coral reef sites already enjoy protected status, while numerous additional sites are being suggested for protection. In many cases, the site selection process is driven from a perception that a particular site is in a stressed or deteriorating condition. This approach focuses limited resources on a series of crisis situations and perpetuates a reactive management posture. Ideally, the scientific community should provide theoretical guidance and taxonomic information regarding appropriate site selection. The conservation community, capable of rapid response and adept at raising public awareness, then should use scientific data to target specific sites or regions for protection. The resource management community does not become established until an administrative structure is created. Every effort should be made to incorporate high-quality science in management approaches that document biodiversity. Coral reef sites that enjoy preexisting protected status should be encouraged to adopt programs that adequately document levels of biodiversity if such information is lacking. These major groups—scientists, conservationists, and resource managers—must work in concert along established guidelines to maximize the application of diffuse resources and funding to help identify and protect coral reefs.

ESTABLISHING PRIORITIES IN MARINE BIODIVERSITY

Priorities vary depending on individuals, agencies, and processes involved. For example, taxonomists place the highest priority for research on reefs that contain the highest biodiversity at the level of species or are the least impaired or impacted by man. Conservation groups may target areas that house species of special interest or support high proportions of endemics, while resource managers might focus on reef areas that seem threatened by human impacts. It must be emphasized that most coral reef systems, except for limited areas in the Caribbean and the Great Barrier Reef in Australia, are claimed by developing countries with little or no scientific or administrative resources. Therefore, any system of establishing priorities must take into account a variety of factors.

Developing Selection Strategies

The following approaches should be employed to develop priorities for marine biodiversity initiatives:

(1) Identify centers of evolutionary diversification, or major centers of genetic biodiversity by surveys and inventories, existing museum collections, and cladistics.

(2) Target areas where geological history indicates a history of vicariant events that might lead to high levels of endemic species or areas of composite biodiversity.

(3) Implement surveys and inventories to determine biodiversity baselines. While entire protected areas must be surveyed initially, ongoing monitoring of biodiversity could be restricted to special protected areas, such as replenishment zones located within a protected area (Bohnsack, 1993). It is important to stress that our understanding of biodiversity in coral reefs is rudimentary; therefore, initial surveys must incorporate a representative cross section of the entire area. Once levels of biodiversity have been determined, decisions can be made as to the location of permanent monitoring sites and schedules.

(4) Publish comprehensive identification guides to biota. Most information regarding marine invertebrates from coral reefs is uneven in terms of taxonomic and geographic coverage. Computer-based identification manuals should be a priority for development and distribution.

(5) Establish parataxonomic training programs to augment survey and inventory capabilities. Basic biodiversity surveys and inventories of many protected areas remain to be initiated and are not currently a primary management concern. Two of the largest marine protected areas in the world, the Florida Keys National Marine Sanctuary and the Great Barrier Reef Marine Park, Australia, have yet to institute systematic biotic inventories, despite stated management concerns to "monitor biodiversity."

A Geological Frame of Reference for Biodiversity

The term "hotspots," a geological term originally used to describe fixed volcanic sources, has recently gained popularity in the biological sense to indicate areas of high biodiversity. This unfortunate amalgamation of terminology confuses the original use of the term. In this chapter, the term "hotspot" is used in a geological context.

Recent efforts to test the hypothesis that geologic hotspots accurately record the passage of lithospheric plates generally have shown it to be valid (Duncan, 1981, 1991; Duncan and Richards, 1991; Fleitout and Moriceau, 1992; Morgan, 1981, 1983). If hotspots indeed are fixed with respect to one another over significant periods of geological time, then they would constitute an accurate frame of reference that could be used to determine precise lithospheric plate movement and allow reconstruction of plate positions back into time. Morgan (1981) estimated that hotspots migrate less than 5 mm per year relative to each other. Hotspot trails have been clearly delineated for the time span of zero to 100 million years, thus allowing reconstruction of southwest Pacific tectonic elements. This is done by first determining linkages between individual rifted continental fragments and island arcs and then by assigning correct plate motion derived from hotspot models, thus providing a "plate circuit" (Yan and Kroenke, 1993). Such approaches to tectonic plate circuitry allow visualization of the complex movements of plates through space and time. Yan and Kroenke (1993) provide a CD-ROM that contains animated sequences of the southwest Pacific plate movements from zero to 100 million years ago in 0.5-million year increments that provides additional insight on how and when marine organisms may have dispersed or been constrained by various geological events over the last 100 million years.

THE MADANG LAGOON, PAPUA NEW GUINEA

The island of New Guinea, composed of Papua New Guinea (PNG) and Irian Jaya, is located on the leading edge of the Australian plate that has been moving rapidly northward. The country of PNG comprises a landmass slightly larger than California (457,000 km²). The north coast of PNG has been formed by collision events along the north edge of the Australian plate and subsequent "docking" of the east Papuan composite and other terranes (a series of continuously related geological formations) in the mid-late Miocene 15-20 million years ago (Pigram and Davies, 1987). I suggest that the composite marine fauna of the Madang Lagoon is the result of the accretionary process along the north coast of PNG, involving rapid volcanic uplift as islands, archipelagos, submerged plateaus, and continental terranes are thrust rapidly upward in the collisional process. This process differs from other areas of the circum-Pacific, where terranes are predominantly of oceanic affinity.

This "docking" process introduced a number of previously discrete biotic assemblages that then intermingled with established floral and faunal elements.

Reefs in the Madang Lagoon illustrate the importance of taxonomic data in identifying areas of scientific concern. At first glance, the reefs in the Madang

Lagoon appear "less spectacular" (in a visual sense) than their southern counterparts on the Great Barrier Reef. However, taxonomic surveys of marine invertebrates suggest that Madang reefs are some of the most biologically diverse reefs yet documented. Sustaining native levels of biodiversity in these reefs as a potential genetic seed source for other South Pacific reefs is important in the larger context of regional biodiversity.

Marine Invertebrates of the Madang Lagoon

Over the past several years, increasing numbers of scientists have focused their research efforts in the Madang Lagoon in an attempt to document the unusual levels of biodiversity found there. The following information summarizes briefly some of the results and trends that have been documented. A more complete discussion can be found in the Proceedings of the Seventh International Coral Reef Symposium (Richmond, 1992).

Scleractinian Corals.   While much attention has been given to the scleractinian corals of eastern Australia (Veron and Pichon, 1976, 1979, 1982; Veron and Wallace, 1984; Veron et al., 1977) the nature of reef corals from northern New Guinea was not well understood due to lack of distributional data. Potts suggests that the Madang Lagoon may prove to be the single most diverse site in the world for scleractinian corals (Potts, personal communication, 1994). Hoeksema's recent treatment of fungiid corals (1992) represents the only reef coral family on which detailed distribution and taxonomic data are available for the Indo-Pacific region. According to Hoeksema (1992), northern New Guinea appears to have the highest fungiid biodiversity (39 species), with a fauna most similar to that of the Philippines and eastern Indonesia (37 species). Diversity in eastern Australia was the next highest (31 species), followed by western Australia and Taiwan (26 species), northwest Java (25 species), southern Papua New Guinea (24 species), and northeast Borneo (19 species). Hoeksema (1992) stresses that the generic diversity of hermatypic corals in the Indo-Pacific region is quite large, and that lists of genera are not as informative as species diversity and monospecific genera. Because all taxonomic categories above the species level are arbitrary, lists that include species and monospecific genera are more likely to reflect evolutionary history, and thus allow for a more precise comparison of biodiversity.

Octocorals.   Winston (1988) estimates that only 50% of the octocoral fauna from the Indo-Pacific region is known at present. Like most other marine invertebrate groups, our basic taxonomic knowledge of the Indo-Pacific octocorals is poorly known (Williams, 1992). A recently described species of octocoral from the Madang Lagoon was unlike any species of the genus previously recorded (Bayer, 1994).

Polyps of an organ pipe coral.

Amphipods. Thomas (1992) found that the amphipod fauna from coral reefs of the Madang Lagoon exhibited exceptional levels of species diversity. The amphipod fauna of Madang reefs is a composite, consisting of approximately 180 species, 60% of which are new to science and exhibit multiple biogeographic affinities. The Madang Lagoon amphipod fauna is taxonomically distinct from other South and Indo-Pacific sites, and the amphipod biodiversity on the north coast of PNG likely exceeds that of any coral reef area studied thus far. However, many coral reef systems in the Indo-Pacific have never been systematically analyzed for smaller crustaceans. The author suggests herein that future biodiversity inventories and surveys be undertaken with regard to selective criteria. Due to the limited dispersion capabilities and habitat specificity of amphipods, amphipods may be of use in biogeography and environmental monitoring in coral reef systems.

Crinoids. Messing (1992) reported 39 species of comatulid crinoids from the Madang Lagoon and, with limited sampling, found the crinoid fauna of the Madang Lagoon comparable to other more intensively studied sites such as Lizard Island and Davies Reef (Australia, Great Barrier Reef), and Palau.

Gastropod Molluscs. Working in the Madang Lagoon, Gosliner (1992) found that the north coast of PNG supports a more diverse fauna of opistobranch gastropods (538 species) than has been reported from any single geographical area studied thus far. The next richest tropical site is Guam (395 species), followed by Hawaii (244 species), the Caribbean (232 species), and Japan (184 species). Gosliner's faunal records are significant because of intensive field efforts in numerous tropical localities using snorkeling and SCUBA that enable comparative studies. Other areas that are known or suspected to house high diversity have not been studied adequately to allow comparisons of opistobranch biodiversity (Gosliner, 1992).

Other Biota and Habitats. Kristian Fauchald (personal communication, 1994) reported that the polychaete fauna of the Madang Lagoon exceeded that of any area yet sampled. Clyde Roper and Mike Sweeney (personal communication, 1994) reported similar findings for cepahlopod molluscs. In a preliminary survey of the marine algae of the Madang Lagoon, Mark and Diane Littler (personal communication, 1994) reported that not only were there more species of algae collected, but the number of undescribed species surpassed that of any region previously sampled.

The geological history that may have contributed to the extraordinary levels of marine invertebrates in the shallow waters of the north coast of PNG also may have influenced the fauna in deeper waters of the region. Studying a collection of deep sea crustaceans from the Bismarck Sea region (1,200 m), Austin Williams of the National Marine Fisheries Service Systematics Laboratory reports unusual levels and types of biodiversity (personal communication, 1994). More investigation of the deep sea component of this region is warranted in light of this preliminary information.

SELECTING CORAL REEF SITES BY ATTRIBUTES

Most current approaches to protecting biodiversity place emphasis on areas or species of special human interest and natural beauty or areas with novel biological features (endemics). Therefore, highest value is placed on unique and unusual biota, with little regard to the nature and quality of biodiversity on a larger scale. It is imperative to develop priorities using the geological past, the nature of the ecosystem, the characteristics of the region, and the available specialists.

Developing Action Strategies

The following tactics can be used to initiate biodiversity inventories in marine areas of particular interest and importance:

(1) Initiate surveys and inventories that document biodiversity in those areas (known or suspected to be) of extraordinary scientific or conservation value as determined by a rigorous selection process.

(2) Establish biologically based monitoring programs using selected groups of organisms that have an established reputation as bioindicators of high quality.

(3) Inventory existing historical collections from reef systems deposited in natural history museums. Many museums house historical collections of organisms from tropical regions. Such collections could serve as a "biodiversity baseline."

(4) Publish primary taxonomic monographs, identification guides, keys, and manuals, especially computerized, graphically based manuals for nonspecialists and resource managers.

(5) Develop parataxonomic training programs targeted for specific taxa and geographic regions.

(6) Implement scientific taxonomic training programs coordinated through a network of natural history museums, academic institutions, government agencies, and other organizations.

SUMMARY

Oceanic islands and their associated coral reefs have provided scientists with a wealth of biogeographic information. Levels of biodiversity on modern coral reefs provide a window on past evolutionary events that detail the correlation between biogeographic pattern and geological history. Reefs of the Madang Lagoon in Papua New Guinea exhibit levels of biodiversity exceeding all other reef systems studied thus far. Because of this, the Madang Lagoon represents a scientific and conservation resource of the highest priority. While biodiversity research on coral reefs is in its infancy, the need for this information is acute.

Within PNG, a complex pattern of land ownership combined with negligible developmental pressures and resource exploitation have allowed the reefs to remain relatively unaffected by anthropogenic impacts. That situation promises to change as the country seeks to modernize its largely subsistence economy and as rich mineral deposits and timber resources are developed. The rugged topography of the interior of the country virtually assures that the majority of this developmental pressure and impact will be in the coastal region, adjacent to reefs.

The reefs of the north coast of PNG provide an unequaled opportunity to study marine biogeography in what is probably a major source of biodiversity for a large area of the South Pacific. Research and conservation efforts must be focused on this invaluable biotic resource before significant impacts occur that will affect yet unstudied and undocumented groups of organisms.

ACKNOWLEDGMENTS

The author wishes to thank the Christensen Research Institute in Madang for research support on coral reef systems in and around the waters of Madang. Former Director Matthew Jebb and current Director Larry Orsak graciously have provided facilities and equipment to marine researchers working in the Madang region.

REFERENCES

Angermeier, P. L., and J. R. Karl. 1994. Biological integrity versus biological diversity as policy directives. BioScience 44:690-697.

Bayer, F. M. 1994. A new species of the Gorgonacean genus *Bebryce* (Coelenterata: Octocoralia) from Papua-New Guinea. Bull. Mar. Sci. 54(2):546-553.

Bohnsack, J. A. 1993. Marine reserves: They enhance fisheries, reduce conflicts, and protect resources. Oceanus 36(3):63-71.

Carey, S. W. 1958. The tectonic approach to continental drift. Pp. 177-355 in Symposium on Continental Drift. University of Tasmania, Hobart.

Carey, S. W. 1976. The Expanding Earth. Elsevier Publishing Company, Amsterdam.

Duncan, R. A. 1981. Hotspots in the Southern Oceans—An absolute frame of reference for motion of the Gondwana continents. Pp. 29-42 in S. C. Solomon, R. Van der Voo, and M. A. Chinnery, eds., Quantitative Methods of Assessing Plate Motions. Tectonophysics 74(1/2):1-208.

Duncan, R. A. 1991. Age distribution of volcanism along aseismic ridges in the eastern Indian Ocean. Pp. 507-518 in J. Weissel, J. Pierce, E. Taylor, and J. Alt, eds., Proceedings of the Ocean Drilling Program, Scientific Results 121. Texas A & M University, College Station.

Duncan, R. A., and M. A. Richards. 1991. Hotspots, mantle plumes, flood basalts, and true polar wander. Rev. Geophys. 29:31-50.

Fleitout, L., and C. Moriceau. 1992. Short-wavelength geoid, bathymetry and the convective pattern beneath the Pacific Ocean. Geophys. J. Int. 110:6-28.

Gosliner, T. M. 1992. Biodiversity of tropical opisthobranch gastropod faunas. Pp. 702-709 in R. Richmond, ed., Proceedings of the Seventh International Coral Reef Symposium, Vol. 2. University of Guam Press, Mangilao.

Grassle, J. F., and N. J. Maciolek. 1992. Deep-sea species richness: Regional and local diversity estimates from quantitative bottom samples. Amer. Nat. 139:313-341.

Hoeksema, B. W. 1992. The position of northern New Guinea in the center of marine benthic diversity: A reef coral perspective. Pp. 710-717 in R. Richmond, ed., Proceedings of the Seventh International Coral Reef Symposium, Vol. 2. University of Guam Press, Mangilao.

Kay, E. A. 1984. Patterns of speciation in the Indo-West Pacific. Pp. 15-31 in F. J. Radovsky, P. Raven, and S. H. Sohmer, eds., Biogeography of the Tropical Pacific. B. P. Bishop Museum Special Publication 72. Bishop Museum Press, Honolulu, Hawaii.

May, R. M. 1994. Biological diversity: Differences between land and sea. Phil. Trans. R. Soc. Lond. (B) 343:105-111.

Messing, C. G. 1992. Diversity and ecology of comatulid crinoids (Echinodermata) at Madang, Papua New Guinea [abstract]. P. 736 in R. Richmond, ed., Proceedings of the Seventh International Coral Reef Symposium, Guam, Vol. 2. University of Guam Press, Mangilao.

Morgan, W. J. 1981. Hotspot tracks and the opening of the Atlantic and Indian oceans. Pp. 443-487 in C. Emiliani, ed., The Sea, Vol. 7: The Oceanic Lithosphere. John Wiley and Sons, N.Y.

Morgan, W. J. 1983. Hotspot tracks and the early rifting of the Atlantic. Tectonophysics 94:123-139.

Nur, A., and Z. Ben-Avrahm. 1977. Lost Pacifica Continent. Nature 270:41-43.

Pandolfi, J. M. 1992. A review of the tectonic history of New Guinea and its significance for marine biogeography. Pp. 718-728 in R. Richmond, ed., Proceedings of the Seventh International Coral Reef Symposium, Guam, Vol. 2. University of Guam Press, Mangilao.

Pigram, C. J., and H. L. Davies. 1987. Terranes and the accretion history of the New Guinea orogen. J. Aust. Geol. Geophys. 10:193-212.

Platnick, N. I. 1992. Patterns of biodiversity. Pp. 15-24 in N. Eldredge, ed., Systematics, Ecology, and the Biodiversity Crisis. Columbia University Press, N.Y.

Ray, G. C., and J. F. Grassle. 1991. Marine biological diversity. BioScience 41(7):453-469.

Richmond, R., ed. 1992. Proceedings of the Seventh International Coral Reef Symposium, Guam, Vols. 1 and 2. University of Guam Press, Mangilao. 1,240 pp.

Roberts, L. 1988. Coral bleaching threatens Atlantic reefs. Science 238:1228-1229.

Springer, V. G. 1982. Pacific Plate biogeography, with special reference to shore-fishes. Smithsonian Contr. Zool. 367:1-182.

Thomas, J. D. 1992. Biodiversity and biogeography of coral reef amphipods from the north coast of New Guinea [abstract]. P. 736 in R. Richmond, ed., Proceedings of the Seventh International Coral Reef Symposium, Guam, Vol. 2. University of Guam Press, Mangilao.

Thomas, J. D. 1993. Biological monitoring and tropical biodiversity in marine environments: A critique with recommendations, and comments on the use of amphipods as bioindicators. J. Nat. Hist. 27:795-806.

Vermeij, G. J. 1990. Tropical Pacific pelecypods and productivity: A hypothesis. Bull. Mar. Sci. 47:62-67.

Veron, J. E. N., and M. Pichon. 1976. Scleractinia of eastern Australia. I. Families Thamnasteriidae, Astrocoeniidae, Pocilloporidae. Aust. Inst. Mar. Sci. Monogr. Ser. 1:1-86.

Veron, J.E.N., and M. Pichon. 1979. Scleractinia of eastern Australia. III. Families Agariciidae, Siderastreidae, Fungiidae, Oculinidae, Merulinidae, Mussidae, Pectiniidae, Cryophyllidae, Dendrophylliidae. Aust. Inst. Mar. Sci. Monogr. Ser. 4:1-422.

Veron, J. E. N., and M. Pichon. 1982. Scleractinia of eastern Australia. IV. Family Poritidae. Aust. Inst. Mar. Sci. Monogr. Ser. 5:1-159.

Veron, J. E. N., and C. C. Wallace. 1984. Scleractinia of eastern Australia. V. Family Acroporidae. Aust. Inst. Mar. Sci. Monogr. Ser. 6:1-485.

Veron, J. E. N., M. Pichon, and M. Wijsman-Best. 1977. Scleractinia of eastern Australia. II. Families Faviidae, Trachyphilliidae. Aust. Inst. Mar. Sci. Monogr. Ser. 3:1-233.

Williams, G. C. 1992. Biotic diversity, biogeography, and phylogeny of pennatulacean octocorals associated with coral reefs in the Indo-Pacific. Pp. 729-735 in R. Richmond, ed., Proceedings of the Seventh International Coral Reef Symposium, Guam, Vol. 2. University of Guam Press, Mangilao.

Williams, L. B., and E. H. Williams. 1990. Global assault on coral reefs. Nat. Hist. 4:47-54.

Winston, J. E. 1988. The systematists' perspective. Pp. 1-16 in D. Fautin, ed., Biomedical Importance of Marine Organisms. Mem. California Acad. Sci. 13:1-157.

Yan, C. Y., and L. W. Kroenke. 1993. A plate tectonic reconstruction of the southwest Pacific, 0-100 Ma. Pp. 697-707 in W. H. Berger, L. W. Kroenke, and L. A. Mayer, eds., Proceedings of the Ocean Drilling Program, Scientific Results 130. Texas A & M University, College Station.

CHAPTER 25

Ecological Restoration and the Conservation of Biodiversity

WILLIAM R. JORDAN, III

Editor and Outreach Officer, University of Wisconsin Arboretum, Madison

Through its effects both on the land and on the human community, restoration has a crucial role to play in the conservation of diversity.

Discussions about the conservation of biological diversity generally concentrate on strategies for preserving existing habitat for native species, but of course preservation of what exists is only part of a comprehensive program of biodiversity conservation. The other, complementary, part is restoration, the active attempt to return an ecological system—whether conceived as an ecosystem, an ecological community, or a landscape—to some previous condition following a period of change or disruption, usually resulting from human activities such as agriculture, development, waste disposal, or mining.

Ecological restoration in its most ambitious and ecologically sophisticated form, involving the attempt to actually recreate ecological systems that closely resemble specific historic systems in composition, structure, dynamics, and function, is a relatively new form of environmental technology, and environmentalists have tended to be wary of it. There are good reasons for this. There are serious questions about our ability, given the present state of technology and understanding of the ecosystems involved, to actually re-create ecologically accurate examples of many kinds of systems. Even when reasonably high-quality restoration is possible on a scale of a few hectares or tens of hectares, this is often labor-intensive and costly, and may not be feasible on an ecologically significant scale. In addition to these technical and economic concerns, environmentalists have tended to regard restored ecosystems, even those of the highest ecological quality, as inherently inferior to, less natural, or even less "real" than the historic systems they are intended to represent. For all these reasons, people concerned about the conservation of biodiversity and about the quality of the environment in general have tended to be skeptical about the promise of restoration and to regard it less as a promising way to reverse environmental damage than as a false promise that may be used to undermine arguments for the preservation and protection of existing natural areas.

These are not unfounded concerns. Restoration is, in fact, an immature discipline, and the success of restorationists' efforts varies widely. Moreover, the promise of restoration frequently has been used to circumvent efforts at preservation, especially of wetlands. This is obviously a matter of deep concern, and it is a concern that is shared by most restorationists, including those who make their living doing restoration to compensate for or "mitigate" environmental damage, as required by law under certain conditions. At the same time, while this may be a legitimate concern at the political level, it should be clear that, at a more fundamental level, the dichotomy between preservation and restoration is a false one. This is true because vast areas of the Earth already have been profoundly altered by human activities, so that in many cases there are valuable opportunities to expand preserves or even create new ones through restoration. Also, and more fundamentally, everything within an ecosystem interacts with everything else so that, whether we choose to regard ourselves as outside of "nature" or as part and parcel with it, we cannot actually disengage ourselves from it or avoid influencing it. In the last analysis, then, preservation of natural landscapes is in the strictest sense impossible—or rather is properly seen not as a conservation *strategy* at all, but rather as a conservation *goal* or objective that can be achieved only through a continual effort to identify novel influences on a given landscape and to find ways of compensating for these influences in ecologically effective ways.

This of course amounts to a continual program of restoration. This may be intensive in situations where influence is severe and has dramatic effects on the landscape—for example, where prairie has been plowed down to grow corn or wheat. Or it may be less intensive in situations where the influence is subtle or indirect—a change in the frequency of fire, for example, or an alteration in a hydrological cycle, or the extirpation of a native plant or animal. Nevertheless, the principle is the same: influence is inevitable, and conservation depends not on eliminating novel (or "external") influences, but on finding ways of compensating for them in such a way that the system resumes behaving—or can continue to behave—*as if* these influences were absent.

Preservation, in other words, *depends* on restoration. Of course, when people talk about restoration, what they generally have in mind is intensive restoration, such as the wholesale replanting of prairie in an abandoned cornfield. There is a tendency to avoid the term "restoration"when referring to attempts to compensate for subtler forms of influence, and to use softer-edged terms such as "management," "maintenance," or "stewardship." The distinc-

Returning fire to fire-dependent ecosystems such as prairie is key to their restoration.

tion, however, is an arbitrary one and can be seriously misleading. Strictly speaking, *all* attempts to compensate for novel influences in order to allow an ecological system to remain on its "natural" or historic trajectory are restorative acts. Restoration in its more dramatic forms differs only in degree from other kinds of management aimed at the conservation of the classic ecosystem. Properly understood, it is not a peculiar activity, distinct from these other forms of management, but is of a piece with them. In fact, precisely because it is dramatic and conspicuous and because it makes such explicit claims and raises so clearly and forcefully all the questions involved (about the role of humans, the quality of the resulting ecosystem, and so forth), restoration is best regarded not as peculiar but as a paradigm for all activities aimed at the conservation of classic ecosystems against the unavoidable pressure of novel influences.

There are several reasons why it is desirable to be explicit about this, and to recognize much of what is called "management" or "stewardship" as "restoration." Doing so clearly establishes the crucial fact, rooted in ecological principle, that preservation is not a strategy but a goal, and that achieving it will in the last analysis always entail some kind of compensatory manipulation. Also, terms such as "management," "stewardship," and even "preservation" imply no commitment to any particular end result. "Management," for example, promises only to manipulate the system, while "preservation," construed as a strategy or means for conservation, promises only to leave it alone, which of course is impossible. Neither represents any clear commitment to actually ensuring the existence or the well-being of the system over the long term.

Only "restoration" does that. And of course this is one reason why restoration has been peculiarly vulnerable to criticism. Environmentalists often point out that "restoration" in the strictest sense is usually impossible, and of course they are right. Restoration in fact may be impossible in a particular situation. But at least the restorationist is committed to a particular objective and is clear and explicit about what he or she is trying to accomplish. In this sense, the very vulnerability of restoration to criticism is one of its great values. Besides this, the very notion of restoration brings into the open a series of fundamental questions that may be overlooked under headings such as "management" or "stewardship." Is restoration (i.e., ecologically effective compensation for novel influences on an ecosystem) actually possible for a given system under a given set of conditions? How are model systems to be selected? Why choose this system—or period—rather than another? And how are standards to be defined and the results of a project evaluated? This process also forces us to acknowledge the unavoidable fact of our influence on the system, the specific ways we have influenced it, and the fact that in many if not all situations the quality of the so-called "natural" landscape will depend at least in part on our understanding of it and our ability to conserve it through an ongoing process of restoration, or what might be called compensatory maintenance. Acknowledging all this is important. It not only clarifies what is going on, it also opens the way to certain experiential and performative benefits that may otherwise remain inaccessible.

In this chapter, I summarize some of my own ideas about the role of restoration in the conservation of biodiversity. In doing so, I take into account not only the direct value of restoration as a technology for conserving classic ecosystems against the pressure of novel influences, but also what might be called its *indirect* value as a way of changing people and bringing them into a positive, mutually beneficial relationship with the classic landscape. So far, evaluations of restoration have concentrated almost exclusively on its products—the restored ecosystem itself—and have overlooked what may well be the most important (and are unquestionably the most immediate and the most *ecological*) aspects of restoration—restoration as an experience and a performance, and as a way of bringing about changes in the practitioner and in his or her audience. As a result of this one-dimensional evaluation, many of the benefits of restoration simply have been overlooked. Indeed, as I try to make clear, *most* of them have been overlooked. Not that the product of the restoration effort is unimportant. Restoration is defined—and distinguished from other forms of agriculture—by its product, and it deserves to be evaluated on the basis of its ability to produce it. At the same time, whenever there is a product, there is also a process. And when humans carry out a process, there are also experience and performance.

Taken together, these four dimensions—of product, process, experience, and performance—define any act. Value is to be found in all of them, not just the first.

This chapter briefly explores each of these dimensions, considering what each has to contribute to the value of ecological restoration as a strategy for conserving biological diversity.

THE PRODUCT: THE RESTORED ECOSYSTEM

I begin with the most obvious benefit of restoration—the product of the restoration effort, the restored ecosystem itself. Clearly, restoration has an important role to play in conserving diversity in the direct, technical, or ecological sense. To begin with, in the case of ecological systems that have been more or less protected from the more dramatic forms of human influence, an ongoing program of low-key restoration generally will be required to prevent the system from drifting ecologically in response to subtle influences and in the process losing native species, picking up exotics, and perhaps suffering the impairment of various ecological functions and processes. This is perhaps not a serious issue among environmental managers, most of whom are fully aware of the pervasiveness of human influence on the natural landscape and accept the responsibility of identifying these influences and finding ways of compensating for them. What needs to be stressed in this regard is simply that even these relatively low-key forms of ecological management are in fact forms of restoration and entail many of the same technical and philosophical problems—and also offer many of the same opportunities—as do the more dramatic forms of restoration. A crucial point here is that the very act of recognizing the necessity for low-key restoration efforts, even in remote areas that may be protected from the more conspicuous forms of human influence, draws attention to the reality of ecological interrelatedness and the extent to which the conservation of diversity actually depends on deliberate human effort to compensate for influences that may be indirect, obscure, or even unintentional.

The ultimate example of this would be global climatic change. If the Earth's climate actually changes significantly as a result of human activities, massive restoration and relocation efforts will be required to maintain existing ecosystems or to move them to areas offering suitable conditions, perhaps at higher latitudes or elevations. Of course, this is an extreme case and may or may not actually transpire. In the meantime, however, there are already many natural areas that are changing more or less irreversibly as a result of changes in the landscape around them. A classic example is the fragments of tallgrass prairie, often less than an acre in size, that still survive in old cemeteries and odd neglected areas in many parts of the Midwest, but that are now tiny islands of native biodiversity in a monocultural ocean of corn. Another example is the tiny, isolated relics of tropical dry forest in Central and South America, which

Killing trees to restore prairie.

ecologist Dan Janzen has referred to as examples of the "ecological walking dead." By this vivid phrase, Janzen means quite simply that, existing under altered conditions, these forests are no longer viable ecological entities. Though biologically alive, they are ecologically dead. No amount of protection will enable them to survive, and their future, like that of the prairies of the Midwest, will depend wholly on an effective, sustained program of restoration and recreation to compensate for their altered circumstances.

This may seem an extreme example, and in some ways it is. But it is at the same time paradigmatic. Though we hesitate to admit it, the same thing is true in some degree of all ecosystems everywhere. This fact may be obscured by the subtlety of influence in some cases and also by the ecological inertia that may enable an ecosystem to persist for a time in the presence of changing conditions (Magnuson, 1990). Nevertheless, it is unavoidable, and it provides the bedrock principle on which all conservation thinking, policy, and practice must be based. With the possible exception of some areas that may be large enough and remote enough to preserve a high level of ecological integrity—and that have no history of human influence—the conservation of natural or classic levels of biodiversity will always depend on some degree of restoration as I have defined it. The question, then, is not whether to restore, or even whether we can restore perfectly (usually we cannot). The question is simply how close we can come to restoring the natural or historic features of a given ecosystem. In the long run, "natural" landscapes everywhere on the planet will necessarily (and we might say naturally) be to some extent artificial landscapes. Indeed, the best and most "natural" of these will be those that have been most skillfully and diligently restored (restoration being taken here to *include* protection to minimize novel influences). It is important that we not only recognize this, but that we learn to celebrate it—indeed to make a celebration of it—rather than to deplore it.

Of course, the question of restorability takes on a special urgency in the case of restoration projects that are carried out to reverse the effects of the most dramatic kinds of environmental damage—plowing, for example, or surface mining—and that entail not just tinkering with key processes or the reintroduction of a few extirpated species, but the wholesale reassembly of the entire system. This is what most people have in mind when they talk about ecological restoration, and the question of the quality of the resulting ecosystems has led to a considerable amount of discussion and debate and, in recent years, to some systematic research. The answer, not surprisingly, is complex and depends on many factors, including the nature of the ecosystem, the nature and degree of disturbance to which it has been subjected, the conditions under which the work is carried out, the resources available for the project and, of course, the skill of the restorationist. Obviously, this question is far too complex to answer in detail here, but there is a growing literature on the ecological quality of restored ecosystems, and it is possible to make a few generalizations based on this emerging body of knowledge (for earlier overviews see also Jordan et al., 1988; MacMahon and Jordan, 1994).

Of all ecosystems that have been the subject of intensive restoration efforts, wetlands are perhaps the most intensively studied, at least in part because the restoration of wetlands is mandated by law under certain conditions, and recent research in this area has led to a considerable amount of new information about the quality of several kinds of restored wetlands, and also to a considerable amount of debate over the meaning and significance of this information. A prime example is an evaluation of a restored coastal marsh near San Diego that recently was carried out by ecologist Joy Zedler and her colleagues at San Diego State University (Zedler and Langis, 1991). Zedler and Langis' report on this project is generally downbeat. They found that the restored wetland differed markedly from a natural reference wetland with respect to a number of indicators, including biomass, plant height, soil organic-matter content, and eight other indicators of structural and functional quality. Of special importance to us here, Zedler and Langis found that, 5 years after restoration began, the wetland did not yet provide the habitat for the endangered light-footed clapper rail, which had been a primary objective of the project. They concluded that the restored wetland resembled the reference wetland by only about 57% (a figure obtained by averaging values for the 11 attributes of the ecosystem) and suggested that, because of this low value and because the restored wetland was not yet a suitable habitat for the rail, the project had been disadvantageous from an environmental point of view.

Welcome as these data were as a contribution to the relatively small body of published information on restored ecosystems, Zedler and Langis' conclusions did not go unchallenged. Restorationist John Rieger, for example, pointed out that the marsh Zedler and Langis chose for study was only 5 years old, so that it was unreasonable to compare it with a natural marsh that is thousands of years old (Rieger, 1991). Rieger also argued that the San Diego project had suffered from several years of unseasonably dry weather, and he called into question the validity of averaging measurements of different features as an index of quality.

In similar situations, restorationists have also questioned the techniques involved in restorations, arguing that projects chosen for evaluation have not been

the best examples of the restorationist's craft. This point, which I have generally encountered in conversation and not in print, is of considerable interest
because it points toward the disparity in outlook between ecologists, who may
have little knowledge of horticulture or other practical aspects of restoration,
and restorationists, who somehow have to combine horticultural skill with ecological understanding into a kind of ecological horticulture.

Overall, ecologists' assessment of the quality of restored wetlands generally
has been guarded. A good example is the comment by Mary Kentula and Jon
Kusler in their executive summary of a recent survey of techniques for restoring
and creating wetlands (1990:xviii):

> "Total duplication of natural wetlands is impossible due to the complexity and
> variation in natural as well as created or restored systems and the subtle rela-
> tionships of hydrology, soils, vegetation, animal life, and nutrients which may
> have developed over thousands of years in natural systems. Nevertheless, ex-
> perience to date suggests that some types of wetlands can be approximated and
> certain wetland functions can be restored, created, or enhanced in particular
> contexts. It is often possible to restore or create a wetland with vegetation
> resembling that of a naturally-occurring wetland. This does not mean, how-
> ever, that it will have habitat or other values equaling those of a natural wet-
> land nor that such a wetland will be a persistent, i.e., long term, feature in the
> landscape, as are many natural wetlands."

As Kentula and Kusler acknowledge, restoration is a complex and uncertain
business, yet projects can have strikingly positive results. Two examples are
appraisals of the attempt to restore the estuary of the Salmon River in Oregon
(Morlan and Frenkel, 1992) and the restoration of Henry Greene Prairie, a classic
project begun at the University of Wisconsin-Madison Arboretum in the mid-
1940s (Kline, 1992). In the first case, success was attributed partly to the fact
that the disturbance involved had been relatively mild—a change in hydrology
that could be reversed by removal of an artificial dike, allowing the ecosystem
to recover more or less on its own. But in the case of Greene Prairie, the situation was quite different. The historic prairie on this site had been virtually
eliminated as a result of three-fourths of a century of intensive farming, and the
entire plant community had to be reassembled virtually plant by plant. Moreover, this project, undertaken in the 1940s, was only the second large-scale prairie restoration ever attempted and so necessarily involved a great deal of trial
and error. Despite this, parts of Greene Prairie now closely resemble natural
prairies in southern Wisconsin, at least with respect to the species, abundance,
and distribution of vascular plants. Less is known about other features of the
system, including animals and ecosystem processes. In fact, the neglect of animals and functional aspects, both of which are often relatively difficult to study
quantitatively, is often a weakness of evaluations of restored ecosystems. Nevertheless, it is fair to point out that restoration projects have at times resulted in
the creation of habitat for rare and often difficult-to-restore species—for ex-

ample, a number of rare and endangered plants now growing in restored prairies such as Greene Prairie, and the least Bell's vireo along the Kern River in California (Baird, 1989). In landscapes now profoundly altered by development, the expansion of habitat, even for relatively common native species, may be an important contribution to the native biodiversity of an area. Indeed, a site like Greene Prairie, harboring several hundred species of native plants and animals, many of which are now rare in the area, represents a biodiversity hotspot of great beauty and inestimable ecological, biological, cultural, and spiritual value. Moreover, while intensive restoration projects have generally been carried out on a modest scale in the past, restorationists are developing techniques that allow them to undertake projects on a much larger scale without compromising ecological quality. An example is the recent development of a technique sometimes called "successional restoration" for restoration of tallgrass prairies in the Midwest (Packard, 1994).

Overall, the answer to the question of the quality of restored ecosystems is both complex and incomplete, but will no doubt become clearer, more comprehensive, and more accurate as the craft of restoration develops and as ecologists learn more about how to evaluate the quality—or assess the health—of an ecological community or ecosystem.

Beyond the essentially technical question of ecological quality or "accuracy," however, there is the larger question of authenticity. Even supposing that a restored ecosystem is a faithful replica of the "natural" or model system, closely resembling it in all technical features, including function and dynamics as well as composition and structure, questions remain as to its authenticity, its *ontological* status or value or, simply, its realness. Is the restored ecosystem as "real" as its natural counterpart? Or is it, as some have suggested, merely a "copy" or even a "fake"?

Most discussions of this issue have taken for granted that restoration does in some sense compromise nature, and that the restored ecosystem is not only less "natural" but even in a sense less real than a natural or historic system that, even if humans played a role in shaping it historically, may be perceived as "given" and therefore fully "natural" (see Elliott, 1994, for example).

On this assumption, the restored ecosystem is unauthentic simply by definition. But the real issue here is not actually whether the system is "authentic" in some final sense. It is rather what we mean by authenticity—what we mean by the word "real," what we take to be the basis for realness, what is the ground or touchstone of reality. Curiously, the assumption that an artificial copy of a thing is inherently less real than the original is rooted in two philosophical traditions, both of which are now widely recognized as being antiecological in their implications. The first of these is the platonic idea that what is most real is the changeless form of a thing. In this view, actual objects such as a prairie are mere representations of the ideal form or idea of "prairie," and are therefore less real or of lower ontological status. By the same reasoning, representations of objects (e.g., a restored prairie), being mere representations of representations, are of even lower status. The second philosophy is the Cartesian idea that being is—or at least can be—known only by the solitary individual, and that relationships of any kind can only be inferred, not experienced, and are therefore of doubtful ontological value. From this perspective, the restored ecosystem is ontologically suspect because it does not exist in and of itself, but is contingent on entities outside itself. In fact, it is the product of relationships. To us, this simply means that it is ecological. But from the Cartesian perspective, it means that its reality has been compromised. Curiously, this kind of thinking is clearly evident in modern environmental thinking, as indicated by the conventional environmental critique of restoration.[1]

In any event, both conceptions of the "real"—the platonic and the Cartesian—result in a conception of authenticity within which a restored ecosystem will be judged unauthentic. But both take for granted a conception of the world and our relationship with it that is inimical to the conception of radical relatedness that is at the heart of an ecological sensibility. In fact, the ecological sensibility takes for granted a quite different idea of authenticity, in which the realness and ontological value of things emerges precisely from the way they interact with, register on, and we might even say "contaminate" each other. This underlies, for example, the conception, common among people of archaic and premodern cultures, that the realness of the world depends on its participation in a higher or sacred reality, mediated by human awareness expressed in ritual (Eliade, 1971). This ecological sensibility is also integral to the conception of "performed being" developed by the poet and philosopher Frederick Turner (1985). In Turner's deeply ecological view, realness is not compromised by relationship, but emerges from it. Realness is, in fact, the *product* of interaction and engagement—of the mutual registration of objects on each other—and the more intense and the more reflexive and self-aware the engagement becomes, the more realness there is. From this perspective, nature is neither compromised, or made less real, by deliberate human participation in it, nor is the restored ecosystem in any way less real because it is partly the result of human effort. It is, rather, *more* real in the precise sense that it is more fully *realized*.

Interestingly, the ontological status of a restored ecosystem in this view bears comparison with that of objects or landscapes made real or sacred through ritual action in certain archaic traditions as interpreted by historian of religion, Mircea Eliade. This seems to me a matter of considerable importance in the evaluation of restoration, in part because our conception of the authenticity of the restored ecosystem will affect profoundly the way we interpret this work and the spirit in which we undertake it, which will in turn largely determine its effectiveness as an occasion for social- and self-transformation. (For further dis-

---

[1] I am indebted to Professor Gene Hargrove of the Department of Philosophy at the University of North Texas for an introduction to these ideas and their implications for restoration.

cussion of this point, see Jordan, 1993.) My point here is simply that there are many ideas of authenticity; some are more "ecological" than others, and they provide a more favorable perspective on restoration than those rooted in either the platonist or the Cartesian/modernist tradition.

The discussion so far pertains to the direct contribution that restorationists can make to the conservation of biodiversity through the restorative maintenance of existing ecosystems, or through the wholesale creation or re-creation of native ecosystems on more or less severely disturbed sites. This is the most obvious contribution of restoration to biodiversity conservation, but it is by no means the only one. The effects of the restorationist's efforts on the landscape are obviously important. Just as important, however, are their effects on those who carry out the work and on those who merely witness it and act as an audience for it. Since the well-being of the natural landscape will ultimately depend to a considerable extent on its relationship to humans and on how people understand and value it, this is an important matter. Indeed, it may well be that what is most important about restoration in the long run will be the way it affects the human community by providing the basis for a repertory of experiences and rituals for negotiating the intellectual, psychological, and spiritual, as well as the purely physical, reentry of nature. This being the case, I conclude with a few comments on the other three dimensions of the act of restoration—restoration as process, as experience, and as performance—and on the implications of these for the conservation of biodiversity.

THE PROCESS: RESTORATION ECOLOGY

One outcome of the act of ecological restoration is the restored ecosystem itself. But restoration is more than a way of creating ecosystems. It is also a powerful way of learning about them. Indeed, the history of both ecology and of restoration clearly illustrate the value of restoration—the attempt to repair, heal, or reassemble a living ecosystem—as a way of raising questions and testing ideas about it. A classic example from the University of Wisconsin Arboretum in Madison was the discovery of the importance of fire in the ecology of prairies that resulted from the Arboretum's early attempts at prairie restoration (Curtis and Partch, 1948). But ecology provides endless examples of the heuristic value of the reassembly of ecological systems, all pointing toward an obvious fact: we can change a complex system without understanding very clearly what we are doing, but we generally cannot restore that system to its former condition without having a clear idea of how we have influenced it and of the ecological implications of that influence.

Reflecting on this early in my career at the Arboretum, I coined the term "restoration ecology" to refer to restoration efforts that were undertaken specifically to raise questions or to test ideas about the system being restored (Jordan et al., 1987). It seemed to me that it was important to introduce this term in order to call attention to this approach to ecological research and to encourage its systematic development. But this is of more than academic or purely intellectual importance. In the long run, the well-being of these ecosystems will depend in part on how well we understand them and are able to care for them. An important value of restoration, then, is its value as a way of booting us up intellectually to become competent stewards of the natural landscape. Indeed, through this process of heuristic reassembly, the ecosystem may be said to become more aware of itself through transcription into human understanding, and in this sense even to acquire a kind of immortality—that is, viability in a landscape dominated by human beings.

THE EXPERIENCE: REENTERING THE FOREST

We all know that we are shaped by experience and that we are to some extent the product of our experiences. The question for us here, then, is what kind of experience is ecological restoration, and in what way might we expect it to affect or shape the person who carries it out? This is obviously a complicated question. But it is not too difficult to sketch at least the broad outlines of an answer. To begin with, restoration provides an opportunity to participate in a constructive way in the ecology of the ecosystem being restored. In this sense, then, it makes us one with other species of animals and plants which, in their various ways, also contribute to the shaping and well-being of the system.

This makes us members of the biotic community in a purely ecological sense. But it does something else as well. If we consider the act of restoration carefully, we can see that it is a form of agriculture, and that, like any form of agriculture, it entails the manipulation of nature. At the same time, it is in some ways a peculiar form of agriculture. While other, more traditional forms of agriculture manipulate nature creatively, the restorationist manipulates it *conservatively*; i.e., his or her objective is not to change nature or improve on it, but rather to maintain it, or we might say to turn

The classic experience of the gatherer.

it back into itself. Thus, while the farmer or gardener attempts to imitate nature, the restorationist attempts something similar but psychologically very different, i.e., not to imitate, but actually to *copy* nature. This entails great humility, and even a measure of self-abnegation, a setting aside of creativity and preference in deference to nature. Restoration is, then, an expression of humility in the exercise of technology. And like the ritual abasement that often accompanies rites of initiation, this helps prepare the restorationist psychologically for initiation into the ecological community.

Finally, it is worth pointing out that the act of restoration is itself complex and rich, drawing on a wide range of human abilities and interests. Specifically, it entails not only the work of the farmer and the gardener, but also of the hunter and gatherer and, as we have seen, of the scientist as well. As a result, it provides an opportunity to explore all these classic ways of experiencing and interacting with nature by reenacting them. In other words, it provides a way of reentering nature without ceasing to be fully ourselves—without abandoning what naturalist Loren Eiseley referred to as the lessons learned on the pathway to the moon. In this way, we may hope to become more fully at home in nature, more, as Thoreau wrote, "a part of herself." This obviously has important implications for the conservation of biodiversity in any area that is subject to any significant amount of human influence. And the fact that the work of restoration draws on and appeals to a wide range of human aptitudes and interests has important political implications as well.

PERFORMANCE: A NEW COMMUNION WITH NATURE

When we think of ecological restoration, we tend to think of it primarily as a more or less effective process, but like any human activity it is more than this. It is also an expressive act—in fact an act in the dramatic, theatrical, or ritual sense of the word: an action that not only accomplishes work, but that also conveys information, meaning, and feeling. This is an aspect of work that a puritan society, with its deep-seated skepticism regarding ritual and the experience of performance, tends to overlook. Nevertheless, there are reasons to believe that it is of immense importance to us in the task of negotiating a healthy relationship between human communities and the larger biotic community.

Relationship, as noted earlier, is crucial to the ecological sensibility—the bedrock of its conception of what is real. Yet, as Frederick Turner and others have pointed out, the relationship between nature and culture, like many kinds of relationships, entails deep tensions that actually cannot be resolved in literal terms, but only in psychological terms, and only by stepping out of the literal dimension and into the dimension of make-believe, of performance and of ritual (Turner, 1991). This is true of many kinds of relationships, among humans, obviously, but also among individuals of other species, and certainly between humans and the rest of nature. A convenient example is the conflict between

Checking Curtis Prairie at the UW-Madison Arboretum, the world's oldest restored prairie.

the urge to mate and the urge to defend territory in a territorial species. Another is the irreducible tension between predator and prey. Yet another is the tension that exists between creatures endowed with different levels of reflexivity or self-awareness. Such tensions, Turner suggests, cannot be resolved literally, but only in performative, often counterfactual, terms. It is at such points of irreducible tension that ritual develops or is invented as a way of negotiating and resolving psychologically (and in the imagination) a tension that cannot be resolved in literal terms—hence the mating rituals of many species of animals, including humans. Hence, too, the rituals human communities commonly provide as ways of negotiating the deeply problematic entry of the individual into the community, or the relationship between the human and the larger biotic community.

Simply put, community depends on ritual. The anthropologist Victor Turner, a pioneer in the study of ritual, argued that what he called *communitas*, or the full *experience* of community, is available only in the central and climactic phase of community-making rituals (Driver, 1991). Though Turner is concerned primarily with the human community, his ideas clearly have profound implications for the nature-culture relationship, and so ultimately for the conservation of biodiversity. Since the development of a relationship between the human community and the biotic community is essentially an act of community-making, it may well be that the failure of modern environmentalism to deal successfully with a wide variety of environmental problems, including the loss of natural habitat, is due in part to its failure to provide rituals suitable for the purpose and, more deeply, to a wariness of ritual experience and a skepticism as to its efficacy that is part of our puritan and modernist heritage.

This being the case, it would seem that we at least should explore the value of ritual as a way of addressing what is really the central problem of environmentalism: the negotiation of a satisfactory relationship between the human community and nature in its wilder, more primitive, or less reflexive forms. Hence the importance of viewing the act of restoration as a kind of performance and as a basis for ritual—specifically rituals for negotiating our relationship with the rest of nature. As I discussed earlier, we tend to regard an activity such as restoration primarily, if not exclusively, as a form of technology—simply a way of fixing the landscape. In other words, we see restoration simply as a more or less effective process, overlooking its value as an experience, an expressive act and a means of self-transformation. In this way, our culture is the obverse of archaic cultures, for which the expressive, ritual value of an act is often taken as being of primary importance, the culture often devoting a large fraction of its resources to activities that have no technical or literal efficacy at all, but that are rich in expressive value and in the power to bring about some desirable inner transformation of individuals and groups.

An example of special significance here is the tradition of world-renewal common to archaic cultures in many parts of the world (LaChapelle, 1988). These characteristically entail elaborate rituals that both inform and transform the participants, but that have little or no direct effect on the "world" or the landscape at all. In fact, what is renewed by rituals of this kind is not the landscape but the community's *idea* of the landscape and its place in it. The crucial observation, however, is that this frequently works, resulting in at least a measure of reconciliation between nature and culture, while literal acts of restoration or management, undertaken in a purely technical spirit, commonly do not work as well, even in the purely literal sense in which they are intended. They may be successful in the short term, and they may meet regulatory requirements and ecological specifications, but, having failed to bring about the inner changes—the renewal of ideas and belief within the human community—on which the well-being of the "natural" landscape typically depends, they will inevitably fail in the long run.

Hence the crucial importance of construing the work of ecological restoration not only as a process and technique but also as an expressive and transformative action and the basis for modern rituals of world-renewal (Jordan, 1994). Indeed, conventional technical activities such as ecological restoration have a crucial role to play in the invention of ritual. According to Victor Turner, ritual commonly emerges from ordinary acts such as mating, the eating of a meal, or the killing of animals for food. Ritual, Turner suggests, is the "quintessence of custom" (Turner, 1968)—an idea that points to the futility of importing rituals from other cultures and attempting to graft them onto one's own. Since ritual—like story, dance, and other art forms—emerges from experience and custom, it follows that we ought to look to custom—the shared work and experience of a particular human community—to provide a foundation or starting point for the invention of the rituals needed to negotiate the relationship between industrial and postindustrial societies and nature in its primitive or classic forms.

Fortunately, restoration is already a "custom" for a growing number of people who are participating in community-based restoration projects. Besides this, restoration incorporates many activities such as gardening, birding, hiking, and even hunting that are conventional activities in our society, and are in fact partly ritualized avocations for millions of people. What remains is the integration of these activities into the task of restoration and the self-conscious development of this work as the basis for festival and other ritual activities needed to bring the human community together and to negotiate its relationship with the biotic community. Construed and developed in this way, restoration no longer will be merely a way of patching up environmental damage but will become in the deepest sense a basis for world-renewal.

In fact, this is already happening in the case of some community-oriented, volunteer-driven restoration projects. In Lake Forest, Illinois, just outside Chicago, the burning of brush piles created by volunteers working on the restoration of oak savannas has become the occasion for an annual community festival, complete with hot-air balloons and a parade of bagpipes reflecting the community's Scottish heritage. In the view of those involved, this seasonal festival plays a vital role in bringing the Lake Forest community together and orienting it toward the reinhabitation of the ancient oak groves (Christy, 1994; Holland, 1994).

What this kind of experience, together with the experience of traditional cultures, suggests is that in the last analysis it is the ritual that matters most. If we get that right, then technique will follow. In fact, in my view, it is not any lack of technical know-how or even ecological understanding that currently is limiting the conservation of biodiversity in most situations—it is the inadequacy of our ritual tradition. If, as Victor Turner and others have argued, ritual is crucial to the process of community-building, then it clearly will be crucial to the process of developing the human community and bringing it into a healthy relationship with the larger biotic community. One value of restoration, then, will be its value as a basis for the creation of new rituals for this purpose. Thus, in my view, the ritualization of restoration, its development as a basis for community-building ritual and festival, is a matter of great urgency, and ought to proceed hand in hand with the development of restoration as a science and an art of environmental healing.

REFERENCES

Baird, K. 1989. High-quality restoration of riparian ecosystems. Restor. Managemt. Notes 7(2):60-64.

Christy, S. F. 1994. A local festival. Restoration and Management Notes 12(2):123.

Curtis, J.T., and M. L. Partch. 1948. Effect of fire on the competition between blue grass and certain prairie plants. Amer. Midl. Nat. 39(2):437-443.

Driver, T. F. 1991. Pp. 160-173 in The Magic of Ritual: Our Need for Liberating Rites that Transform Our Lives and Our Communities. Harper, San Francisco.

Eliade, M. 1971. The Myth of the Eternal Return. Princeton University Press, N. J.

Elliott, R. 1994. Extinction, restoration, naturalness. Envir. Ethics 16(2):135-144.

Holland, K. 1994. Restoration rituals: Transforming workday tasks into inspirational rites. Restor. Managemt. Notes 12(2):121-125.

Jordan, W. R., III. 1993. The ghosts in the forest. Restor. Managemt. Notes 11(1):3-4.

Jordan, W. R., III. 1994. Sunflower forest. Pp. 17-34 in A. D. Baldwin Jr., J. De Luce, and C. Pletsch, eds., Beyond Preservation: Restoring and Inventing Landscapes. University of Minnesota Press, Minneapolis.

Jordan, W. R., III, M. E. Gilpin, and J. D. Aber. 1987. Restoration Ecology: A Synthetic Approach to Ecological Research. Cambridge University Press, N.Y.

Jordan, W. R., III, R. L. Peters, II, and E. B. Allen. 1988. Ecological restoration as a strategy for conserving biological diversity. Envir. Managemt. 12(1):55-72.

Kentula, M. E., and J. A. Kusler. 1990. Executive summary. Pp. xvii-xxv in J. A. Kusler and M. E. Kentula, eds., Wetland Creation and Restoration: The Status of the Science. Island Press, Washington, D.C.

Kline V. M. 1992. How well can we do? Henry Greene's Prairie. Restor. Managemt. Notes 10(1):36-37.

LaChapelle, D. 1988. Sacred Land. Sacred Sex. Rapture of the Deep. Kivaki Press, Durango, Colo. Pp. 239-249.

MacMahon, J., and W. R. Jordan, III. 1994. Ecological restoration. Pp. 409-438 in G. K. Meffe and C. R. Carroll, eds., Principles of Conservation Biology. Sinauer Associates, Sunderland, Mass.

Magnuson, J. 1990. Ecological research in the invisible present. BioScience 40(7):495-501.

Morlan, J. C., and R. E. Frenkel. 1992. How well can we do? The Salmon River Estuary. Restor. Managemt. Notes 10(1):21-23.

Packard, S. 1994. Successional restoration: Thinking like a prairie. Restor. Managemt. Notes 12(1)32-39.

Rieger, J. 1991. San Diego Bay mitigation study: A response. Restor. Managemt. Notes 9(2):65-66.

Turner, F. 1985. Performed being: Word art as a human inheritance. Pp. 3-58 in Natural Classicism: Essays in Literature and Science. Paragon Press, N.Y.

Turner, F. 1991. Beauty: The Value of Values. University of Virginia Press, Charlottesville.

Turner, V. 1968. Introduction. P. 21 in The Drums of Affliction: A Study of Religious Processes among the Ndembu of Zambia. Clarendon Press, Oxford, England.

Zedler, J. B., and R. Langis. 1991. Comparison of constructed and natural salt marshes of San Diego Bay. Restor. Managemt. Notes 9(1):21-25.

CHAPTER
26

Tropical Sustainable Development and Biodiversity

PATRICK KANGAS

Coordinator, Natural Resources Management Program,
University of Maryland, College Park

Sustainable development is one of the great hopes for the conservation of biodiversity. It is a complicated concept that has arisen to synthesize ideas that simultaneously address environmental impacts and the needs of people. The idea of sustainability, or long-term renewable use, is old in terms of fisheries and wildlife harvest management. However, recently it has been applied more broadly to cover a variety of development activities. Although much work is now being classed as sustainable development, the term did not even appear less than a decade ago in the 1986 BioDiversity Forum (Wilson and Peter, 1988). The purpose of this chapter is to review the concept of sustainable development and its relationship with biodiversity. The focus of the chapter is on the tropics, where biodiversity is high and environmental impacts are increasing.

Sustainable development commonly is defined as the process of meeting the needs of the present generation without compromising the ability of future generations to meet their needs (Goodland and Ledec, 1987). This concept arose in the late 1980s as an approach to balance economics and environment. Sustainable development embodies several advancements over other models that have characterized the development process (Norgaard, 1990). One advancement is drawing distinctions between the growth component and the efficiency component of development (Costanza and Daly, 1992; Goodland and Daly, 1992). The growth component refers to an increase in size or dimension of an economy. Assuming that an economy draws on a finite resource base, growth cannot continue indefinitely, i.e., it cannot be sustained (Daly, 1990). The efficiency component refers to the "realization of potential," which is a more qualitative sense of development that may be able to be sustained on a finite resource base. Thus, the recognition of the different components of development can lead to more effective policies for resource use that focus on efficiency rather than growth. These ideas, along with recognition of the importance of future generations and the value of the environment, make sustainable development an attractive philosophy for planning. Although sustainable development is still just an alternative to more conventional economic development models, it is beginning to be adopted at the national level, as seen in the sustainable biosphere initiative of the Ecological Society of America (Lubchenco et al., 1991), and at the international level, as seen in Agenda 21 (Piel, 1992).

Sustainable development is tied indirectly to biodiversity through the need to maintain overall environmental values. Its purpose is to sustain both man and nature, and thus it is directly related to conservation and the wise use of resources, including biodiversity. In the tropics, conservation has focused on controlling the deforestation process (Table 26-1; see also Gradwohl and Greenberg, 1988). All of the strategies listed in Table 26-1 can be thought of as forms of sustainable development, but the extraction of products and values from existing forest may be the approach that most directly links biodiversity to sustainable development. This is the "use it or lose it" approach that is receiving much recent attention (Bawa, 1992; Dobson and Absher, 1991; Janzen, 1992). The idea is that forests or other habitats can be maintained if they can be shown

TABLE 26-1 Techniques for Mitigating Deforestation and Maintaining Tropical Forest Habitat

Strategy	Mechanism	Examples
Preservation	Deforestation is eliminated as a land-use option	National parks Community sanctuaries Biosphere reserve
Conversion from extensive to intensive forms of agriculture	Less land is used to produce the same amount of yield, thus freeing other land for forest preservation	Polycropping Intercropping Fertilization Agroforestry Intensive pasture
Extraction of products and values from existing forest	Forest is maintained because of demonstrated market values	Selective logging Ecotourism Harvest of nontimber products
Reforestation	Restored forest directly replaces deforested land	Managed succession Tree planting

Rainforest canopy.

to provide direct value to people, especially rural people who coexist with the forests. Direct market values of forests are derived from the ecosystem as a whole for ecotourism (e.g., Tobias and Mendelsohn, 1991) and from harvest of individual species for timber or nontimber resources. Of course, the value of logging or the harvest of trees for timber is well known, but assessments of value of nontimber products are very recent. The much-cited paper by Peters et al. (1989) led the way in showing that harvest of nontimber products (i.e., latex and fruits) from intact forests can generate more revenue than other forms of forest land-use that are more destructive, such as logging or conversion to pasture or tree plantations. Although the estimate of the value of the forest given by Peters et al. (1989) may not be representative of tropical forests in general (Godoy et al., 1993), their study has led to a new vision of rain forest economics. Harvest of medicinal plants is another important example of this approach (Akerele et al., 1991; Balick and Mendelsohn, 1992; Elisabetsky and Nunes, 1990), but many other nontimber resources potentially can be sustainably harvested from intact tropical forests (Anderson, 1990; Anderson et al., 1991; Durning, 1993; Nepstad and Schwartzman, 1992; Panayotou and Ashton, 1992; Plotkin and Famolare, 1992; Robinson and Redford, 1991).

THE MCKELVEY BOX CLASSIFICATION OF BIODIVERSITY

In viewing extraction as a form of sustainable development, biodiversity is a resource to be utilized by people. In fact, each species is a potential resource.

To clarify the role of species in sustainable development, an analogy can be made with a classification used for mineral deposits (McKelvey, 1972; U.S. Geological Survey, 1980). The McKelvey box separates mineral deposits based on the knowledge of their existence and their economic concentration (Figure 26-1). The term "reserves" is applied only to those deposits that have been identified and are in sufficiently high concentrations to be mined economically. All other deposits are termed "resources." This classification provides a status assessment or inventory that is useful for decision-makers. Hypothetical examples for two forms of biodiversity (higher plants and insects) are given in Figures 26-2 and 26-3. Obviously, the numbers shown in these McKelvey boxes are tentative, but several patterns are evident. First, there are relatively few species that currently and directly are utilized by people and therefore fall into the "reserves" category. Also, there is a high proportion of undiscovered species, especially insects. Many and probably most of these undiscovered species occur in tropical forests, most of which are undergoing deforestation. Although these undiscovered species present conceptual problems (Kangas, 1992), their loss to extinction certainly would be significant.

It is important to recognize that the McKelvey box represents a single time frame and that over time the numbers inside the box change. As taxonomic exploration continues, the number of undiscovered species will go down. As technology advances, the number of economically useful species will increase. New mechanisms are developing to stimulate these changes. Biodiversity prospecting (Reid, 1993; Reid et al., 1993; Rubin and Fish, 1994) is being advocated,

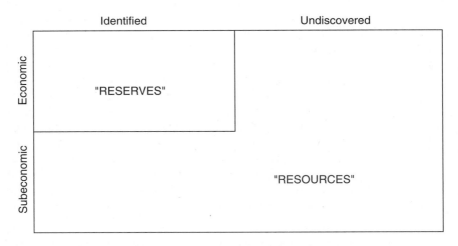

FIGURE 26-1 The McKelvey box classification of mineral deposits from geology (McKelvey, 1972; U.S. Geological Survey, 1980).

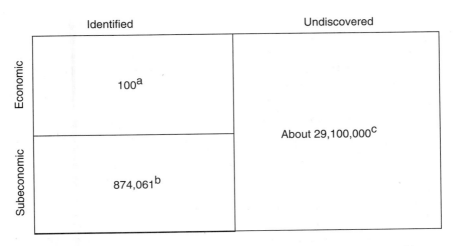

FIGURE 26-2 An hypothetical McKelvey box classification of biodiversity of insects and other species of arthropods. [a]Assumption. [b]Found by subtraction of known species (Wolf, 1987) and economically important species from a. [c]Found by subtraction of estimated total species and known species (Wolf, 1987).

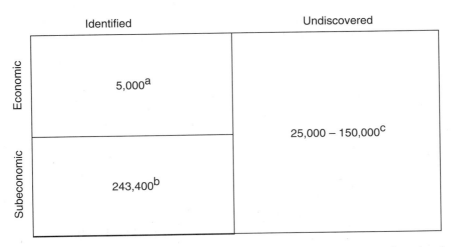

FIGURE 26-3 An hypothetical McKelvey box classification of biodiversity of species of higher plants. [a]National Research Council (1982). [b]Found by subtraction of known species (Wolf, 1987) and economically important species from a. [c]Found by subtraction of estimated total species and known species (Wolf, 1987).

and mechanisms for distributing income that is generated from biodiversity are being developed. These mechanisms include the work of national institutions such as the Instituto Nacional de Biodiversidad de Costa Rica (INBio) (Gamez et al., 1993; Tangley, 1990) and, at the international scale, the Biodiversity Treaty (Broadus, 1992; Stone, 1992).

IMPEDIMENTS TO SUSTAINABLE DEVELOPMENT

The idea of sustainable development is attractive but there are a number of problems or impediments in practice (Table 26-2). This should not be surprising, since the system of man and nature that is to be sustained is complex. Some of the problems with sustainable development are described below. To a greater or lesser extent, these and other problems must be addressed for success to be achieved.

Overharvesting or overuse is a critical issue in most resource systems (e.g., Bodmer et al., 1990; Vasquez and Gentry, 1989). Because income often is perceived to be directly proportional to harvest, users of resources easily can be enticed into overharvest. Thus, attention must be given to harvest and use schedules. This requires the best scientific knowledge (Ewel, 1993; Hall and Bawa, 1993; Levin, 1993), but it also requires that users of resources face the issue of limitation. In some cases, short-term sacrifices in harvest may be needed to achieve long-term benefits of sustainability. In this regard, management of people is just as important as management of resources in sustainable development (Fiske, 1990).

One approach to supplement income of rural people is to develop new markets, either for existing products or products of new resource-use systems, such as harvest of nontimber products from forests (Jukofsky, 1993). The hope is that, by providing additional income derived from forests, there will be less incentive to cut the forests. An interesting example of this strategy is the opening of markets in the developed countries for tropical forest products. This is

TABLE 26-2   Summary of Impediments to Sustainable Development

---

Overharvesting of renewable resources

Lack of markets for products from sustainably developed operations

Short-sighted political economies that do not properly value sustainability or the contributions of nature to economies

Land-tenure problems and the uneven distribution of land-holdings

Government subsidization of counterproductive land-use programs

Political backlash caused by the influence of developed countries on land-use and on conservation decisions of lesser-developed countries

Violent conflicts, especially over natural resources

---

difficult to achieve (e.g., Pinedo-Vasquez et al., 1992), but strategies are being developed (Clay, 1992a,b,c).

Existing political economies in the tropics can promote environmental impacts (Hecht, 1993; Schmink, 1987). The recognition of this problem has led to an exciting new approach, termed "political ecology" (Schmink and Wood, 1987; Thrupp, 1990). This approach goes beyond combining economic and ecological perspectives to add the political context within which land-use decisions are made. Political ecology is helping to identify the complexity of tropical problems, but new political economies are needed. Accounting systems are being developed that may fill these needs (Ahmad et al., 1989; Costanza and Daly, 1992; Godoy, 1992; Odum, 1984; Repetto, 1992), but they are difficult to implement. Areas that are undergoing economic upheaval, such as Brazil and the Eastern European countries, may be the most likely candidates for the implementation of these new approaches to accounting and valuation.

The typical pattern of land-holdings for a tropical country is where a large portion of land is controlled by relatively few wealthy people and the remaining small portion of land is divided among a large number of poor people. This uneven distribution leads to social problems that can inhibit wise land-use strategies and encourage deforestation (Eckholm, 1979; Rudel, 1993). Approaches for redistributing wealth, such as the development of new markets for rural people described earlier, may help mitigate this problem. The concept of extractive reserves was developed in part to deal with this problem. Rural people would have access to these lands to generate income through harvest of non-timber products (Fearnside, 1989). However, controversy exists as to whether or not sufficient income can be derived from extractive reserves, at least under conditions in the Amazon (Browder, 1990; 1992a,b; Vantomme, 1990).

The problem with government subsidy of bad land-use has been demonstrated by analyses given in Repetto (1988) and Repetto and Gillis (1988). These authors provide many examples, perhaps the best known of which is the tax incentive program that was provided to ranchers in the Amazon during the 1970s and 1980s. This program led to much deforestation and subsequent declines in productivity of pastures. These subsidy programs can be difficult to overcome because they have support for political reasons, which can supersede the wise use of land.

Political backlash can occur when one country interferes with the sovereignty of another. Several examples of this phenomenon are known from Brazil, whose government has been criticized for deforestation by conservation groups in developed countries (Guimaraes, 1991). Stimulated by a World Bank publication, a major controversy arose concerning rates of deforestation in the Amazon (Fearnside, 1990; Golden, 1989; Mahar, 1989; Neto, 1989a). As a consequence, the President of Brazil even stopped accepting financial support from foreign conservation organizations because he feared that they were having too much influence on Brazilian land-use decisions. This position was soon reversed, but it illustrates the potential magnitude of political backlash (Radulovich, 1990). Other Amazonian examples of backlash have involved a benefit rock concert (Neto, 1989b) and the advocacy actions of a foreign anthropologist (Anonymous, 1989).

War and violent conflict can severely inhibit conservation and sustainable development (Homer-Dixon et al., 1993). During such times, priorities change and environmental values frequently are overlooked as conflicts are resolved. An example from the Neotropics was the "Soccer War" between El Salvador and Honduras in 1969 (Durham, 1979). The "war on drugs" in tropical countries also leads to both social and environmental problems (Goodman, 1993a; Kangas, 1990).

HISTORY LESSONS FOR TROPICAL SUSTAINABLE DEVELOPMENT

One way to gain perspective on the problems associated with sustainable development is to look to history. Analogous examples of conservation strategies from Table 26-1 and impediments to sustainable development from Table 26-2 have occurred in the past. Comparative analysis of these examples potentially can provide inspiration and insight for present-day issues. Historical failures or problems can be lessons to learn from, and historical successes can be models to emulate. Thus, using Deevey's metaphor for historical ecology (Deevey, 1969:40), we "coax history to conduct experiments" in sustainable development with 10 examples.

The Titanic Effect and the Collapse of the Maya

The Maya civilization of Middle America reached remarkable achievements in art, architecture, science, and social organization from approximately A.D. 300-900. This civilization developed a sophisticated culture in a rain forest environment that supported a high population density with intensive agriculture. However, in spite of all of their achievements, the Maya civilization collapsed after A.D. 900, with depopulation of urban sites, loss of cultural knowledge, and dramatic social transformations (Culbert, 1973; Lowe, 1985). A number of causes have been suggested for the collapse, including malnutrition, disease, foreign invasion, internal civil uprisings, and agricultural decline. Environmental degradation certainly occurred and played a role in the collapse (Abrams and Rue, 1988; Deevey et al., 1979). The dimensions of the collapse of Maya civilization remain a mystery, but the lesson seems clear (Rice and Rice, 1984). Sustainability is not assured, not even for a high civilization such as that of the Maya. This seems to be an example of what Watt (1974) has termed "the Titanic effect." The Titanic was considered to be so unsinkable that it was not even equipped with an adequate supply of life preservers. When it struck an iceberg and sank, there was considerable loss of life. During their classic period, the elite of the Maya, surrounded by their beautiful cities and volumes of books, must have thought their society and economy was without limit. They were wrong, and we must learn from their failure to develop systems that were sustainable.

Henry Ford's Sustainable Villages

Henry Ford was a dynamic man who revolutionized America through his contributions to the auto industry in the early 1900s (Wik, 1972). He developed an effective line of automobiles and an industry that could mass-produce them. In developing this industry, he also pursued related social agendas that included worker benefits, international relations, and the state of agriculture. One of his agendas was the decentralization of his industry and the back-to-the-farm movement. This philosophy became embodied in his "village industry" network in southeastern Michigan. In each small town, he built a factory for automobile parts, usually manned by part-time farmers and supported by a hydroelectric power plant. Each community had an element of self-sufficiency, but they also were connected to each other through the parts of the automobile that they produced. This was a type of utopian plan that was fairly successful until the region became more diversified with the growth of the country. Perhaps Ford's village industry philosophy could be a model for tropical landscapes with sustainable mixtures of wilderness, agriculture, and towns (Janzen, 1990; Lugo, 1991).

The American Civil War and the North-South Dichotomy

The Civil War in the United States was the most costly war in American history in terms of loss of life. It originated from a North-South dichotomy that is reminiscent of the current global dichotomy between developed ("North") and less-developed ("South") countries. The American North-South dichotomy arose early in colonial times from differences in cultural background of the settlers and differences in land-use as determined by geography and climate. The dichotomy solidified after the revolutionary war, with the North becoming primarily industrial and commercial and the South remaining primarily agrarian. At the beginning of the Civil War, these differences were dramatic. The North had five times as many factories and more than double the population of the South, which was dominated by monoculture crops such as cotton, tobacco, and rice, as well as pastures for beef cattle production (Flato, 1960). Slavery was common to both the North and the South before the American Revolution and was not the primary factor that originally distinguished the two regions. However, slavery quickly declined in the North, and by the early 1800s became an important issue in distinguishing the North and the South.

The end of the Civil War (1861-1865) was followed by the Reconstruction

Era, which lasted 10-15 years and consisted of political attempts to manage the process of reuniting the divided country (Franklin, 1961; Lynd, 1967). In general, Reconstruction was dominated by corrupt politicians and shrewd businessmen (i.e., carpetbaggers) who failed to generate economic and social development.

The obvious lesson of the American Civil War is that dichotomies, such as that between the North and South, which are characterized by extremely un-equal distributions of wealth and industrial development and strong relations of dependency, are not stable and can result in violent conflict. (The slavery issue is not a critical qualification to this analogy. Brazil was able to emancipate slaves without violent conflict, in part because there was no major regional dichotomy underlaying slavery.)

The Reconstruction Era also may provide lessons on how not to stimulate development in a depressed region. There are similarities with recent develop-ment efforts in Africa, which have been described by the metaphor of "tropical gangsters" (Klitgaard, 1990). Thus, some nations in the tropics now are facing some of the same problems of development that the South had to face after the Civil War.

Tropical Partnerships for Transportation and Development

An important aspect of tropical development has involved the actions of large foreign companies. The American-owned fruit companies of Central America are good examples: Standard Fruit (Karnes, 1978) and United Fruit (Adams, 1914; May and Plaza, 1958). These companies began operations in the Atlantic lowlands of Central America in the late 1800s with the primary focus of production being bananas. Towards this objective, thousands of hectares of rain forest were converted to banana plantations in an export business which was established to supply the demands of a rapidly growing market in the United States. As time passed, the negative aspects of this form of land-use became obvious: loss of watershed protection, reduction in diversity, local economic dependence on a monoculture, exportation of the crop, and degradation of wa-ter quality. Certainly these are social and environmental impacts that perhaps could and should have been avoided. However, at least in some cases, there also were positive aspects of the presence of foreign companies in the tropics, includ-ing improvements in health care and infrastructure. In particular, the construc-tion of roads and railroads was a benefit in the sense that capital would not have been available for these developments without foreign input. Often the trans-portation networks were designed to meet the needs of the company, i.e., move-ment of bananas to coastal ports, but they also were used for the benefit of local populations. Transportation infrastructure, especially roads, has been recog-nized as leading to deforestation and nonoptimal land-use by opening terrain that previously was not available to land-poor farmers. The tropical road has become "the symbol of modern deforestation" (Forsyth and Miyata, 1984). Perhaps this negative view needs to be balanced by the positive aspects of transportation networks. One example of the positive aspect has been noted by Clay and Clement (1993:21): "The development of markets for sustainably harvested commodities and the destruction of the rainforests both depend, ironically, on the same thing: improved transport systems . . . Forest residents have long realized that roads are both their salvation and their demise." Although alternative transportation systems need to be researched, roads will continue to play an important role in tropical development. In this regard, partnerships between conservationists and foreign companies, such as the old United Fruit Company, may provide the means for this planning.

The Columbian Exchange and Marketing of Nontimber Products

Although Christopher Columbus may have been disappointed that his voyages did not discover the trade route to Asia, the "Columbian Exchange" of materials, species, and cultures that he initiated was dramatic (Crosby, 1972; Viola and Margolis, 1991). One of these exchanges (New World foods) may represent a model for current efforts to develop markets for rain forest products. The New World provided many species of food for the Old World, including maize, potatoes, cacao, and others (Foster and Cordell, 1992). Some of these were similar to species used for food in the Old World and diffused easily. Others, such as potatoes and tomatoes, were unlike anything in the Old World and, thus, took longer to be accepted. Davidson (1992) describes some reasons why people were reluctant to take up certain New World foods and why other foods were accepted quickly. The process of acceptance of a foreign food item that occurred with these species from the New World may provide ideas and interesting contrasts with efforts being made to market rain forest products in the United States and Europe. Clay (1992a,b,c) discussed this new marketing challenge, drawing heavily on his experience with the organization called Cultural Survival. This effort requires information on biology, ethnology, food science, international trade, and business. Several nongovernmental conservation organizations are working on these exchanges, which involve paying producers a fair world price or sending a portion of the profits back to the tropical cultures that produced the item, as with other forms of biodiversity prospecting. The marketing of rain forest products is not limited to food items, and the New World to Old World "Columbian Exchanges" also may provide useful insights for facilitating other efforts in present-day marketing of products related to conservation.

Ideas From the Origin of Agriculture

The origin and spread of agriculture was one of the most important developments in human history (Anderson, 1956; MacNeish, 1991). It occurred between 7,000 and 12,000 years ago and marked the transition between the old Stone Age (Paleolithic) and the new Stone Age (Neolithic) periods. The advent of agriculture and the related sedantism that more or less occurred simultaneously changed the cultural capacity of humans and led to urban civilization. This process has been called the Agricultural Revolution, and it embodied a number of changes resulting in new forms of land-use. As we now search for new and sustainable land-uses, the literature on the original Agricultural Revolution may represent a source of ideas. For example, the idea of cultivation of crop species is thought to have spread from centers of origin to prospective farmers in surrounding regions through a diffusion process (Harlan, 1971; Sauer, 1952). This process must have involved communication of the benefits and risks of new crops or cultivation methods. The current need in the tropics is similar to some extent in that we must spread new forms of sustainable agriculture and forest-use to rural peoples. Perhaps the theories for how agriculture originally spread can help us now devise new forms of technology transfer. Kangas and Rivera (1991) suggested that existing agricultural extension services may provide a mechanism for the flow of information about sustainable land-use. Existing models of the diffusion process (Fliegel, 1993; Rogers, 1983) can readily be adopted for this purpose.

Another example involves the role of women in the origin of agriculture. Women probably played a critical role in all aspects of the Agricultural Revolution, including selection, domestication, harvest, storage, and preparation of new crops and food products (Ehrenberg, 1989). Women recently have been seen to be a significant element in planning for sustainable development (Abramovitz and Nichols, 1992; Braidotti et al., 1994), and perhaps they can be targeted to play important roles in the adoption of sustainable land-use options, as they did in the original Agricultural Revolution.

Amazonia and the Louisiana Purchase

The vastness of the Amazon basin long has been a stimulus to the imagination of people who wish to exploit its resources. This has led to explorations from the time of Orellana (Gheerbrant, 1992) to recent times (Cousteau and Richards, 1984). These explorations resulted in both a romantic literature (Preto-Rodas, 1974) and assessments of resources that have been used as a basis for public policies (Tambs, 1974). Unfortunately, the images of the Amazon have not always been accurate and have lead to unrealistic expectations. Jordan (1982) describes one example in relation to agricultural potentials, which never were realized.

A somewhat similar situation existed in American history with the Louisiana Purchase of 1803. The United States acquired land from France that, in one stroke, more than doubled the size of the country. This land extended from the colonial frontier through the Mississippi and Missouri River watersheds. It con-

An aerial view of the Amazon River.

tained unknown resources that were the basis for ambitious images of rich and rapid development (Allen, 1975). Thomas Jefferson commissioned the Lewis and Clark expedition to explore the northwest, and the work of this expedition both qualified and expanded the development images. The lesson here is that images are an important part of the development process, but that they also can be detrimental to the extent that they mislead development (Lugo and Brown, 1981). Perhaps careful analysis of historical examples, such as the development of the American northwest, can provide useful information for the development of tropical areas such as the Amazon.

Export Agriculture in Central America and the Chesapeake Tidewater

Central America's economy has been characterized by a reliance on export agriculture. Originally, the focus was the "dessert economies" or luxury crops of coffee, sugar, and bananas. After World War II, cotton and beef gained in production. These exports primarily have gone to Europe and the United States, where the crops either cannot be grown or, at least, cannot be grown as cheaply as in Central America. The resulting relationship of dependency has negative effects on many aspects of the Central American economy, environment, and society (Boucher et al., 1983; Pelupessy, 1991; Williams, 1986).

A similar situation has characterized the southern region of the United States. In fact, Persky (1992) uses the Latin American example as a model for southern economic history. One example was the colonial tobacco agricultural system of the Chesapeake Tidewater areas of Maryland and Virginia. From the late 1600s until after the Revolutionary War, large plantations of tobacco were developed in the tidewater area for export to England (Breen, 1985; Goodman, 1993b; Kulikoff, 1986). A strong dependence arose, with the American planters exchanging tobacco for clothes, farm machinery, and other supplies from English merchants. This was a classic case of exchange of raw materials for manufactured goods with a trade deficit. Accordingly, the planters fell seriously into debt over time. Many went bankrupt and were thrown into debtors' prison, but a few were able to switch to more balanced agricultural production that served a domestic market. The best example was George Washington, who gradually switched from production of tobacco to wheat and corn, which he exchanged with merchants in nearby Philadelphia for American-made goods (Mee, 1987). Other examples are described by Clemens (1980) for Maryland's Eastern Shore. Thus, some of the colonial planters were able to diversify their production and reach new markets. This is a strategy that has been recommended for Central America (Tucker, 1992). The transition will be difficult, but helpful insight might come from analysis of the colonial Americans who were able to escape dependence on England some 200 years ago.

Puerto Rican Deforestation and Reforestation

The history of Puerto Rico provides an interesting microcosm of tropical land-use dynamics and development (Dietz, 1986). When Columbus visited the island in 1494, people of the Taino culture occupied the landscape with a relatively low population density. Over the following 400 years, the Spanish colonized the island and converted the landscape from tropical forest to a mix of export crops (primarily sugarcane, coffee, and tobacco) and subsistence farms. Puerto Rico was ceded to the United States after the Spanish American War of 1898, but the agricultural focus continued until the 1940s, when industrialization became dominant. During this period, there was a shift in social structure, with a decline in agriculture and an increase in industrial employment. The island became more strongly tied to the mainland United States and, in a sense, the people became uncoupled from the land. Although there were problems with this development process (Sanchez-Cardona et al., 1975; Weisskopf, 1985), a byproduct has been a net conversion of land-use from agriculture to forest. This has been a dramatic change; forest area has increased from a low of about 9% of the island in the early 1900s to nearly 35% in the 1980s (Birdsey and Weaver, 1982, 1987).

Although this is a special case involving a unique set of sociopolitical circumstances, it does demonstrate that tropical forests are resilient and that deforestation is not irreversible. In this case, there was no active reforestation, and the increase in forest area was due to natural succession on abandoned crop land

and pastures. If active programs of reforestation had been employed, the results could have been larger. Thus, the historical case of Puerto Rico provides incentives to develop programs that will allow restoration of tropical forests on degraded landscapes.

The Civilian Conservation Corps and Tijuca Forest

The Civilian Conservation Corps (CCC) probably was the most massive program of human and natural resource management in American history (Lacy, 1976). From 1933-1937, more than a million young men were employed by the federal government to work on a variety of natural resource projects. The CCC program was devised by Franklin D. Roosevelt as part of the New Deal, primarily as a way to increase employment during the Depression. The program was restricted to unemployed and unmarried young men between 18-25 years of age who signed up for tours of between 6 months and 2 years. During work tours, men were clothed, housed, fed, and paid a modest wage. Men lived in simple, rustic camps located near work projects, which included such activities as forestry, prevention of soil erosion, and flood control. The program involved a rapid, large-scale mobilization of both manpower and facilities and was accomplished by incorporating the War Department in a supervisory role. The U.S. Army played a major role in organizing the men, while other departments, including Interior, Agriculture, and Education, supervised the actual work in the field. The program also had the goal of training the men so that they could find employment after they left the Corps. Thus, the CCC was both an ambitious social program that provided employment and training to primarily unskilled men and a major capital works program that resulted in the following significant accomplishments, which occurred after just the first year: ". . . construction of 25,000 miles of truck trails; 15,000 miles of telephone lines; 420,000 erosion check dams; disease and insect control on 3 million acres of forest; 98 million seedlings planted; forest stand improvement on a million acres; and 687,000 man-days of fire fighting" (Lacy, 1976:38).

Another example of this type of large-scale deployment of manpower occurred with the reforestation of Tijuca Forest in Rio de Janeiro during the late 1800s (Por, 1992). The Emperor of Brazil had the coffee plantations converted to forests in order to save the water supply of the city. It was a massive program, organized by the army and carried out by slaves, that involved the planting of more that 100,000 trees. Today, Tijuca Forest is a national park with a magnificent rain forest that is surrounded by the city of Rio de Janeiro.

In both historical cases described above, a large-scale, labor-intensive public works program in natural resource management was conducted in response to a crisis. Perhaps similar programs can be initiated now in response to the current crisis of tropical deforestation. If properly managed, they could have multiple positive effects, as did the historical examples of the CCC and

Tijuca Forest. Perhaps some aspects of the historical examples could be used in the new programs, such as the use of the military as a positive, organizing authority.

CONCLUSIONS

The 1986 National Forum on BioDiversity (Wilson and Peter, 1988) focused interest on biodiversity and helped to bring the subject to the attention of both policy-makers and the general public. Sustainable development is needed to improve the standard of living of tropical people, but it is especially needed to conserve biodiversity (National Research Council, 1992). In fact, as Janzen (1990) noted, these goals are not independent, but instead they must be tied closely together if conservation is to be successful. Biodiversity is needed to help maintain the global life-support system of humanity and as a direct source of products for the economy of humans. Thus, a symbiotic relationship between sustainable development and biodiversity in the tropics must be designed.

The main value of the history lessons described in this chapter (Table 26-3) may be to illustrate that at least some of the current problems of development in the tropics are not new. Although the tropics present unique problems (Kamarck, 1976), many ideas for sustainable development are being tested, and history can provide more models. However, there is an urgency for tropical sustainable development that is new.

Sustainable development presents multidisciplinary and interdisciplinary challenges and has led to hybrid approaches such as political ecology, conservation biology, and ecological economics. Thus, biologists are learning how to market rain forest products and economists are learning the importance of species diversity in ecosystems. One hopes that there is time to learn these lessons before tropical biodiversity becomes seriously degraded.

TABLE 26-3 Summary of Historical Lessons for Sustainable Development

Sustainability should not be taken for granted
Self-sufficiency should be a development goal, but it should be balanced with linkages to other communities
North-south dichotomies are a liability
Cooperation between conservationists, foreign companies, and other large businesses should be encouraged rather than focusing on negative aspects
New markets for exotic items, such as nontimber rain forest products, should be encouraged
Technology transfer is necessary for communicating ideas about sustainable development
There may be unique roles for women in sustainable development
Images and expectations of development should be explicitly analyzed and should be realistic
A regional agricultural system of exportation with strong dependency can be diversified
Deforestation is not necessarily irreversible
Large-scale public works projects can provide employment and achieve environmental goals

ACKNOWLEDGMENTS

Campbell Plowden and the editors provided many useful comments. Staff at the National Colonial Farm and the Accokeek Foundation in southern Maryland provided inspiration for many of the historical analogies presented here.

REFERENCES

Abramovitz, J. N., and R. Nichols. 1992. Women and biodiversity: Ancient reality, modern imperative. Development 1992:85-90.

Abrams, E. M., and D. J. Rue. 1988. The causes and consequences of deforestation among the prehistoric Maya. Human Ecol. 16:377-395.

Adams, F. U. 1914. Conquest of the Tropics. Doubleday, Page and Company, Garden City, N.Y.

Ahmad, Y. J., S. El Serafy, and E. Lutz, eds. 1988. Environmental Accounting for Sustainable Development. World Bank, Washington, D.C.

Akerele, O., V. Heywood, and H. Synge. 1991. The Conservation of Medicinal Plants. Cambridge University Press, Cambridge, England.

Allen, J. L. 1975. Lewis and Clark and the Image of the American Northwest. Dover Publications, N.Y.

Anderson, A. B., ed. 1990. Alternatives to Deforestation: Steps Toward Sustainable Use of the Amazon Rain Forest. Columbia University Press, N.Y.

Anderson, A. B., P. H. May, and M. J. Balick. 1991. The Subsidy from Nature. Columbia University Press, N.Y.

Anderson, E. 1956. Man as a maker of new plants and new plant communities. Pp. 763-777 in W. L. Thomas, ed., Man's Role in Changing the Face of the Earth. University of Chicago Press, Chicago.

Anonymous. 1989. The Kayapo bring their case to the United States. Cultural Survival Quart. 13(1):18-19.

Balick, M. J., and R. Mendelsohn. 1992. Assessing the economic value of traditional medicines from tropical rain forests. Conserv. Biol. 6:128-130.

Bawa, K. S. 1992. The riches of tropical forests: Non-timber products. Trends Ecol. Evol. 7:361-363.

Birdsey, R. A., and P. L. Weaver. 1982. The Forest Resources of Puerto Rico. Resource Bull. SO-85. U.S. Department of Agriculture, Forest Service, Southern Forest Experiment Station, New Orleans, La.

Birdsey, R. A., and P. L. Weaver. 1987. Forest area trends in Puerto Rico. Research Note SO-331. U.S. Department of Agriculture, Forest Service, Southern Forest Experiment Station, New Orleans, La.

Bodmer, R. E., T. G. Fang, and L. Moya. 1990. Fruits of the forest. Nature 343:109.

Boucher, D. H., M. Hansen, S. Risch, and J. H. Vandermeer. 1983. Agriculture. Pp. 66-73 in D. H. Janzen, ed., Costa Rican Natural History. University of Chicago Press, Chicago.

Braidotti, R., E. Charkiewicz, S. Hausler, and S. Wieringa. 1994. Women, the Environment and Sustainable Development. Zed Books, London.

Breen, T. H. 1985. Tobacco Culture. Princeton University Press, N.J.

Broadus, J. M. 1992. Biodiversity, Riodiversity. Oceanus 35(3):6-9.

Browder, J. O. 1990. Extractive reserves will not save tropics. BioScience 40:626.

Browder, J. O. 1992a. The limits of extractivism. BioScience 42:174-182.

Browder, J. O. 1992b. Social and economic constraints on the development of market-oriented extractive reserves in Amazon rain forests. Adv. Econ. Bot. 9:33-41.

Clay, J. 1992a. Some general principles and strategies for developing markets in North America and Europe for nontimber forest products. Pp. 302-309 in M. Plotkin and L. Famolare, eds., Sustainable Harvest and Marketing of Rain Forest Products. Island Press, Washington, D.C.

Clay, J. 1992b. Buying in the forests: A new program to market sustainably collected tropical forest products protects forests and forest residents. Pp. 400-414 in K. H. Redford and C. Padoch, eds., Conservation of Neotropical Forests. Columbia University Press, N.Y.

Clay, J. 1992c. Some general principles and strategies for developing markets in North America and Europe for non-timber forest products: Lessons from Cultural Survival Enterprises, 1989-1990. Adv. Econ. Bot. 9:101-106.

Clay, J. W., and C. R. Clement. 1993. Selected Species and Strategies to Enhance Income Generation from Amazonian Forests. FAO Forestry Paper. Food and Agriculture Organization, Rome, Italy.

Clemens, P. G. E. 1980. The Atlantic Economy and Colonial Maryland's Eastern Shore. Cornell University Press, Ithaca, N.Y.

Costanza, R., and H. Daly. 1992. Natural capital and sustainable development. Conserv. Biol. 6:37-46.

Cousteau, J.-Y., and M. Richards. 1984. Jacques Cousteau's Amazon Journey. Harry N. Abrams, Inc., N.Y.

Crosby, A. W., Jr. 1972. The Columbian Exchange. Greenwood Publishing Company, Westport, CN.

Culbert, T. P., ed. 1973. The Classic Maya Collapse. University of New Mexico Press, Albuquerque.

Daly, H. E. 1990. Sustainable growth: An impossibility theorem. Development 1990:45-47.

Davidson, A. 1992. Europeans' wary encounter with tomatoes, potatoes, and other New World foods. Pp. 1-14 in N. Foster and L. S. Cordell, eds., Chilies to Chocolate. University of Arizona Press, Tucson.

Deevey, E. S. 1969. Coaxing history to conduct experiments. BioScience 19:40-43.

Deevey, E. S., D. S. Rice, P. M. Rice, H. H. Vaughan, M. Brenner, and M. S. Flannery. 1979. Mayan urbanism: Impact on a tropical karst environment. Science 206:298-306.

Dietz, J. L. 1986. Economic History of Puerto Rico. Princeton University Press, N.J.

Dobson, A., and R. Absher. 1991. How to pay for tropical rain forests. Trends Ecol. Evol. 6:348-351.

Durham, W. H. 1979. Scarcity and Survival in Central America. Stanford University Press, Stanford, Calif.

Durning, A. T. 1993. Saving the Forests: What Will it Take? Worldwatch Paper 117. Worldwatch Institute, Washington, D.C.

Eckholm, E. 1979. The disposed of the Earth: Land reform and sustainable development. Worldwatch Paper 30. Worldwatch Institute, Washington, D.C.

Ehrenberg, M. 1989. Women in Prehistory. University of Oklahoma Press, Norman.

Elisabetsky, E., and D. S. Nunes. 1990. Ethnopharmacology and its role in third world countries. Ambio 19:419-421.

Ewel, J. J. 1993. The power of biology in the sustainable land use equation. Biotropica 25:250-251.

Fearnside, P. M. 1989. Extractive reserves in Brazilian Amazonia. BioScience 39:387-393.

Fearnside, P. M. 1990. Deforestation in Brazilian Amazonia. Conserv. Biol. 4:459-460.

Fiske, S. J. 1990. Resource management as people management: Anthropology and renewable resources. Renewable Res. J. 8(4):16-20.

Flato, C. 1960. The Golden Book of the Civil War. Golden Press, N.Y.

Fliegel, F. C. 1993. Diffusion Research in Rural Sociology. Greenwood Press, Westport, CN.

Forsyth, A., and K. Miyata. 1984. Tropical Nature. Charles Scribner's Sons, N.Y.

Foster, N., and L. S. Cordell, eds. 1992. Chilies to Chocolate. University of Arizona Press, Tucson.

Franklin, J. H. 1961. Reconstruction After the Civil War. University of Chicago Press, Chicago.

Gamez, R., A. Piva, A. Sittenfeld, E. Leon, J. Jimenez, and G. Mirabelli. 1993. Costa Rica's conservation program and National Biodiversity Institute (INBio). Pp. 53-68 in W. V. Reid, S. A. Laird, C. A. Meyer, A. Sittenfeld, D. H. Janzen, M. A. Gollin and C. Juma, eds., Biodiversity Prospecting. World Resources Institute, Washington, D.C.

Gheerbrant, A. 1992. The Amazon: Past, Present, and Future. Harry N. Abrams, Inc., N.Y.

Godoy, R. 1992. Some organizing principles in the valuation of tropical forests. Forest Ecol. Management 50:171-180.

Godoy, R., R. Lubowski, and A. Markandya. 1993. A method for the economic valuation of non-timber tropical forest products. Econ. Bot. 47:220-233.

Golden, F. 1989. A catbird's seat on Amazon Destruction. Science 246:201-202.

Goodland, R., and H. H. Daly. 1992. Three steps towards global environmental sustainability (Part I). Development 1992:35-41.

Goodland, R., and G. Ledec. 1987. Neoclassical economics and principles of sustainable development. Ecol. Modeling 38:19-46.

Goodman, B. 1993a. Drugs and people threaten diversity in Andean forests. Science 261:293.

Goodman, J. 1993b. Tobacco in History. Routledge Publishers, London.

Gradwohl, J., and R. Greenberg, eds. 1988. Saving the Tropical Forests. Smithsonian Institution Press, Washington, D.C.

Guimaraes, R. P. 1991. Ecopolitics of Development in the Third World. Lynne Rienner, Boulder, Colo.

Hall, P., and K. Bawa. 1993. Methods to assess the impact of extraction of non-timber tropical forest products on plant populations. Econ. Bot. 47:234-247.

Harlan, J. R. 1971. Agricultural origins: Centers and noncenters. Science 174:468-474.

Hecht, S. B. 1993. The logic of livestock and deforestation in Amazonia. BioScience 43:687-695.

Homer-Dixon, T. F., J. H. Boutwell, and G. W. Rathjens. 1993. Environmental change and violent conflict. Sci. Amer. 270(2):38-45.

Janzen, D. H. 1990. Sustainable society through applied ecology: The reinvention of the village. Pp. xi-xiv in R. Goodland, ed., Race to Save the Tropics. Island Press, Washington, D.C.

Janzen, D. H. 1992. A south-north perspective on science in the management, use, and economic development of biodiversity. Pp. 27-52 in O. T. Sandlund, K. Hindar, and A. H. D. Brown, eds., Conservation of Biodiversity for Sustainable Development. Scandinavian University Press, Oslo, Norway.

Jordan, C. F. 1982. Amazon rain forests. Amer. Sci. 70:394-401.

Jukofsky, D. 1993. Can marketing save the rainforest? Envir. Magazine 4(4):33-39.

Kamarck, A. M. 1976. The Tropics and Economic Development. World Bank, Washington, D.C.

Kangas, P. 1990. Ecology and the war on drugs. Bull. Ecol. Soc. Amer. 71:105-111.

Kangas, P. 1992. Undiscovered species and the falsifiability of the tropical mass extinction hypothesis. Bull. Ecol. Soc. Amer. 73:124-125.

Kangas, P., and W. M. Rivera. 1991. Mitigating tropical deforestation and the role of extension. Pp. 79-88 in W. M. Rivera and D. J. Gustafson, eds., Agricultural Extension: World-wide Institutional Evolution and Forces for Change. Elsevier Publishing Company, Amsterdam.

Karnes, T. L. 1978. Tropical Enterprise. Louisiana State University Press, Baton Rouge.

Klitgaard, R. 1990. Tropical Gangsters. Basic Books, N.Y.

Kulikoff, A. 1986. Tobacco and Slaves. University of North Carolina Press, Chapel Hill.

Lacy, L. A. 1976. The Soil Soldiers. Chilton Book Company, Radnor, Pa.

Levin, S. A., ed. 1993. Perspectives on sustainability. Ecol. Appl. 3:545-589.

Lowe, J. W. G. 1985. The Dynamics of Apocalypse. University of New Mexico Press, Albuquerque.

Lubchenco, J., A. M. Olson, L. B. Brubaker, S. R. Carpenter, M. M. Holland, S. P. Hubbell, S. A. Levin, J. A. MacMahon, P. A. Matson, J. M. Melillo, H. A. Mooney, C. H. Peterson, H. R. Pulliam, L. A. Real, P. J. Regal and P. G. Risser. 1991. The sustainable biosphere initiative: An ecological research agenda. Ecology 72:371-412.

Lugo, A. E. 1991. Cities in the sustainable development of tropical landscapes. Nature Resources 27:27-35.

Lugo, A. E., and S. Brown. 1981. Tropical lands: Popular misconceptions. Mazingira 5(2):10-19.

Lynd, S., ed. 1967. Reconstruction. Harper and Row, N.Y.

MacNeish, R. S. 1991. The Origins of Agriculture and Settled Life. University of Oklahoma Press, Norman.

Mahar, D. J. 1989. Government Policies and Deforestation in Brazil's Amazon Region. World Bank, Washington, D.C.

May, S., and G. Plaza. 1958. The United Fruit Company in Latin America. National Planning Association, N.Y.

McKelvey, V. E. 1972. Mineral resource estimates and public policy. Amer. Sci. 60:32-40.

Mee, C. L., Jr. 1987. The Genius of the People. Harper and Row, N.Y.

National Research Council. 1982. Ecological Aspects of Development in the Humid Tropics. National Academy Press, Washington, D.C.

National Research Council. 1992. Conserving Biodiversity. National Academy Press, Washington, D.C.

Nepstad, D. C., and S. Schwartzman, eds. 1992. Non-timber Products from Tropical Forests. Advances in Economic Botany, Vol. 9. New York Botanical Garden, N.Y.

Neto, R. B. 1989a. Disputes about destruction. Nature 338:531.

Neto, R. B. 1989b. Brazilian general stung by Sting. Nature 339:7.

Norgaard, R. B. 1990. The development of tropical rainforest economics. Pp. 171-183 in S. Head and R. Heinzman, eds., Lessons of the Rainforest. Sierra Club Books, San Francisco.

Odum, H. T. 1984. Embodied energy, foreign trade, and welfare of nations. Pp. 185-199 in A-M. Jansson, ed., Integration of Economy and Ecology, An Outlook for the Eighties. Proceedings of the Wallenberg Symposium. Asko Laboratory, University of Stockholm, Stockholm.

Panayotou, T., and P. S. Ashton. 1992. Not by Timber Alone. Island Press, Washington, D.C.

Pelupessy, W., ed. 1991. Perspectives on the Agro-Export Economy in Central America. University of Pittsburgh Press, Pittsburg, Pa.

Persky, J. J. 1992. The Burden of Dependency. Johns Hopkins University Press, Baltimore, Md.

Peters, C. M., A. H. Gentry, and R. O. Mendelsohn. 1989. Valuation of an Amazonian rainforest. Nature 339:655-656.

Piel, G. 1992. Agenda 21: Sustainable development. Sci. Amer. 269(10):128.

Pinedo-Vasquez, M., D. Zarin, and P. Jipp. 1992. Economic returns from forest conversion in the Peruvian Amazon. Ecol. Econ. 6:163-173.

Plotkin, M., and L. Famolare, eds. 1992. Sustainable Harvest and Marketing of Rain Forest Products. Island Press, Washington, D.C.

Por, F. D. 1992. Sooretama, The Atlantic Rain Forest of Brazil. SPB Academic Publishing, The Hague, Netherlands.

Preto-Rodas, R. 1974. Amazonia in literature: Themes and changing perspectives. Pp. 181-198 in C. Wagley, ed., Man in the Amazon. University Presses of Florida, Gainesville.

Radulovich, R. 1990. A view on tropical deforestation. Nature 346:214.

Reid, W. V. 1993. Bioprospecting: A force for sustainable development. Envir. Sci. Technol. 27:1730-1732.

Reid, W. V., S. A. Laird, C. A. Meyer, R. Gamez, A. Sittenfeld, D. H. Janzen, M. A. Gollin, and C. Juma, eds. 1993. Biodiversity Prospecting. World Resources Institute, Washington, D.C.

Repetto, R. 1988. The Forest for the Trees? World Resources Institute, Washington, D.C.

Repetto, R. 1992. Accounting for environmental assets. Sci. Amer. 269(6):94-100.

Repetto, R., and M. Gillis, eds. 1988. Public Policies and the Misuse of Forest Resources. Cambridge University Press, N.Y.

Rice, D. S., and P. M. Rice. 1984. Lessons from the Maya. Latin Amer. Res. Rev. 19(3):7-34.

Robinson, J. G., and K. H. Redford, eds. 1991. Neotropical Wildlife Use and Conservation. University of Chicago Press, Chicago.

Rogers, E. M. 1983. Diffusion Innovations. Collier MacMillan, London.

Rubin, S. M., and S. C. Fish. 1994. Biodiversity prospecting: Using innovative contractual provisions to foster ethnobotanical knowledge, technology, and conservation. Pp. 23-58 in Endangered Peoples. University Press of Colorado, Niwot.

Rudel, T. K. 1993. Tropical Deforestation. Columbia University Press, N.Y.

Sanchez-Cardona, V., T. Morales-Cardona, and P. L. Caldari. 1975. The struggle for Puerto Rico. Environment 17:34-40.

Sauer, C. O. 1952. Agricultural Origins and Dispersals. American Geographical Society, N.Y.

Schmink, M. 1987. The rationality of forest destruction. Pp. 11-30 in J. C. Figueroa, F. H. Wadsworth, and S. Branham, Management of Forests of Tropical America: Prospects and Technologies. Institute of Tropical Forestry, Rio Piedras, Puerto Rico.

Schmink, M. and C. H. Wood. 1987. The "political ecology" of Amazonia. Pp. 38-57 in P. D. Little, M. M. Horowitz, and A. E. Nyerges, eds., Lands at Risk in the Third World. Westview Press, Boulder, Colo.

Stone, R. 1992. The biodiversity treaty: Pandora's box or fair deal? Science 256:1624.

Tambs, L. A. 1974. Geopolitics of the Amazon. Pp. 45-87 in C. Wagley, ed., Man in the Amazon. University Presses of Florida, Gainesville.

Tangley, L. 1990. Cataloging Costa Rica's diversity. BioScience 40:633-636.

Thrupp, L. A. 1990. Environmental initiatives in Costa Rica: A political ecology perspective. Soc. Nat. Res. 3:243-256.

Tobias, D., and R. Mendelsohn. 1991. Valuing ecotourism in a tropical rain-forest reserve. Ambio 20:91-93.

Tucker, S. K. 1992. Equity and the environment in the promotion of nontraditional agricultural exports. Pp. 109-141 in Poverty, Natural Resources, and Public Policy in Central America. Transaction Publishers, New Brunswick, N.J.

U.S. Geological Survey. 1980. Principles of a resource/reserve classification for minerals. Circular 831. U.S. Government Printing Office, Washington, D.C.

Vantomme, P. 1990. Forest extractivism in the Amazon: Is it a sustainable economical viable activity? Pp. 105-112 in Annals of the First International Symposium on Environmental Studies on Tropical Rain Forests. FOREST '90, Rio de Janeiro, Brazil.

Vasquez, R., and A. H. Gentry. 1989. Use and misuse of forest-harvested fruits in the Iquitos area. Conserv. Biol. 3:350-361.

Viola, H. J., and C. Margolis, eds. 1991. Seeds of Change. Smithsonian Institution Press, Washington, D.C.

Watt, K. E. F. 1974. The Titanic Effect. Sinauer Associates, Stamford, CN.

Weisskopf, R. 1985. Factories and Food Stamps—the Puerto Rico Model of Development. Johns Hopkins University Press, Baltimore, Md.

Wik, R. M. 1972. Henry Ford and Grass-Roots America. University of Michigan Press, Ann Arbor.

Williams, R. G. 1986. Export Agriculture and the Crisis in Central America. University of North Carolina Press, Chapel Hill.

Wilson, E. O., and F. M. Peter, eds. 1988. BioDiversity. National Academy Press, Washington, D.C. 521 pp.

Wolf, E. C. 1987. On the brink of extinction: Conserving the diversity of life. Worldwatch Paper 78. Worldwatch Institute, Washington, D.C.

CHAPTER

27

Wildland Biodiversity Management in the Tropics[1]

DANIEL H. JANZEN

Professor, Department of Biology, University of Pennsylvania, Philadelphia

Humans have been studying the biodiversity of wildlands as long as there have been humans. The goal was extirpating, eating, avoiding, inhaling, domesticating, controlling, and predicting. We have sought simplification and homogenization of the natural world to facilitate these activities.

The outcome is that today any given tropical nation or large multinational region has three basic kinds of land-use: urban, ever more intensively managed agroscape, and ever dwindling wildlands. The latter are generally patches of comparatively biodiverse habitats on socially or physically inaccessible sites or on "poor" agricultural soils.

The urban habitat is viewed as productive even if restive. The agroscape is productive, with largely pacific and homogenized biodiversity. The wildlands largely are viewed as removable, conservable, or conserved; i.e., they have been set aside by someone "else" for strip-mining of their natural products or for social fossilization, outside of the national economy. They are like cash in a shoebox under the bed, neither earning interest nor circulating, but of value to someone.

This perception of tropical wildlands is unfortunate, and fortunately it is waning in popularity. There are encouraging nuclei of voices dotted across the tropical (and extra-tropical) landscape arguing that conservable and conserved tropical wildlands are a category of highly productive land-use. Conserved wildlands are a different kind of field, just as ecotourists are a better kind of

---

[1]Derived from D. H. Janzen. 1994. Wildland biodiversity management in the tropics: Where are we now and where are we going? *Vida Silvestre Neotropical* 3:3-15.

411

cow, just as drug precursors are another kind of cotton, just as literacy in biodiversity is another kind of rice. In contrast to pastures, fields, and paddies, all three products of biodiversity—and many more—can come from the same hectare.

Such a shift in social and economic attitudes demands that a conserved wildland be blessed with the level of planning, knowledge, investment, oversight, budget, technology, and political attention that long has been characteristic of the more productive sectors of the agroscape, and also of a nation's institutions—highway systems, hospitals, education, and communication. Traditional tropical conserved wildland management—"fence it and put a guard on it"—is to such a blessing as a guard at the bank's front door is to the stock market, Federal Reserve, free market economy, taxes, and trade barriers all rolled into one.

We may anticipate a new edition of "potential land-use" maps for tropical countries. This is really what the "thou shalt inventory thy biodiversity" component of the Biodiversity Treaty is all about. No longer will there be a soil and contour map marked "apt for agriculture," "apt for forestry," and "apt for conservation," with conservation meaning "useless" and therefore to be assigned to the national park service or its equivalent. Rather, these maps will show what has been explicitly designated as agroscape and wildlands conserved for their biodiversity and its nondestructive use, with awareness that any hectare of a nation can be developed as either, depending on society and history rather than on soil type, rainfall, slope, and distance from a road or border war. The overall goal will be to render both types of land-use to be sustainably productive, high quality, and much valued by a nation and a region.

Up to the present, relatively nondamaging consumption from wildlands—humanity's hallmark during the first 99% of human evolution—gradually has lost out in competition with the agroscape. Today's wildlands appear to be substantially less productive than are many kinds of agroscapes. Humanity has cleared the way for its domesticates—including humans that function as urban or rural draft animals—and invested huge amounts in domestication. However, as the agroscape becomes ubiquitous across the tropics, the value of conserved wildland that is multiply used increases for society as a whole, and for nations specifically, because of its scarcity. Simultaneously, as the desires of humanity become more diverse and more perceptive, the value of a unit of wild biodiversity increases. Finally, as the knowledge base of humanity increases in bulk and interconnectivity, the intrinsic potential for multiple use of a unit of biodiversity increases. All of these increases are proportional to our investment in them.

The outcome is that a smart, modern, tropical government explicitly farms and ranches the information in an explicitly designated portion of its wildland biodiversity, just as a smart government resists pulping its national library during a newsprint shortage or using Internet cables to construct fences. It uses the income generated, in many currencies, to support the costs of managing the conserved wildland, further its development, and meet its costs of opportunity. This builds employment and capacity. This is sustainable development—living off the interest rather than consuming the capital.

WHERE ARE WE NOW, AND WHERE ARE WE GOING?

We are at a crossroads. Do we allow the progression of tropical land-allocation to conserved wildland biodiversity to continue as has been the case for the past 2,000 years? If so, 10-30% of tropical terrestrial biodiversity eventually will be conserved in 1-2% of the tropics. The locations of this remnant will be the serendipitous outcome of a multitude of social and economic forces acting largely irrespective of biodiversity's traits. One example of this process is that there are no unambiguously conserved large tropical wildlands on "good" agricultural soils. This "happy accident" strategy for conservation of tropical biodiversity will continue unabated if there is no major shift in social attitudes and economic processes. This strategy is quite comfortable for the majority of individual, national, and institutional agendas in the contemporary tropics.

The "use it or lose it" strategy is the other road. Less comfortable, it envisions 80-90% of tropical terrestrial biodiversity conserved in 5-15% of the tropics. The locations will be the serendipitous and planned outcome of a multitude of social forces acting irrespective of, and with respect to, the traits of biodiversity.

The major shift in social attitude and economic forces that are required by the "use it or lose it" strategy is that tropical conserved wildlands are conserved for nondamaging use by all sectors of society rather than because they are wastelands, for our grandchildren, for the sake of conservation, crown jewels, biodiversity-prospecting pits, observation posts for bar-coded horses, or to fill the agenda of any other single social sector. Each of these seven sacred and reasonable cows, and a whole herd more, become byproducts and ingredients, rather than *the* goal. This needs to be true even if each today is of major importance somewhere in the tropics. This attitude is somewhat akin to recognizing that the value of good agricultural soil or quality roadworks is not in the specific crop or the specific truck, but rather in being a platform on which society carries out a multitude of diverse activities.

The unhidden agenda is to move tropical wildlands into that social category of "so useful to society that no matter what form a society or nation takes, tropical wildland biodiversity will be woven into and through it"—as is the case with health, education, welfare, market economics, and communication.

Should someone mistake this essay as an argument for the simple commercialization of tropical wildland biodiversity, please note that humanity has in fact won the basic battle against terrestrial nature. It is not if, but how. Wildlands are rapidly becoming historic events. We are no longer afraid of the dark, spirits are no longer The Cause. We are in fact polishing the globe clean of most wild biodiversity that weighs more than a gram through species-specific harvest, habitat destruction, and contamination. Even the little things—fungi, bacteria, insects, and their brethren—are being removed or thoroughly impacted by these processes.

If we do indeed sweep the battlefield of the wild things, if we do reduce our globe to the playground of domesticates, we consign humanity to the doldrums of just those things that humans can imagine, invent, and control. We as thoroughly deprive ourselves as if we excise our color vision, our sense of smell but for frying chicken, our taste but for salt and sugar, our hearing but for high, low, and middle C. The brain is a computer with tens of thousands of applications invented to deal with nonhuman nature. By the removal of tropical wildland biodiversity, we are permanently relegating it to word processing. But the other side of the coin is that our appreciation for superlative architecture does not demand that we have only those buildings that will win international prizes. There is a place on the landscape for a healthy agroecosystem as well as the wildland crop.

Because the emphasis throughout this chapter is on "use it or lose it," there are several caveats, all of which are traditional in other social sectors but have been slow to be applied to conservation. They boil down to several equivalent expressions. The frontier is gone. You are always in someone's living room. Tropical biodiversity must escape the Tragedy of the Commons. There is no free lunch. The only sure things are death and taxes. Applying these age-old concepts to the case at hand:

- The more we know about wild biodiversity, the more we can use it without destroying it.
- Not all persons can use wild biodiversity as much as they would like.
- The use of wild biodiversity must be scheduled and monitored.
- There are all sorts of users.
- Users pay in all sorts of currency.

Tropical wildland biodiversity needs detailed, knowledgeable and dedicated management as much as does any other social sector. Ironically, since it is so underdeveloped, the returns on an additional unit of investment in fact are likely to be often substantially greater than is the case with many well-developed sectors.

THE MORE WE KNOW ABOUT WILD TROPICAL BIODIVERSITY, THE MORE WE CAN USE IT WITHOUT DESTROYING IT

What Do We Need To Know?

In order to begin to use wildland biodiversity, we must come to know:

- What it is: Identification and taxonomy.
- Where it is: Microgeography.
- How to get it in hand: Trappers' tricks and husbandry.
- What it does: Basic natural history.

All of this must come to be in the electronic public domain, because hard copy is not functionally public, is out of date the minute it is published, and does not have the opportunities for massage and recombination that is so easy with electronic media.

Identification and Taxonomy. We need to know what biodiversity is so that we can:

- know its parts when we see them,
- communicate about it today,
- pool and massage our information about it, and
- link what we find out with what others have found out.

The latter is of particular importance with tropical biodiversity, much of which has geographic distributions spreading across many countries. With an accurate identification, the user in one country has potential access to all the information that has been accumulated about that species across its range, rather than being dependent solely on local knowledge.

There is a second and equally powerful reason to know what are the component parts of biodiversity. By knowing what it is—that is, by putting a Latin binomial on it—the species is placed within the purview of taxonomy's enormous power of inference. This inference is based on relatedness as expressed through grouping into genera and higher taxa and is derived from gene to whole-organism similarity. Placement of a species in a higher taxon is more than for filing convenience. It tells one what to expect of that species, based on what we know of the others in its taxon. Full realization of this kind of knowing demands an electronic database based on observations and specimens. These must back up the derivative information bases and knowledge bases that are derived from information on species.

Microgeography. We need to know where biodiversity resides, or at least where a portion of it resides, so that we can get to its location "on call." A catalog of the books in a Library of Congress is of severely reduced use if there is no knowledge as to where the books are to be found, even if they can be recognized and read once in hand.

Trappers' Tricks and Husbandry. We need to know how to get biodiversity to hand or eye so as to get the information that we seek or can use from it. All hunter-gatherers and their field-biologist counterparts long have experienced the circumstance where a species is known and appears to be absent, yet—with the appropriate collection method—the species appears in droves.

We also need to know how to get biodiversity to hand so that we can care for it and multiply it, so that with this husbandry we can introduce it to the agroscape—rural or urban.

Basic Natural History. We need to know what biodiversity does—its natural history in the broadest sense—so as to:

- give us clues as to what it offers by itself and through its interactions;
- suggest how to farm it elsewhere;
- allow us to know the impact of our presence, studies, and sampling; and
- allow us to know when it is in trouble and what to do about it.

The first three of these four needs tend to be open-ended and cumulatively solved and, for a given site, require progressively less investment per species across time. However, understanding of natural history is ever expanding and peaks much later, if ever, in the cycle of involvement.

How Do We Get This Information?

Taken in collaboration, the four activities above represent genuine biodiversity inventorying, and they constitute the real base on which biodiversity management is constructed. They also fully recognize that management for a given site can be built on one or more inventorying activities as the others are being developed.

These four activities can be, and will be, carried out by a diversity of persons for a diversity of agendas in a diversity of wildland sites. However, a tropical nation with species-rich conserved wildlands may well be fortunate enough to have >100,000 hectare blocks containing 100,000 to a million species and all their interactions. In such cases, a major strategy for biodiversity management and development is to select—in the context of a nation's full gambit of users and managers—a site, and rapidly inventory all of its species. This is an All Taxa Biodiversity Inventory (ATBI). The function of an ATBI is to set up a major block of a nation's wildland biodiversity for all users. It:

- projects a massive block of diverse raw materials onto society's table,
- enjoys substantial economies of scale,
- foments mutualistic gains among executors as well as among users, and

elevates biodiversity inventory far beyond being a taxonomist's tool or a conservationist's listing.

When ATBIs occur in various countries and are firmly networked, a global network of these four advantages of an ATBI can and should be achievable. For example, this has been envisioned through the DIVERSITAS network as visualized by the United Nations Educational, Scientific, and Cultural Organization (UNESCO), and will be realized in the biodiversity clearinghouses of the Biodiversity Convention. An ATBI is a major advance over the diffuse and dilute approach currently in play, an approach clearly rooted in the time-honored traditions of curiosity-driven and highly individualistic field biology as performed by taxonomists and ecologists.

Why must all this activity be in the electronic public domain? First, in contrast to the past centuries of "public" publication of information on wildland biodiversity, which was aimed almost entirely at the very specialized audience of the scientific community, we now have the technical opportunity and ability to put information about tropical biodiversity truly in the national and global public domain through world-level electronic networks. Second, the goal of tropical wildland biodiversity management is to imbed it in society—all sectors of society and not just those with access to scientific journals and preprints. Third, the greater part of tropical terrestrial biodiversity is international; biodiversity represents a global effort even if a nation is the primary custodian. What we know, and will come to know, of the wild silkmoth *Rothschildia lebeau* is based on the aggregation of information from studies in Texas, Mexico, Costa Rica, Venezuela, and Colombia, among others, and conducted for a multitude of reasons—schoolyard exercises, pharmaceutical prospecting, ecotourism guiding, silk research, insect disease transmission, and religious symbolism. Questions of the ownership of information, costs, and charges—such unfamiliar ground for the community of taxonomists, ecologists, and conservationists—have very much in common with the well-worked terrain of ownership, costs, and charges for other social sectors such as trails, roads, highways, waterways, and airports.

The other side of public domain is the responsibility to actually conduct ATBIs and other kinds of inventories and the subsequent management and development of wildland biodiversity. Much of what commonly has been the social responsibility of the traditional academic/museum community on the one hand, and "park guards" or distant government offices in the capital city on the other hand, can be passed most profitably to parataxonomists, paraecologists, ecologists, educators, ecotourism guides, administrators of biodiversity, and other forms of site-based paraprofessionals. There is huge potential in training residents that neighbor the conserved wildland or live in it. They can accept a major portion of the responsibility to carry out nondamaging management and user-processes. This transfer of power and decentralization is essential for moving beyond what is today largely management of tropical wildlands by absentee-landlords.

This transfer, however, does meet with two major classes of social resistance. First, the scientific community understandably is reluctant to invest the energy and modification of tradition that will bring this about without compensation by senior administrators of science and by society at large. There are widespread benefits that could result from such a leveling of the playing field, and some of these need to feed back to its contributors. Second, such a transfer of political and economic power to rural areas—in modern parlance, decentralization and management horizontality—is theoretically attractive but very difficult to bring about in the face of contemporary vertically organized society. A large and properly managed conserved wildland dances dangerously close to secession from the federal state in virtually all tropical countries.

What Do We Not Need To Know?

I have tried to stay away from counterproductive commentary on alternate trade routes to the same new world. However, to put the above schema for site management and ATBIs in clearer perspective, I would like to suggest ever so gently a few areas in biodiversity biology where energy might be more productive for long-term conservation of biodiversity if spent elsewhere. These comments are likely to receive mixed reviews from the ranks of biodiversity biologists. These comments run in direct conflict with the very human behavior of attempting to mold the newly emerging activity of conservation of biodiversity so that its energy feeds one's entrenched agenda rather than targets the goal that elicited the activity—conservation of biodiversity.

• We do not need to know how many species there are in the world, in a country, in a large conserved wildland. We already know that there are hundreds of thousands to millions of species, and most are unknown in most respects. That is enough information to get on with knowing biodiversity and setting it up for nondestructive use. It is not shameful that "science" does not know whether there are 10 million, 30 million or 100 million species of organisms, and it is a waste of precious time and human resources to focus on refining this estimate. Would the conservation of biodiversity be aided for us to know that there are 502,451 species in Costa Rica? Would the Library of Congress be substantially more effective if someone counted all the books or the kinds of books? All of these estimates of biodiversity seem to have forgotten the existence of bacteria and the oceans. The count of species in a large aggregate of biodiversity is a by-product, not a goal. One can use a telephone directory to count approximately how many telephone numbers there are in a city, but that is not why one makes a telephone directory. What does need our attention is our ignorance of biodiversity in organisms and processes. This ignorance does not imply that the numbers of species per se are of importance, though numerical relationships do play their usual role in sampling community properties.

• We do not need to know the world-level or even national-level detailed geographic distributions of all butterflies, birds, or trees (or some other conspicuous taxon). The world is simply not a sandbox offered to scientists to reorganize as they wish so as to save their favorite higher taxon. The function of detailed biodiversity inventory is *not* to choose sites for conservation. One invests inventory attention on an area that already has been seriously designated for conservation status, with the goal of ensuring that status through understanding. The bulk of the significant blocks of conserved or conservable biodiversity in the Earth's terrestrial tropics already are known and largely delimited. Where this is not the case, there already exist knowledgeable field biologists and conservationists—national and international—who can quickly set the majority of those limits through rapid ecological assessments and other protocols. What is needed is not many international "choose your favorite site to conserve" exercises, but rather a focus of the world's scientific, conservation, and user energy on making those 5-15% of the world's tropics into places that society really wants to keep.

• We do not need to know about one more set of data on traditional wildlife management about this or that turtle, macaw, tiger, or deer. Yes, there are some large conspicuous tropical organisms that need a close look—vis à vis the real biological and social context of where they live. But, in general, it is the other 98% of biodiversity that needs much of our attention in natural history. The attention should be in the context of the society that surrounds and infuses the biodiversity of the focal site and not in the context of the time-honored initiation rituals of academic titles and institutions. Let us stop making conservation science be the science of trying to figure out how to get more money for biological research by piggybacking on the biodiversity crisis.

• We do not need to know where each individual of every species is (or was) over the surface of the tropical landscape. Far more than in extra-tropical habitats, the living dead and the population fragments sprinkled across the tropical agroscape are slated for the dust bin. We long have been deceived by the ability of extra-tropical species to persist as the sum of minute fragments in severely impacted landscapes. Sixty to 80% of North America's biodiversity probably can survive in a scattered and porous network of many small reserves and on marginal farmland or ranchland. The analogous lifeboat in the tropics would do well to save 30%. Our tropical resources, always in short supply, should be directed toward saving the bulk of biodiversity in a few large and well-distributed blocks, hopefully robust in the face of climatic change through their elevational and multihabitat diversity. This comment is not meant to denigrate the small patches and fragments of widespread tropical species and the occasional isolated endemic—at times of high value to their immediate owners and neighbors irrespective of their eventual demise or impoverishment of biodiversity. Rather, it is meant to suggest that we coldly practice a biologically realistic triage so as to bring conserved wildland biodiversity into peace with tropical society at large and, more specifically, the agroscape. I would argue that it is a much wiser investment to assure the survival of a few large blocks than to continue to harass the agroscape over the unsustainable survival of tiny remnants.

WHAT DO WE DO WITH BIODIVERSITY
ONCE WE KNOW SOMETHING ABOUT IT?

When one begins to be familiar with biodiversity in a large conserved tropical wildland, what does one do with it? This is somewhat akin to asking what does society do with the Library of Congress, Internet, a university, the Missouri Botanical Garden, a supermarket, and Disney World all tied up in one. The answer is "everything." This reply stresses the importance of conducting an ATBI on a very large area, an area large enough that it can be used for multiple purposes without the users destroying it. It is of restricted use to know a huge amount about an area so small that its biodiversity must be treated like the rare book section of the public library. The following begin to answer the question:

• All conserved tropical wildlands are islands (or shortly will be), and an ATBI is the beginning of providing a ground zero for asking how its biodiversity reacts to such things as global climatic change, pesticide contamination, use, and insularization. Tropical wildlands, fractured into habitat islands with distinctive conditions, and perhaps subject to severe global change, will be the sites of a global speciation event that will dwarf anything generated by, for example, Pleistocene drying. An ATBI sets up a gigantic canary in the mine for these reactions.

• The known universe of an ATBI site can be a standard for calibration (and development) of any and all kinds of inventorying, sampling, and monitoring technologies and protocols that can be applied in lesser known circumstances for a multitude of reasons. Virtually all of today's methods and protocols for inventorying biodiversity have been developed in the process of inventorying the unknown, rather than the known, universe.

• The known universe of each ATBI site will serve as the foundation on which to construct an extremely diverse array of question-driven ecological, behavioral, demographic, and ecosystem studies. The possibilities are mind-boggling with respect to what sorts of ecological work can be done with a complex habitat where virtually every species can be identified and categorized ecologically.

• An ATBI site is, in effect, a well-organized wild zoo, greenhouse, and culture facility in which to prospect for wildland organisms, genes, and their products. An ATBI will remove the single largest obstacle to serious examination of large arrays of wild biodiversity for human use. This information has enormous applied value for the further development of the millions of hectares of serendipitously developing agroscape throughout the tropics. A happy agroscape is far more likely to live in peace with its neighboring wildland crop than is an agroscape at war with itself, a war brought on by monocultures, synthetic product substitution, and bored human draft animals.

- An ATBI site is a national museum, national library, art gallery, concert hall, national zoo, national botanical garden, and national university for ecotourists and other forms of students ranging from grade-school children to senior citizens. How can we hope to even begin to develop biological literacy for the tropics when wild nature appears as a homogeneous green wall, when its incredible array of solutions, questions, and examples are illegible and undecipherable?

- An ATBI site can be a major provider of ecosystem services, especially if the site is chosen with that as an additional criterion, and simultaneously can be known well enough to study the internal mechanics of ecosystem services. Carbon sequestration can be a boring monoculture, or it can have complex value-added properties.

The Limiting Resource: Knowledge Itself

Today we are in the throes of determining how to record, manage, and transmit information about wildland biodiversity—the structure of databases, networks, distribution of databases, image transmission, authority files, authorship attribution, clearinghouses, bar-coding, retroactive data-capture, and computer-capture of literature—but tomorrow these technologies and protocols will have been resolved for the most part. Then, and for centuries thereafter, the resource in short supply will be the information about biodiversity itself. Who eats what, what breeds when, why is this pond green and that blue, when will the mushrooms bloom, when are the birth peaks? What genes code for magnesium resistance, for morphine synthesis, for dry season dormancy, for sex? What does a complex tropical ecosystem do when the annual rainfall declines by 40%? How ironic that just as the great bulk of tropical humanity flees the countryside or polishes it clean, humanity is coming to have the wherewithal to record forever what some grandparents knew, and the grandchildren will want to find out, about the vaporizing wildlands. How ironic that tidbits of natural history gleaned from local naturalist's publications, birdwatchers' notes, and schoolyard exercises—properly collated through the Internet—may turn out to be as valuable as the information on specimen labels in the world's natural history museums.

Are we going to shed our distorted visions of tropical biodiversity gained from centuries of touristic field biology, and really begin to offer society an understanding of biodiversity in its heartlands? What is a keystone species? It is a species that is so well known that we can recognize the ecological ripples that occur when it is removed. All species are keystone species on some scale, though not necessarily on the scale and ruler of a 1.6 m-tall diurnal vertebrate. We need to look at more than our big woolly relatives. What is a redundant species? It is a species that does not yield what you want. This is not a biological trait. What is an indicator species? Any species can be a miner's canary in the right circumstances. Please let us leave the Holy Grail for other social sectors.

Computerization and The Transfer of Knowledge

The all-invasive wave of computerization represents a quantum and qualitative change in the acquisition, massage, distribution, and archiving of information about biodiversity. It will change humanity's relationship with biodiversity more than has the printing press, the camera, or the chainsaw. Computerization is a great part of what allows the realization of all the prognosis mentioned or alluded to here and elsewhere in biodiversity management. For the first time in human history, there is the opportunity of open and massive intra- and intersociety flows of information about biodiversity, something that was alluded to through "publication" but in fact has not been achieved even minimally compared with what is to come.

For the first time, it is possible for an individual and a site to acquire, massage, distribute, and archive the unimaginably large quantity of highly particulate information—images, specimen descriptors, species descriptors, habitat descriptors, circumstances, previous knowledge—that is pertinent to the management and use of a conserved wildland that contains hundreds of thousands of species and has been or is being studied by tens to thousands of observers over the years or even at one time. The essentiality of bar-coded uniquely tagged vouchers and specimen-based information becomes self-evident. The primal necessity of attributing authorship and evaluating input for all these data is written in stone. The real art is how to massage information and put it in a multitude of formats for a multitude of users. The real question in biodiversity then becomes whether the developed world is willing to accept the leveling of the global playing field that all this implies. Most evident of all, the last thing biodiversity management needs is new hard-copy journals, more hard-copy books (except as temporary reports for some kinds of convenience), and a continuation of the stultifying hard-copy traditions of biodiversity information management from the past several centuries. What tropical wildland biodiversity management needs is for all the holders of information about biodiversity to get that information as fast as possible into the Internet, rather than waiting decades (if ever) to see it frozen onto thin sheets of wood.

Taxonomy and the Taxasphere

Taxonomy is the basic philosophical and technological infrastructure for wildland biodiversity management. Without taxonomy, there is no inventory, no collation and distribution of information about biodiversity in space and time, no inference among species. But taxonomy, like conservation, ecology, and other specialities in science, has evolved to its own drumbeat. Most encouragingly, taxonomy currently is reexamining its mission through efforts such as *Systematics Agenda 2000* and a multitude of international symposia. Government agencies of the United States are beginning to undertake and support a global responsibility in taxonomy, and taxonomy once again is coming to be supported as a form of national development.

Some things are evident in the changes that biodiversity management asks of taxonomy. No more turgid keys, please. Expert systems, picture keys, Intkey, and the like are a major step forward. Give top priority to the coming together of standards for taxonomic and specimen data, data models, and user-friendliness in computerization. We all need to be headed toward identification guides that allow us to flip through an electronic (or hard-copy) picture book, with centralized or networked processors where an image or discussion of a doubtful organism can be sent for taxonomic confirmation. Close on our heels is the magic box into which a piece of a bug is dropped, sequenced, sequences compared with a library, and a name spit out if it matches. Then, with the name in hand, one calls up what the greater global network already knows about the biology and biodiversity of that species or population—a global field guide in a pocket. Once again, it is the information about genetics and biodiversity from the field that becomes the resource in short supply.

Where is most of that information pool today? It is in the heads of retiring taxonomists. Speaking quite coldly, these most honorable systems should be data-based, information-based, and knowledge-based—the brain dump—to say nothing of put diligently into mentorships for the next generation of those who will manage this (to date) highly personal tradition. The new PEET initiative (Partnership for Enhancing Expertise and Taxonomy) of the National Science Foundation is a major step in this direction. This information-capture might well be done in conjunction with the retroactive data-capture that is possible in the world's large museums, but, if not, the highly perishable should be given priority over the pinned and dried.

Taxonomy is really a taxasphere with nodes of specialists, collections, knowledge-bases, and hard-copy data—all strung together on an Internet lattice and variably plugged into the world's biodiverse sites. The nodes interconnect as much for taxonomy's own work as for all the other users of biodiversity. There is a major question as to whether, and to what degree, it is worthwhile to retroactively capture the information in museums. Ironically, museums were, on the one hand, the expressions of interest in species as manifest through specimens. Information contained in museums therefore often was not gathered in a manner conducive to its maximum use in biodiversity management. On the other hand, museums are the depositories of the raw material on which taxonomists have largely built their science. To someone concerned with a given conserved wildland, the international and national distribution of a species (as recorded on selectively and serendipitously collected museum specimens over previous decades) may be of limited interest. What may be of greater interest is whether and where that species occurs today within one's local or national gambit of interaction.

At the very time when extant museums are rethinking the value of their collections, they are the logical recipients of the new and enormous responsibility of curating the mass of voucher specimens that will appear from inventories and other kinds of biodiversity management. While many of these specimens are perhaps of lesser direct taxonomic interest, they are of huge importance in underpinning the current mass of information about biodiversity, a base on which much more information will be built. We find ourselves in the ticklish position of explaining to the tropical world at large that the specimen is of little or no value per se, and thus should not be the focus of nationalistic possessiveness, while at the same time it may be a voucher specimen or source of genetic information that merits long-term maintenance costs. The more the biotechnologists tell us, the closer that specimen comes to being a legible cookbook for many of the things that it did in nature.

The taxasphere has long run on the engine of personal interest in organisms by taxonomists and other kinds of field biologists, rather than on a true economic and social recognition of the critical nature of the taxonomic underpinning of the use of biodiversity (though an impressive amount of research on wildland biodiversity was conducted in previous centuries in the name of economic interests). To the degree that society neglectfully accepts that taxonomy is run by such a volunteer work force, we are confronted with the advantages and disadvantages of trying to run an army or national park staffed with unsalaried volunteers, even very competent ones. While the taxasphere needs to reach out with joy for the finances and responsibility that should come with a reversal of this trend, this same taxasphere then is confronted with an increased accountability to the funder, a kind of accountability not usually associated with those who operate in a free-spirited and artistic social sector. It will be most helpful if the taxasphere can manifest some self-directed willingness to spread responsibility to those taxa and technologies previously unconsidered, as a response to society's willingness to put resources behind this action.

Targeting the Small Stuff

The bulk of biodiversity comprises very small organisms—easily 80% of a conserved wildland's biodiversity weighs less than a few grams even as an adult. However, many of the traditions of information management, field ecology, species-use, conservation, wildland education, and evaluation of biodiversity have been little affected by the biology of the small stuff. On the other hand, the enormous biodiversity of small species constitutes much of the potential for the use of biodiversity and offers a huge part of the complexity of managing this biodiversity.

This means that finding out which aspects of biodiversity reside in a site and getting it in order for society will involve a very large number of field taxonomists and ecologists spending their time getting their (easily inventoried) big organisms into situations where they can be poked and searched by the people who work with viruses, bacteria, fungi, mites, small insects, protozoans, parasites, algae, and other little things. This means that the quality of on-site laboratory facilities will need to take a megastep upward to complement the old tent and machete. This means that the conserved wildlands will be brought yet closer to society.

MANAGING WILDLAND BIODIVERSITY FOR SUSTAINABLE USE

As I mentioned at the outset, the frontier is gone. Sustainability is eating the interest, not the principal. The use of wildland biodiversity must be scheduled, planned, and monitored. There will be all sorts of users, and they will compensate for their impact through payment in a very wide range of currencies.

It is no secret that tropical wildland biodiversity currently is threatened by the nearly invisible symphony of a multitude of threats that are exponentially gaining force from unseen and unexpected directions. They impact simultaneously in different countries, and the well-established lack of intercountry communication renders them even yet more invisible. That little farmer in the forest with his chainsaw is now unexpectedly given a huge boost by the fall of trade barriers, by pharmaceuticals abruptly rendering yet another major tropical disease less of a barrier to wildland clearing, by the introduction of newly genetically engineered domesticates, and by the speeding of the process of domestication through biotechnology. Knockout punches are gathering silently in the wings. Yet the left hand needs to be doing something quite noticeable before the right hand takes note. Conserved wildlands require aggressive and eager succor from society at large if they are to survive the very onslaught that often is generated quite innocently by that same society.

But do extinction rates really matter? Does it matter if this or that species goes extinct? The fact is, we will lose 10-20% of them over the coming century, important or not. So let's get busy delimiting the areas that will be conserved tropical wildlands, largely forget about those things that live outside, and get on with making high-quality conservation areas out of the 10-15% of the Earth's surface that contains the remaining 80-90% of the species. And let's make very high-quality agroscapes and urbanized habitats in which these areas are imbedded. All the mental energy and all the funds put into anguishing over the losses could be spent far better on quality survival of the survivors.

It might be useful to note that the current extinction differs from the Cretaceous extinction in the following ways:

- We are not about to give the terrestrial world back to biodiversity to reevolve after this is all over.
- We will have reduced terrestrial biodiversity to tens of thousands of analogs of the Galapagos Islands and New Guinea, where speciation and higher taxon evolution will work apace (and generate a plethora of endemics).

- The surviving subset of species will be defined by area and habitat, rather than being ecological groupings such as those small vertebrates that could aestivate or stay warm (microreptiles, micromammals, and feathered hotblooded microdinosaurs) and survive on a diet of plant and animal carrion, seeds and other dormant organisms, and the insects and fungi that feed on the same, during a winter induced by meteor impact.

So the first self-denial that must become characteristic of users of biodiversity and practitioners of biodiversity management is the temptation to have different government agencies and nongovernment organizations (NGOs) harvest their own information about biodiversity and wield it to their own end. We hopefully are about to enter an era of interagency and inter-NGO cooperation. The Oaxaca Declaration, the U.S. National Biological Service, much intermuseum collaboration in the development of databases, the information clearinghouse of the Biodiversity Convention, and databases moving onto the Internet are recent examples. ATBIs, institutions like INBio, national biological surveys, and the Internet itself are all manifestations of this process of collaboration among users.

Terrestrial conserved wildlands are habitat islands and will become more so. They are habitat islands joined only by a selected few (largely) aerially mobile organisms, islands positioned in an ocean of intensely managed domesticates. This insularity means that no matter how large and how well planned and inventoried, each conserved wildland will have a different and heterogenous ceiling for intensity of impact by on-site users. We even have the irony that established conserved wildlands can render the concept of "endangered species" an anachronism. If species are in truly conserved wildlands, they survive in those habitats at their naturally achievable densities or they go extinct. Outside of the conserved wildland, they are basically forgotten. Yes, some will survive as society's pet trees and animals, or as domesticates and weeds, but these are not the focus here. The question is not whether we can bustle across the agroscape feeling valiant in the protection of the living dead, but whether we can design rules for maximum nondamaging use of significantly large conserved wildlands and tempt society to live by these rules. Let us use our energy *now* to make the wildlands into better islands, rather than dream that we are being effective conservationists by saving a noble tree left standing in a tropical bean field.

Introduced organisms would seem to be a sort of unconscious use of conserved wildlands and marginal farmland. First, please stop the introductions until the sink as a whole has been taken into account. No matter how many firewood trees have been cut down in Africa or India, the solution is not the introduction of new species of firewood trees from the Neotropics. No matter how little water can be allocated to Hawaiian home ornamentals, the solution is not the massive introduction of drought-resistant plants from Costa Rican dry forest to Hawaiian gardens.

Second, recognize the extreme contradiction inherent in biodiversity. At least in mainland habitats of broad extent, virtually every organism in the natural community evolved as a little population somewhere else. It is an immigrant in most of where it is found today, and wildland biodiversity programs on mainlands are put together mostly through ecological interactions rather than through on-site evolutionary fine-tuning. The horse is an instructive example. The horse is a New World native and could be argued to be a proper part of Mesoamerican conserved wildlands, albeit a machine gun is needed to substitute for the sabertoothed cats. The survival of the horse in the Old World, while we extinguished it in the New World and then reintroduced it, really is not that different from reducing the North American bison to a tiny herd and then building it back up (and partly domesticating it as well). Use should be measured by real impact, not by the license plate on the "immigrant." All wildlands are strongly impacted by humans already—extinction of the Pleistocene megafauna, extinction of contemporary vertebrates, global warming and other change, hunting, roadside secondary succession, and introduction of bacteria, fungi, algae, mites, and herbs. There is no "pristine" nature, free of "introduced species" and human influence, to conserve.

Once designated as conserved wildland for the nondamaging use of its biodiversity, this land-use categorization needs to be inviolate. In this respect, wildlands conserved for their biodiversity are qualitatively different from other kinds of land-use, and are not easily interchangeable with other kinds of land-use. The agroscape easily can move from peanuts to sorghum to cows to peanuts over the years, but moving a given hectare from rice to forest to rice to forest requires considerably more cost and long-range structure, and often is not biologically possible. Restoration is a limited tool, not a panacea. However, restoration does offer enormous potential in the siting of conserved wildlands throughout the tropics.

A conserved wildland is far more sensitive to context than is an equally-sized portion of the agroscape, and a conserved wildland cannot afford to go bankrupt—unless society also is willing to then leave it in peace until wildland production starts up again. In the same vein, we must come to recognize that a conserved wildland is no more or less responsible for contributing to a country's national budget and the solution of its social ills than is any other kind of land-use. A successful rice farm is not held accountable for the social welfare of all of its neighbors' ills except through some variety of national income tax distribution. There is no reason to expect that the earnings of a conserved wildland will provide the solution to all of its neighbors' ills except through the same kind of distribution of earnings, employment opportunities, and taxes as for other forms of land-use.

Ironically, the direct use of a conserved site by people is perhaps one of the easiest of all facets of biodiversity-use to manage. Society at large, and specific individuals, are very good at using or visiting a conserved wildland area to the level of intensity allocated—if they are clearly informed of the limits and if the method of explanation is clear and cast in a socially perceptible format. This communication requires more or less direct human presence and interpretation, depending on the society and circumstance. The more specific harvesters—researchers, staff, biodiversity-prospectors, inventoriers, ecosystem service personnel—likewise are proving themselves to be highly responsible socially in conserved tropical wildlands if they find themselves cast in a responsibly managed and forward-directed interaction between society and biodiversity. But we can never forget that the finest farm or ranch easily can be destroyed through overgrazing of pastures, improper irrigation, failure to rotate crops, poor selection of varieties, or sloppy agrochemical application. Wildland biodiversity is another kind of farm or ranch.

Throughout the tropics, lured by the ecotourism dollar, there has been a very strong tendency to use the dollar as the primary currency in valuating conserved wildlands. While this has its valid points, what seems to be forgotten—largely through the inconvenience of leveling the national social playing field—is that the "poor" national user of a conserved wildland pays in votes (as well as through some decentralization of cash flow) and in emotional attachment to the conserved wildland. When the fourth grade schoolchild is voting on the irrigation district board as a 55-year-old adult, that person will remember what was learned in the conserved wildland 46 years ago, what experiences were had there, and visualize the grandchildren as doing the same. This phenomenon is reinforced when the conserved wildland and its associated processes constitute a major local employer, spends millions of dollars per year locally in operations costs, and uses its income to establish its own management endowment.

Equally revealing, and long-term, is the biodiversity-prospecting loop. When a conserved wildland or its facilitators bring home the first contract for biodiversity-prospecting, the returns seem very large when set against the background of tropical conserved wildlands—viewed as all cost and no visible income other than piddling ecotourism entrance fees. However, a ministry of natural resources will take notice when the first actual royalties from a drug discovery flow into the national budget for conserved wildlands or, better yet, into the endowment fund of the conserved wildland from which the raw materials were collected. But even then, a ministry of the economy will not take notice. That will occur when the pharmaceutical company decides to move some substantial portion of the development process—more than $200 million per successful drug—into the source country. Once again the leveling of the playing field reappears, with all its advantages and impediments. The North American Free Trade Agreement (NAFTA) and the General Agreement on Tariffs and Trade (GATT) relate directly to conservation of biodiversity.

The art of valuation of nondestructive use of biodiversity rests heavily on being able to work in many currencies, to recognize the market value of information to many different sectors. A field guide to the birds of a tropical country is not just "a bird book." It is essential technology in the ecotourism industry. It is fertilizer for the ecotourism crop. Yet also, without a conserved wildland in which to observe the birds and the things they do, it becomes just a bird book. The value of information about biodiversity is extremely dependent on context. A country may "have" (really, may be custodian for) the most marvelous set of endemic species or bizarre habitats, but if the information they contain and display is not put into various social currencies, those species and habitats will contribute little or nothing to their very survival in a world dominated by humans.

All of this can do nothing but reemphasize the critical need for institutions and processes that accept the responsibility and challenge of the specific task of gathering, collating, massaging, and distributing information from and about a nation's conserved wildlands. This essential process must occur at the level of each specific wildland and at the level of the national synthesis, at the least. You, society, hardly can be expected to value that which is invisible to you. Ironically, the very salvation of biodiversity—its valuation by society—is a multiedged sword.

First, if the area is conserved for its value on just one or a few axes, then it is in the same risk zone as the country that depends on a monocultural agroscape—coffee, bananas, and Costa Rica are close to mind. Fortunately, wildland biodiversity is in fact far more diverse than is the agroscape and, as such, diversification of crops as well as diversification of markets is very feasible (though hardly developed).

Second, information differs from agricultural produce in that one consumes produce today and needs more tomorrow. Once consumed, information is public domain and continually widely available, and is even more so in the electronic and computerized age. Therefore, a given piece of new information is unlikely to have nearly the same value in next year's market as in this year's market. There is a high premium on rapid development of products, almost as one encounters in the newspaper business. However, as in the news business, naive consumers of information about biodiversity continually do appear through human biological processes (birth, forgetting, nostalgia), and the amount of absolutely new information about biodiversity to be gathered and developed certainly is limitless for many decades to come. Information about biodiversity also can become "new" through the appearance of a new use.

Finally, it is no secret that a nation's conserved wildlands are its package of local varieties. All that a nation does to both share and profit from the varieties in its agroscape is pertinent by analog to the treatment of the breeding stock and genes from its conserved wildlands. Just like petroleum, which occurs in a multitude of countries, the value of any one of these species depends on what the country constructs on top of its national supply of this basic raw material.

Any conserved wildland will need to struggle with the question of physical use and the impact of sampling, observing, studying, experimenting, and visiting. Given that all conserved wildlands are in fact impacted already by humanity, and always will be, the question is basically what level of use falls within the "natural" ups and downs and expansions and contractions of behavior, demography, and interactions. What level of use is "nondamaging"? Any user does leave a footprint or a beer can if one knows enough biology to see it. However, just as the tapir-nibble out of the top of a bush blurs into biological "noise" within a few days to weeks, the biodiversity-prospecting sample taken from that bush blurs as well over time. Just as the loss of the annual baby agouti to a boa constrictor changes the mother's foraging pattern for a year, the monkey-watcher's trail changes the sleeping site of the local peccary herd. But the next year, both perturbations are indistinguishable from the multitude of other nonanthropogenic changes.

At present, perhaps the largest single near-sighted user of detailed tropical biodiversity is the international academic and museum community. In what currency will they pay for and value their use? Long we have cast our graduate students in our own image, and now and then we have done the same to a student from a tropical country. But it is not at all clear that this is the kind of payment we would make if we really were to think out what a tropical resident needs, for example, to be part of the managerial cadre of biodiversity. Even more basic is whether we should be expending so much energy on producing yet more graduate students in a steady-state system, or expending that energy in collaborating with the tropics as it comes up to speed. We are letting the lifestyles in our developed world define the way that we examine and study tropical biodiversity. That is okay, more or less, if the biodiversity is in our backyards in Minnesota or California, but it definitely is not if the biodiversity is in Madagascar or Colombia and the training is in England or Illinois.

The upcoming presidents of tropical countries often will have advanced degrees from universities in the developed world as well as from those in their home countries. Will they have learned about biodiversity around those northern universities? Or will they have learned how to deal with the biodiversity in their home countries, a situation that desperately needs their political attention? Costa Rica's new President, José María Figueres (1994-1998), has accepted the challenge of steering his country in the direction of sustainable development and management of conserved biodiversity for society's nondamaging use. Has his university training, and that of his advisors, prepared him for this?

IN CLOSING

A peculiarity of taxonomy and natural history—those pivotal professions for the management of biodiversity—raises its hand here. Taxonomy and natural history represent some of the very few subsectors of science that strongly depend on people—amateurs and professionals alike—who really love the actual objects of their research as well as being intensely curious about their traits.

High-quality managers of biodiversity—wildland and urban, taxonomists and many other kinds—are largely born and then facilitated, just as are musicians, politicians, scientists, basketball players, and others. The expression of their genes requires an accepting society, and the facilitation of their abilities costs money and job security. Retooling people and institutions from other areas into biodiversity management has the usual advantages and drawbacks, with institutions being the most difficult. Who is going to crawl around in the hot tropical sun doing natural history without being in love with the organisms? Who is going to spend 50 years of their life peering intently at some organism that is 1 mm long in return for just salary and prestige? We would do far better to feed and support those who are by nature inclined in this direction, as we do with musicians and many other professions, than to issue a call for all good people to come and be good taxonomists and natural historians. They will not be. But we can reinforce those with a propensity in this direction, and draw out the best in them.

ACKNOWLEDGMENTS

This manuscript is the result of many discussions with many people, but I would like to particularly acknowledge the time and attention of F. C. Thompson, K. Krishtalka, J. Edwards, W. Hallwachs, R. Gámez, D. E. Wilson, P. Jutro, I. D. Gauld, F. Chavarría, A. M. Solis, P. Rauch, J. Busby, A. Chapman, J. Croft, P. Raven, T. Duncan, E. Hoagland, D. Brooks, A. Rossman, M. M. Chavarría, J. Jiménez, A. Sittenfeld, W. Reid, R. Curtis, J. Burns, D. Miller, G. Barnard, D. Brooks, M. Carvajal, J. Croft, R. Espinosa, R. Moraga, J. Hester, M. Ivie, N. Zamora, D. Schindel, F. Harris, N. Chalmers, K. MacKinnon, F. Talbot, E. Nielsen, M. Boza, C. Quintela, J. Sarukhan, L. Tilney, S. Ulfstrand, A. Ugalde, J. Coddington, A. Umaña, S. Marin, A. Pescador, R. Blanco, M. Zumbado, A. Solis, C. Wille, J. Corrales, J. Ugalde, R. Castro, J. Phillips, E. Sancho, A. Piva, M. Koberg, M. Molina, M. Segnestam, J. Burley, R. Poole, R. Hodges, Q. Wheeler, T. Erwin, H. Daly, T. Eisner, E. O. Wilson, J. Soberon, H. Kidono, J. Jensen, M. Bystrom, P. Froberg, G. Hubendick, R. Rimel, D. Martin, N. Meyers, D. Johnson, S. Viederman, O. T. Sandlund, F. Joyce, J. M. Figueres, and the INBio parataxonomists.

GETTING THE JOB DONE: INSTITUTIONAL, HUMAN, AND INFORMATIONAL INFRASTRUCTURE

Inventory, study, and wise use:
three steps to conserving biodiversity.

28

Taxonomic Preparedness: Are We Ready to Meet the Biodiversity Challenge?

QUENTIN D. WHEELER
Professor, Department of Entomology and L. H. Bailey Hortorium, Cornell University, Ithaca, N.Y.

JOEL CRACRAFT
Curator, Department of Ornithology, American Museum of Natural History, Central Park West at 79th Street, N.Y.

Taxonomic and genetic diversity—essential for human health, agriculture, and ecosystem function—are being eroded as wildlands are converted to other uses. Many species already have become extinct as a result, and thousands, perhaps millions, of others soon face a similar fate. Equally alarming, the rate of habitat conversion is accelerating, along with—by implication—extinction of species. Consider, for example, the approximate doubling of the estimated annual rate of tropical deforestation in the 1980s (Myers, 1991), which has led some to predict that one-fourth or more of our planet's species may disappear within a few decades (National Science Board, 1989).

Humans use tens of thousands of species in their daily lives for food, shelter, medicines, and diverse forms of commerce. As population pressures increase and more and more people must make use of other species for subsistence, managing these resources in a sustainable manner will become increasingly difficult. Modern forms of transportation have reached all corners of the globe so that people, whatever their economic circumstances, are now more mobile than at any other time in human history. As a consequence, thousands of species are being transported around the world, both intentionally and unintentionally. New diseases are emerging; agricultural systems are being exposed to a multitude of new crop pests; the natural structure and function of ecological communities are being torn apart by the introduction of exotic species; and intricate distributions of species produced over thousands or millions of years are rapidly being shuffled. Laissez-faire public policies regarding the management of natural ecosystems that once were considered benign now are seen to carry serious societal consequences that seldom can be reversed, and then only at substantial expense.

Discovering new biological resources and managing existing ones depend on access to reliable scientific knowledge about biodiversity. Yet as implied above, the challenges to effective management of the world's species are multiplying at a rate that far outstrips our acquisition of the information needed to confront them (Cracraft, 1995; Systematics Agenda 2000, 1994a,b). Ironically, the biodiversity crisis has emerged as a global issue at the same time that support for basic research and training in the biodiversity sciences has declined sharply (Holden, 1989; House of Lords, 1991; Nash, 1989; Schrock, 1989; Wheeler, 1995a). The absence of adequate scientific infrastructure in most countries, especially in those that are species-rich, constitutes a major impediment to an international response by the scientific community. Even those countries with substantial scientific resources cannot meet their management needs (National Research Council, 1993). In these countries, for example, systematic collections are not funded at a level that is capable of keeping up with the existing rate of specimen acquisition, let alone at a level appropriate for the biodiversity crisis. Existing data in herbaria and museums remain largely inaccessible by modern technologies for data management. Funds available for investigating fundamental questions about biological diversity are severely limited relative to the task at hand. And, finally, the numbers of students trained in systematics and organismal biology have diminished, contributing to what many, including the DIVERSITAS program of the International Union of Biological Sciences, the Scientific Committee on Problems of the Environment, and the United Nations Education, Scientific, and Cultural Organization, have called the "taxonomic impediment."

Of all the biological information that is needed to manage the world's species, the most fundamental is that provided by the discipline of systematic biology. The four primary components of systematics—discovery and description of species, phylogenetic analysis, classification, and biogeography—provide basic biological information about species, including their name, characterization, relationships to other species, and geographic distribution, thus establishing the foundation for all the other biodiversity sciences, such as ecology, population biology, genetics, and behavior. Taken in aggregate, these components support the ultimate aim of systematics to know and understand the taxonomic and phylogenetic diversity of life on Earth.

The need for systematics has never been greater. Despite having accumulated significant knowledge about the world's species over the past 2 centuries, we still cannot provide accurate answers to the simplest of all questions about biodiversity. How many species are there? Estimates vary from 3 to 100 million species. What are the relationships among species? Except for a small number of taxa, the pattern of life's history remains an enigma. Where among the myriad of Earth's habitats are these species distributed? Detailed answers exist for no more than a few species; indeed, the distributions of many of the most thoroughly studied vertebrates, including groups such as birds, remain imprecisely known.

The continuing loss and degradation of the world's biological resources compromises the ability of nations to create a sustainable future for their citizens. Managing these resources will necessitate an increased commitment on the part of the world community to support the biodiversity sciences, especially in the species-rich countries where scientific capacity is least developed. This should include programs to build new or to improve existing infrastructure, enhance human resources, and establish a world-wide biodiversity information network.

The systematics community, through its initiative *Systematics Agenda 2000*, has established a framework that can be used to develop the science of systematics world-wide (Systematics Agenda 2000, 1994a,b). Advances in the theory and methods of systematics, computer management of vast collections of specimens, and existing descriptions for more than a million of Earth's species provide a context and starting point for creating the knowledge-base in systematics that will be required to confront the challenges of managing species and their ecosystems.

DEVELOPING A SYSTEMATIC UNDERSTANDING OF THE WORLD: LIMITATIONS OF THE ECOLOGICAL APPROACH

Maintaining global ecological systems and achieving sustainability of biological resources are now widely recognized goals of the world's nations. For this to happen, however, each nation must adopt management policies that are based on, and consistent with, credible scientific knowledge. A key component of this knowledge is derived from systematic biology. Unfortunately, sufficient systematic information neither exists today in the quality or quantity required, nor does the scientific community have the capacity to acquire it rapidly.

Given that systematics is of clear importance for successfully managing global biodiversity, and that resources to support systematics science will need to be increased, we may ask "What is the most cost-effective and efficient means of developing a systematic understanding of biodiversity?" The broad, comparative scope of systematics research is unique among the biological sciences (Nelson and Platnick, 1981), suggesting that the best returns will be gained from strategies that meet its special needs for research resources.

Perhaps no more than 5% of the world's species have been discovered, described, or classified. There is broad agreement that human societies would benefit immensely from having knowledge of this unknown diversity, yet until recently no coherent research programs have been proposed to address this problem.

One approach to providing a taxonomic understanding of global biodiversity has been developed largely from an ecological perspective on diversity. It begins with the concept of an All Taxa Biodiversity Inventory (ATBI) at a single, geographically localized site, in which an effort is made to collect and identify "all" the species at that site (Janzen and Hallwachs, 1994; Janzen, Chapter 27, this volume). The overriding purpose of an ATBI is to contribute to the sustainable development activities of the country in which the site is located, and for that purpose an ATBI has obvious potential. Because of the substantial systematic activity necessary to undertake and support an ATBI, global knowledge of biodiversity seems certain to increase, and the results of such an exercise are likely to have important benefits for basic and applied biology, as well as for conservation and economic development. But are ATBIs the most logical and efficient way to approach an inventory of Earth's biological diversity (Wheeler, 1995b)?

In addition to their function in the development of nations, ATBIs also are being proposed as a model for achieving a systematic understanding of global biodiversity (Janzen, 1993; Langreth, 1994). It has been suggested that a dozen ATBIs, carefully sited and successfully completed, would sample as much as 40-50% of the world's species diversity and could be undertaken for about $1 billion (Janzen, quoted in Langreth, 1994:81). Leaving aside the fact that we currently do not have the necessary collections-based infrastructure or systematic expertise to complete a single ATBI, the application of what is essentially an ecological approach to sampling will result, over the long term, in an ineffective and cost-inefficient program, whether the goal of that program is to develop a systematic understanding of the world's species or a credible and predictive understanding of the local biodiversity.

The reasons for this are several. First, a dozen ATBIs barely would begin to provide the geographic coverage necessary to discover a sizable percentage of the world's estimated species diversity. This follows from the fact that when groups are investigated in detail, it is found that most species are rather narrowly distributed (see also Reaka-Kudla, Chapter 7, this volume). To sample the Earth's diversity adequately, therefore, would require an ATBI in each of the major areas of endemism in the world. A dozen ATBIs, in fact, would be a small number for some continents. In Australia, for example, the vertebrate fauna is partitioned into many areas of endemism, and an ATBI in only one of them would recover a very small portion of the Australian biota (Cracraft, 1991). As a consequence, an ATBI in one country will have limited utility for representing the diversity of other countries, including in many cases those in relatively close proximity to the ATBI site. An ATBI in one country, moreover, often will have limited relevance for the development process in others, except perhaps in the sense that they contribute to global systematic knowledge. Because adequate knowledge for any given taxon may dictate studies in a unique number and combination of geographic locations and habitats, it is doubtful that any finite number of ATBI sites ever would prove to be fully adequate for a detailed systematic understanding of more than a few taxa. While periodic collecting at known sites is a prerequisite for documentation of the status and trends of biodiversity, the very notion of long-term study at a few anointed sites is inherently an ecological approach while the resolution of fundamental questions about biodiversity require answers grounded in a systematic biological approach.

Ironically, were most species found to have broad geographic ranges, in contradiction to 2 centuries of experience and observation, then comparative studies throughout their ranges would be of critical importance. Widespread species often can be recognized definitively only after their full range of genetic variation has been studied (Mayr, 1963). The history of taxonomy suggests that the isolated description of local floras and faunas frequently results in confusion about species identities and a proliferation of redundant names. If most species actually were narrowly distributed, perhaps this problem could be avoided and provincial ATBIs could succeed in recognizing endemic species without creating such confusion. But then the assumption that species are widespread would be violated, and the ATBI model would fail to capture a significant representation of the world's biological diversity.

Second, ATBIs are particularly inefficient for setting global priorities for systematic research or for facilitating the growth of systematic knowledge, being biased as they necessarily are by the geographic location of the site and by the development and conservation needs of that particular country. Systematics can contribute far more to meeting the international need for knowledge by working with all nations to set priorities for revisionary studies, such as on those groups that are important for agriculture, human health, ecosystem function, and growth of scientific knowledge. Designing a global program to target such groups and to share the resulting discoveries would more efficiently meet diverse international needs for knowledge of biodiversity than investment of human and financial resources in documenting the biodiversity at a limited number of sites. For taxa of demonstrated economic or societal relevance, it is unlikely that either the majority of species or the most promising underutilized species will be found in a handful of ATBI sites, regardless of how carefully they are identified. For taxa of unsuspected or unappreciated importance, a worldwide perspective is even more essential. No one yet can predict where valuable discoveries will take place. Opportunities are enhanced by assuring an exploration of the most disparate branches of the evolutionary tree of life, rather than restricting our discoveries to the species and clades that happen to live in a dozen or so places on Earth.

Third, collecting organisms is typically the easiest, least labor-intensive aspect of systematic activity. Specimens cannot be identified reliably to species without access to synoptic collections of previously described species, specialized expertise, and extensive comparative analysis. This takes considerable time and access to resources for systematic study. Thus, current conceptual schemes for ATBIs significantly underestimate the time and scientific resources that will be required to identify and classify specimens collected during the ATBI process. With the exception of a few well-known groups, or species that are widespread or "weedy," identifications frequently will be a slow process, if they can be done at all. As is well known, there is exceedingly little taxonomic expertise or background knowledge for many groups of organisms.

Finally, ATBIs will not fulfill their potential for many development purposes until the relationships of the species are understood, inasmuch as it is the knowledge provided by an understanding of the close relatives of species and their biological characteristics that makes newly discovered species useful to society. Unless studies of phylogenetic relationships go hand-in-hand with the inventory process, we will neither make maximal use of these discoveries, nor will we be able to construct maximally efficient and useful information systems or predict where research, development, or conservation dollars are most profitably spent. Thus, systematic research is a prerequisite for the successful completion of any ATBI and as a consequence should have priority when building scientific capacity.

Drawing on extensive experience in the discovery, description, and classification of biodiversity and a remarkable recent theoretical revolution within the discipline, systematists are uniquely qualified to propose an alternative strategy for inventorying biodiversity that draws on what we already know and takes full advantage of research expertise and resources.

REALIZING *SYSTEMATICS AGENDA 2000:* AN ACTION PLAN FOR MEETING THE BIODIVERSITY CHALLENGE

In contrast to a world-wide network of ATBIs, the most efficient and cost-effective mechanism for developing a systematic understanding of the world's biodiversity is to promote capability in systematics for all nations (Systematics Agenda 2000, 1994a,b). This entails an integrated research program of global inventories, revisionary taxonomy and phylogenetic analysis, and electronic access to this knowledge via a biodiversity information system. It also involves setting priorities for research that benefit all nations, not just a few, which can be accomplished within a time-line that meets the needs of countries over the short term.

It is impossible to identify, classify, or understand the biota of one nation without also knowing and studying the same or related taxa from other countries. Assessing variation within species or the diversity of characters among species necessitates access to correctly identified specimens of all species of the group under study world-wide.

Because it is impractical to house a complete collection of the world's species in each country, comparative systematic research will rely on a limited number of institutions focused on particular groups and networked electronically and programmatically. By creating such centers with special expertise on specific taxa, resources for research can be concentrated at that location. Such resources would include collections, systematic experts, specially trained support staff, and databases. This results in cost-effectiveness for original research as well as training new generations of experts, validating and updating data included in vast databases, and providing a source for taxonomic expertise and services wherever they are needed around the globe. Such laboratories typically would be located within museums and universities that at the same time support reference collections of the species indigenous to their own countries. Because no country can support experts on all groups living within its borders, each country must depend on a network of international scientists in order to have duplicate, accurately identified specimens of as many of their species as possible.

For many branches of science, such global vision and international cooperation are taken for granted. Imagine astronomical, seismological, or global change research without data gathered world-wide or without relatively open exchange of data and scientists among nation states. In each case, data must be gathered at many sites around the globe and integrated in order to make sense of the local data. At the same time, budgets for such "big science" are assumed to be a shared, multinational responsibility. Similarly, an understanding of the Earth's species diversity will require no less than a globally conceived research effort.

Acceptance of the premise that research should be organized around particular groups affects virtually every aspect of research in biodiversity. Research centers should be structured and staffed so as to maximize and take full advantage of accumulated knowledge of a taxon. Field inventory work must be organized to document all species of a group throughout its geographic range rather than to concentrate efforts at one or a few study sites. Ultimately, this information needs to be summarized, interpreted, and communicated—in either printed or electronic form—through comprehensive monographs. Databases should be kept current and accurate by locating them where experts, libraries, and voucher specimens exist. In each case, research, funds, and personnel would be organized around the unique requirements for the study of specific taxa.

What kinds of research infrastructure are necessary in order to mount an effective scientific response to the biodiversity challenge? While existing numbers of specialists, institutions, and funding sources are grossly inadequate to address the biodiversity challenge, there is little doubt about what needs to be done or that the organizational components of an effective scientific program exist. Theoretical advances in systematic biology over the past 3 decades have revolutionized and rejuvenated the field, arming it with the ideas and methods appropriate for the exploration and analysis of biodiversity (e.g., Eldredge and Cracraft, 1980; Forey et al., 1992; Nelson and Platnick, 1981; Schoch, 1986; Wiley, 1981). Collection and data management practices have incorporated modern computational capabilities so that they are prepared for the acquisition of large numbers of specimens and vast quantities of data. The tragic erosion of systematic expertise, moreover, could be reversed in a single generation of scholars, given adequate resources to recruit, train, and employ such scientists.

PLANETARY EXPLORATION: TAXON INVENTORIES

The process of discovering, discriminating, and describing the world's species is an enormous undertaking, amounting to the scientific exploration of all forms of life on an entire planet (Raven and Wilson, 1992). In order to accomplish this ambitious undertaking efficiently and rapidly, it is critical that such work be approached so as to take full advantage of current taxonomic expertise and research resources. What does this mean?

Just as systematic research is focused on taxa, as opposed to geopolitical areas or ecosystems, so too an efficient scientific program must be aimed at generating systematic knowledge of Earth's species. Scientists studying a particular group must collect specimens from wherever on the globe such organisms live. Short-term or intermittent study sites in diverse geographic locations are most appropriate for meeting the global needs for systematic knowledge. In essence, systematics views biodiversity differently than experimental or functional biology by making broad comparisons across all species of a clade. The following, then, represent top priorities:

• **Action Item 1:** Provide for diverse world-wide inventorying efforts, each of which is directed at one (or several related) taxa, making full use of specialized expertise and research resources.

• **Action Item 2:** Increase the effectiveness of the inventorying effort, and the systematic research derived from it, by expanding support for the construction, improvement, management, and growth of natural history collections in museums, botanical gardens, and universities. These should include institutions that build comprehensive, global research collections as well as reference collections that hold representatives of the species indigenous to each nation.

• **Action Item 3:** Establish international networks of taxonomic experts with the goal of ensuring that there is open access to expertise for every group of organisms somewhere in the world. Each country should have sufficient systematics expertise to coordinate cooperative international research and to link with and interpret existing data.

DOCUMENTING DIVERSITY: MONOGRAPHY, REVISIONARY TAXONOMY, AND PHYLOGENETIC ANALYSES

The scientific documentation of biodiversity goes far beyond the description of species, and ultimately extends to providing access to all that is known about related species in larger taxa as well as enabling a logical classificatory scheme to be constructed from which predictions about their properties, distributions, and attributes can be made.

Drawing on powerfully predictive phylogenetic analyses, monographers pull together, analyze, and interpret all that is known about the species of a taxon, elucidate the clade's evolutionary history, and produce a classification that makes data easily retrieved and understood. Consequently, the following is imperative:

- **Action Item 4:** Competitive research funds should be made available to support comprehensive, comparative revisions and monographs that are focused on taxa at all levels from individual genera to phyla.

ACCESS TO KNOWLEDGE ABOUT BIODIVERSITY

Because more and more nations, managers of natural resources, and scientists will make ever-increasing demands for accurate and detailed knowledge about biodiversity, it is critical that what is learned from this global inventory effort and subsequent systematic analysis be accessible as part of the world's emerging information highway. Much of the information on biodiversity that is housed in the world's collections remains inaccessible electronically. Furthermore, descriptions of the approximately 1.5 million existing species are widely scattered in the technical literature, making an accurate count of species, not to mention retrieval of the descriptions themselves, problematic. Two hundred million specimens exist in natural history collections in the United States alone (Edwards et al., 1985), yet databases permitting access to their data do not yet exist. Therefore, the following must receive priority attention:

- **Action Item 5:** A network of world-wide taxon databases should be created which make specimen-based knowledge accessible and guarantee that those data are maintained by institutions possessing the collections, libraries, and taxonomic specialists necessary for ensuring their integrity over time.
- **Action Item 6:** An international effort should be made to electronically capture the specimen-based information in major natural history collections of the world and to make that information freely available to all nations for their use and benefit.

HUMAN RESOURCES AND TRAINING:
MEETING NEEDS FOR TAXONOMIC EXPERTISE

A major challenge facing the study of biodiversity is the creation, reestablishment, and expansion of taxonomic expertise on neglected taxa, thus confronting the "taxonomic impediment." Evidence from around the world indicates that taxonomic expertise has declined and continues to do so at just the time when the world is expressing its desire for credible taxonomic knowledge (e.g., Edwards et al., 1985; National Science Foundation, 1990). In the United Kingdom, for example, about one-half of the botanical taxonomists teaching in leading universities have disappeared in recent years (House of Lords, 1991). A recent survey of universities in the United States which offer doctoral degrees in systematic entomology shows that 60% of existing faculty have an average of one Ph.D. student and the remaining 40% have none (Daly, 1995). In many universities, faculty positions in systematics have been replaced by non-taxonomic, often molecular, ones (Holden, 1989; Nash, 1989). And even in museums, efforts to be "modern" and compete for limited research dollars have resulted in the de-emphasis of traditional taxonomic research, particularly monography.

Steps must be taken immediately to train and support experts on diverse taxa, to clarify the unique responsibility of museums, herbaria, and universities to conduct taxonomic research, and to educate administrators to understand that good biology involves asking and answering questions at many levels of organization using a wide range of techniques. Each nation and its systematics institutions must accept responsibility for facilitating taxonomic work that is not only relevant to its own interests and objectives but which also contributes to the global knowledge-base.

Despite the urgency for additional systematic expertise, the contributions of the limited number of taxonomists working today can be increased immediately and significantly to meet the demand for knowledge about biodiversity if given the requisite support. Support personnel who are knowledgeable about the group being studied can multiply the productivity of single or teams of scientists. Collectors, preparators, curatorial assistants, illustrators, database managers, laboratory technicians, and others can perform tasks that today occupy a significant proportion of systematists' time. Such personnel, familiar with common species, proper collecting protocols, and procedures for handling specimens, also could extend the field activities of a taxonomist to many sites simultaneously.

EPILOGUE: AN OPTIMISTIC FUTURE

Systematics Agenda 2000 must be seen for what it is: big science. It is nothing short of planetary exploration, requiring the resolve to understand the biota of Earth for the future benefit of humanity and the conservation of the natural world. The use of biodiversity already contributes trillions of dollars to the world economy through the goods and services it provides. Thus, it is not only a good investment to have scientific knowledge about the world around us, it is a matter of survival.

Because many aspects of taxonomic research are not perceived as being expensive or technologically intensive, it has been falsely assumed that the missions of systematics do not meet the contemporary criteria for status as "big science." Yet the missions of systematic biology constitute an immense and complex scientific enterprise, ultimately accounting for the biodiversity of the present and past world (see Systematics Agenda 2000, 1994a,b). The dimensions of the biodiversity crisis are immense, and the extinction of several million species over the next few decades seems certain unless nations respond forcefully. This, and the lack of sufficient scientific information to confront the biodiversity crisis with effective policies, contributes to a sense of doom. We are, nonetheless, cautiously optimistic. Whereas taxonomic expertise has declined sharply, the field is more intellectually vibrant and exciting than ever, and existing scientists are eager to train a new generation of experts. With adequate resources for the kind of research taxonomists do, positions would open, and students would gravitate to the field. Given successful examples of the knowledge and understanding that would emerge from world-wide inventories of targeted taxa, benefits to nations participating in this global exploration of biodiversity would become evident.

In the industrialized countries, at least, construction of a massive scientific infrastructure from scratch to meet the biodiversity challenge would not be necessary. A first step would be to expand, amplify, and support more fully those scientists and infrastructure that are already in place and to make accessible data from the hundreds of millions of specimens already housed in our museums. Recent advances in the theory of systematics have been rapid and profound, giving a promising conceptual context for understanding the origin and diversification of organisms on Earth. New and diverse sources of comparative data have been developed, extending and enhancing our means of critically testing phylogenetic hypotheses. The expanding capacities of computational tools provide hope for managing facts about tens of millions of species around the globe. Systematic biologists thus know what must be done in response to the biodiversity challenge and have the conceptual and practical tools to accomplish it. Society now needs the courage and foresight to invest in the growth of the fundamental taxonomic knowledge that will make scientifically informed decisions about resource management and conservation possible in the future.

REFERENCES

Cracraft, J. 1991. Patterns of diversification within continental biotas: Hierarchical congruence among the areas of endemism of Australian vertebrates. Aust. Syst. Bot. 4:211-227.

Cracraft, J. 1995. The urgency of building global capacity for biodiversity science. Biodiv. Conserv. 4:463-475.

Daly, H. V. 1995. Endangered species: Doctoral students in systematic entomology. Amer. Entomol. 41:55-59.

Edwards, S. R., G. M. Davis, and L. I. Nevling. 1985. The Systematics Community. Association of Systematics Collections, Lawrence, Kans.

Eldredge, N., and J. Cracraft. 1980. Phylogenetic Patterns and the Evolutionary Process. Columbia University Press, N.Y.

Forey, P. L., C. J. Humphries, I. L. Kitching, R. W. Scotland, D. J. Siebert, and D. M. Williams. 1992. Cladistics: A Practical Course in Systematics. Clarendon Press, Oxford, England.

Holden, C. 1989. Entomologists wane as insects wax. Science 246:754-756.

House of Lords. 1991. Systematic Biology Research. Select Committee on Science and Technology, First Report. HL Paper 22-I. HMSO, London.

Janzen, D. H. 1993. Taxonomy: Universal and essential infrastructure for development and management of tropical wildland biodiversity. Pp. 100-113 in O. T. Sandlund and P. J. Schei, eds., Proceedings of the Norway/UNEP Expert Conference on Biodiversity. NINA, Trondheim, Norway. 190 pp.

Janzen, D. H., and W. Hallwachs. 1994. All Taxa Biodiversity Inventory (ATBI) of Terrestrial Systems. A Generic Protocol for Preparing Wildland Biodiversity for Non-damaging Use. Report of National Science Foundation Workshop, 16-18 April, 1993, in Philadelphia. National Science Foundation, Washington, D.C. 132 pp.

Langreth, R. 1994. The world according to Dan Janzen. Pop. Sci. 245:78-82, 112, 114-115.

Mayr, E. 1963. Animal Species and Evolution. Harvard University Press, Cambridge, Mass.

Myers, N. 1991. Tropical forests: Present status and future outlook. Climate Change 19:3-32.

Nash, S. 1989. The plight of systematists: Are they an endangered species? Scientist (16 Oct):7.

National Research Council. 1993. A Biological Survey for the Nation. National Academy Press, Washington, D.C.

National Science Board. 1989. Loss of Biological Diversity: A Global Crisis Requiring International Solutions. National Science Foundation, Washington, D.C.

National Science Foundation. 1990. Systematic Biology Training and Personnel. U.S. Higher Education Survey, No. 10. National Science Foundation, Washington, D.C.

Nelson, G., and N. Platnick. 1981. Systematics and Biogeography: Cladistics and Vicariance. Columbia University Press, N.Y.

Raven, P. H., and E. O. Wilson. 1992. A fifty-year plan for biodiversity studies. Science 258:1099-1100.

Schoch, R. M. 1986. Phylogeny Reconstruction in Paleontology. Van Nostrand Reinhold, N.Y.

Schrock, J. R. 1989. Pre-graduate education in systematics and organismic biology. Assoc. Syst. Coll. Newsletter 17:53-55.

Systematics Agenda 2000. 1994a. Systematics Agenda 2000: Charting the Biosphere. Technical Report. Systematics Agenda 2000, a Consortium of the American Society of Plant Taxonomists, the Society of Systematic Biologists, and the Willi Hennig Society, in cooperation with the Association of Systematics Collections, N.Y. 34 pp.

Systematics Agenda 2000. 1994b. Systematics Agenda 2000: Charting the Biosphere. Systematics Agenda 2000, a Consortium of the American Society of Plant Taxonomists, the Society of Systematic Biologists, and the Willi Hennig Society, in cooperation with the Association of Systematics Collections, N.Y. 20 pp.

Wheeler, Q. D. 1995a. The "old systematics": Phylogeny and classification. Pp. 31-62 in J. Pakaluk and S. A. Slipinski, eds., Biology, Phylogeny, and Classification of Coleoptera: Papers Celebrating the Eightieth Birthday of Roy A. Crowson. Muzeum i Instytut Zoologii PAN, Warszawa, Poland.

Wheeler, Q. D. 1995b. Systematics, the scientific basis for inventories of biodiversity. Biodiv. Conserv. 4:476-489.

Wiley, E. O. 1981. Phylogenetics. John Wiley and Sons, N.Y.

CHAPTER
29

Museums, Research Collections, and the Biodiversity Challenge

LESLIE J. MEHRHOFF

*Supervising Biologist, Connecticut Geological and Natural History Survey;
and G. Safford Torrey Herbarium, Department of Ecology and
Evolutionary Biology, University of Connecticut, Storrs*

A few years ago a young student asked a very straightforward question in response to my statement that there were about 1.5 million species currently known to science and that there could be anywhere from 5 to 30 million species on this planet. He wanted to know who counted them. I explained how Wilson (1988) arrived at the number of 1.5 million and that seemed to suffice. In retrospect, I missed a chance to explain the importance of museum collections in documenting global biodiversity.

Museums are important tools for inventorying our planet's biological diversity, second only to the discerning eyes, love of field work, and innate senses of the naturalists[1] who collect the specimens. Natural history collections throughout the world house over 2 billion specimens (Duckworth et al., 1993). Many of these specimens represent the 1.5 million species currently known to science. Most museums have specimens that await processing and some of these may be new to science. Since it is possible that there may be as many as 100 million species of organisms in the world (Wilson, 1992), the specimens already in mu-

---

[1] The term "naturalist" is used here in the classical sense of a field biologist who is knowledgeable about the systematics and ecology of organisms and their environment and who gathers data in order to speculate on scientific problems. Examples such as Charles Darwin, William Beebe, William Morton Wheeler, or Edward O. Wilson quickly come to mind. I do not mean naturalist in the sense of park naturalists (who would be better called interpreters) or people who simply enjoy the outdoors and who are really nature-lovers. If "naturalist" be thought of as an archaic or atavistic, if not somewhat derogatory, distinction, I suggest Mehrhoff (in press), Wheeler (1923), or Wilson (1992:243, 1994b) be consulted. Perhaps the late Raymond Fosberg (1972:633) said it best: "The better scientist may be the one who can validly claim to be both systematist and ecologist, better called a naturalist."

seum collections may represent only a portion of the Earth's biological diversity. Inventories must be conducted, specimens collected, identified, and—if new to science—named and classified, and the current museum backlog must be processed if we are going to have a complete understanding of biodiversity. There are daunting tasks ahead for natural history museums, research collections, and the curators, systematists, and staff who are responsible for them.

There are two types of natural history museums. To most biologists, the word "museum" means research collections. These museums are the subject of this chapter. The kind of natural history museum with which most people are familiar, however, emphasizes public exhibits. The importance of this kind of natural history museum in inspiring young naturalists or in educating the millions of people who visit them annually should not be overlooked.

Outreach and education are components of most modern research museums. Museum exhibits depict biological diversity and serve as a valuable tool for stimulating an interest in natural history as well as educating people about many things. These are important and necessary services in the name of biological diversity. Not the least of the important messages is the challenge of maintaining global biodiversity.

No one can forget his or her first visit to a natural history museum. Most of us were fortunate to visit a large museum when we were young and impressionable. For some it was the dinosaurs that were the most memorable, for others the cases of beautiful butterflies. Still others marvelled at the birds, or beetles, or turtles. The diversity of interests were bounded only by the breadth of the museum's collections and exhibits.

The natural history museum was a place to which the many youngsters who grew up with an interest in natural history could return to see specimens of animals that fascinated them. For those of us who grew up in northern, temperate climates where much of the biota becomes dormant each winter, the museum represented a mental sanctuary where we could continue to absorb more natural history until the spring thaw once again brought an abundance of life to the fields and woods.

Museums shaped the lives of countless young naturalists whose interests have withstood the test of time and the rigors of education. Unfortunately, many budding careers were derailed later in life for any of a thousand different reasons. I suspect, however, that memories of those early visits to museums are still in the subconscious of those who chose different roads, and can be brought out, dusted off, and the interests rekindled.

Research natural history museums are collections of specimens that document the diversity of organisms that exist, or have existed, on this planet[2]. These usually are associated with research institutions (e.g., the Academy of Natural

---

[2]These collections may be botanical (usually referred to as herbaria), including vascular or non-vascular plants, or zoological, including vertebrates and invertebrates. Other kingdoms of organisms also can be included.

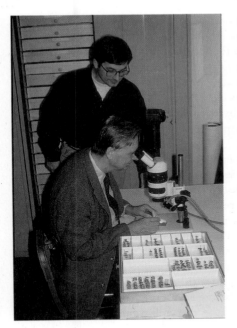

Professor Edward O. Wilson with museum staff Stefan P. Cover in the Entomology Division of the Museum of Comparative Zoology, Harvard University.

Sciences of Philadelphia, the Smithsonian Institution, the Missouri Botanical Garden, and the American Museum of Natural History) or major universities (e.g., Harvard University's Museum of Comparative Zoology and Gray Herbarium, the University of Kansas' Museum of Natural History, and University of California at Berkeley's Jepson Herbarium and Museum of Vertebrate Zoology). Often the lines between research museums and universities are not clear, with curators from the museums serving joint appointments on the university faculty and vice versa.

The building blocks of any museum are its specimens. Specimens can be of whole organisms or parts, such as bones, fur, fruits, or spores. Specimens are mounted on microscope slides, kept in alcohol, or preserved in more esoteric ways. Museums have a diversity of types of specimens and have developed sophisticated ways of preparing and maintaining the specimens. There are many helpful books, pamphlets, or other aids for the preparation of specimens. Much time and effort usually has gone into the preparation and curation of the final specimens.

Some of the larger research museums have millions of specimens. In addition, there are many smaller collections, usually at colleges and universities, whose numbers of specimens are not as large but are also important. These smaller collections frequently specialize in particular taxonomic groups or geographic areas and have curators or faculty who are specialists in these areas.

The unique value of collections to conservation is that they are an irreplaceable library of knowledge on the diversity of life. Each specimen is part of a composite picture. They can be used to document biodiversity over both time and space. If maintained properly, specimens in every collection provide a permanent record of life on Earth.

FROM CABINETS OF CURIOSITIES TO MODERN NATURAL HISTORY MUSEUMS: A BRIEF HISTORY

In Europe, the collection of natural history specimens for the purpose of displaying the variety of living organisms in museums began in earnest in the sixteenth century. The first natural history museums belonged to wealthy individuals who maintained them both as a hobby and as an indication of social standing. Wealthy patrons supported these collections because it was fashionable to do so or because of *noblesse oblige*, not to facilitate the dissemination of knowledge. Even in the sixteenth century, knowledge of the living world represented power. Many early collectors were doctors, clerics, or teachers. Their specimens either were commissioned by the wealthy or sold by subscriptions. Social interests provided economic support for naturalists and their cabinets of curiosities (Bowler, 1992).

The utilitarian basis for these museums was to show what God had placed on Earth to benefit mankind. There were strong religious overtones in the display of God's creations (Bowler, 1992). To study nature was to worship God's work that was manifest on Earth. There was no need to collect or display multiple specimens because of the fixity of nature in God's creations. For the most part, there was little or no recognition of diversity within a species. In addition, the fact that many collections had few individuals of the same species was probably a practical consideration of space and costs of maintenance as well.

Exploration in the seventeenth, eighteenth, and nineteenth centuries produced many plants and animals never before seen by Europeans. Many of the early museums in Europe were built from the vast collections brought back from abroad by voyages of exploration (Mayr, 1982). By the last half of the eighteenth century, the number of known species was becoming cumbersome for naturalists. The establishment of a species concept and binomial nomenclature were of paramount importance to natural history (Mayr, 1946). Specimens had to be described and classified to be of use to naturalists (Bowler, 1992). The naturalists in Europe who described and classified species had to rely on specimens in museums because they had little or no field experience with these organisms. The best known of the early naturalists, Carolus Linnaeus[3] (1701-1778) built a highly respected collection representing the world's known biota. His students and correspondents sent back specimens from all around the globe (Blunt, 1971).

The Procrustean explanations of some religious doctrines were beginning to be challenged. Even the religious Linnaeus questioned how the 5,600 species of animals that he had named, not to mention all others then known, could have been saved in Noah's Ark. In addition, distributions of animals could no longer be made to fit with Biblical interpretations. The diversity of vertebrates alone made a literal translation of Genesis and the account of the "Great Flood" untenable. The Noachian story was abandoned for scientific reasons (Browne, 1983).

---

[3] When Linnaeus died, his personal herbarium and library were offered to Sir Joseph Banks, but later were sold to James Smith. His specimens and books were used to found Britain's oldest natural history society in 1788, the Linnean Society of London. After Smith left London, the Linnean Society purchased the collection from him. The fact that anyone was willing to pay for a collection of dried plants was indicative of the value that naturalists placed on specimens (Bowler, 1992).

In the United States, Charles Willson Peale's natural history museum opened in Philadelphia in 1786. It attracted the public, who wished to see wonders from the natural world, more than naturalists, who wished to study organisms. Many of the early museums had a decidedly "circus side-show" flair (Barber, 1980).

Natural history museums prospered in the nineteenth century because the people who came to see them were interested in the bizarre, strange, unfamiliar, or unknown (Barber, 1980). Exploratory voyages returned from far-off lands loaded with specimens that differed from anything previously known. Many novelties were put on public display long before they were interpreted by naturalists. These were the harvest days of museums (Goode, 1901a).

While museum visitors viewed the public displays, naturalists relied heavily on museums to show similarities and differences between species. Taxonomic research on the diversity of organisms was done in museum collections. Naturalists could no longer be expected to be familiar with all known species (Huxley, 1861). As the number of species grew, naturalists were forced to specialize (Mayr, 1946).

Many natural history museums were started in larger cities of the United States during the nineteenth century. Philadelphia's Academy of Natural Sciences was established in 1812 (Bennett, 1983). The Boston Society of Natural History began in 1830 (Creed, 1930). The Smithsonian Institution was created in 1846 and by 1850 included 6,000 specimens (Rivinus and Youssef, 1992). The American Museum of Natural History opened in 1877 (Preston, 1986).

The National Museum of Natural History, Smithsonian Institution, Washington, D.C.

The Smithsonian Institution's first Secretary was Joseph Henry, who was not overly enthusiastic about collections (Dupree, 1957). His assistant secretary, Spencer F. Baird, however, was a collector, and was good at persuading others to collect for him (Yochelson, 1985). He was the champion of a national museum and did much to catapult it into international prominence (Rivinus and Youssef, 1992). The term "National Museum" was not used for the collections until 1851 (Goode, 1901b). The first collections came to the Smithsonian from the Patent Office's "National Cabinet of Curiosities," where they had been moldering for years (Dupree, 1957). The original natural history museum was housed in the Smithsonian "Castle" in Washington, D.C., but a new building, the present National Museum of Natural History, was opened across from the Castle in 1909 (Yochelson, 1985).

Swiss-born Louis Agassiz came to the United States in 1846 to deliver a series of lectures at the Lowell Institute in Boston. His charismatic nature and his reputation as a naturalist so impressed the people of Boston that they encouraged him to stay. He remained and taught at Harvard University from 1848 until his death in 1873 (Lurie, 1960). Agassiz was a collector who envisioned a natural history museum in Cambridge, Massachusetts, that would resemble the best European museums. His museum would illustrate patterns of similarities in nature. He would have complete control over this museum, allowing him to teach zoology the way he thought it should be taught. Agassiz's method of studying nature was to amass a large collection of individual specimens in order to make comparisons. Agassiz reasoned that many specimens of the same taxon were needed in order to truly interpret natural history (Winsor, 1991). The Museum of Comparative Zoölogy, "Agassiz's Museum," opened in 1859. It is a monument to his far-sighted vision.

By the end of the nineteenth century, many large collections of natural history specimens existed in the United States, rivaling the holdings in European museums. Many universities had large research museums that were used by biologists—the name biologist having replaced the earlier professional name of naturalist—to study and classify organisms. These collections formed the basis for an increasing knowledge of global biological diversity.

Interest in museums, biological research, and the value to education of their collections was apparent by the beginning of the twentieth century. In 1903, the New York State Museum published a catalog of the natural history museums of the United States and Canada (F. Merrill, 1903). This catalog resulted from a survey of both large and small museums in order to fill the "lack of general and specific information concerning the natural history museums of this country and their collections." Two hundred sixty-four museums from the United States and Canada responded to the survey. The catalog attempted to inventory the holdings of each museum as to kinds of specimens they maintained (botanical, vertebrate-skin, skull, liquid, etc.), numbers of each kind, and important historical collections contained in the museum.

The beginning of the twentieth century also saw an expansion of interest in the world's tropical regions. Many universities and museums in temperate countries began detailed floristic and faunistic surveys of tropical natural history. These yielded numerous publications and many thousands of specimens to be identified, classified, and curated. World War II brought another resurgence of interest in the biology of tropical countries. During the war, many biologists were sent to tropical countries to collect new or better sources of medicines. Biologists collecting in tropical countries brought back specimens to major museums (E. Merrill, 1946).

Today, museums that house research collections are an invaluable and irreplaceable source of knowledge on the flora and fauna of the world. New collections are being added while the older, historically significant specimens are being maintained for research. This represents a large scientific and financial responsibility for the museums. These collections are the very basis of our knowledge of biodiversity.

GOVERNMENT BIOLOGICAL SURVEYS: STATE NATURAL HISTORY SURVEYS

An important component for inventorying the biological diversity of the United States are the state and national biological surveys. These surveys have collected many specimens that help document biological diversity. Although their current roles are expanded greatly beyond museum collections, they still maintain an active interest in their collections. Their specimens increasingly are used in conservation work.

Many states established state natural history surveys to inventory their respective biotas. Some of the early surveys were combined with geological surveys of the state's mineral resources (Socolow, 1988). The earliest of these was the New York State Geological and Natural History Survey. The New York Survey was established in 1836 and later was split into a biological survey and a geological survey (Miller, 1986). Some state geological surveys provided much biological information. A well-known example was the early California Geological Survey, which began in 1860 (Brewer, 1966). Other state natural history surveys were disbanded, reorganized, or shifted to other programs. The Illinois Natural History Survey, by far the biggest and most active of remaining state natural history surveys, began in 1858. Connecticut's State Geological and Natural History Survey, begun in 1903, is one of two active surveys that includes both biology and geology. The youngest state survey is the Rhode Island Natural History Survey. It was initiated in 1993 and incorporated in May 1994. Unfortunately, there are less than a dozen active state biological surveys currently in existence.

Most of the remaining state biological surveys perform a variety of activities within the state (Risser, 1986). They conduct inventories of species and

TABLE 29-1   Consortium of Biological Surveys

Name	Year Established	Government Affiliation
Connecticut Geological and Natural History Survey	1903	Department of Environmental Protection
Hawaii Biological Survey	1992	Bernice P. Bishop Museum, and State Museum of Natural and Cultural History
Illinois Natural History Survey	1858	Department of Energy and Natural Resources
Kansas Biological Survey	1959[a]	University of Kansas
New York State Biological Survey	1836	New York State Museum, and State Education Department
North Carolina Biological Survey	1976	North Carolina State Museum of Natural Resources
Ohio Biological Survey	1912	The Ohio State University
Oklahoma Biological Survey	1988[b]	University of Oklahoma
Pennsylvania Biological Survey	1988[c]	Private Consortium
Rhode Island Natural History Survey	1993	Private Consortium

[a]Initiated in 1911.
[b]Initiated in 1927.
[c]Initiated in 1979.

habitats and publish the results. Most surveys function as clearinghouses for information for other state agencies, academia, and the public. Publishing has been a major part of many state biological surveys. A few surveys are involved in environmental review or provide management recommendations. The State Heritage Program is part of the state survey in Connecticut, Kansas, and Oklahoma. Most states maintain collections that document their state's biological heritage. Some of the state surveys, such as those of Hawaii and New York, are associated with state natural history museums and share collections. Others, such as those in Kansas, Oklahoma, Ohio, and Illinois, are associated with the state university and deposit their specimens in the university's research collections.

In 1993, 10 state biological surveys formed a Consortium of State Biological Surveys (see Table 29-1). These surveys have a unifying interest in collections and common goals of inventory, research, and dissemination of information. At its initial meeting held in Columbus, Ohio, in December 1993, the Consortium passed a resolution to support the goals of the National Biological Survey[4] and to work with the National Biological Survey to make the best use of their existing information and collections. It was exciting that both the oldest and the newest state surveys were founding members of the Consortium.

---

[4]The Secretary of the Interior, Bruce Babbitt, recently has changed the name to the National Biological Service.

GOVERNMENT BIOLOGICAL SURVEYS:
THE NATIONAL BIOLOGICAL SURVEY

Biologists and conservationists were encouraged by the announcement of the Secretary of the Interior, Bruce Babbitt, of the formation of a National Biological Survey in 1993. His intention was to unite the biological research that currently is under way within the eight agencies of the Department: the Bureau of Land Management, Bureau of Mines, Bureau of Reclamation, Fish and Wildlife Service, Geological Survey, Minerals Management Service, National Park Service, and Office of Surface Mining (Corn, 1993). One of the primary objectives of the National Biological Survey is "to develop comprehensive ecosystem management strategies . . ." (Corn, 1993:CRS-1). An additional objective is to "give land and resource managers more timely, objective scientific information essential for decision-making . . ." (Corn, 1993:CRS-1).

This was really the second beginning of a National Biological Survey. The earlier national survey mostly has been forgotten, even by the scientific community. Although it had a convoluted nomenclatural history, it played an important role in attempting to understand the natural history of the United States.

In 1885, Congress created a Section of Ornithology within the U.S. Department of Agriculture's Division of Entomology. Within a year, this became the Division of Ornithology and, by 1888, it was officially called the Division of Economic Ornithology and Mammalogy (Osgood, 1943). "Economic" usually was dropped from the title, simplifying it to the Division of Ornithology and Mammalogy. This was probably at the insistence of its first director, C. Hart Merriam, who favored the subordination of the economic nature of the Division's work to the scientific (Cameron, 1929). By 1889, Merriam was advocating the establishment of a "systematic Biological Survey." Finally, in 1896, Congress transformed his agency into the Division of Biological Survey. This gave official recognition to what Merriam and his staff had been doing for almost a decade (Sterling, 1974).

In 1905, the title of the Division again was changed, this time to the Bureau of Biological Survey. In 1940, the Bureau was transferred to the U.S. Department of the Interior, where it was joined with the former Fish Commission to become the Fish and Wildlife Service (Osgood, 1943).

The Bureau of Biological Survey was under the direction of Merriam from 1885 until 1910, when he stepped down to pursue independent research. He felt that the Bureau should be primarily a research program aimed at gathering information on the biology of the United States. He was best known for his pioneering work on "life zones." Merriam led by example, much preferring to spend his summers in the field and winters in Washington, D.C. (Osgood, 1943). Merriam was a prodigious collector of natural history specimens, first of birds and eggs and later—mostly during his work with the Bureau of Biological Survey—of mammals. He began collecting in 1870 when he was 15 years old, and by 1940 his mammal collection was reported to contain 136,613 specimens with full data and in prime condition (Osgood, 1943).

While there seemed not to be a direct governmental mandate to collect specimens, Merriam's field method was to collect mammals, birds, reptiles, and amphibians from a study site. Representative plant specimens were taken, as were photographs when possible. All these served as the basis for the reports he was compiling. This method later was adopted by other members of the Bureau (Sterling, 1974). Many specimens collected under the auspices of the Bureau of Biological Survey are in the Smithsonian's Natural History Museum (Hampton, personal communication, 1994).

The current National Biological Survey hopes to increase the degree of communication and collaboration between federal agencies and the museum community (National Biological Survey fact sheet, no date). The U.S. Department of the Interior signed a memorandum of understanding with the Association of Systematic Collections in February 1994. The memorandum recognizes the common mission and mutual interest in collections and biological inventory. The National Biological Survey and the Association of Systematics Collections, which represents member institutions, societies, and individuals, have formed a working group to determine the policies of the National Biological Survey regarding museums and collections (Hoagland, personal communication, 1994). Furthermore, they hope to ensure that specimens gathered during the work of the National Biological Survey will be accessioned and curated efficiently (National Biological Survey fact sheet, no date). The National Biological Survey may be able to encourage inventory of the biotic resources of the United States and support basic taxonomic work on priority groups.

SYSTEMATISTS AND THE USE OF COLLECTIONS

The two primary roles of the research collection are education and documentation. I have divided these roles into five greatly overlapping functions: systematics research, documentation in space and time, identification, education, and specimen maintenance (Mehrhoff, in press). Every museum that maintains research collections, large or small, must serve these roles and functions if it is to continue to play an important part in the preservation of biological diversity.

Each specimen in a museum is a data set of useful information. Information can be gleaned on identification, classification, morphology, phenology, and distribution. Collections also meet the needs of applied biology, such as health sciences, agriculture, resource management, and biotechnology (Systematics Agenda 2000, 1994).

In addition to the value of the specimen for systematics and scientific research, there are other, less frequently considered, values of collections (Mayr and Goodwin, no date). Museum collections are the only place to see examples

A researcher sorting mammal specimens.

of extinct species. If we are to understand where these taxa fit in classification schemes, or reevaluate phylogenetic relationships, we must have the specimens. Given the backlog of material waiting to be curated in many collections, one wonders how many of Wilson's "centinelan extinctions[5]" will come to light. Specimens also document a previously existing biota. This documentation also can work for recent arrivals (Mehrhoff, in press). Specimens represent expenditures in obtaining them, especially if they were collected in inaccessible areas (Mayr and Goodwin, no date).

Specimens must be properly curated and maintained for all of these reasons. If specimens are lost, damaged, or discarded, the value of the information they represent will be severely decreased. Stewardship of collections should be a primary concern for everyone in charge of natural history collections. This is very nicely addressed in a project report of the National Institute for the Conservation of Cultural Property (Duckworth et al., 1993). Every curator should have a copy of this report.

---

[5]Silent, undocumented extinctions—named after a cloud-forested ridge in the western foothills of the Andes that was cleared for farming, causing the extinction of a large number of rare, endemic species within a few months. The extinctions of Centinela were observed only through an accident in timing, when Alwyn Gentry and Calaway Dodson of the Missouri Botanical Gardens happened to visit the site immediately before its demise (Wilson, 1992:243).

Museum collections represent cumulative knowledge, with each museum representing a piece of that knowledge. No museum exactly duplicates the holdings of any other collection. A network of museums can make their specimens available for research through national and international loan programs. It is much more efficient to send specimens to researchers at their home institutions rather than expect the researchers to visit each museum where material is available (Quicke, 1993). The means by which electronic data can be transferred to expedite certain aspects of research should be explored (Alberch, 1993).

The most important way for a museum to address the roles of their collections is to maintain an active research and education program for students interested in careers in systematics. This would solve many of the problems revolving around the needed biological inventories. Systematists do systematics research so that the classification of known organisms can be improved and the proper placement of newly discovered taxa can be accomplished. Specimens for this systematics research, which also document biodiversity, will come into museums from naturalists in the field as well as from the systematists themselves. Students can be trained to take on parts of larger projects.

This is a simplistic solution, especially when considering the current problems facing museums. Things are always more complex than they seem. The two problems produced by this simple recommendation are space for the new specimens and curatorial costs. Some scientists are calling for a global inventory of biodiversity. Additional trained field people will be necessary. Museums will be forced to expand their holdings. New material for mounting specimens, cabinets, space, and curation will be necessary. Ultimately these all lead to the other major problem: cost. Who will pay for this education and expansion?

Space is not a small issue by itself, let alone in conjunction with cost. A few years ago, three Australian "botanical taxonomists" suggested that only type specimens and a few important vouchers were really necessary to maintain in herbaria (Clifford et al., 1990). Their suggestion was to save these limited collections, and all other material could have the data recorded into a database and then the specimens could be "pulped." There was an outpouring of disagreement in a subsequent issue of the same journal (West et al., 1990) and, fortunately, no curators seem to have followed their suggestion.

Just at a time when the world is losing species at an unprecedented rate, collections are being forced to be selective about acquisitions (Quicke, 1993). If collections swell from the influx of new specimens, then both space and curatorial costs will increase concomitantly. Clearly, these two issues may become important limiting factors in completing global inventories of biodiversity.

The curators of many collections are forced to consider unofficial policies on what to collect or accession. I formulated guidelines to help determine what to collect for the Connecticut Geological and Natural History Survey's project on state flora. I am attempting to use these same guidelines in determining which student collections should be accessioned into the G. Safford Torrey Herbarium at the University of Connecticut. While funding previously may not have been a primary consideration for curators, it now has become one of paramount importance. At the very least, major granting sources such as the National Science Foundation must increase their grants to systematists and systematic collections. Grants from other less traditional sources also must be sought. It is conceivable that curators might find new sources of money if their museum's specimens were viewed as part of our natural heritage.

Systematists must see as part of their responsibility the funding of collections holding the specimens that are necessary for their work. Ecologists and other scientists who routinely deposit voucher specimens to document their research also should support collections financially to ensure their long-term maintenance (Quicke, 1993).

Scientists must consider museums when budgeting for research work. This becomes especially important for museums in other countries. When working in developing countries, scientists should budget for extra time to be spent in those countries visiting national museums and universities in order to annotate specimens, give short-courses, or present seminars on their research. They should hire field assistants from those museums and universities. Biologists should allow for costs of extra publications resulting from their research so that they can be made available at no cost to scientists and conservationists in developing countries.

Additional sources of funds for both systematics research and for collection maintenance must come from the conservation community. In the past, conservation programs—both in government and private nonprofit organizations—have taken museum work for granted, especially when it comes to funding and support (Mehrhoff, in press). All too often, they use the collections due to the good-will of the curators and institutions. This has to change. Collections are basic not only to every scientific discipline but to the majority of conservation issues as well (Lutz, 1994). We must realize that the limited funds for conservation also must be used to assist collections and their long-term maintenance.

Interest in systematics seems to be waning just at a time when it is most necessary (Kosztarab and Schaefer, 1990; Mehrhoff, in press; Parnell, 1993; Quicke, 1993; Systematics Agenda 2000, 1994). If museums are to function as research centers, then they must be staffed by competent curators and systematists who are using their specimens. I suspect that there are few deans or department heads who could support the long-term existence of unused research collections. As we face a paucity of trained systematists, it becomes increasingly difficult to imagine who will use existing collections. Biodiversity inventories can create problems of increased needs for space and increased expenses, but they also can suggest a solution. Museum staff must attempt to meet the challenging task of preserving global biodiversity.

CONCLUSIONS

Scientists and conservationists acknowledge that we must have accurate information in order to protect biological diversity (Committee on the Formation of the National Biological Survey, 1993; McNeely et al., 1990; Norton, 1987; Wilson, 1992; World Resources Institute et al., 1992). Basic to this need is a global inventory of the world's biota (Wilson, 1992). No single person, agency, program, or institution is capable of this task. It must be a collaborative effort of Herculean proportion. While some scientists are calling for a full-scale inventory of the world's biota, others are advocating more realistic approaches (Raven and Wilson, 1992; Wilson, 1994a).

Given the magnitude and estimated costs of a global inventory, there is no room for duplication of effort. We are fortunate that research museums have been involved in biotic inventories for centuries. The specimens from these museums can tell us not only what organisms exist or have existed, but they also can give us an indication over time of distribution and, in some cases, abundance (Mehrhoff, in press).

Naturalists are needed to participate in these inventories. No matter how sophisticated equipment and techniques become, they cannot produce results without dedicated, trained scientists to conduct the field work. Naturalists of this caliber are becoming scarce. This is why the loss of Al Gentry and Ted Parker (Stevens, 1993) was so keenly felt by the scientific community (Forsyth, 1994; Hurlbert, 1994).

Systematists also are necessary to deal with the increase of specimens that will come into existing museums (Systematics Agenda 2000, 1994). The decrease in trained systematists to deal with the increased specimens obtained by this renewed interest in inventorying biodiversity is not only paradoxical; it is both noticeable and lamentable (see Lutz, 1994).

The current picture of systematics research is bleak. There presently are only about 4,000 specialists in the United States and Canada capable of classifying the organisms occurring in those countries. According to Kosztarab and Schaefer (1990), there are only a few hundred specialists who are competent to identify or conduct systematic research on some of the insect and arachnid groups in the United States. It is important to remember that most of these specialists only work part-time on the taxonomy of their group because they teach at universities (Kosztarab and Schaefer, 1990). Some have additional administrative responsibilities. Add to this the loss of positions in taxonomy at universities and in government work. If positions vacated by taxonomists are not being filled by taxonomists, who will teach the next generation of taxonomists? If we do not train students to become systematists, who will direct the inventories necessary to assess biodiversity or name the plethora of new species that will come from the specimens collected? The magnitude of the problem quickly becomes sobering.

The situation is even bleaker for the species-rich tropics. There are probably no more than 1,500 professional systematists capable of dealing with the myriad of tropical organisms (Wilson, 1992). Most of the museums that have large holdings of tropical organisms are in Europe, North America, and Australia (Alberch, 1993), adding to the difficulties in studying biological diversity in developing countries. There are more practicing plant taxonomists in Europe (about 1,000) and the United States (about 650) than in the tropics (about 520) (Parnell, 1993). The area involved is small, yet tropical species represent over 70% of the world's biodiversity.

Larger museums have a responsibility to developing countries. They must make their holdings available to researchers working in tropical groups or in tropical countries. This, however, places the financial burden on the museums to support their collections, curate their material, pay their staff, and process loans to researchers in other museums. These are all increasingly expensive in today's financial environment. Who will pay for these services? The financial conundrum becomes more complex if you consider the alternative. It would be almost prohibitively expensive for developing countries to build a comparable reference collection (Alberch, 1993). It is likely that existing museums will remain the centers for taxonomic research, so we must find ways to support their continued existence. Funding for museums, systematics, and for training future systematists must be increased. Federal and state governments and the conservation community all must do their share. The loss of any major museum or research collection is unacceptable.

One program beyond the scope of this chapter deserves special mention. Costa Rica's Instituto Nacional de Biodiversidad (INBio) should be used as a model for other countries who wish to develop an inventory of their nation's biological diversity (Valerio Gutierrez, 1992). This highly publicized program, started in 1989, employs local people as parataxonomists who are trained to collect museum specimens. These specimens later can be sorted and identified at INBio's main office and museum on the outskirts of San José. INBio is expanding its services to meet the needs for data by government agencies, scientists, educators, planners, and industry (World Resources Institute et al., 1992).

We have a long and venerable history of biological inventory that should not be overlooked. More importantly, these collections represent a composite picture of everything we currently know about the world's biodiversity. We need to increase support for systematics and collections. This increased support has to be in services, such as training new field naturalists, systematists, and curators, as well as in financial backing. The value of collections must be acknowledged by government and private conservation programs.

Museum collections must be made "user-friendly." Museum staff must do their part to encourage students, faculty, and conservationists to use the collections. In addition to the roles and functions of museums already addressed, curators must try to meet the realistic needs of conservationists.

Curators and systematists, those most familiar with the value of collections, must convince the public of their importance. Museums should increase the number of exhibits that tie their collections to the preservation of biodiversity. Too often, collections and those who work with them are taken for granted. A worthwhile effort would be to publicize the value of collections via the popular press.

Curators and systematists must take time from their research to spread the word about biological diversity. Increased support for collections and systematic research will come only with increased understanding of the roles that collections play in protecting biological diversity.

The word "enthusiasm" comes from the Greek word "*enthousiamos,*" meaning "to inspire." What is education if it is not to inspire? Enthusiasm for natural history can excite people. Many biologists became interested in science by studying plants or animals. The only place better than in a museum to kindle an interest in natural history is the field. Museum and field research are necessary to renew interest in systematics. I suspect it was not thoughts of gene sequencing, electrophoresis, or cladistics that got many of today's systematists interested in their work. The excitement of youth that pushed many of us toward careers in systematics, inventory, or museums must be encouraged.

Most curators and systematists are enthusiastic about their work. Most people who work in museums find them exciting places, not the dark, musty images that some would paint. Since most systematists and curators spend a portion of their time collecting, students can be hired as field assistants who will learn by direct experience, especially from contagious enthusiasm. Future systematists should be encouraged by example. If more students take an interest in systematics, either from a pure perspective or as an approach to the global biodiversity challenge, then support likely will follow.

In our zeal to protect biodiversity, we cannot afford to ignore our collections. Collections of natural history specimens represent a priceless heritage as well as an increasingly important database. It is of paramount importance that no bits of this collective knowledge be lost.

ACKNOWLEDGMENTS

I wish to thank the many biologists and conservationists who have knowingly or unknowingly contributed to this paper by discussing with me over the years the roles and functions of research collections. I have acknowledged many of these in another paper on this subject and will not repeat them here. In addition, I wish to thank those who have contributed information or helped me find obscure references for this paper: Stefan P. Cover, Joseph J. Dowhan, Craig Freeman, Michelle Hampton, K. Elaine Hoagland, John Ketchum, Gordon L. Kirkland, Julie Lungren, Norton G. Miller, Robert M. Peck, Robert S. Ridgely, Gary D. Schnell, David L. Wagner, and Alfred G. Wheeler.

Jane E. O'Donnell critically proofread the manuscript and made many helpful and much-appreciated suggestions. Susan Carsten helped with word processing.

REFERENCES

Alberch, P. 1993. Museums, collections, and biodiversity inventories. Trends Ecol. Evol. 8:372-375.

Barber, L. 1980. The Heyday of Natural History, 1820-1870. Doubleday and Company, Garden City, N.Y. 320 pp.

Bennett, T. P. 1983. The history of the Academy of Natural Sciences of Philadelphia. Pp. 1-14 in A. Wheeler, ed., Contributions to the History of North American Natural History. Special Publication No. 2. Society for the Bibliography of Natural History, London.

Blunt, W. 1971. The Compleat Naturalist. A Life of Linnaeus. The Viking Press, N.Y. 256 pp.

Bowler, P. J. 1992. The Norton History of the Environmental Sciences. W. W. Norton and Company, N.Y. 634 pp.

Brewer, W. H. 1966 (1930). Up and Down California in 1860-1864: The Journal of William H. Brewer. University of California Press, Berkeley. 583 pp.

Browne, J. 1983. The Secular Ark: Studies in the History of Biogeography. Yale University Press, New Haven, CN. 273 pp.

Cameron, J. 1929. The Bureau of Biological Survey: Its History, Activities and Organization. Institute for Government Research, Service Monographs of the U.S. Government, No. 54. Johns Hopkins University Press, Baltimore, Md. 339 pp.

Clifford, H. T., R. W. Rogers, and M. E. Dettmann. 1990. Where now for taxonomy? Nature 346:602.

Committee on the Formation of the National Biological Survey. 1993. A Biological Survey for the Nation. National Academy Press, Washington, D.C. 205 pp.

Corn, M. L. 1993. Biological Surveys: Current Proposals and Past Practices. Congressional Research Service, Washington, D.C. 6 pp.

Creed, P. R. 1930. The Boston Society of Natural History, 1830-1930. The Merrymount Press, Boston. 117 pp.

Duckworth, W. D., H. H. Genoways, and C. L. Rose. 1993. Preserving Natural Science Collections: Chronicle of Our Environmental Heritage. National Institute for the Conservation of Cultural Property, Washington, D.C. 140 pp.

Dupree, A. H. 1957. Science in the Federal Government: A History of Policies and Activities to 1940. Harvard University Press, Cambridge, Mass. 460 pp.

Forsyth, A. 1994. Ted Parker: In memoriam. Conserv. Biol. 8:293-294.

Fosberg, H. R. 1972. The value of systematics in the environmental crisis. Taxon 21:631-634.

Goode, G. B. 1901a (1888). The beginnings of American science. The third century. Pp. 409-466 in A Memorial of George Browne Goode. Annual Report of the Board of Regents of the Smithsonian Institution. Report of the U.S. National Museum, Part II, Washington, D.C.

Goode, G. B. 1901b (1893). The genesis of the United States National Museum. Pp. 83-191 in A Memorial of George Browne Goode. Annual Report of the Board of Regents of the Smithsonian Institution. Report of the U.S. National Museum, Part II, Washington, D.C.

Hurlbert, K. J. 1994. A tribute to Alwyn H. Gentry. Conserv. Biol. 8:291-292.

Huxley, T. H. 1861. On the study of zoology. Pp. 94-119 in T.H. Huxley (1903), Lay Sermons, Addresses, and Reviews. D. Appleton and Company, N.Y.

Kosztarab, M., and C. W. Schaefer. 1990. Conclusions. Pp. 241-247 in M. Kosztarab and C. W. Schaefer, eds., Systematics of the North American Insects and Arachnids: Status and Needs. Virginia Agricultural Experiment Station Information Series 90-1. Virginia Polytechnic Institute and State University, Blacksburg.

Lurie, E. 1960. Louis Agassiz: A Life in Science. The University of Chicago Press, Chicago. 449 pp.

Lutz, D. 1994. More than stamp collecting. Amer. Sci. 82:120-121.

Mayr, E. 1946. The naturalist in Leidy's time and today. Proc. Acad. Nat. Sci. Philadelphia 98:271-276.

Mayr, E. 1982. The Growth of Biological Thought: Diversity, Evolution, and Inheritance. Harvard University Press, Cambridge, Mass. 974 pp.

Mayr, E., and R. Goodwin. No date. Biological Materials, Part 1: Preserved Materials and Museum Collections. National Research Council Pub. 399. National Academy of Sciences, Washington, D.C. 20 pp.

McNeely, J. A., K. R. Miller, W. V. Reid, R. A. Mittermeier, and T. B. Werner. 1990. Conserving the World's Biological Diversity. International Union for the Conservation of Nature and Natural Resources, World Resources Institute, Conservation International, World Wildlife Fund-US, and the World Bank, Washington, D.C. 193 pp.

Mehrhoff, L. J. The roles of collections in conservation biology: A pressing concern. Rhodora (in press).

Merrill, E. D. 1946. Merrilleana: A selection from general writings of Elmer Drew Merrill, Sc.D., LL.D. Chronic Botanica 10:127-394.

Merrill, F. J. H. 1903. Natural History Museums of the United States and Canada. Bull. N. Y. State Mus. 62, Misc. 1. 233 pp.

Miller, N. G. 1986. New York State Museum: Sesquicentennial of the biological and geological surveys. Assoc. Syst. Coll. Newsletter 14:25-29.

National Biological Survey. No date. Initiative: NBS Museum Initiative. National Biological Survey fact sheet, Department of the Interior, Washington, D.C.

Norton, B. G. 1987. Why Preserve Natural Variety? Princeton University Press, N.J. 281 pp.

Osgood, W. H. 1943. Clinton Hart Merriam—1855-1942. J. Mammal. 24:421-436.

Parnell, J. 1993. Plant taxonomic research, with special reference to the Tropics: Problems and potential solutions. Conserv. Biol. 7:809-814.

Preston, D. J. 1986. Dinosaurs in the Attic: An Excursion into The American Museum of Natural History. St. Martin's Press, N.Y. 244 pp.

Quicke, D. L. J. 1993. Principles and Techniques of Contemporary Taxonomy. Blackie Academic and Professional, N.Y. 311 pp.

Raven, P. H., and E. O Wilson. 1992. A fifty-year plan for biodiversity surveys. Science 258:1099-1100.

Risser, P. G. 1986. State and private legislative and historical perspectives, with comments on the formation of a National Biological Survey. Pp. 177-182 in K. C. Kim, and L. Knutson, eds., Foundations for a National Biological Survey. Association of Systematics Collections, Lawrence, Kans.

Rivinus, E. F., and E. M. Youssef. 1992. Spencer Baird of the Smithsonian. Smithsonian Institution Press, Washington, D.C. 228 pp.

Socolow, A. A., ed. 1988. The State Geological Surveys: A History. American Association of State Geologists, Washington, D.C. 499 pp.

Sterling, K. B. 1974. Last of the Naturalists: The Career of C. Hart Merriam. Arno Press, New York. 478 pp.

Stevens, W. K. 1993. Biologists' deaths set back plan to assess tropical forests. New York Times (17 August):C1,C3.

Systematics Agenda 2000. 1994. Systematics Agenda 2000: Charting the Biosphere. Technical Report. Systematics Agenda 2000, a Consortium of the American Society of Plant Taxonomists, the Society of Systematic Biologists, and the Willi Hennig Society, in cooperation with the Association of Systematics Collections, N.Y. 34 pp.

Valerio Gutierrez, C. E. 1992. INBio: A pilot project in biodiversity. Assoc. Syst. Coll. Newsletter 20:101,104-105.

West, J. G., B. J. Conn, E. A. Jarzembowski, P. F. Stevens, R. M. Harley, C. H. Stirton, L. Boulos, T.

D. MacFarlane, N. P. Singh, A. Nicholas, N. Chalmers, D. Hawksworth, D. S. Ingram, G. Long, G. T. Prance, P. H. Raven, and L. E. Skog. 1990. In defense of taxonomy. Nature 347:222-224.

Wheeler, W. M. 1923. The dry-rot of our academic biology. Science 57:61-71.

Wilson, E. O. 1988. The current state of biological diversity. Pp. 3-18 in E.O. Wilson and F. M. Peter, eds., BioDiversity. National Academy Press, Washington, D.C.

Wilson, E. O. 1992. The Diversity of Life. Belknap Press, Cambridge, Mass. 424 pp.

Wilson, E. O. 1994a. Biodiversity: Challenge, science, and opportunity. Amer. Zool. 34:5-11.

Wilson, E. O. 1994b. Naturalist. Island Press, Washington, D.C. 380 pp.

Winsor, M. P. 1991. Reading the Shape of Nature: Comparative Zoology at the Agassiz Museum. The University of Chicago Press, Chicago. 324 pp.

World Resources Institute, the World Conservation Union, and the United Nations Environmental Programme. 1992. Global Biodiversity Strategy. World Resources Institute, N.Y. 244 pp.

Yochelson, E. L. 1985. The National Museum of Natural History: 75 Years in the Natural History Building. Smithsonian Institution Press, Washington, D.C. 216 pp.

CHAPTER

30

Resources for Biodiversity in Living Collections and the Challenges of Assessing Microbial Biodiversity

RICHARD O. ROBLIN

*Associate Director for Science, American Type
Culture Collection, Rockville, Maryland*

Although not generally visible to the naked eye, ubiquitous microbes are a crucial element in the biological infrastructure of our planet. They break down complex organic materials for reuse by new generations of plant and animal cells, are central to the proper functioning of vital geochemical cycles, and their genes code for the synthesis of a wide variety of drugs that are essential to the treatment of human diseases (Bull et al., 1992). Because they spread easily, they are found in air, water, and soil, as well as on the exterior and interior surfaces of essentially all plants, birds, insects, and other animals. Microorganisms may make up as much as 25-50% of the biomass of the Earth. Because microbes are ubiquitous and functionally important, they clearly belong on the biodiversity research agenda.

In this chapter, I first review some of the authenticated microorganisms that living collections, such as the American Type Culture Collection (ATCC), can provide for use as reference standards in the assessment of microbial biodiversity. Second, I highlight some of the substantial challenges that will be encountered in attempts to even partially describe the extent of microbial diversity in even the smallest areas of the Earth. Finally, I stress the importance of collaborative transnational efforts among scientific institutions in the development of programs for measuring and developing the riches of the microbial cornucopia.

Like macroorganisms, microorganisms are classified using genus and species names. However, because the concept of a species is less definitive for microorganisms than for higher organisms, there are multiple definitions for this term in different subfields of microbiology. For bacteria, a microorganism where the concept of species is perhaps the most well-defined, I use the following definition:

> "A bacterial species may be regarded as a collection of strains that share many features in common and differ considerably from other strains. (A strain is made up of the descendants of a single isolation in pure culture, and usually is made up of a succession of cultures ultimately derived from a single colony.) One strain of a species is designated as the *type strain*; this strain serves as the name-bearer strain of the species and is the permanent example of the species, i.e., the *reference specimen for the name*. The type strain has great importance for classification at the species level, because a species consists of the type strain and all other strains that are considered sufficiently similar to it to warrant inclusion with it in the species" (Staley and Krieg, 1984:1).

As microbiologists isolated single-colony microbial strains from different habitats around the world during the last 120 years, they described the organisms and their properties in the scientific literature. More and more strains accumulated, and the field of microbial systematics developed through attempts to determine whether new microbial isolates were the same as or different from those previously described. Frozen storage of microorganismal samples was introduced as a way of minimizing the changes in properties observed when microorganisms are cultured continuously for long periods of time. As new fields of science developed, they were successively applied to determine the degree of relatedness among different microbial isolates.

Removal of frozen samples from a liquid nitrogen container.

Founded in 1925 by scientists desiring to preserve and distribute useful microbial strains, the ATCC serves as a repository of authenticated type strains, the fundamental reference materials of the system of microbial classification. As of early 1994, the ATCC had approximately 13,500 strains of bacteria (including about 2,600 type strains representing different species), approximately 26,000 strains of fungi and yeasts (including about 3,800 type cultures), and approximately 1,200 protozoan strains (including about 60 "authenticated"[1] strains). These cultures are available to the scientific community of the world for use as reference materials in studies of microbial diversity. Their availability is limited only by the requirements to obtain permits from the U.S. Department of Agriculture and U.S. Public Health Service for shipment of agricultural, human, or animal pathogens, respectively, and the requirement for a valid export license for shipment of materials outside the United States.

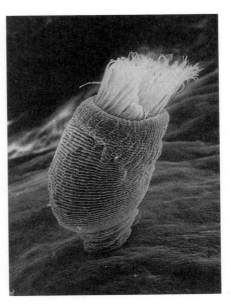

A peritrich ciliate growing on a hydroid.

Although the numbers of microbial strains (and species) in collections like those of the ATCC are large, one estimate puts the number of known bacterial species at 4,800; known fungal species at 69,000; and known protozoan species at 30,800 (Wilson, 1992). Using this estimate and the ATCC as examples, the collection contains about 54% of the type strains for the known bacterial species, about 5.5% of the type cultures for the known fungal species, and only 0.2% of the "authenticated" strains for the known protozoan species.

While this situation may appear somewhat reassuring (at least for bacteria), many bacteriologists think that the number of known bacterial species is only a small fraction of the total number of different bacterial species that currently exist on Earth. For example, Palleroni (1994:538) writes:

> ". . . the four volumes of *Bergey's Manual of Systematic Bacteriology* describe about 3,000 species, a number now considered to represent from a fraction of one percent to a few percent of the total bacterial species in nature."

With fungi and protozoans, the estimates are even less reassuring. One recent well-documented estimate puts the total number of different species of fungi on Earth at 1.5 million (Hawksworth, 1991). Thus, living culture collections such as the ATCC contain only a small fraction of the bacterial, fungal, and protozoan species thought to exist on Earth. We therefore are likely to encounter many new microbial species through projects that assess microbial diversity.

---

[1]Protozoan type specimens generally are nonliving preparations on microscope slides. "Authenticated" protozoan cultures are those obtained from the original describer of the species and subsequently grown and cryopreserved by ATCC.

At the current level of resources, new bacterial type strains are being added to the ATCC at an annual rate of about 100-125 per year, while the total number of new bacterial strains added is 250-300 per year. A total of 500-700 new fungal and yeast strains are added to the mycology collection each year. There are usually more useful strains identified than the ATCC can accession each year. Thus, the ATCC would not be able, at the current level of resources, to absorb a major increase in the number of new bacterial, fungal, yeast, and protozoan strains accessioned per year.

The ATCC provides information on each strain in the collections in several different media formats. The most recent versions of the written catalogs that list each bacterial, fungal (including yeast), and protist strain total about 1,700 pages. This information is also available in CD-ROM and diskette form for personal computers and on the Internet via the World Wide Web (http://www.atcc.org). Together with specialized monographs on the distinguishing features of different microbial species, the information and the cryopreserved living specimens housed by collections such as the ATCC provide the framework of previously known microbial species that is used to assess the novelty of new isolates from the environment.

A broad variety of other general and specialized microbial collections can provide information and examples of microbial type cultures. The Microbial Germplasm Data Net (Moore, 1993) provides data on many microbial research collections that is searchable over the Internet. The *World Directory of Collections of Cultures of Microorganisms* (Sugawara et al., 1993) provides data on 481 collections in 51 countries and includes the scientific names of 334,312 strains of bacteria and 351,263 strains of fungi and yeasts. It also can be searched over the Internet. These resources should be useful to researchers studying biodiversity as points of contact and centers of expertise on the properties of a wide variety of microorganisms.

Can this network of culture collections absorb the increase in new microbial species that likely would result from implementation of an All Taxa Biodiversity Inventory (ATBI)? Probably not, in my view, without a substantial additional investment in this critical component of infrastructure. Many of the collections are small and specialized, and primarily reflect the research interests of single investigators or a small group of researchers. Even at current levels of support, important specialized microbial collections periodically become "endangered" through loss of financial support or retirement of the interested scientist.

THE CHALLENGES OF MEASURING MICROBIAL DIVERSITY

Recent molecular biological measurements estimate that about 4,000 different bacterial species reside in a gram of Norwegian forest soil (Torsvik et al., 1990; Wilson, 1992:143-144). Assuming that this is a reasonable measure of the bacterial species complexity in such an environment, one begins to see the magnitude of the task that is required to enumerate even the bacterial species (much less the fungal and protozoan species) present in the air, soil, and water, and on the outer and inner surfaces of all the plants, insects, birds, and other animals in a chosen area of tropical rain forest.

Part of the difficult nature of the task comes from my assumption that many of the bacterial species isolated from the aforementioned different habitats of the tropical rain forest will be very similar to those encountered in adjacent environments. Thus, much work may be expended to identify what ultimately will turn out to be very similar strains of the same microbial species. While this initially might seem like "wasted effort," it will aid in answering fundamental questions about the lateral spread of some microorganisms in a specific environment.

Research on microbial diversity also will continue to isolate many organisms that apparently are previously unknown. Only through more detailed biochemical, immunological, or molecular biological tests will it become clear, however, whether the new isolates are sufficiently novel to be called new microbial species. Depending on which combination of techniques is chosen, a reasonably large amount of laboratory research will be required to answer the question of novel species even for a single isolate. As the number of apparently novel isolates increases, the amount of effort to characterize them quickly will become more than any one institution can handle.

For this and other reasons, the ATCC joined forces with four other institutions in the Washington, D.C., area (the Smithsonian Institution's National Museum of Natural History, the U.S. Department of Agriculture's Agricultural Research Service, the University of Maryland at College Park, and the Maryland Biotechnology Institute) to form the Consortium for Systematics and Biodiversity. The application of systematics, ecology, and evolution to biodiversity will be a primary emphasis for Consortium members, who have agreed to cooperate in developing and enhancing research and training programs and facilities for systematics and biodiversity. In this context, the ATCC will provide both expertise in systematics and living examples of the type specimens that form the current framework of microbial systematics.

Visionary proposals to carry out an ATBI (Janzen and Hallwachs, 1994) are exciting in their promise to provide an integrated estimate of the total species diversity in a selected area. However, if "all taxa" includes the microbes, as I believe it should, some differences between the objectives for microorganisms and macroorganisms will have to be introduced into the program in order to develop practical plans. In the February 1994 draft of the ATBI planning document, Janzen and Hallwachs (1994) discuss the microbes and several other groups of organisms under the heading "Group G3. Species-rich taxa that are sufficiently problematical that it may not be feasible to inventory a noticeable fraction of their species in a 5 year period."

The ATBI proposal nevertheless stimulates thinking about how large-scale

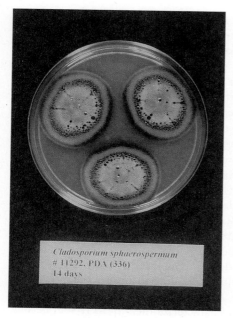

Cladosporium sphaerospermum
11292, PDA (336)
14 days

Growth of microorganisms in a petri dish.

investigation of microbial diversity might be done. For example, Tiedje (1995) recently has described the outlines of an experimental approach. This approach emphasizes reduction in the number of organisms analyzed through identification of those that have unique value for intensive study. It advocates use of rapid, semi-automated methods that generate results in a form that can be placed immediately into large computer databases. Finally, it can serve as a useful starting point for further discussion among microbiologists about the most fruitful and cost-effective methods for characterizing microbial diversity in the context of an ATBI.

Living culture collections should study and selectively implement actions that will increase the utility of their resource materials for the assessment of microbial diversity. First, they can make detailed information about their microorganisms available on a world-wide basis over the Internet. Second, they can enhance the usefulness of the descriptive information about their collection materials via multifaceted databases that include pictorial as well as text descriptions. Finally, they can implement ways to increase their rate of accessions so that, as new microorganisms are identified through ATBIs and other initiatives to describe microbial diversity, more strains can be authenticated and cryopreserved for further study and use by the scientific community of the world.

INTERNATIONAL COOPERATION

The days when one could walk into any country with interesting habitats for microbial diversity and walk out with one's pockets full of interesting samples appear to be over. Countries containing such habitats now are aware that they may harbor microorganisms with commercial potential. Following the Rio Conference in 1992, the *United Nations Convention on Biological Diversity* (Reid et al., 1993:303-324) is in the process of being ratified by many nations. This Convention sets out a framework for cooperation between nations in sustaining and developing biodiversity (including microbial diversity), but will require implementing legislation in many nations before it is broadly effective.

If an international consortium of scientists were to work on the identification and characterization of microbial samples identified through an ATBI, then some new mechanisms may have to be put in place to delineate the rights and responsibilities of the cooperating parties. For the next few years, before national legislation implementing the U.N. Convention on Biological Diversity is passed, specific contractual arrangements may have to be used as a substitute. In the context of an ATBI, these should spell out the rights and responsibilities of funding agencies, incountry institutions, and international collaborators. The experience of the Instituto Nacional de Biodiversad (INBio) in Costa Rica can provide useful models for some of the necessary arrangements (Reid et al., 1993:53-67).

New modes of international scientific interaction may help manage the costs and increase the feasibility of such studies. The Internet is being used increasingly for international collaborative work on scientific projects. Videoconferencing by small groups of international collaborators may be useful in bringing taxonomic expertise in close proximity to ATBI sites. Videoconferencing and the Internet also may be able to facilitate transnational training in microbial systematics.

CONCLUSION

There are clearly many technical, financial, and administrative problems to be solved on the way to implementing an ATBI. Nevertheless, the allure of participating in a large, international, collaborative project to describe in detail a small part of the Earth's species-scape remains. Living microbial culture collections can play multiple constructive roles in such projects as suppliers of information and standard specimens of currently known microbes, as sources of taxonomic and educational expertise, and as eventual repositories of some of the new microbes discovered.

REFERENCES

Bull, A. T., M. Goodfellow, and J. H. Slater. 1992. Biodiversity as a source of innovation in biotechnology. Ann. Rev. Microbiol. 46:219-52.

Hawksworth, D. L. 1991. The fungal dimension of biodiversity: Magnitude, significance and conservation. Mycol. Res. 95: 641-655.

Janzen, D. H., and W. Hallwachs. 1994. All Taxa Biodiversity Inventory (ATBI) of Terrestrial Systems. A Generic Protocol for Preparing Wildland Biodiversity for Non-damaging Use. Report of National Science Foundation Workshop, 16-18 April, 1993, in Philadelphia. National Science Foundation, Washington, D.C. 132 pp. (Available over the Internet via Gopher to huh.harvard.edu; see Biodiversity Information Resources for All Taxa Biological Inventory.)

Moore, L. 1993. Microbial Germplasm Data Net, Vol. 3, No. 4. (Internet address mgd-feedback@bcc.orst.edu.)

Palleroni, N. J. 1994. Some reflections on bacterial diversity. Amer. Soc. Microbiol. News 60:537-540.

Reid, W. V., S. A. Laird, C. A. Meyer, R. Gamez, A. Sittenfeld, D. H. Janzen, M. A. Gollin, and C. Juma. 1993. Biodiversity Prospecting. World Resources Institute, Washington, D.C. 341 pp.

Staley, J. T., and N. R. Krieg. 1984. Classification of prokaryotic microorganisms: An overview. Pp. 1-4 in N. R. Krieg and J. G. Holt, eds., Bergey's Manual of Systematic Bacteriology, Vol. 1. Williams and Wilkins, Baltimore, Md.

Sugawara, H., J. Ma, S. Miyazaki, J. Shimura, and Y. Takashima. 1993. World Directory of Collections of Cultures of Microorganisms, fourth ed. World Federation of Culture Collections Data Center on Microorganisms, Saitama, Japan. 1185 pp. (Internet address http://www.wdcm.riken.go.jp.)

Tiedje, J. M. 1995. Approaches to the comprehensive evaluation of prokaryote diversity of a habitat. Pp. 73-87 in D. Allsopp, D. L. Hawksworth, and R. L. Colwell, eds., Microbial Diversity and Ecosystem Function. CAB International, Wallingford, England.

Torsvik, V., J. Goksoyr, and F. L. Daae. 1990. High diversity in DNA of soil bacteria. Appl. Envir. Microbiol. 56:782-787.

Wilson, E. O. 1992. The Diversity of Life. Belknap Press, Cambridge, Mass. 424 pp.

31

Integration of Data for Biodiversity Initiatives

DAVID F. FARR
Research Mycologist

AMY Y. ROSSMAN
Research Leader, Systematic Botany and Mycology Laboratory

Plant Sciences Institute, Agricultural Research Service,
U.S. Department of Agriculture, Beltsville, Maryland

"Science is built up with facts, as a house is with stones. But a collection of facts is no more a science than a heap of stones is a house" (taken from Jules Henri Poincaré, La Science et L'Hypothése [1908], in Beck, 1980:673).

Vast amounts of information about biological diversity exist in forms ranging from systematic monographs and regional checklists to the data associated with the millions of specimens held in the nations' collections. This wealth of information should be organized and integrated into a readily accessible, comprehensive knowledge base. Such a knowledge base is needed for global land-use planning to ensure the long-term economic and environmental benefits of these vital biological resources.

Although some data about biological diversity are available electronically, most are scattered in the literature, associated with specimens in museum collections, and contained in other resources of the biological disciplines. In addition, existing data often must be edited laboriously before they can be integrated with other data in electronic form. Yet integration of data about biological diversity is the key to preservation of these biological resources. Capturing this information electronically and integrating it into an accessible knowledge base is a significant step toward cataloging, developing, and preserving the world's biodiversity.

In order to integrate databases of biological information, two types of activities must be undertaken: (1) the actual process of obtaining and integrating data in electronic form, and (2) reviewing and reconciling differences in consistency of data. The first activity is computer-oriented and is thus generally easier to complete than the second. Although the second activity should be the purview of the publishing scientist, in reality, those who attempt to integrate data must wrestle with and resolve problems of inconsistencies.

Two projects are discussed as examples in which biological data have been successfully integrated to provide a source of information that is useful in managing biological diversity. This is not to imply that these projects are the only examples of such activity or that they are necessarily the most sophisticated. However, the projects discussed below have integrated large data sets successfully and can be used as a frame of reference for developing additional approaches to the integration of data for biodiversity initiatives.

FUNGI ON PLANTS AND PLANT PRODUCTS: A SPECIALIZED PROJECT MANAGED BY A SINGLE INSTITUTION

This project illustrates the activities and procedures required to provide a diverse community of users with a well-researched, authoritative, and comprehensive body of specialized biological information. The goal of the project was to develop a database of current information on the occurrence and distribution of fungi associated with vascular plants in the United States. The immediate objective of the project, and major result to date, has been a book entitled *Fungi on Plants and Plant Products in the United States* (Farr et al., 1989). Because this product was intended to meet the needs of plant regulatory officials and extension agents as well as scientists from a variety of disciplines, considerable effort was devoted to reviewing and consolidating data and presenting it in a format that was easy to retrieve. The book includes two major sections (Figures 31-1 and 31-2). The first lists species of vascular plant hosts that are arranged by family of plants and the fungi that are associated with each species of vascular plant. The distribution of each fungal species is listed state by state, based primarily on reports from the literature (Figure 31-1). The second section presents the accepted scientific name of the 13,000 fungal species with their basionym, important synonyms, the names of alternate life forms, comments on world-wide geographic and host distribution, citations to relevant systematic literature, and the genera of vascular plants upon which that fungus is reported in the United States (Figure 31-2).

The amount of effort required to complete a project such as this depends on the group of organisms involved. In the case of fungi associated with plants, it was a major effort because standard references, such as lists of authoritative scientific names, did not exist, and the number of both fungal and vascular plant host names was high—much higher than expected at the outset of the project. As a result of this project, authoritative information on over 24,000 scientific names of fungi is now available. Four major steps were involved: (1) compiling the information; (2) entering information into the computer or obtaining data in some electronic form; (3) analyzing the new data for accuracy and

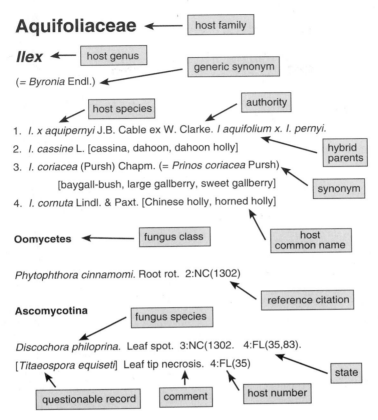

FIGURE 31-1   Components of a record in the host-fungus list (Farr et al., 1989).

consistency, particularly of the scientific names and reports of the host-fungus association; and (4) producing the desired output, in this case, as both hard-copy and on-line information.

Compilation

By far the most important source of data in this project was the literature, although other sources were included. Ease of access to sources of information is important and confines such projects to organizations that have comprehensive literature or herbarium resources. A project such as this one must be thoroughly documented and based on primary literature, rather than extracting information from other databases (e.g., from a library reference database). Such databases rarely are cataloged in sufficient detail to satisfy the needs of a narrowly defined project. In addition, the information derived from the primary

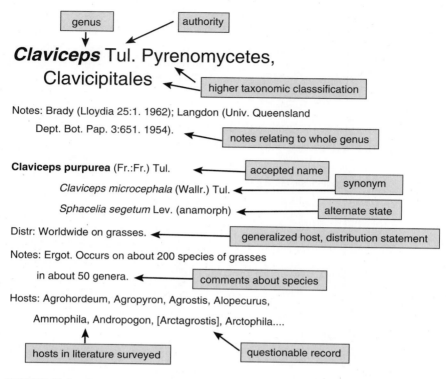

FIGURE 31-2   Components of a record in the fungus list (Farr et al., 1989).

resource may be incomplete or erroneous. For this project, several relevant journals were examined page by page by a knowledgeable individual. A second consideration is that primary resources must be readily available to the project coordinators because rarely is a piece of literature passed through the system once and then not needed again. Questions arise that require returning to the original piece of literature. For example, in reviewing the accumulated information on a particular host-fungus distribution, one record may appear anomalous, as in the questionable record shown in Figure 31-1. Reexamination of the original literature is required to resolve this suspected erroneous information. Lack of documentation of the primary source of a report, such as in the *Index of Plant Diseases in the United States* (Anonymous, 1960), the predecessor to Farr et al. (1989), creates confusion and devalues the information because questionable reports cannot be verified through either the literature or a specimen.

Herbarium specimens provided an additional source of data for this project. The primary obstacle to incorporating a large quantity of specimen information, assuming that the specimen has been correctly identified, concerns problems that arise in computerizing data from specimens, namely inconsistencies in and legibility of labels. Handwritten labels often represent a challenge to personnel who enter the data. The specimens themselves also may require special handling that can introduce labor-intensive steps in the process of entering data. If specimen labels are photocopied, they then can be handled similarly to a piece of literature.

Regardless of the source of information, a rarely discussed but time-consuming activity is locating and marking specialized data in the primary literature. This step is required to make the information comprehensible to those who enter the data. In the case of this project, the name of the fungus and the corresponding name of the vascular plant host were marked lightly in pencil to be read by the personnel entering the data. Although editing can be coordinated with entry of the data, editing data from primary resources as an initial separate step affords greater efficiency by matching skills to job requirements. For example, the knowledgeable individual who scans the journals for data readily can indicate whether a scientific name is that of the fungus or the vascular plant host. Such information greatly assists those who enter the data but lack a background in biology. In addition, accurate preliminary editing reduces errors during the entry of data and saves time in the later editing phases of the project.

Data Input

Electronic validation by the computer of the information entered is a common feature in most data-entry programs and greatly increases the accuracy and consistency of entering the data. Simple types of validation include checking that the entered data is in the correct format. More powerful validation includes comparing entered data against data that is in a separate authority file (data dictionaries). A common use of this technique is to check a country name against a previously established computer file of accepted country names. This type of validation not only assures accurate entry of data but also assumes an editing role in that it requries consistency of abbreviations and format. Since only certain place names are used, the information is edited to meet these requirements. Although this type of validation requires that the entered data match those in the authority file, such editing does not necessarily ensure that the entered data reflects that in the primary source. Often some interpretation and editing of the primary data are necessary.

Verifying the accuracy of entered data is a laborious process. Although it is tempting to bypass this step, proofreading against the original source is necessary during the initial stages of a project both to uncover errors and to determine if information was skipped during the entry of data. Specialized editing printouts such as alphabetical listings of certain fields can be an important and helpful component of the proofreading process. For this project, printouts were proofread on a daily basis during the data-entry stage. At the conclusion of the project, the database included 120,000 reports of fungi derived from over 4,000 sources of literature.

Analysis of Raw Data

The analysis and consolidation of data obtained from various sources are important steps in producing a database that is of value to a range of potential users. Users of the database on hosts and fungi generally lack the background, expertise, or the time needed to properly analyze the original unedited information. This is particularly true for outdated scientific names that are now synonyms. Without thorough analysis and consolidation, the raw data are likely to be underutilized or possibly used in an unreliable manner. In the case of fungi associated with plants in the United States, and probably for many projects that integrate data on biological diversity, the inconsistencies due to differing taxonomic concepts and changing nomenclature must be reconciled.

Conversion of the primary data to reflect, for example, accurate scientific names of taxa, can be handled by the computer in at least two ways. The simplest involves altering the data in the original records to reflect the current concepts, i.e., using flat files. Although straightforward logistically, this approach has severe disadvantages. One significant drawback is that the original entry no longer can be verified against the primary source. On several important occasions, it has been necessary to refer back to the original entry and primary resource, e.g., when the validity of a report has been questioned. Second, errors cannot be traced that are introduced during the original process of extracting data from the primary sources and during entry of data. Because considerable labor is involved in these initial steps, there is significant opportunity for the introduction of errors, which cannot be traced if the original record has been changed.

The second approach, the one that we followed—building relational databases—, is more complicated but produces a reliable, flexible result. In order to convert the "raw" data extracted from the primary sources, authority files were built that were used to compare entered data with the "correct" data and convert the data into an acceptable form. In this case, a file of accepted scientific names of fungi and a file of accepted scientific names of vascular plant hosts were used to relate the data entered from the primary resource to the final data-output, i.e., fungi on vascular plant hosts (Fig. 31-2). This method requires increased software support, but has the significant advantage of leaving the original record intact, thus providing an important resource for future reference. Questions arising about the original information are easier to answer, and the original records do not have to be changed, thereby avoiding the continuous introduction of errors.

The process of reviewing the scientific names to produce an authoritative database must be accurate and represent a consensus from the information available in the literature (see Thompson, Chapter 13, this volume). Professionally trained systematists have the required expertise to complete such work, but it is rarely feasible to add this work to the activities of those already conducting active research programs. Despite a need for authoritative information about biological resources, little professional recognition is attached to the consolidation and dissemination of these most fundamental data. The issue was resolved by employing knowledgeable individuals trained in systematic mycology, usually with a Ph.D., with some background in plant pathology. Other resources were essential in reviewing the fungal names, particularly the continuously updated databases of taxonomic literature.

When this project was initiated, authoritative lists of scientific names of fungi were simply not available, and, even for vascular plants in the United States, an agreed-upon computerized list of scientific names was lacking. Therefore, for the fungi, each name from the literature was reviewed to determine if it was an accepted taxon, was a synonym, or belonged in some other category. Thus, a database of fungal names was built that contained the following core information:

> Accepted scientific name
> > basionym
> > synonyms
> > alternate state names (e.g., anamorph, teleomorph).

For the vascular plant hosts, scientific names derived from the literature were compared electronically with several authoritative lists. If the name from the literature agreed with those on the lists, that name was accepted. Discrepancies between the names of the vascular plants in the data-source (i.e., the report of a fungus on a host) and the authoritative lists (used to verify the names of the vascular plants) were resolved by consulting additional literature. The literature used to verify the scientific name of the vascular plant was recorded in the database.

Two databases of scientific names were constructed, one for the names of the fungi and one for names of the vascular plants. Every name extracted from the literature was included in one of these databases. Many anomalies were encountered that were difficult to resolve, including errors in the primary literature. In all cases, an electronic record was maintained of which authority was followed and how anomalies were resolved.

Producing the Output

In order to produce an output, either electronically or as hard-copy, it was necessary to integrate the original data derived from the literature, including geographic information, with the reviewed scientific names of both the fungi and the vascular plants. An application was developed that converted records from the literature to output based on the currently accepted scientific name. In other words, a new checklist was synthesized that reflected the most recent taxonomy and nomenclature of the fungi and their hosts (Figure 31-3). Following the review of the taxonomic literature and the development of an authoritative database of scientific names for fungi, specialized lists were printed and sent for review to systematic experts. Such reviews by experts throughout the world served to include the scientific community in this project and were not considered a burdensome task. The taxonomic literature on vascular plants is more comprehensive, thus the reviewing and updating of vascular plants was based solely on the literature. In the printed version, accurate scientific names of fungi and their synonyms are listed with a summary of their hosts. For both the hard-copy and on-line output, synonymous scientific names on which the report may be based are converted to the accurate scientific name. For example,

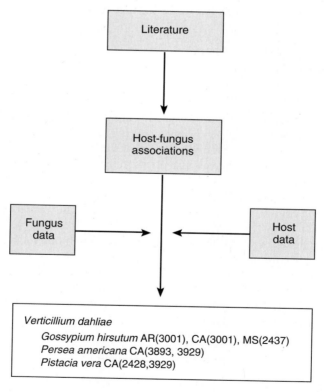

FIGURE 31-3 Data source and relational databases used to generate output.

if data were requested about the occurrence of *Verticillium ovatum*, the following would be displayed:

> Hosts and distribution for the fungus *Verticillium ovatum*, a synonym of *Verticillium dahliae*.
>
> *Verticillium dahliae*
> *Gossypium hirsutum* AR(3001), CA(3001), MO(3001), MS(2437), NE(3090), TX(3001)
> *Persea americana* CA(3893,3929)
> *Pistacia vera* CA(2428,3929)
> *Prunus armeniaca* CA(3929)
> *Prunus cerasus* CA(3929)
> *Prunus dulcis* CA(3929)
> *Prunus salicina* CA(3929)
> *Rubus idaeus* CA(3929)
> *Solanum melongena* var. *esculentum* MA(94), NY(94), WA(94)

Because *V. ovatum* is now considered to be a synonym of *V. dahliae*, the information for both of these names is included as one entry under the currently accepted name, *V. dahliae*. The entry for Massachusetts (MA) under *S. melongena* var. *esculentum* is actually a report of the fungus in the literature in which the name *V. ovatum* is used, a name that now is considered a synonym of *V. dahliae*.

Although this was a project strongly driven by computers, the primary objective was the publication of a hard-copy book. Without a book as a product, the project probably would not have been funded. Although this project technically was completed with the publication of Farr et al. (1989), several additional products and services have resulted from the project. A second hard-copy and on-line product derived from this project was a book, *Scientific and Common Names of 7,000 Vascular Plants in the United States* (Brako et al., 1995), with emphasis on those plants that are reported to be hosts for fungi. Over Internet, outside users have on-line access to data in *Fungi on Plants and Plant Products in the United States* (Farr et al., 1989), in addition to the other electronic resources developed at the Systematic Botany and Mycology Laboratory (Anonymous, 1994). Ad hoc queries (Clay, 1994), construction of look-up files for other projects, and use of subsets of data in other database projects (Alfieri et al., 1994; French, 1989) are some of the additional products that have resulted from the database. Fortunately, the files were constructed with the flexibility required to accommodate such unanticipated uses. Once data exist in electronic form, there is nothing more frustrating than not being able to use them fully. This inability can come from restrictions on software, lack of appropriate restrictions of software, or restrictions of institutional systems (e.g., inability to provide access to data to those outside the institution). In general, computer databases do not exist for a single purpose. They are meant to be manipulated in new ways, divided, updated, and modified with new and additional information. They become part of the universe of computer databases, waiting to be integrated with other data to provide information to solve unanticipated problems.

ENVIRONMENTAL RESOURCES INFORMATION NETWORK: A COMPREHENSIVE PROJECT INVOLVING MULTIPLE INSTITUTIONS

The Australian Biological Resources project, called the Environmental Resources Information Network (ERIN), is an example of integration of data at a national level. The mission of this project is to provide geographically related environmental information of the extent, quality, and availability that is required for planning and making decisions (Slater, 1993). As governmental bodies in Australia were making decisions about issues that affected the environment, they recognized that they lacked adequate information on the environmental attributes and characteristics of the areas under discussion. To address this problem, ERIN was developed to produce various kinds of computer maps that integrate biological and geographical data.

To develop a comprehensive database that would meet the goals of this project, rapid access to accurate information was needed about the distribution of targeted groups of organisms. These biological data then could be combined with geological and cultural features to present a graphical overview of a particular area of land. The organizational approach used for this project was to build a computer network that linked the sources of information (e.g., databases of the localities from which herbarium specimens were collected) to a central node that coordinates information (Figure 31-4). One node on the network is the ERIN database, which contains data ("point data") that are derived from specimens, information on taxa, and information on the sets of data that are available on the network. At this centralized unit, various kinds of data were integrated and made available to those requesting information, such as government officials making decisions about land-use.

Compilation

How were the point data obtained? In the case of the localities for vascular plants, information was based on the collections in various Australian herbaria. ERIN initiated a contract with each herbarium to supply the required locality data. Each contract was specific and included details about the fields and format of the data that were needed. Other kinds of data included in the ERIN system were data on specimens for mammals, amphibians, insects, and fungi, as well as geologic, climatologic, and cultural data that were derived from both governmental and nongovernmental sources.

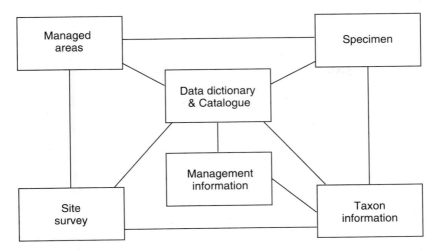

FIGURE 31-4 Interactions in ERIN database between data sources, validation points, and user-interface (Slater and Noble, 1991).

Data Input and Analysis of Raw Data

How were these data integrated? After passing through two validation points, the data were integrated continuously at the central node of the ERIN computer. The first validation point was a check on the scientific name of an organism. Reports with unacceptable scientific names were rejected for resolution at the source database. After the scientific name had been accepted by the central node, information about the locality that was associated with the record was validated by comparison with other data. Anomalous data were readily apparent when integrated with other data (i.e., the report of a plant specimen that occurred outside its range or on an unusual type of soil). When both criteria were satisfied, the data were loaded into the Geographic Information System—the common reference through which decision-makers can extract point data for analysis. The ERIN computer maintains, for example, lists of acceptable scientific names of vascular plants with associated information on localities that can be integrated with other kinds of data.

Producing the Output

Core data are provided to a centralized database via a network so that information is immediately available to a potential user. The detailed data from which the core data are extracted are not stored centrally, but rather exist in separate nodes where they are under the control of the custodian. This ensures that the data will be updated at the point of origin and thus maintained by those best able to do so. All data, regardless of source, are available through an interface that is easy to use, incorporating a comprehensive directory. Analytical and modeling tools are available through the same user-interface as the data. Priority is given to the acquisition of primary point data, rather than aggregated or interpreted information. This ensures that conclusions based on those data can be reviewed rigorously, alternative analyses can be performed, and baselines for monitoring can be established. Using these centralized data, computerized maps can be produced in response to queries and specialized requests.

The ERIN project is a functioning database system that addresses issues that arise when discussing the integration of biological information. This information resource is the result of a national effort to provide environmental information related to geography using a centralized database. As a result, planning, research, development, and management is based on environmental information provided through a well-coordinated interdisciplinary and multi-institutional collaboration. Data are readily accessible, both at reasonable cost and without encumbrances that otherwise might impede responsible environmental decision-making.

COMPARISONS OF THE TWO PROJECTS

The project on associations between fungi and plants in the United States and ERIN in Australia share two important features. Both use a taxon-based list of accepted scientific names to allow the integration of data from diverse sources. The importance of a list of accurate scientific names of taxa cannot be overestimated. These lists are the linchpin on which all attempts to integrate biological information are dependent. Integration of biological information will be possible only with the availability of such lists. In addition, these two projects demonstrate the need for validation of data if information from diverse sources (either from the literature or from collections of specimens) is to be integrated meaningfully.

The important differences between the projects are the means by which data are handled and the number of institutions that are involved. For the database on fungi, all of the data were generated, edited, and manipulated within one computer at a single institution. The ERIN project has developed a system that uses distributed sets of data at several institutions. Most of the resources for the project on fungi were devoted to verifying and updating the information. For ERIN, the responsibility for verifying and updating the information was contracted out to multiple institutions. Project resources were directed to developing the network with a central node and a user-interface that could provide the data. Quality-control was ensured by validating incoming data as well as contracting with institutions that had strong systematic expertise to serve as the custodian for the entry and review of specimen-based data. The ERIN database demonstrates that multiple institutions can be involved in the development and maintenance of integrated sets of data.

AVAILABILITY OF INFORMATION ON BIODIVERSITY

Information on biodiversity is available from a number of sources, including those used in the examples cited above—reports from the literature and data from herbarium or museum specimens. The most important and authoritative sources are comprehensive taxonomic monographs. Such monographs reflect the state of knowledge about particular groups of organisms based on years of accumulated experience, detailed scrutiny of specimens from throughout the world, and the synthesis and analysis of many kinds of data. Unfortunately, the data accumulated in monographs often are not stored in well-developed databases. Integration of the wealth of data in systematic monographs with other sources of information requires orientation toward the databases during the collection and collation of the information. Currently, the software program DELTA (DEscriptive Language for TAxonomy) provides this capability (Askevold and O'Brien, 1994; Dallwitz et al., 1993; Fortuner, 1993). All of the character states for an organism are treated as distinct entries in the database. The recent monograph of *Cucumis*, a genus of vascular plants that includes melons, squashes, and gherkins, was produced using DELTA (Kirkbride, 1993). Although published as a book, it includes a disk with data in DELTA format. Ad hoc queries based on any of the types of data used in the monograph are possible. In addition to providing a user-friendly synoptic means of identification, the data can be used for secondary analyses, e.g., PAUP programs. Subsets of data for specialized purposes are created easily. For example, one can select only those species found in a certain geographic area and produce a key to those species for a local user-community. Integration of data sets that have used the DELTA format is easily accomplished, and thus the treatment of two related genera could be combined into a single data set. Most importantly, the DELTA format allows for the manipulation, extraction, and display of data to meet the needs of a large and diverse user-community, and these data can be made available. Users of biological information in DELTA format range from plant-explorers and plant-breeders to phylogeneticists and molecular biologists.

Large-scale integration of data about biological diversity requires that much of the primary information be in an electronic form. How can this electronic information be located and accessed? Currently, Internet provides the best means of accessing such information. Internet is a series of connections between computers that allows for the transfer of data. Having been built from the ground up, there is no overall structure to the Internet. For example, when a computer that is part of our Laboratory system at Beltsville communicates via Internet with a computer at Harvard University, it connects with 19 other computers along the way. The recent surge in users attached to Internet means that millions of individuals now have access to computerized information throughout the world. Through Internet, electronic systematic data can be made available to everyone.

Internet facilitates finding and disseminating data in several ways (Krol, 1992). One is the use of Gopher software to locate subject-specific files throughout the world. For example, this program can be used to gain access to information about biodiversity located on the Biodiversity and Biological Collections Gopher at Cornell University, the Harvard University Gray Index of vascular plants described from North America, or the Register of Type Specimens in the Botany Department available through the Smithsonian Gopher. Another means of accessing data is through Telnet, the Internet remote login application. Using Telnet, the user actually logs onto another computer and then can work with the databases on that computer. Our Systematic Botany and Mycology Laboratory computer in Beltsville, Maryland, has such a capability, which allows our major databases on fungi to be accessed by the outside world (Anonymous, 1994). File Transfer Protocol (FTP) is a procedure available over Internet that allows the rapid transfer of computer files. FTP can be used to attach to another computer, select the file of interest, and copy it in a matter of seconds. A final method of communication through Internet is the use of mail groups. These communication tools permit groups of individuals in specific disciplines to ask questions and discuss current subjects of interest to that group. Most recently, the use of Mosaic and the World Wide Web server has made locating relevant data that are available through Internet increasingly easy. Clearly, the suite of tools available over Internet has greatly improved progress toward the integration of data.

FUTURE CHALLENGES

Abundant data about biological diversity exist in various forms; the information and technology needed to integrate these data are also readily available. In fact, no significant hardware or software problems constrain the development of systems for integrating data. The projects discussed in this chapter provide examples of procedures that can be used to develop the integration of data on a large scale. Two developing large-scale projects that integrate taxonomic and distributional information are the International Organization of Plant Information (IOPI) project (Bisby et al., 1993) for vascular plants, and the Biosystematic Information on Terrestrial Arthropods (BIOTA) project for insects (Hodges, 1993). Such projects always require a strong motivating force backed by fiscal and intellectual resources.

Several factors have impeded progress in large projects on biodiversity. One is a lack of clear goals and objectives. What are the goals of research on biodiversity? The ultimate goal is global planning of land-use that would yield long-term economic and environmental benefits. Intermediate goals must be defined with objectives that can be achieved using available or obtainable financial and personnel resources. Two of the most urgent attainable objectives for biodiversity initiatives are: (1) the development of authoritative databases of scientific names of organisms that are available over Internet (efforts are under way to provide and continuously update such data as part of the U.S. Department of the Interior's National Biological Survey[1]; National Research Council, 1993), and (2) electronic access to baseline data on the presence/absence of species, particularly of speciose groups such as insects and fungi that are sensitive to environmental pollution and habitat alteration. Baseline data exist in collections of specimens as well as in the literature, but they must be made available and reviewed prior to integration with climatological and other abiotic data.

A second factor hindering the integration of information for biological diversity is a lack of stable resources. Long-term commitments from funding agencies and research organizations are needed to foster long-term mega-projects. The Australian government recognizes the economic value of making judicious decisions on land-use based on data from ERIN, and thus has made a long-term financial commitment to this project. Long-term financial commitments from national and international science and environmental funding agencies would strongly motivate projects to integrate information about biological diversity.

CONCLUSION

The multidisciplinary integration of biological and nonbiological data requires a coordinating organization to spearhead initiatives in biological diversity from *talk to action*. The Australian plan, with its centralized decision-making, is an extremely successful example of a centrally coordinated program that should be emulated around the world. An organization is needed that will set goals, promulgate standards, consolidate information about available databases, develop procedures for integrating data, and make available a comprehensive database of information relevant to biological diversity. Such an organization for the United States has been proposed in the U.S. Department of the Interior's National Biological Service[1] (National Research Council, 1993). At the international level, the United Nations Environmental Programme could provide the leadership to bring into reality the information resources that are needed to wisely direct international biological diversity initiatives. Within the next 10-20 years, it is essential that data on the world's biological resources be integrated in order to fully explore, sustainably utilize, and preserve these vital resources.

---

[1] The National Biological Survey subsequently was renamed the National Biological Service by Secretary Babbitt.

REFERENCES

Alfieri, S. A., Jr., K. R. Langdon, J. W. Kimbrough, N. E. El-Gholl, and C. Wehlburg. 1994. Diseases and disorders of plants in Florida. Florida Dept. Agri. Conserv. Serv., Bull. No. 14:1-1114.

Anonymous. 1960. Index of plant diseases in the United States. Agri. Handbook 165:1-531.

Anonymous. 1994. On-line information available about U.S. National Fungus Collections. Phytopathol. News 28:159-160, 163.

Askevold, I. S., and C. W. O'Brien. 1994. DELTA, an invaluable computer program for generation of taxonomic monographs. Ann. Entomol. Soc. Amer. 87:1-16.

Beck, E. M., ed. 1980. Familiar Quotations. A Collection of Passages, Phrases and Proverbs Traced to Their Sources in Ancient and Modern Literature, by John Bartlett, fifteenth ed. Little, Brown and Company, Boston. 1540 pp.

Bisby, F. A., G. F. Russell, and R. J. Pankhurst, eds. 1993. Designs for a Global Plant Species Information System. Clarendon Press, Oxford, England. 346 pp.

Brako, L., A. Y. Rossman, and D. F. Farr. 1995. Scientific and Common Names of 7,000 Vascular Plants in the United States. American Phytopathological Society, St. Paul, Minn. 304 pp.

Clay, K. 1994. Patterns of biodiversity in plant pathogens (abstract). P. 39 in Fifth International Mycological Congress, Vancouver, British Columbia, Canada.

Dallwitz, M. J., T. A. Paine, and E. J. Zurcher. 1993. User's Guide to the DELTA System: A General System for Processing Taxonomic Descriptions, fourth ed. CSIRO Division of Entomology, Canberra, Australia. 142 pp.

Farr, D. F., G. F. Bills, G. P. Chamuris, and A. Y. Rossman. 1989. Fungi on Plants and Plant Products in the United States. American Phytopathological Society, St. Paul, Minn. 1259 pp.

Fortuner, R. 1993. Advances in Computer Methods for Systematic Biology: Artificial Intelligence, Databases, Computer Vision. Johns Hopkins University, Baltimore, Md. 560 pp.

French, A. M. 1989. California Plant Disease Host Index. California Department of Food and Agriculture, Sacramento. 394 pp.

Hodges, R. W. 1993. Biosystematic Information on Terrestrial Arthropods. Assoc. Syst. Coll. Newsletter 21:72.

Kirkbride, J. H., Jr. 1993. Biosystematic Monograph of the Genus *Cucumis* (Cucurbitaceae). Parkway, Boone, N.C. 159 pp.

Krol, E. 1992. The Whole Internet. User's Guide and Catalog. O'Reilly and Associates, Sebastopol, Calif. 376 pp.

National Research Council. 1993. A Biological Survey for the Nation. National Academy Press, Washington, D.C. 205 pp.

Slater, W. 1993. Introductory remarks. International Workshop, Designing Spatial Information Systems to Manage Biodiversity Information, March 1-5, Canberra, Australia. 5 pp.

Slater, W. R., and S. J. Noble, eds. 1991. ERIN Program Brief, June, 1991. Commonwealth of Australia, Canberra.

CHAPTER

32

Information Management for Biodiversity: A Proposed U.S. National Biodiversity Information Center

BRUCE L. UMMINGER
Senior Advisor on Biodiversity

STEVE YOUNG
Advisor on Biodiversity Information

Office of the Assistant Secretary for Environmental and External Affairs, Smithsonian Institution, Washington, D.C.

To explore the essential role that information management must play in biodiversity conservation, consider the arenas of finance, weather, and sports. For many of us, our quality of life is affected by one's financial posture—more and diversified assets are preferable, and it is desirable that income exceed spending. Try to imagine a world in which you lack reliable information about your assets, income, and expenditures. Imagine no payroll statements, no bank statements, no records of checks written and charges made, and so on. Life would be a perilous joust with bankruptcy.

Fortunately, most of us do not face this problem. We receive ample information about assets, income, and liabilities. We merely have to put the information to use; guesswork should not be necessary. In most cases, we get the information we need to empower us to manage our personal finances, whether or not we proceed to manage them well.

Consider weather information. Mark Twain pointed out that everyone talks about the weather, but no one does anything about it. Strictly speaking, that is not true. When we hear that rain is in the forecast, we may pick up an umbrella or raincoat. If the weather is predicted to be cold and windy, we wear a warm coat. We respond to current and forecasted weather conditions. We cannot change the weather, but we do something about it by reacting, by adapting.

The United States and other nations have worked together to build a marvelous weather information infrastructure. Consider how pervasive weather information is. We hear about the weather on the radio, see it in the newspaper, and now television provides spectacular satellite images and movies, Doppler radar displays, animated graphics, and lively human commentary to explain current conditions and forecasts. Weather is part of every local news show, and in the United States we have the Weather Channel on cable. We have spent billions of dollars on capabilities to gather weather data and generate and disseminate forecasts. Weather satellites and information networks provide global coverage. By most reckoning, thousands of lives have been saved by improved forecasting and warning capabilities, and large economic losses have been avoided through timely responses to better predictions.

Finally, consider sports, another area of pervasive information flows. For most of us, professional sporting events lack the immediate, direct impacts on our lives that financial outcomes and weather events have. Nonetheless, millions of people are keenly interested in sports, and another elaborate information infrastructure has been built to satisfy that interest. We can speculate about which came first: the overwhelming interest in professional sports, or the vast information infrastructure that stimulates and satisfies that interest. Sports information captures large sections of the printed media and is inescapable in broadcast media. Sports has its own publications and radio and television networks. And the wide availability of sports information influences behavior. How many people would follow their favorite sports teams fanatically if they could not get information about team standings and catch some of the games on television and radio? We might say that sports information "programs" people to take a greater interest in sports and protect sporting interests.

Now consider biodiversity, which affects our lives in the most fundamental ways possible. People are extraordinarily interested in the environment and biodiversity. Recent polls indicate that approximately 80% of Americans call themselves "environmentalists." That interest extends to the natural environment. More Americans visit museums and zoos every year than attend professional sporting events. Tens of millions of Americans enjoy various activities outdoors in ecosystems: hiking, birding, fishing, hunting, camping, and so on. When they are exposed to good information about biodiversity, young children show great interest—comparable to the levels of interest that they later demonstrate for sports.

Not only is interest keen in biodiversity and the environment, but, unlike the weather, humans can do something to change the situation; and we do a great deal. We eliminate species like the ivory-billed woodpecker and passenger pigeon. We introduce species like the gypsy moth and zebra mussel. We alter and destroy habitat. Sometimes we restore ecosystems and reintroduce native species that had been extirpated. In short, unlike the weather, we can do much to change the environment for the better or worse. Each of us takes actions that to various degrees help or harm.

The present administration of the United States is committed to an ecosystem management approach to the environment and the economy. The concept of ecosystem management explicitly recognizes the linkages between ecology and economy and rejects the assumption that there is an inherent conflict between jobs and the environment. The United States also is committed to a policy of sustainable development; conservation of biodiversity and management of ecosystems are vital approaches toward sustainability.

There are strong parallels between financial management and conservation of biodiversity. If we do not know what our ecological assets are, we are poorly equipped to safeguard those assets through wise use and management. If we "spend" too much of the natural production and capital of an ecosystem, we see symptoms of impending bankruptcy, e.g., the collapse of fisheries, endangered species, and loss of jobs. Managing ecosystems by deficit-spending results in what Secretary of the U.S. Department of the Interior, Bruce Babbitt, calls "train wrecks," with examples like the forest and fish crises in the Pacific Northwest.

Information is equally critical to financial and ecosystem management. Without information, we cannot manage finances at the personal, corporate, or governmental levels; without information, we cannot manage personal "backyard," local, or regional ecosystems so as to conserve biodiversity.

THE U.S. NATIONAL INFORMATION INFRASTRUCTURE: ROOM FOR THE ENVIRONMENT?

We have looked at the pervasiveness of financial, weather, and sports information. Now, the United States is building its national information infrastructure (NII), a national system of information superhighways. It is a given that the NII will offer all the financial, weather, and sports information that we could possibly want, in fact more than we can possibly absorb. Through a National Biodiversity Information Center (NBIC), the NII also could be a rich source of information on biodiversity for the public.

Is information on biodiversity pervasive in our lives today? How much information is readily available about the species and ecosystems in your area? Have they been inventoried? Are their populations, health, and trends known? Regrettably, in the United States, information on biodiversity receives a minuscule fraction of the attention that is paid to finance, weather, and sports.

Lack of information is the silent partner to the more prominent agents of environmental destruction. If the destruction happens gradually enough and invisibly enough, without galvanizing incidents like the catastrophes of Bhopal and the *Exxon Valdez*, by the time a threshold is crossed and the train wreck occurs, it is too late. We need to make the gradual, seemingly invisible changes visible to people. Financiers and meteorologists know this lesson. Financiers build indicators to track changes that otherwise would be hidden, and meteorologists use sensing technologies to "see" events that cannot be seen by the eyes of observers on the ground. And in sports, instant replays help us see what happened when we missed it during the live action.

The U.S. National Performance Review, Vice President Albert Gore's initiative for "reinventing government," suggests that the federal government should place greater reliance on strategies of empowerment that help people do the right thing. Recall that 80% of the public say they are environmentalists. Are there ways to empower them to do more for the environment, to act on what they say?

Information is the key to an effective strategy of empowerment for the environment. If we were to build a robust environmental information infrastructure, comparable in capabilities to those of finance, weather, and sports, we could provide people with environmental information to empower them. This kind of empowerment would work the same way that financial information empowers us to make informed choices in our financial management, weather forecasts empower us to prepare for the weather, and sports coverage empowers us to feel a strong emotional involvement with sporting events.

Empowerment through environmental information requires a new way of thinking. Our robust weather information infrastructure follows an implicit "observe-forecast-warn-react" model, as shown in Figure 32-1. This works quite well for the weather, since we cannot change it and can only react to it. But this model is too limited for the environment, where the relationship is more complex—the environment changes us, but we also change the environment.

We need to follow a "feedback" model that acknowledges the feedbacks between humans, information, and our environment. Figure 32-2 illustrates

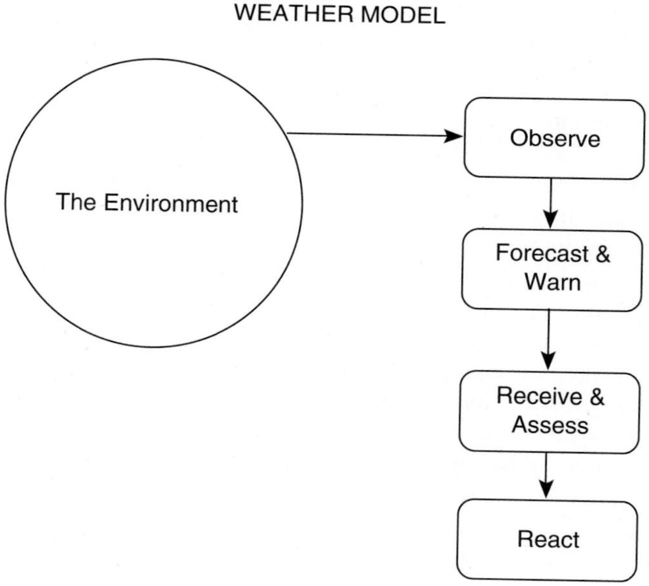

WEATHER MODEL

FIGURE 32-1   Model for using information about the weather.

such a model. It shows that the flow of information can influence human behavior, thus changing our impacts on the environment—one hopes for the better. Financiers understand that this model also applies to financial markets. Consider the well-known impact that information has on the stock market. Responses to information about actual events and rumors can cause waves of buying or selling, moving prices up and down. Analysts know that there is a highly dynamic feedback process between information and markets. We need to apply this same insight to information and ecosystems. We need to put information to work as a tool that will empower us to protect, restore, and manage our ecosystems.

If we adopt this feedback model, we see that information is the fundamental mechanism that influences human behavior toward the environment. Initiatives for the conservation of biodiversity can succeed only to the extent that we put information to work as a stratagem, resource, and tool.

Unfortunately, today it is not clear that information on biodiversity will be a prominent part of the NII. The weather model has been the dominant model in thinking about the application of the information infrastructure to the environment. If the weather model is the controlling paradigm, then it seems that there is not much to observe, forecast, warn of, and react to, because most environmental changes take place, seemingly invisibly, on a longer time scale than

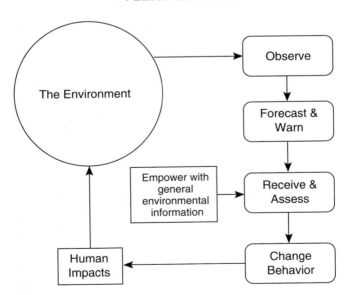

FEEDBACK MODEL

FIGURE 32-2 A model that acknowledges feedbacks between humans, information, and the environment.

weather events, which happen on scales of minutes to a few days and pose immediate and obvious threats to life and property.

On the other hand, if the feedback model is the controlling paradigm, it is imperative that we create an "environmental channel" to provide people with the information they need for empowerment and to create feedback loops that change behavior and result in positive changes for the environment.

Key components of a vision that is both possible and necessary for an effective environmental information infrastructure have been demonstrated technically in other contexts (the financial, weather, and sports realms referenced above). Also, the Environmental Resources Information Network (ERIN) in Australia, the Right-to-Know Network (RTKNet) in the United States, and other advanced practitioners already have demonstrated some of these capabilities in the environmental arena.

Imagine the time, not very distant in the future, when virtually every home is connected to the information highway. It is a given that you will be able to "surf" through video entertainment and engage in interactive home shopping. It is also a given that you will have access to a tremendous range of financial, weather, and sports information. But, in our vision, you also will have interactive access to a vast range of environmental information—in effect, an "environmental channel" on the information highway. The proposed U.S. NBIC may function as a major resource center for the channel.

An environmental channel will offer six major capabilities. First, it will allow its users to explore past, present, and predicted future environmental conditions at local, state, regional, national, and global scales, using ecosystem or political boundaries. Second, it will provide capabilities for "what-if?" modeling, visualization, and simulation to explore alternative future scenarios. Third, it will provide information about biodiversity and other natural resources in local areas and about stresses on those resources. Fourth, it will provide information oriented toward solutions that will empower individuals to identify what can be done to minimize harm and improve the environment. Fifth, it will allow individuals to "publish" observations, ideas, and questions about the environment. Sixth, you will be able to find information, other resources, and collaborators for environmental efforts.

For example, an environmental channel will allow you to view detailed computer maps and select an ecosystem of interest. For that ecosystem, you will be able to review current environmental data and view animations showing changes over time, e.g., changes in land-use and land-cover over the last 200 years. You will be able to forecast the continuation of such changes into the future and use simulation models to explore alternative futures. You will be able to review the known species within the ecosystem; their status, trends, and natural histories; and the known threats to the health of each species. You will be able to access information about recommended management practices to conserve biodiversity within the ecosystem. You will be able to publish data that you collect and ideas that you generate for conservation. And finally, you will be able to find others who are interested in the same ecosystem and develop collaborations with them.

INTERNATIONAL MODELS

ERIN

The Australian Environmental Resources Information Network (ERIN), a unit of the Department of Environment, Sport, and Territories, provides an excellent model for the sophisticated application of information tools, technologies, and stratagems to the challenges of conservation of biodiversity and ecologically sustainable development. ERIN demonstrates attributes of the needed capabilities of an environmental channel and of a biodiversity information center.

ERIN's mission is to provide geographically-related environmental information of the extent, quality, and availability that is required for planning and decision-making, fulfilling an "infrastructure support" role. With a relatively small staff (approximately 24) and budget (approximately $2.5 million per year in Australian dollars), ERIN has been able to make a disproportionately large impact on environmental decision-making in Australia.

ERIN takes a distributed network approach, in which it identifies the custodians of high-priority data, such as standard spatial data sets, and works with those custodians to make their data available over the information network. The agencies retain custodial responsibilities. ERIN seeks to ensure that primary data are available, so that users are free to make their own interpretations rather than being forced to rely on derivative data such as classification schemes for land-use or land-cover. Finally, ERIN emphasizes the need for standards to facilitate sharing, synthesis, and understanding of data and information.

ERIN operates a World Wide Web server, "Australian Environment Online," that is accessible internationally over Internet through Web browser software like the popular Netscape and Mosaic packages. Its uniform resource locator (URL) is http://www.bdt.org.br/bin21/bin21.html. The ERIN web server demonstrates advanced information technology that brings together a wide range of environmental information and presents it to users, along with tools for spatial display of geographical information, modeling, and viewing.

BIN21

The Biodiversity Information Network (BIN21) initiative, managed by the Base de Dados Tropical in Brazil, was created to provide informational support that would further the purposes of the Convention on Biological Diversity and Agenda 21. BIN21 is developing a distributed, international network of sources of information to support research on biodiversity and conservation. Like ERIN, BIN21 now operates a World Wide Web server (URL http://www.ftpt.br).

BIN21 is working to create international "nodes" on the Internet and foster international exchange of information and collaboration. A U.S. National Biodiversity Information Center logically might join BIN21 as a node in the United States.

WCMC

The World Conservation Monitoring Centre (WCMC) offers yet another useful model. WCMC's Web server is found at URL http://www.wcmc.org.uk/. Also, WCMC operates the pilot Clearing House Mechanism for the Convention of Biological Diversity at URL http://www.wcmc.org.uk/~chm.

A PROPOSED U.S. NATIONAL BIODIVERSITY INFORMATION CENTER

Background

The concept of a focal point for efforts to understand and sustainably manage biodiversity in the United States has been discussed for over 20 years. In 1974, The Nature Conservancy pioneered the first of its State Natural Heritage Inventories in South Carolina. Natural Heritage Data Centers now exist in almost every state of the United States, using methodology that has been improved continuously since 1974 (Jenkins, 1988).

In 1986, a report of the Association of Systematics Collections (ASC), *Foundations for a National Biological Survey*, and a background paper of the Office of Technology Assessment (OTA), *Assessing Biodiversity in the United States: Data Considerations*, brought increased attention to this area. In 1987, the OTA made a modest recommendation for "a small clearinghouse for biological data" in its report, *Technologies to Maintain Biological Diversity*.

More recent examples of the continuing interest in the management of information about biodiversity include the 1991 policy dialogue report of the Keystone Center, *Biological Diversity on Federal Lands*, the 1993 report to the Smithsonian Institution on the National Center for Biodiversity by the ASC (Hoagland, 1993), and the 1993 report from the Council on Environmental Quality, *Incorporating Biodiversity Considerations Into Environmental Impact Analysis Under the National Environmental Policy Act*. The discussions in these reports dealt with a broad array of activities designed to improve the management of biodiversity, including regulation, conduct of government activities, and improvements in the amount and availability of information.

A Proposal for a National Institute for the Environment, published by the Committee for the National Institute for the Environment (1993), recommended that the proposed agency include a National Library for the Environment with modern information services and electronic technologies. Both the 1993 report from the National Research Council, *A Biological Survey for the Nation*, and the

1994 *Systematics Agenda 2000: Charting the Biosphere* cited decentralized computer networks of databases as absolutely central to the study and preservation of biodiversity.

From 1987-1992, several Congresses (with 1 Senate and 10 House of Representatives hearings on four versions of legislation on biodiversity) considered bills that would have instituted various regulatory programs, required consideration of biodiversity in environmental impact statements under the National Environmental Policy Act (NEPA), and established a national center of biodiversity. The Bush administration resisted these efforts, primarily due to concerns over NEPA and other regulatory and quasiregulatory provisions.

In these bills, the concept of a center was quite broad. A center would be the home of fairly ambitious efforts to (1) locate information that is available on the nation's biological resources; (2) assess the completeness of that base of information; (3) catalyze and, in some cases, actually fund efforts to fill gaps in information; (4) serve as a clearinghouse, or "database of databases," that would facilitate improved access to information by users; and (5) generate its own products, such as maps, flora, and other types of publications and media.

At the Earth Summit in Rio, in June 1992, the Bush administration embraced the idea of a national center of biodiversity, based on a more limited conception of its scope and role. A center was envisioned to include three general functions: (1) developing methods to access widely scattered information on biodiversity, (2) assessing the state of knowledge and identifying gaps, and (3) providing leadership in understanding and communicating about biodiversity. Plans were developed to establish by Executive Order a center at the Smithsonian Institution, but the order was not signed before President Clinton took office.

In 1993, the U.S. Department of the Interior established the National Biological Survey (later renamed the National Biological Service, NBS), with information management as an integral component. A report from the National Research Council, *A Biological Survey for the Nation* (1993:103), recommended that "Under the leadership of the NBS, the National Partnership for Biological Survey[1] should develop a National Biotic Resource Information System . . . (The NBS) should also participate in interagency initiatives to coordinate collection and management of biodiversity data by the federal government."

Recent Planning

With the establishment of a National Biological Survey, the Environmental Protection Agency provided funds to the Smithsonian Institution to lead an

---

[1] A term the National Research Council uses to refer to a broad spectrum of federal and nonfederal agencies and organizations that are carrying out activities which, in effect, constitute the National Biological Survey as broadly defined.

interagency planning study, beginning in the fall of 1993, to clarify the concept of a National Biodiversity Information Center (NBIC). A small ad hoc working group of staff that represented several federal agencies developed background materials to prepare for the establishment of a center. The materials included a "strawman" description of what might be included in a National Biodiversity Information Center and questions to be addressed in the planning process.

The Office of Management and Budget shared the results of this working group with the assistant secretaries of over 10 federal agencies involved in biodiversity research at a meeting in January, 1994. After receiving advice on policy from the assistant secretaries, the Smithsonian Institution began the next phase of planning with representatives of federal agencies, state agencies, academia, museums, nongovernmental organizations, and industry. An Advisory Planning Board chaired by Dr. Thomas E. Lovejoy, Smithsonian Assistant Secretary for Environmental and External Affairs, guided the process with representation from the organizations mentioned above. The Board held two meetings, March 21-22 and November 15, 1994, and published a consensus paper, *The National Biodiversity Information Center*, in December 1994.

The Board was responsible for setting the specific charge for and overseeing the work of the Drafting Committee, which was chaired by Dr. Robert S. Hoffmann, Assistant Secretary for Science at the Smithsonian Institution, and included members from the full spectrum of stakeholder organizations. The Drafting Committee produced a draft report that was taken to the meeting of the Board on November 15 for adoption. The report provided both a conceptual framework and a detailed account of the structure, location, and institutional setting of the Center and its principal tasks. Relationships between the Center and other organizations (including federal agencies) also were addressed.

Changes to the draft report, suggested at the Board meeting of November 15, were incorporated into the final consensus paper published in December 1994. The full text of the final report, minutes of the two meetings of the Advisory Planning Board, and a list of Board members can be obtained on the World Wide Web through the Biodiversity and Ecosystems NEtwork (BENE), URL http://straylight.tamu.edu/bene/nbic/nbic.html; via anonymous FTP from keck.tamu.edu/pub/bene/bene_texts; or from the NBIC area on the Smithsonian Institution's National Museum of Natural History Web and Gopher servers. The principal concepts coming from this report are presented below.

Mission and Objectives

The idea of the National Biodiversity Information Center is simple: it provides a focal point where the many parties that generate, manage, or use data on biological resources can collaborate and make decisions leading to broader access to that information. The Center will direct users to sources of the data they seek, while working with funding agencies to encourage development of tools and strategies to make data more accessible. The Center will not duplicate existing databases or information, but will provide directory services for the large array of information that is available. It also will identify gaps where new databases are needed, and assist the development, transfer, and application of new technologies. Further, the Center could coordinate access to data outside the usual realm of the biological sciences.

NBIC's mission will be to provide leadership and a neutral venue in which to facilitate collaborative discussions about the availability of data and information on biodiversity. It also will be a clearinghouse to provide knowledge of, enable access to, and facilitate the use and exchange of data and information on biodiversity. The Center's objectives will be to promote and encourage the use of well-documented data and information on biodiversity, address the full scope of biodiversity from molecular data through ecosystems, connect those seeking information and data on biodiversity to those having custody of such data, and facilitate structured identification of and access to data that is pertinent to a user's needs. This will be accomplished through an interactive computer system that uses metadata (data about the data) on geographic location, species, ecosystem, or other keywords to sort, aggregate, or integrate data sets, identify gaps in existing data and knowledge, and provide a forum for collaborative approaches to issues in biodiversity information.

Guiding Principles

The Center must be responsive to the needs of users, providing both data and information services that are tailored for different audiences. NBIC also must be responsive to the needs of providers, and must offer incentives and encouragement for them to provide their data through NBIC. The Center will facilitate the development of standards for metadata (minimum criteria for the documentation and format of data) and the establishment and provision of protocols for collecting and reporting data. Guidance on appropriate uses of data or information also will be provided. NBIC will facilitate the improvement in the quality of data sets with a feedback system that allows comments on the quality and utility of data. Custody of data will reside largely with primary collectors and producers of data, and users will be referred to original sources of data. Therefore, data holdings by NBIC will be reduced. NBIC will use appropriate technologies for the integration and analysis of information and will promote the adoption and use of appropriate standards of information.

Structure and Location

The Center will have a distributed structure that will function on a democratic, consensus-building, and partnership approach. NBIC will serve as a convener, facilitator, and host. Center experts will move discussions along and involve key constituencies. An Advisory or Governing Board from the broad community of contributors and users will provide general direction. A Technical Advisory Committee will ensure that NBIC receives computer and technical advice. NBIC must establish partnerships with the other organizations, such as the NBS, whose activities include collection of data and information and assessment of biodiversity issues.

NBIC's location should be decided by open competition, with a review after 3 years and recompetition every 5 years if so recommended by the review. NBIC should be designed to be moved if needed. Individual institutions or consortia of institutions should be invited to apply for selection as the Center host. Desirable characteristics of the host institution should include strong support for computational and information management services; a creative and active program in biological sciences, especially involving the use of computers in biodiversity information management; broadly based expertise and strong links to systematics, ecological research, and information management of collections; understanding of modern and historical North American collections; history of service and support to users; reasonable access to national and international transportation; and comfortable, modern facilities for conferences and the staff of the Center.

The National Biodiversity Information Center (NBIC), the National Biological Service (NBS), and the National Biological Information Infrastructure (NBII)

Both the NBS and the report of the National Research Council, *A Biological Survey for the Nation*, conceived of the establishment of a highly distributed information resource that would involve participants from federal, state, private, academic, and other institutions at scales ranging from local to national and global. Participants in this National Biological Information Infrastructure (NBII)[2] would work to develop tools, strategies, and collaborative activities that will bring coherence to the collection, storage, and management of data and increase access to biological information.

The newly created NBS has as a significant component of its mission the augmentation of access to biological information, relying upon a "distributed federation of databases" in accordance with the recommendations of the National Research Council. To achieve this goal, NBS plans to gather and make available information on sources of biological information (an electronic "card catalog" will be available in the near future), encourage the development of standards for metadata, and make investments in both its own and its partners'

---

[2]A term used by NBS to reflect a parallel with the existing National Spatial Data Infrastructure and equivalent to the National Biotic Resource Information System described in the report of the National Research Council.

hardware and software capabilities to move toward a highly distributed communications and access environment.

These activities and objectives closely complement those of the NBIC. The Center will assume the clearinghouse function of the NBII as envisioned by the NBS. NBS views the Center as a powerful tool for achieving key NBS objectives. The development of the "National Partnership for Biological Survey" requires that a wide range of partners meet regularly to work toward common strategies for achieving specific partnership goals (in this case, to increase the accessibility of information). NBIC provides the "neutral forum" around which those common strategies can be developed. The relationship between NBS and NBIC will evolve, with a shared goal of improving the accessibility of data and information on biodiversity.

In addition, the Center—by representing not solely a single government agency or the federal government but the entire biological information community in the United States—can provide important international linkages and make an important statement regarding the need for broad collaboration. The Convention on Biological Diversity encourages nations to develop such integrated information networks. Collaboration with the nongovernmental sector provides a powerful example for other nations.

It should be noted that, as this article is being written, the NBIC, NBS, and NBII and their interrelationships are still in a stage of formulation and flux. For this reason, the descriptions, roles, and interactions of these entities should be viewed as preliminary and are likely to change as technology advances and policies evolve.

CONCLUSION

The United States is beginning to recognize the critical role that information performs in supporting conservation of biodiversity. The Nature Conservancy's Natural Heritage data program pioneered more aggressive and systematic approaches to collect, process, and share information about the occurrences of biodiversity (Jenkins, 1988). The calls for creation of a new U.S. National Biological Survey have been answered, and the National Biological Service (NBS) now exists. The NBS will be leading a National Partnership for Biological Survey and helping to build a National Biological Information Infrastructure for the United States. International efforts such as ERIN and BIN21 are improving the global biological information infrastructure. A vision for a global "environmental channel" is emerging. The proposed U.S. National Biodiversity Information Center likely will play a key part in helping a wide range of users to gain awareness of and access to information on biodiversity in the United States. Interested readers may wish to examine the Biodiversity and Ecosystems NEtwork (BENE) on the World Wide Web to see a demonstration of some of the capabilities proposed for the US NBIC. BENE's URL is http://straylight.tamu.edu/bene/.

ACKNOWLEDGMENTS

Many of the ideas in this paper have been expressed previously by Jonathan Z. Cannon, Assistant Administrator for Administration and Resource Management, U.S. Environmental Protection Agency, in his presentation on "Information—The Key to Ecosystem Management" to the Environmental Information and Computing Technologies Conference on June 9, 1994. We thank Dr. David Blockstein of the Committee on the National Institute for the Environment and Dr. K. Elaine Hoagland of the Association of Systematics Collections for background information on the history of activities relating to the establishment of a National Biodiversity Information Center. The final concept for a National Biodiversity Information Center is taken from the report of the National Biodiversity Information Center Advisory Planning Board. We also acknowledge the ideas and other contributions of Dr. John R. Busby and Jacques Kapuscinski.

REFERENCES

Association of Systematics Collections. 1986. Foundations for a National Biological Survey. Association of Systematics Collections, Lawrence, Kans. 215 pp.

Committee for the National Institute for the Environment. 1993. A Proposal for a National Institute for the Environment: Need, Rationale, and Structure. Committee for the National Institute for the Environment, Washington, D.C. 99 pp.

Council on Environmental Quality. 1993. Incorporating Biodiversity Considerations Into Environmental Impact Analysis Under the National Environmental Policy Act. U.S. Government Printing Office, Washington, D.C. 29 pp.

Hoagland, K. E. 1993. A National Center for Biodiversity: A Report to the Smithsonian Institution. Abridged version for the National Research Council Committee on the National Biological Survey. Association of Systematics Collections, Washington, D.C. 16 pp.

Jenkins, R. E., Jr. 1988. Information management for the conservation of biodiversity. Pp. 231-239 in E. O. Wilson and F. M. Peter, eds., BioDiversity. National Academy Press, Washington, D.C.

Keystone Center. 1991. Biological Diversity on Federal Lands. Report of a Keystone Policy Dialogue. Keystone, Colo. 96 pp.

National Biodiversity Information Center Advisory Planning Board. 1994. The National Biodiversity Information Center. Washington, D.C. 13 pp.

National Research Council. 1993. A Biological Survey for the Nation. National Academy Press, Washington, D.C. 205 pp.

Office of Technology Assessment. 1986. Assessing Biodiversity in the United States: Data Considerations—Background Paper #2. OTA-BP-F-39. [NTIS-PB-86-181-989/AS] U.S. Government Printing Office, Washington, D.C. 72 pp.

Office of Technology Assessment. 1987. Technologies to Maintain Biological Diversity. OTA-F-330. U.S. Government Printing Office, Washington, D.C. 334 pp.

Systematics Agenda 2000. 1994. Systematics Agenda 2000: Charting the Biosphere. Technical Report. Systematics Agenda 2000, a Consortium of the American Society of Plant Taxonomists, the Society of Systematic Biologists, and the Willi Hennig Society, in cooperation with the Association of Systematics Collections, N.Y. 34 pp.

PART
VII

CONCLUSIONS

Declining amphibians are a worldwide warning of a biodiversity crisis.

CHAPTER
33

Santa Rosalia, the Turning of the Century, and a New Age of Exploration

MARJORIE L. REAKA-KUDLA
Professor, Department of Zoology, University of Maryland, College Park

DON E. WILSON
Director, Biodiversity Program, National Museum of Natural History, Smithsonian Institution, Washington, D.C.

EDWARD O. WILSON
Pellegrino University Professor, Museum of Comparative Zoology, Harvard University, Cambridge, Massachusetts

Although a number of scientists (including several contributors to this volume) previously had targeted the origin and maintenance of diversity in biological communities as a central issue in biology, the scientific community and the public began to be generally aware of declining biodiversity—and the tragically coincidental decline in numbers of scientists trained to analyze the diversity of the world's organisms—only within the last 10 years. One of the primary means by which these dual crises became known to the scientific community and the public was through the publication of *BioDiversity* by the National Academy Press (Wilson and Peter, 1988).

Since that time, national and international interest in the issue of biodiversity has risen to an all-time high. The public has become informed about the natural diversity of communities such as rain forests, prairies, wetlands, and coral reefs, and alarmed at the rate at which these natural wonders are being lost. Policy-makers have advocated or argued about our responsibility for saving endangered species and the costs and benefits of preserving them. Major strides have been made by the scientific community in understanding the processes that regulate biodiversity, predicting how much will be lost if habitats are degraded, and developing new technologies for the conservation and sustainable use of biodiversity. There has been a pervasive realization, however, that the number of trained specialists is low and the job of understanding the world's biodiversity is vast; and vigorous discussions have raged about how to approach the all-important task of documenting and conserving the world's biodiversity. Thus, some might think that the speed of destruction is too fast and that even to survey the world's resources in biodiversity is a task too overwhelming and expensive to undertake. The fact remains that far-reaching changes in the infrastructure of our knowledge have occurred since the seminal publication of *BioDiversity*.

The goal of the present volume is to summarize important conceptual and technological developments that have occurred in the field of biodiversity since the publication of the original volume, with a view to whether or not and how we can cost-effectively assess, understand, and manage our total global biodiversity. It is critically important to present this information in a format that is accessible to scientists, students, policy-makers, and the public, for only with their collective concurrence and support will the effort succeed.

THE HISTORICAL CONTEXT OF BIODIVERSITY IN WESTERN CULTURE

Humans always have been fascinated by biodiversity. As pointed out by Patrick (1983), an appreciation of the diversity of animals in included in the book of Genesis. Furthermore, the history of the modern science of biodiversity can be traced from the writings of the Greek philosopher, Aristotle, in the fourth century B.C., through medieval myths and manuscripts, which were ornamented with all manner of plants and animals, including invertebrates (Hutchinson, 1974; Waddell, 1934). In his *Systema Naturae* (1758), Carolus Linnaeus consolidated western knowledge of the names, pedigrees, and descriptions of all of the kinds of animals that were known at the time (4,379 species, of which 1,937 were insects).

The academic appreciation of the diversity of life continued into the nineteenth century, as represented especially in the work of Charles Darwin and Alfred Wallace, and experienced a resurgence during the synthesis of genetics, paleontology, comparative anatomy, and modern evolutionary theory in the "new systematics" of the 1940s and 1950s. A flurry of papers (mostly published in the young journal *Systematic Zoology*) raised questions about how many of the world's species from different major taxa and different environmental realms were known (Hyman, 1955; Muller and Campbell, 1954; Sabrosky, 1953a; Thorson, 1957). Citing how little still was known and how variable were the estimates of the numbers of species of different animal groups, Sabrosky (1953b) called for the Society of Systematic Zoologists to sponsor a global census of the animal kingdom that would be completed in 1958—the two-hundredth anniversary of the tenth edition of *Systema Naturae*.

Building upon earlier traditions in ecology (Elton, 1930; Hutchinson, 1957), G. Evelyn Hutchinson (1959) provided an icon for the synthesis of evolution, systematics, and ecology in his presidential address to the American Society of Naturalists, "Homage to Santa Rosalia, or why are there so many kinds of animals?" A stalactite-encrusted skeleton of Santa Rosalia, a twelfth century Sicilian saint, had been found in a cave near a hillside pond where Hutchinson found two species of water boatmen (Corixidae) living in abundance. This aquatic community stimulated Hutchinson to muse about how so few or so many species can coexist in any given environment. The query led Hutchinson, his students, and coworkers to formulate theories of the niche and species packing—and thus to address the number and functional assembly of species in natural communities—through their investigations of food webs, the structure of environmental mosaics, body size relationships among components of the community, and the evolutionary biogeography of island biotas. Thus, Santa Rosalia was erected as the "patroness of evolutionary studies" (Hutchinson, 1959:146).

Although it is clear how Hutchinson's question was connected to Santa Rosalia, an exploration of medieval manuscripts, which recognized the importance of human harmony with the world's biodiversity even then, suggests that some other saints might have been more appropriately tied to ecological and evolutionary studies of biodiversity than Santa Rosalia. For example, the medieval St. Kevin of Ireland lived among "the wild things of the mountains and the woods" that "came and kept him company, and would drink water . . . from his hands," and "there were times that the boughs and the leaves of the trees would sing sweet songs to St. Kevin" (Waddell, 1934:129; Figure 33-1). A huntsman

FIGURE 33-1 St. Kevin surrounded by beasts of the woods (from the translation of medieval legends from Latin by H. Waddell, with woodcuts by R. Gibbings; Waddell, 1934).

FIGURE 33-2 St. Brendan and his men above innumerable beasts of the sea (Waddell, 1934).

and his hounds who chased a wild boar into the vicinity were so impressed with the diversity of animals flitting around St. Kevin and perched on his shoulders that the hounds lay down, refusing to attack the boar, and the hunter returned to tell the King of Leinster. Similarly for the marine realm, another medieval tale tells of St. Brendan and his men, who were celebrating the Feast of St. Paul in a boat on the Irish Sea when the men said "Sing lower, Master, or we shall be shipwrecked. For the water is so clear that we can see to the bottom, and we see innumerable fishes great and fierce, such as never were discovered to human eye before . . . The creatures rose on all sides, making merry for joy of the Feast, followed after the boat for the day, and then returned to the deep" (Waddell, 1934:111; Figure 33-2). An additional legend chronicles an appreciation for the utility of biodiversity. St. Colman eschewed earthly possessions, but kept a ·mouse, a cock, and a fly as companions. The rooster crowed to wake him for his prayers, the mouse nibbled at him to remind him of the time for his holy vows, and the fly treaded up and down his codex to sit on the line where St. Colman had halted his reading so that he could refind his place (Figure 33-3). The legend describes his sorrow when his small companions died and he was left alone (Waddell, 1934).

Consequently, an appreciation of biodiversity runs deep in western culture, and an awareness of how much of the abundance and variety of animal life has declined, even since medieval times, can be gained from this literature. While the academic tradition of studying the diversity of life continued into the nineteenth century and flourished beyond the first half of the twentieth century during the development of the "new systematics," the science of biodiversity was overshadowed in the 1960s, 1970s, and early 1980s by the spectacular growth of molecular biology (Wilson, 1994), during which time the numbers of systematists contributing to the "new synthesis" declined unnoticed by most academics. Interest in the field by academicians, policy-makers, and the public was reawakened by publication of *BioDiversity*.

Aside from the academic tradition of biodiversity, another powerful influence, related to biodiversity, brought our culture to its current level of technological development: the exploration of the New World. From the thirteenth to the nineteenth centuries, technological developments in navigation allowed European voyagers to embark on an unprecedented exploration of the globe.

FIGURE 33-3   The utility of beastly companions to St. Colman (Waddell, 1934).

These expeditions revolutionized knowledge of the geography, human culture, and biology of the world at the time. This ultimately led to a reevaluation of human society's place in the world and an understanding of the evolution of all living things. But the explorations also allowed the acquisition of untold wealth in living and nonliving natural resources (see also Kangas, Chapter 25, and Mehrhoff, Chapter 29, this volume), which was brought back from the New World and invested in the culture of western Europe.

In North America, the Louisiana Purchase in 1803 was followed by the Lewis and Clark Expedition (1804-1806), which culminated Thomas Jefferson's long-planned (and congressionally approved) effort to chronicle the natural wealth of the continent. This widely celebrated geographic and scientific expedition mapped the natural history of the central and northwestern part of the continent and brought back many plant and animal specimens that were unknown previously (including the prairie dog, jackrabbit, black-tailed deer, pronghorned antelope, and mountain sheep), providing perhaps the first "national biological survey" (also see Mehrhoff, Chapter 29 of this volume, for the historical role of museums and surveys). The subsequent acquisition of Florida, Texas, and California from France and Spain by the middle of the nineteenth century and the expansion of settlers into these regions widened knowledge of the natural resources of the continent. Continuing the earlier European explorations, the knowledge that was gained and the use of the newly discovered natural resources from the exploration of the American continent changed western civilization forever, bringing us to our current level of technological development (but which, in turn, has drawn us to the current brink of environmental disaster).

THE NEED FOR A NEW EXPLORATION OF THE BIOSPHERE

Five centuries beyond the beginning of New World explorations and 150 years after the opening of the American West, we now have the technological knowledge to embark on a new age of exploration of the globe that is of comparable importance for human culture and knowledge to those historical events. For the first time, we have the potential capability (with a skeletal infrastructure of scientific personnel and institutions) to undertake a thoughtful inventory of global biotic resources (Raven and Wilson, 1992); to understand and predict the processes that govern the amount, location, and sustainability of these resources; and to provide the information that is necessary to protect as much of the remaining global biota as possible so that humans and the biosphere may persist. What is essential now is a coordinated effort—comparable in scale and idealism (and very likely of greater importance for our long-term survival) to the Human Genome Project—to explore and describe (and only thusly to understand, predict, and manage) the biota of planet Earth.

PROGRESS SINCE 1986, PRESENT AND EMERGING INFRASTRUCTURE, AND COST-EFFECTIVE SOLUTIONS

Status of the Field

This volume, an outgrowth of the Inaugural Symposium of the Consortium for Systematics and Biodiversity (comprised of the University of Maryland at College Park, the Smithsonian Institution Natural History Museum, the U.S. Department of Agriculture Systematic Laboratories, the University of Maryland Biotechnology Institute, and the American Type Culture Collection), is organized developmentally. Edward O. Wilson opens the book with a discussion of the historical context of biodiversity and demonstrates how biodiversity includes all facets of biology and even all fields of science. It is a field of synthesis, forged by necessity. As when any society faces a threat to survival, the challenge of declining biodiversity offers unification. All scientific fields can (and do) contribute to the advancing front of knowledge in biodiversity; we are limited only by our vision of how to employ their contributions most effectively to conserve biodiversity. This introductory chapter addresses the opportunity and the need to explore the biosphere.

The Meaning and Value of Biodiversity

Part I of this book is devoted to an exposition of biodiversity by two early explorers of the field, Ruth Patrick and Thomas Lovejoy. The purpose of this volume is to educate students, teachers, scientists, policy-makers, and the public about what biodiversity is and how important it is to understand it. These authors eloquently explain to this broad audience why biodiversity has become a central issue at local, national, and international levels. Chapters elsewhere in the volume also address the meaning and value of biodiversity. For example, Miller and Rossman, as well as Solis, stress the critical importance of the science of biodiversity for agriculture (Part IV). Janzen champions the use of biodiversity as a primary means of making society aware of its value so that it can be conserved (Part V). Jordan describes the process of restoration of biodiversity as meaningful in itself to human society (Part V).

Patterns of Biodiversity

Part II assesses the question of how much and where biodiversity occurs on the globe—a fundamental question that, astonishingly, is still unknown, although, as the chapters and their citations demonstrate, phenomenal progress has been made on this issue during the last decade. Erwin, pioneer of the revolutionary figures for the global diversity of insects that spurred debate on how many millions of species exist on Earth, provides an update to the long-standing question of why the greatest proportion of species on Earth are insects (Erwin,

1982; Gaston, 1991; Hutchinson, 1959; Labandeira and Sepkoski, 1993; Linnaeus, 1758; May, 1978; Muller and Campbell, 1954; Sabrosky, 1953a; Stork, 1988; Wilson, 1988). Focusing on beetles in rain forest canopies, Erwin shows how concentrated studies in local geographic areas can improve our understanding of which and how many species occur in natural communities; importantly, he delineates the practical steps that make the acquisition of this knowledge feasible within a reasonable time frame. Stork carefully evaluates the progress made over the last decade in our estimates of global biodiversity. He concludes that 5-15 million (perhaps near 12 million) species probably occur on Earth. His chapter culminates in a sobering, carefully considered discussion of current extinction rates and prospects. Robbins and Opler also address the most speciose group of organisms on Earth, insects. They provide the important result that biogeographic patterns of butterfly diversity are similar to those of birds, so that conservation efforts targeted at particular regions will protect both components of the biological community. However, mammals show a different pattern, heralding caution in conservation policies (also see chapters by Myers, Scott and Csuti, Dietz, and Thomas in Parts III and IV of this volume for elucidation of emerging technology and recommendations on how to prioritize biodiversity efforts among different areas, habitats, and organisms).

Moving to the little-known aquatic realm, Reaka-Kudla employs biogeographic theory to provide the first quantified estimate of the number of species that inhabit the second pinnacle of diversity and extravagant adaptation on Earth, coral reefs. The results show that only about 93,000 species of described species live on global coral reefs, a paltry figure compared to the number of described species that inhabit rain forests (less than a million, although at least 2 million described and undescribed species almost certainly occur there). However, this chapter determines that global coral reefs occupy 20 times less area than global rain forests, and the data suggest that less than 10% of the species on coral reefs are known. A conservative estimate indicates that about a million species exist on global coral reefs. Moreover, Reaka-Kudla's results suggest that the possibility of current and future extinction of coral reef species (which are predominantly small in body size with restricted geographic ranges) is high, rather than low, as some workers have suggested for species in marine environments. All of the authors in this section and several authors elsewhere in the volume stress that the greatest undocumented diversity is found in organisms of small body size and in tropical regions, and that many species in these groups and regions are likely to become extinct before they are even known if swift conservation measures are not adopted.

Sogin and Hinkle provide a fascinating view of the evolution of microbial diversity (another enigma whose international profile has been elevated by increasing interest in biodiversity over the last decade). Their results are garnered with new molecular techniques that are only beginning to sketch the true dimensions of the biosphere. This approach elucidates the antiquity of genetically distinct lineages in the global biota. Instead of five organismal king-doms (bacteria [prokaryotes], protistans [protozoans and relatives], fungi, plants, and animals), these authors show three lines of descent in the evolution of living organisms: the Archaebacteria (Archaea), Bacteria, and Eukarya (eu-karyotes [protistans and multicellular organisms]). Startlingly, the Eukarya are nearly as ancient as the other two groups, and the diversity and antiquity of relationships among the various microbial organisms is no less than astounding. Colwell also addresses the significance of microbial diversity for biotechnology in a later section of the volume (Part V). Other chapters (Part IV) provide addi-tional contributions to our understanding of patterns in biodiversity, such as the studies of the evolution of bower-building in birds (Borgia), hyperdiversity in an economically important group of moths (Solis), and coevolutionary pat-terns in several groups of parasites and their hosts (Hoberg).

Threats to Biodiversity

The topic of Part III of the volume is as ominous as it is imperative to scruti-nize: the threat of extinction (Wilson, 1989). Myers provides a riveting examina-tion of several issues related to the decline of biodiversity. He notes that, although lakes and rivers contain a quarter of the known species in 0.01% of the planet's water, these freshwater ecosystems are being degraded faster than any other biome. In the United States, only 2% of the rivers are free-flowing and 20-55% of their major taxa are endangered or extinct. In the spectacular species swarms of cichlid fish in the African rift lakes (where 44-95% of the species are endemic), introduced predators and other disturbances have reduced species by 67%, and these losses are likely to accelerate. Discussing the decline of biodiversity over the last 30,000 years (during which time 70% of North American mammals have been lost) as well as the accelerating current impacts of humans, Myers notes the importance of the loss of populations and genetic variability for the future survival of spe-cies. He concludes that, because of the scale of anthropogenic changes in the environment, we must conserve the biosphere rather than only parks and re-serves, and that, if recovery takes 5 million years, this decision will affect more people (an estimated 500 trillion) than any other decision in human history.

Steadman examines human-caused extinctions in birds, which began long before the impact of Europeans. Whereas prehistoric extinctions caused by humans were greatest on islands (especially in the Pacific), population declines and extinctions now have begun in earnest on the continents. Steadman's stud-ies identify what regions and countries (e.g., Indonesia, Solomon Islands) are most at risk of losing species of birds. The most significant agent of bird extinc-tion, habitat destruction, is greatest in the tropics (see also Stork and Reaka-Kudla in Part II). Later in the volume (Part IV), Borgia also points out the dangers that accelerating contemporary extinctions in birds pose for our under-standing of biodiversity. Comparative information from many species is necessary to understand the evolution of complex or extreme traits such as building bowers, elaborate plumage, and nuptial dances in bowerbirds; this window to knowledge soon may be closed. Now is not a good time to reduce funding for basic research.

Wing provides a longer-term perspective on the risks of extinction in his examination of the causes and consequences of the rapid global warming event that occurred during the late Paleocene and early Eocene. This chapter describes different mechanisms that caused extinctions in marine versus terrestrial organisms during this period (which was warmer than now), shows that differential capability for migration may cause different patterns of declining diversity in paleontological data (e.g., in mammals versus plants), and cautions that models in current use may not accurately predict the trajectory of global climate. Forseth also cautions that plant assemblages are likely to change in new ways in response to climatic change, and that we need to understand how species respond to multiple environmental stresses—and how these responses translate to ecosystem responses, biogeography, and potential extinction—in order to sustain biodiversity in the future.

Other chapters in the volume echo the thought that we need to evaluate the responses of organisms to changing environments through historical analysis. For example, Hoberg (Part IV) uses historical reconstruction of the phylogenies of parasites and their hosts to evaluate how patterns of biodiversity are influenced by climatic, geological, and biotic factors. Studies of freshwater stingrays and their parasites in the Amazon and tapeworms and their marsupial hosts in South America and Australia indicate that these ecological associations have been stable for 60 million years. On the other hand, phylogenetic relationships of pinnipeds and seabirds and their relatively recent tapeworm parasites suggest that climatic changes associated with the Pliocene-Pleistocene glaciations promoted isolation and speciation rather than extinction in the North Atlantic and North Pacific regions. Comparing behavior patterns with phylogenies derived from mitochondrial DNA, Borgia (Part IV) evaluates the history of displays in bowerbirds and examines alternative hypotheses for the selective pressures that shaped the diversity of bowers and species in this group. Friedlander et al. (Part V) show how molecular data can be used to estimate times of divergence among major taxa. From such data, one can estimate how many millions of years will be required to replace lineages lost through extinction, and thus how grave the threat of current extinctions is.

Understanding and Using Biodiversity

Part IV addresses the means we use to comprehend and use biodiversity. Patrick explains why the study of systematics is central to the biodiversity agenda. With the description of a systematics training program that was chartered by several leading institutions to provide mentoring for critically needed young systematists, she provides an inspiring example of how relatively small programmatic efforts can have a significant impact. Wheeler and Cracraft also provide an eloquent plea for the training of young systematists in a later section of the volume (Part VI). They indicate that a single generation of young scholars could reverse the erosion of systematic expertise that has occurred over the last several decades.

Embellishing a tradition begun with Linnaeus, Thompson shows how important precise names for organisms are for the effectiveness of the current biodiversity agenda (also see Farr and Rossman, Part VI of this volume). Miller and Rossman and Solis demonstrate the overwhelming economic importance of research in systematics and biodiversity. They show that, as we progress beyond a period in which agriculture was dominated by the use of pesticides, we need to know the world-wide taxonomy and distribution of pests and their natural enemies in order to implement effective mechanisms of biological control and integrated pest management. For example, inaccurate identification of a mealybug—which was costing African cassava farmers $1.4 billion per year—led to the introduction of ineffective natural enemies from the wrong geographic region and delayed control of the pest for several years (at which time a systematist studying the entire group identified the pest as a new species from another region; only then were successful natural enemies obtained). Miller and Rossman document the economic importance of introduced pests into the United States— over $196 trillion for the >450,000 species that had been detected and identified up until 1991. The systematics of these pests must be accurately known before the U.S. Department of Agriculture APHIS quarantine and pest exclusion system can be effective (e.g., a near-international incident was averted when a taxonomist found that smut on Canadian wheat was a contaminant from the storage of rice rather than a new species that afflicted wheat; incorrect identification could have caused a ban on all wheat from Canada). Focusing on a group of hyperdiverse moths that are major pests of crops and stored grains as well as important agents of biological control for weedy plants, Solis' chapter is an excellent example of how to use systematics and education to rapidly generate and transfer essential information about biodiversity in a tropical country (e.g., in 10 versus 30 years). She demonstrates that, in order to increase the taxonomic self-sufficiency of tropical countries and to be effective in pest control, taxonomic studies in individual countries must be framed in a global context—incorporating information from the literature, collections, and taxonomists throughout the world—and placed at the fingertips of ecologists, conservationists, quarantine officers, and farmers in usable form.

Building Toward a Solution

Part V focuses on new technology and new directions that offer hope that the biodiversity crisis can be effectively addressed in a timely fashion. Colwell describes the current state of the biotechnology industry, the promise it offers for agricultural and medical developments that would improve human conditions, and its critical dependence on the preservation of biodiversity in the wild. Showcasing research that emanates from a laboratory intimately involved with the Human Genome Project, Bult and her coauthors describe revolutionary new genetic technology that can be used to identify and quantify genetic diversity at the levels of populations, species, and lineages more rapidly than ever before. Combined with the techniques for data management that have been developed in the authors' institute (see also Farr and Rossman, Umminger and Young, Part VI) and the interdisciplinary collaboration (involving systematists, evolutionary biologists, molecular biologists, computational biologists, computer scientists) that is modeled in Bult et al.'s chapter, these technological breakthroughs offer real hope for an assessment and understanding of global biodiversity that was not possible even 5 years ago. Similarly, the chapter by Friedlander and his coauthors represents rapid advancement in conceptual and molecular analyses of phylogenetic relationships—and hence our understanding of the patterns and processes that generate diversity within lineages. In an earlier section of the volume (Part IV), Thompson also discusses how taxonomic identifications in biodiversity inventories can be made more rapidly with computerized technology.

In quite another new direction of the field, Scott and Csuti describe their Gap Analysis Program, which uses remote-sensing from satellites, geographic information systems, and sophisticated integration of data to assess the status and distribution of several important elements of biodiversity on a landscape scale. Gap analysis provides a layered framework for integrating data on species (abundance, distribution), types of habitats, topography, soils, climate, and human components (land-use, zoning, development, population density) so that critical areas or habitats that contain species or assemblages at risk can be identified, monitored, and their trajectories predicted on computer-generated maps. This type of analysis has shown that endangered assemblages often are not included in protected areas or management plans. For example, only 5% of Hawaii's endangered forest birds live in forest preserves (Scott et al., 1986). Gap analysis also can be used to identify how widespread or abundant species must be to avoid precipitous declines in population, and can identify which of several combinations of contiguous areas would provide the most overall diversity if conserved. The latter allows for some flexibility in conservation policies in response to constraints imposed by human needs or costs.

Building upon a foundation of knowledge about the systematics, geographic distribution, ecology, genetics, and behavior of golden lion tamarins and other neotropical primates, Dietz also employs images derived from satellite-borne instruments, geographic information systems, and sophisticated technology to integrate large sets of data and construct conservation strategies. He outlines several specific approaches—including application of the scientific method, information management, captive breeding and reintroduction, and educational and collaborative activites—to provide hope that these endangered primates will be sustained.

Thomas advocates selecting areas of evolutionary diversification for conservation (also see Myers, Part III of this volume) and indicates that interdisciplinary approaches (including plate tectonics, paleogeography, cladistics, ecology, and anthropology) are necessary to understand and pinpoint critical areas of biodiversity. Using the exceedingly diverse coral reefs of Papua New Guinea as an example, he points out that small, spatially inconspicuous organisms may be better indicators of environmental change than large, conspicuous, and less abundant species. The technological and conceptual developments presented in this section of the volume suggest that assessment and continued monitoring of global natural resources is a feasible goal within the next decade or two.

In an evocative treatment of emerging ideas in the relatively new field of ecological restoration, Jordan emphasizes that "restoration" is not the complete rebuilding of a damaged environment, but is any attempt at conservation that ensures the existence of the system in the long run. Rather than focusing on the products of restored ecosystems, he suggests that the experience and performance of restoration—getting people involved with the environment—is the most immediate and valuable payoff from restoration activities.

Similarly, the concept of sustainable development—the long-term renewable use of biotic resources—is undergoing a renaissance in biodiversity studies from its prior use in fisheries and related harvestables. In a humanistic treatment of sustainable development in tropical forest lands, Kangas discusses the practical interdisciplinary aspects of how to make conservation work. He provides a suite of illustrative historical examples (Mayan culture, Henry Ford's sustainable villages, the roots of the American Civil War, extranational companies, the biotic and economic exchanges fostered by Christopher Columbus, the origin of agriculture, the Louisiana Purchase, the American colonial tobacco agricultural system, Puerto Rican deforestation, the Civilian Conservation Corps, and a Brazilian forest) from which we can learn about interactions between humans, their economic activities, and the environment. These examples of what can be done, of approaches that worked (and some that did not), are very welcome. However, even this chapter cites a new urgency—not seen in the examples from prior centuries—in the need for sustainable development that protects the long-term function of biotic communities.

Janzen indicates that we are at the crossroads of biodiversity and wildland management. One route is to preserve biodiversity without heavy use, which he argues ultimately will result in less biota being conserved in a smaller total area. The other avenue is to use the biodiversity. Because of the attendant valuation of biodiversity, he suggests that a larger percentage of the remaining biota and larger amounts of land can be conserved if the second route is taken.

Due to the urgency of declining biodiversity, Janzen indicates that we need to streamline collection of knowledge about biodiversity and gather information only on the taxonomy, biogeography, and natural history of organisms. This emphasizes the need to reinstate these traditional and important fields in our university curricula and national priorities.

The architect of All Taxa Biodiversity Inventories (ATBIs), Janzen suggests that these intensive field samples provide the best and most cost-effective way to sample biodiversity and assess its trajectory over time, one that is most tailored to the needs of local and national peoples. Others have questioned how many ATBIs would be necessary and whether they would sample global biodiversity adequately. Some scientists think that intensive local or regional programs can sample biodiversity adequately if located properly (chapters by Erwin, Myers, Scott and Csuti, and Thomas in Parts II, III, and V in this volume address aspects of this issue), but other scientists worry that the patterns of species richness and endemism may vary for different taxa (chapters by Wheeler and Cracraft, Robbins and Opler in Parts VI and II) or that too few sites could be sampled and species with restricted ranges would be missed (e.g., see chapters by Wheeler and Cracraft, Reaka-Kudla in Parts VI and II).

Whereas Janzen's vision (Yoon, 1993) contrasts to some extent with that of some systematists (Wheeler and Cracraft, Part VI) and has been termed an "ecological" approach, Janzen's agenda does emphasize the validity and importance of focusing on taxonomy and on genetic variants, populations, species, and lineages of organisms, as opposed to geochemical cycles, nutrient flow, or other ecological attributes that are not directly tied to evolutionary units. In contrast to this view, some researchers have suggested that the function and productivity of ecosystems should be the essential targets of research and conservation as environments deteriorate, and that the number of species present is not relevant (see discussions in Baskin, 1994; Walker, 1992). The latter position ignores the present and future value of all the agricultural, medical, genetic, and other products contained within the species of the global biota (as discussed by Lovejoy, Patrick, Miller and Rossman, Solis, and Colwell in Parts I, IV, and V of this volume; see also "The Fundamental Unit" in Wilson, 1992). Exclusive concentration on the function of ecosystems, rather than on genetic and species units, also ignores the aesthetic and moral values of biodiversity (see Jordan, this section; and Wilson, 1984, 1992). We cannot afford to lose the historical legacy of lineage diversity or the prisms that allow its comprehension and use—systematics, taxonomy, and natural history.

Getting the Job Done

Part VI concentrates on the infrastructure that is in place for assessing and understanding biodiversity. In a chapter that touches on many of the issues addressed throughout the volume, Wheeler and Cracraft document the present and needed infrastructure for undertaking a global systematic survey of biota, as proposed by the *Systematic Agenda 2000*. Mehrhoff documents the historical role that biological surveys and museums—centerpieces in the struggle to document global biodiversity—have played in understanding and ultimately protecting our biota. Their contemporary role may be even more important. By fostering knowledge of systematics and taxonomy, surveys and museums offer the only hope of understanding how our current living resources evolved and how natural communities function. Discussing a technological realm about which most people have little knowledge, Roblin also indicates the promise and the challenge of maintaining living collections (especially of the microorganisms of which we know so little) to facilitate our stewardship of biodiversity. These papers emphasize the need for changes in our approach to elementary, secondary, and higher education. The interest of many well-known contemporary biologists was kindled first in museums or by their experiences in natural history. Will there be a comparable generation 50 years from now? We need more hands-on laboratory experiences and courses that expose students to the emotional thrill and the intellectual challenge of biodiversity that is seen in museums and in the field (see also Wilson, 1994).

The enormously complex task of managing data in an endeavor the magnitude of a global biodiversity assessment is one of the greatest logistical problems faced by scientists, but this problem can and is being solved. Farr and Rossman provide a detailed examination of two approaches to the management and interpretation of data that are effective, accurate, and in place (also see chapters by Bult et al., Scott and Csuti, and Dietz in Part V). Umminger and Young describe the explosion of national and international electronic systems that allow summarization and communication of environmental data. Stressing that information management is critical for the success of the international agenda in biodiversity, these authors describe the development, status, and goals of a U.S. National Biodiversity Information Center related to the National Biological Survey.

CONCLUSIONS

The most important message in this volume is also a hopeful one: the institutional infrastructure (museums and their collections; state, national and global biological surveys and their associated data banks; universities, institutes, and governmental and nongovernmental agencies that support research, training, and conservation policy) is already in place or in developmental stages. The infrastructure to solve the biodiversity crisis does not need to be built from the ground up; but current institutions must increase their linkages, and they must have enhanced support. The human resources are in place: a small but expert and committed community of systematists already is active. It will be essential to augment funding, especially for training young people in systematics and collection management, in order to make this infrastructure more effective for the great task at hand. As is well indicated in this volume, the knowledge-base about biodiversity and how to manage it is substantial, is growing rapidly, and is still very deficient for our needs. We must focus it, enlarge it, and fuse it inextricably with systematics.

Interinstitutional links (maximizing shared human, financial, and institutional resources between museums, universities, and governmental and nongovernmental agencies) will provide the key to cost-effective accomplishment of a global biodiversity agenda. Representing a sea change that already is under way, 10 of the 33 chapters (30.3%) of the present volume are multiply authored, in comparison to 2 of the 57 chapters (3.5%) that were presented in the first biodiversity symposium (Wilson and Peters, 1988). Larger, more interdisciplinary research groups—often representing linkages among several institutions—are increasingly common. Of the 10 chapters with multiple authors in this volume, half result from collaborations among researchers from more than one institution.

Interdisciplinary sharing of knowledge and resources is evident in several chapters of this volume. The Consortium for Systematics and Biodiversity of the greater Washington, D.C., area, one of the stimuli for the present volume, is an example of sharing physical, institutional, and human resources in order to maximize our potential impact on global biodiversity issues. Collectively, this group of institutions includes one of the largest groups of systematists, evolutionists, ecologists, and biotechnologists, and one of the largest biological collections (from molecular to organismal, all kingdoms) in the world. Each institution brings a unique research or training capability to the Consortium. Another example of highly effective collaboration is represented by a group of scientists from British universities and museums who, through individual and joint efforts, have made many significant contributions to understanding biodiversity. The collaborative training program in systematics between a number of major American universities and the Smithsonian Institution described by Patrick (Part IV, this volume) represents an example of productive collaboration among institutions and scientists. Mehrhoff (Part VI, this volume) describes the effective union of 10 state biological surveys into a Consortium. The widely known publications and biodiversity agenda of *Systematics Agenda 2000* resulted from the joint efforts of a consortium of professional associations in systematics (Wheeler and Cracraft, Part VI, this volume). The emerging Informational Center of the National Biological Service (Umminger and Young, Part VI, this volume) will involve collaborations among government agencies, universities, museums, research institutes, and individual scientists. We need more and better-funded cooperative efforts to maximize the ability of museums, universities, governmental and nongovernmental organizations, and the systematic community to implement the biodiversity agenda.

The contributions to this volume suggest that understanding and managing global biodiversity is an attainable, cost-effective, and vitally important goal for human society, and that increased public and governmental support for museums and associated institutions is justified and essential. This support should be directed toward enhancing an existing physical, human, and informational infrastructure in systematics and biodiversity, and especially toward training new young specialists in this area. One of the objectives of the volume has been to educate students who may be stimulated to enter the field, teachers, scientists, policy-makers, and the public about the infrastructure that already exists for understanding biodiversity, and about the urgent need to build upon this strength. We hope to have instilled in this broad audience the idea that we can and must act now.

ACKNOWLEDGMENTS

We wish to express our special appreciation to Valerie Cappola, Jocelyn Kasow, and Mitchell Tartt for their dedication and assistance during the final stages of preparation of this volume.

REFERENCES

Baskin, Y. 1994. Ecosystem function of biodiversity. BioScience 144:657-660.

Elton, C. 1930. Animal Ecology and Evolution. Clarendon Press, London. 296 pp.

Erwin, T. L. 1982. Tropical forests: Their richness in Coleoptera and other Arthropod species. Coleopt. Bull. 36:74-75.

Gaston, K. J. 1991. The magnitude of global insect species richness. Conserv. Biol. 5:283-296.

Hutchinson, G. E. 1957. Concluding remarks. Cold Spring Harbor Symp. Quant. Biol. 22:415-427.

Hutchinson, G. E. 1959. Homage to Santa Rosalia, or why are there so many kinds of animals? Amer. Nat. 93:145-159.

Hutchinson, G. E. 1974. Aposematic insects and the Masters of the Brussels Initials. Amer. Sci. 62:161-171.

Hyman, L. H. 1955. How many species? Syst. Zool. 4:142-143.

Labandeira, C. C., and J. J. Sepkoski, Jr. 1993. Insect diversity in the fossil record. Science 261:310-315.

Linnaeus, C. 1758. Systema Naturae, tenth ed. L. Salvii, Stockholm.

May, R. M. 1978. The dynamics and diversity of insect faunas. Pp. 188-204 in L. A. Mound and N. Waloff, eds., Diversity of Insect Faunas. Blackwell Scientific Publications, Oxford, England.

Muller, S. W., and A. Campbell. 1954. The relative number of living and fossil species of animals. Syst. Zool. 3:168-170.

Patrick, R. 1983. Introduction. Pp. 1-5 in R. Patrick, ed., Diversity. Benchmark Papers in Ecology, Vol. 13. Hutchinson Ross Publishing Company, Stroudsburg, Pa.

Raven, P. H., and E. O. Wilson. 1992. A fifty-year plan for biodiversity surveys. Science 258:1099-1100.

Sabrosky, C. W. 1953a. How many insects are there? Syst. Zool. 2:31-36.

Sabrosky, C. W. 1953b. An animal census for 1958? Syst. Zool. 2:142-143.

Scott, J. M., S. Mountainspring, F. L. Ramsey, and C. B. Kepler. 1986. Forest bird communities of the Hawaiian Islands: Their dynamics, ecology, and conservation. Stud. Avian Biol. 9:1-431.

Stork, N. E. 1988. Insect diversity: Facts, fiction and speculation. Biol. J. Linn. Soc. 35:321-337.

Thorson, G. 1957. Bottom communities. In J. W. Hedgepeth, ed., Treatise on Marine Ecology and Paleoecology, Vol. 1. Mem. Geol. Soc. Amer. 67:461-534.

Waddell, H. 1934. Beasts and Saints. Translations by H. Waddell, woodcuts by R. Gibbings. Henry Holt and Company, N.Y. 151 pp.

Walker, B. H. 1992. Biodiversity and ecological redundancy. Conserv. Biol. 6:18-23.

Wilson, E. O. 1984. Biophilia. The Human Bond with Other Species. Harvard University Press, Cambridge, Mass. 157 pp.

Wilson, E. O. 1988. The current state of biological diversity. Pp. 3-18 in E. O. Wilson and F. M. Peter, eds., BioDiversity. National Academy Press, Washington, D.C.

Wilson, E. O. 1989. Threats to biodiversity. Sci. Amer. (September):108-116.

Wilson, E. O. 1992. The Diversity of Life. Belknap Press, Cambridge, Mass. 424 pp.

Wilson, E. O. 1994. The Naturalist. Island Press, Washington, D.C. 380 pp.

Wilson, E. O., and F. M. Peter, eds. 1988. BioDiversity. National Academy Press, Washington, D.C. 521 pp.

Yoon, C. K. 1993. Counting creatures great and small. Science 260:620-622.

Photo Credits

Index

Germplasm collections, 131. *See also* Research collections, living
Glaciers, 179, 181
Glaphyriinae, 234, 237–238
Gliocladium sp. (fungus), 226
Global climate change
  carbon dioxide, 188, 195
  carbon isotopes, 173
  disturbance regimes, 189
  ecological restoration, 375–376
  effects, 12, 84, 134, 135, 173–175, 187, 195, 255, 257, 258
  environmental stresses, 190–192, 516
  Eocene epoch, 167–170, 171
  equable climate paradox, 167–170
  extinctions, 172–173, 176–181, 182, 516
  general circulation models, 163, 167–168, 181, 187–188, 516
  greenhouse gases, 12, 187, 195, 283
  Holocene deglaciation, 163, 179
  host-parasite systems, 255, 257
  introduced species, 194–195
  leaf physiognomy, 164–167, 168, 169, 171
  nearest living relative, 164
  oxygen isotope studies, 166, 171
  Paleocene-Eocene boundary, 163–164, 172, 176–181, 516
  paleoclimate reconstructions, 164–167, 170, 181–182, 188–189
  plant response, 188–189, 194–195
  Pliocene-Pleistocene glaciations, 255, 256, 257, 516
  proxy data, 163, 164, 166, 188
  research approach, 182
  response of species, 143, 179, 188–189, 257, 258, 516
  Terminal Paleocene Event, 170–176, 179
Global Environment Facility, 126, 134
Glyphyriinae, 233
Golden-cheeked warbler, 146
Golden Lion Tamarin Conservation Program, 349
Gomphonema olivaceum (diatom), 22
Gray birch, 192
Great auk, 145
Great Barrier Reef, Australia, 87, 284, 362, 364
Great Lakes, 127
Greater prairie chicken, 146
Greece, 131
Greenhouse gases, 12, 169, 187, 195, 283
Greenland, 78

Guadalupe River, 18, 19, 24
Guam, 149, 365–366
Guanacaste Conservation Area, Costa Rica, 32–33
Guatemala, 49, 128
Gulls, 148, 149
Guyana, deforestation, 11
Gygis microrhyncha (tern), 151
Gymnogyps californianus (California condor), 144, 145
Gymnorhina tibicen (Australian magpie), 267

H

Habitat loss. *See also specific habitats*
  biodiversity, 11
  birds, 11, 129–130, 142, 144, 154
  deforestation, 1–2
  extinctions, 61, 142, 144, 154
  fragmentation, 11, 102, 135, 143, 146, 375
  rates, 86
Habitat monitoring, primate, 345–347
Haemonchus contortus (nematode), 310
Hakalau Forest National Wildlife Refuge, 322
Halcyon sp. (kingfisher), 152, 153, 155
Hamsters, 306, 307–309
Hanta virus, 285–286
Hapalia machaeralis (snout moth), 235
Haptophytes, 111–113, 114
Hargrove, Gene, 380 n.1
Hawaii Biological Survey, 454
Hawaiian honeycreepers, 7, 148
Hawaiian islands, 11, 86, 147, 148, 322, 366, 518
Hawks, 148, 155
Health, human, biodiversity losses and, 285
Hedylepta accepta Butler (sugarcane leafroller), 235
Heliconius sp. (butterfly), 53–54
Heliothis virescens (tobacco budworm), 219
Helminths. *See* Parasitic helminths
Hemiptera, 55, 97
Henderson Island, 149
Henry Greene Prairie, 378–379
Henry, Joseph, 452
Herons, 148, 150, 151, 155
Hesperiidae, 70, 79
Heterotrophs, 111–113, 114, 115
Hinkle, Gregory, 109–122, 514–515
Histoplasmosis, 113
Historical contexts of biodiversity, 509–512

Measurement of biodiversity. *See also* Models/
modeling
  extinction predictions, 60–64
  extrapolation from samples, 50–57
  geographic ranges, 48–49, 54, 99–100, 326–
  328
  island biogeography concepts, 88, 90–91, 93
  Linneaus, 42–44, 508
  molecular techniques, 109–120, 267, 291–
  292, 317
  number of species, 47–60, 70–76, 89–93
  ratios of known-to-unknown fauna, 47–50,
  96–97
  scaling, 38, 91
Medicines. *See* Drugs and pharmaceuticals
Mediterranean Sea, 145
Mehrhoff, Leslie J., 447–465, 512, 521, 522
Melanesia, 145, 146, 147, 149
Meleagris crassipes (turkey), 145
Meliphaga lewinii (Lewin's honeyeater), 269
Melosira varians (diatom), 22
Megamouth shark, 95
Megapodes, 148, 155
Megapodius sp. (megapode), 155
Merriam, C. Hart, 13
Mesocricetus auratus (hamster), 311
Methane, 283
Mexico
  birds, 141, 143
  corn, 8–9, 143, 224
  deforestation, 143
  insects, 48, 49, 74
  wetlands, 129
Mice, 306, 307–309
Microbial diversity. *See also* Bacteria; Fungi;
  Viruses
  and biotechnology, 279, 284–285, 515
  ecological importance, 282–284
  international cooperation, 472–473
  measurement, 470–472
  number of species, 282
  relation to macroorganismal biodiversity,
  285–286
  repositories, 467–470, 521
Micronesia, 147
Microorganisms. *See* Algae; Bacteria; Microbial
  diversity; Viruses and procaryotes;
  specific organisms and species
Micropygia heana, 151
Microsporidians, 111

Midilinae, 233, 234
Miller, Douglass R., 217–229, 513, 517, 520
Mining, 130
Miocene epoch, 251, 255
Mississippi River drainage area, butterflies, 73,
  74
Mites, 58, 96–97
Mitter, Charles, 301–320, 516, 518
Mittermeier, Russell, 341
Moas, 148
Models/modeling
  bowerbird evolution, 267–268, 273–274
  climate change, 163, 167–168, 181, 187–188,
  516
  species estimates, 57–60
Molecular clock, 113 n.1
Molecular phylogeny, 109–120, 342, 514–515,
  518. *See also* Nuclear-gene-sequence
  character asessment
Mollusca, 64, 86, 87, 94, 96, 101, 102, 365–366
Moluccas, northern, 142
Monarchs, 152, 155
Montastrea annularis (reef-building coral), 95
Moonseed, 178
Moorea, 87
Mosquitos, 78
Moth borers, 221
Moths. *See also* Snout moths
  attributes for biodiversity studies, 231, 234
  concordance studies, 307–309, 318
  number of species, 59
  pests, 218, 219, 221
Mount Desert Island, Maine, 15
Mouse opossum, 253
Mrakia frigida (ant), 117
Mullis, Kary, 13
Mus musculus (domestic house mouse), 293,
  310, 311
Museum of Comparative Zoology, Harvard, 452
Museums. *See* Natural history museums;
  Research collections; *specific museums*
Musotiminae, 237
Mussels, 126
Mutations, 113 n.1
Mycotoxins, 226
Myers, Norman, 125–138, 514, 515, 519, 520
Myiagra sp. (monarch flycatcher), 152, 155
Myiarchus cinerascens (ash-throated flycatcher),
  331
Myzomela cardinalis (honeyeater), 156

O

Oak (*Quercus* sp.), 54
Oaxala Declaration, 426
Oceania, 147–148, 154
Oceans and seas. *See also* Coastal; Deep sea;
    Marine; Open ocean; *individual oceans
    and seas*
    surface area, 88
Octocorals, 364
Oculina arbuscula (coral), 87
Odontiinae, 234
Ohio Biological Survey, 454
Oil extraction, 128
Okavango Swamp, Botswana, 129
Okechobee, Lake, 13
Okhotsk, Sea of, 255
Okinawa, 149
Oklahoma Biological Survey, 454
Oleander, 178
Olson, Storrs, 11
Omiodes Guenée (snout moth), 238
Onchocera volvulus (nematode), 311
Oomycetes, 111–112, 114, 115
Open ocean, bacteria and viruses, 119, 282
Ophyrys sp. (orchids), 16
Opler, Paul A., 69–82, 514, 520
Opuntia Mill (prickly pear cactus), 235
Oranges, 235
Oregon
    butterflies, 73, 74
    gap analysis of biodiversity, 322
Orinoco River, 251–252
Ornithologists, 157
Oryctolagus cuniculus (rabbit), 311
Oryza sativa (rice), 293
Ostrinia nubilalis (European corn borer), 223,
    235
Owlet-nightjars, 148
Owls, 150
Oysters, 10
Ozone, stratospheric, 192

P

Pacific islands, 141, 145, 148, 154
Pacific Ocean, temperatures, 172
Pakistan, 129
Pakitza, Peru, 33, 75, 76
Palau, 87
Palearctic realm, 48
Paleocene epoch, 170–176

Paleocene/Eocene boundary, 176–181, 516
Paleozoic era, 101
Palms, 164, 165
Panama
    birds, 78, 142
    coral reef taxa, 87, 91
    insects, 54, 71–72, 75, 78
Panama Canal, 11
Pantanal Swamp, Brazil, 129–130
Pantoja, Loreto, Peru, 76
Papasula abbotti (Abbott's booby), 149, 150,
    151
Papilionidae, 70, 75, 79
Papua New Guinea, 48, 49, 141, 363–366
Papyrus, 130
Paraguay, 129
Paraguai River, 129–130
Parana River, 251–252, 258
Parasitic helminths
    described species, 244
    food webs, 249
    in freshwater rays, 249, 251–253
    historical reconstruction, 239–258
    host specificity, 244
    in marsupials, 253–254
    in pinnipeds, 255–257
    in seabirds, 255–257
Parasitic wasps, 220, 222
Parasites. *See also* Host-parasite systems
    historical probes, 244
    of mammals, 113, 223–224
Parrots, 148, 149, 150, 152, 153, 155
Partnership for Enhancing Expertise and
    Taxonomy, 423
Parus rufescens (chestnut-backed chickadee),
    329
Passenger pigeon, 145
Passerherbulus henslowii (Henslow's sparrow),
    146
Passerines, 46, 148
Passifloraceae, 53–54
Patrick, Ruth, 15–24, 213–216, 513, 516–517,
    520, 522
Peale, Charles Wilson, 451
Pearl Islands, Panama, 87
Pelicans, 148
Penicillium mold, 9
Pennsylvania Biological Survey, 454
People's Republic of China, 48
Permian/Triassic boundary, 181
Peromyscus sp. (deer mouse), 286